To Ole Rim, with regards

Bro Jaffro
Columbia 7 Dec 71

$3

CRITICAL PHENOMENA IN ALLOYS, MAGNETS, AND SUPERCONDUCTORS

McGraw-Hill Series in Materials Science
and Engineering

Editorial Board

MICHAEL B. BEVER
M. E. SHANK
CHARLES A. WERT
ROBERT F. MEHL, *Honorary Senior Advisory Editor*

AVITZUR *Metal Forming: Processes and Analysis*
AZÁROFF *Introduction to Solids*
BARRETT AND MASSALSKI *Structure of Metals*
BLATT *Physics of Electronic Conduction in Solids*
BRICK, GORDON, AND PHILLIPS *Structure and Properties of Alloys*
BUERGER *Contemporary Crystallography*
BUERGER *Introduction to Crystal Geometry*
DE HOFF AND RHINES *Quantitative Microscopy*
DRAUGLIS, GRETZ, AND JAFFEE *Molecular Processes on Solid Surfaces*
ELLIOTT *Constitution of Binary Alloys, First Supplement*
GILMAN *Micromechanics of Flow in Solids*
GORDON *Principles of Phase Diagrams in Materials Systems*
GUY *Introduction to Materials Science*
HIRTH AND LOTHE *Theory of Dislocations*
KANNINEN, ADLER, ROSENFIELD, AND JAFFEE *Inelastic Behavior of Solids*
MILLS, ASCHER, AND JAFFEE *Critical Phenomena in Alloys, Magnets, and Superconductors*
MURR *Electron Optical Applications in Materials Science*
PAUL AND WARSCHAUER *Solids under Pressure*
ROSENFIELD, HAHN, BEMENT, AND JAFFEE *Dislocation Dynamics*
RUDMAN, STRINGER, AND JAFFEE *Phase Stability in Metals and Alloys*
SHEWMON *Diffusion in Solids*
SHEWMON *Transformations in Solids*
SHUNK *Constitution of Binary Alloys, Second Supplement*
WERT AND THOMSON *Physics of Solids*

CRITICAL PHENOMENA IN ALLOYS, MAGNETS, AND SUPERCONDUCTORS

Edited by

ROGER E. MILLS
Department of Physics, University of Louisville

EDGAR ASCHER
Advanced Studies Center, Battelle Geneva Laboratories

ROBERT I. JAFFEE
Department of Physics and Metallurgy,
Battelle Columbus Laboratories

BATTELLE INSTITUTE
MATERIALS SCIENCE COLLOQUIA

*Geneva and Gstaad, Switzerland
September 7-12, 1970*

Robert I. Jaffee, Chairman

McGRAW-HILL BOOK COMPANY

*New York   St. Louis   San Francisco   Düsseldorf   Johannesburg
Kuala Lumpur   London   Mexico   Montreal   New Delhi
Panama   Rio de Janeiro   Singapore   Sydney   Toronto*

CRITICAL PHENOMENA IN ALLOYS,
MAGNETS, AND SUPERCONDUCTORS

Copyright © 1971 by McGraw-Hill, Inc. All Rights Reserved. Printed in the United States of America. No part of this publication may be reproduced, stored in a retrieval system, or transmitted, in any form or by any means, electronic, mechanical, photocopying, recording, or otherwise, without the prior written permission of the publisher. *Library of Congress Catalog Card Number* 79-158059

07-042365-2

1234567890 MAMM 754321

To **PROFESSOR LARS ONSAGER**

whose exact solution of the two-dimensional Ising model led the way to quantitative treatment of critical phenomena described in these proceedings

CONTENTS

Participants xi
Preface xv
Autobiographical Commentary of Lars Onsager xix
After-Dinner Address: "The Changing Face of Physics" L. VAN HOVE   xxv

Part One   INTRODUCTORY LECTURES

1. *The Ising Model in Two Dimensions*   L. ONSAGER   3
2. *One-dimensional Delta Function Interaction*   C. N. YANG   13
   DISCUSSION   20
3. *The Role of Models in Understanding Phase Transitions*   M. KAC   23
   DISCUSSION   38
4. *Metastability, Instability, and the Dynamics of Unmixing in Binary Critical Systems*   J. W. CAHN   41
   DISCUSSION   64

Part Two   GENERAL THEORY

A. *Thermodynamics and Scaling*

Rapporteur's Introduction: Thermodynamics and Scaling   B. WIDOM . . .   69
1. *Asymmetry of Critical Exponents and the Asymptotic Scaling for Critical Exponents*   B. D. JOSEPHSON   73
   DISCUSSION   75
2. *Generalized Scaling and the Parametric Equation of State in the Critical Region*   M. J. COOPER   77
   DISCUSSION   86
3. *Self-avoiding Walks and Scaling*   C. DOMB   89
   DISCUSSION   103
4. *Secondary Variables in Critical Phenomena, with Application to* λ *Transition in Liquid Helium*   O. K. RICE AND D.-R. CHANG   105
   DISCUSSION   123

viii CONTENTS

5. *Thermodynamics of a Magnetic Model Exhibiting a First-order Phase Transition* H. STENSCHKE 125
   DISCUSSION 133
6. *Critical States of Ferroelectrics* G. FALK 135
   DISCUSSION 137

B. *Ising Systems and Field Theory*

   Rapporteur's Introduction: Ising Systems and Field Theory
   P. C. MARTIN 141
1. *Random Impurities in the Two-dimensional Ising Model* B. M. McCOY 145
   DISCUSSION 153
2. *Randomly Dilute Ising and Heisenberg Systems* G. S. RUSHBROOKE .. 155
   DISCUSSION 163
3. *Analytical Properties of a Class of Planar Ferroelectric Models*
   M. L. GLASSER 165
   DISCUSSION 174
4. *Contributions of Broken-symmetry Modes in Second-order Phase Transitions* R. E. MILLS 177
5. *Renormalization Group and Theory of Phase Transitions*
   G. JONA-LASINIO 189
   DISCUSSION 200
6. *Series Expansions and the Universality Hypothesis* H. S. STANLEY ... 203
   DISCUSSION 217
7. *Inequalities Among the Ising-Heisenberg Model Critical Indices* G. A. BAKER, JR. 221
   DISCUSSION 229
8. *Theory of Critical Slowing Down* M. SUZUKI 231
   DISCUSSION 239

Part Three DEVELOPMENT OF SPATIAL ORDERING

1. *Spatial Order* L. GUTTMAN 243
   DISCUSSION 254
2. *Early-stage Clustering and Ordering Kinetics in Binary Solid Solutions*
   D. DE FONTAINE AND H. E. COOK 257
3. *Bose-Einstein Condensation of Concentration Fluctuations in Binary Solid Solutions* D. DE FONTAINE 277
   DISCUSSION 287
4. *Long-range Order in Beta Brass* C. B. WALKER AND D. R. CHIPMAN 289
   DISCUSSION 297
5. *Multisite Cluster Probabilities in Ising Lattices* P. C. CLAPP 299
   DISCUSSION 307
6. *The "Gestalt" of Local Order* J. E. GRAGG, JR., P. BARDHAN, AND J. B. COHEN 309
   DISCUSSION 336
7. *Ferroelasticity Due to Hydrogen in Metals* G. ALEFELD 339
   DISCUSSION 361

Agenda Discussion: Development of Spatial Ordering Chairman, S. C. Moss 363

CONTENTS ix

Part Four MAGNETS AND SUPERCONDUCTORS

1. *Critical Points Dependent on Parameters* R. B. GRIFFITHS 377
   DISCUSSION 389
2. *Critical Properties of Ferromagnets and Liquid Crystals* J. D. LITSTER 393
   DISCUSSION 404
3. *Scaled Equation of State and Derived Properties* M. VICENTINI-MISSONI 407
   DISCUSSION 421
4. *Critical Relaxation in Isotropic Ferro- and Antiferromagnets Below the Transition Point* J. VILLAIN 423
   DISCUSSION 436
5. *Critical Magnetic Behavior of Palladium-Iron Alloys* J. S. KOUVEL AND J. B. COMLY 437
   DISCUSSION 448
6. *Fluctuations Near the Phase Transition in "One-dimensional" Superconductors* R. J. WARBURTON AND W. W. WEBB 451
   DISCUSSION 468

   Agenda Discussion: Magnets and Superconductors Chairman, R. A. Ferrell.. 471

Part Five TRANSPORT PROPERTIES

1. *NMR and Neutron Scattering Studies of Critical Fluctuations in Uniaxial Antiferromagnets* P. HELLER 483
   DISCUSSION 488
2. *Dynamical Theory of Critical Fluctuations with Applications to Transport Properties in the Critical Region* K. KAWASAKI 489
   DISCUSSION 500
3. *On the Spectrum and the Intensity of Light Scattered from a Fluid Near its Critical Point* G. B. BENEDEK, J. B. LASTOVKA, M. GIGLIO, AND D. CANNELL....................................... 503
4. *Experimental Study of the Dynamics of Critical Fluctuations in a Binary Mixture* P. BERGE AND B. VOLOCHINE 529
5. *Ultrasonic Measurements Near the Liquid-Gas Critical Point* M. BARMATZ 541
   DISCUSSION OF PAPERS 3, 4, AND 5 559

   Agenda Discussion: Transport Properties Chairman, L. P. Kadanoff.. 561

6. *Electronic Phase Transitions* D. ADLER 567
7. *Some Aspects of the Theory of the Mott Transition* T. M. RICE AND W. F. BRINKMAN 593
   DISCUSSION OF PAPERS 6 AND 7 608
8. *Recent Developments of the Lee-Yang Circle Theorem and Analyticity of the Free Energy* M. SUZUKI............................ 613
   DISCUSSION 620

   Summary Agenda Discussion: Critical Phenomena Chairman, E. Montroll.. 623

   Index 647

PARTICIPANTS

DAVID ADLER *Department of Electrical Engineering, Massachusetts Institute of Technology, Cambridge, Massachusetts, U.S.A.*
GEORG ALEFELD *Institut für Festkörper- und Neutronenphysik, Kernforschungszentrum, Jülich, Germany*
EDGAR ASCHER *Battelle Institute, Advanced Studies Center, Geneva, Switzerland*
G. A. BAKER, JR. *Applied Mathematics Department, Brookhaven National Laboratory, Upton, New York, U.S.A.*
MARTIN BARMATZ *Bell Telephone Laboratories, Murray Hill, New Jersey, U.S.A.*
G. B. BENEDEK, JR. *Physics Department, Massachusetts Institute of Technology, Cambridge, Massachusetts, U.S.A.*
H. BÜTTNER *Battelle Institute, Frankfurt/Main, West Germany*
J. W. CAHN *Department of Metallurgy and Materials, Massachusetts Institute of Technology, Cambridge, Massachusetts, U.S.A.*
DO-REN CHANG *Department of Chemistry, University of North Carolina, Chapel Hill, North Carolina, U.S.A.*
PHILIPPE CHOQUARD *Ecole Polytechnique Fédérale, Laboratoire de Physique Théorique, Lausanne, Switzerland*
P. C. CLAPP *Ledgemont Laboratory, Kennecott Copper Corporation, Lexington, Massachusetts, U.S.A.*
J. B. COHEN *Department of Materials Science, Northwestern University, Evanston, Illinois, U.S.A.*
H. E. COOK *Product Development Group, Ford Motor Company, Dearborn, Michigan, U.S.A.*
M. J. COOPER *Statistical Physics Section, National Bureau of Standards, Washington, D.C., U.S.A.*
J. M. COWLEY *Department of Physics, Arizona State University, Tempe, Arizona, U.S.A.*

PARTICIPANTS

DIDIER de FONTAINE *School of Engineering and Applied Science, University of California at Los Angeles, Los Angeles, California, U.S.A.*

CYRIL DOMB *Physics Department, King's College, University of London, London, England*

GEORG FALK *Institut für Mathematische Physik, Universität Karlsruhe, Karlsruhe, West Germany*

R. A. FERRELL *Department of Physics and Astronomy, University of Maryland, College Park, Maryland, U.S.A.*

P. C. GEHLEN *Metal Science Group, Battelle Memorial Institute, Columbus, Ohio, U.S.A.*

M. L. GLASSER *Advanced Study Center, Battelle Memorial Institute, Columbus, Ohio, U.S.A.*

R. B. GRIFFITHS *Department of Physics, Carnegie-Mellon University, Pittsburgh, Pennsylvania, U.S.A.*

CHRISTIAN GRUBER *Ecole Polytechnique Fédérale, Laboratoire de Physique Théorique, Lausanne, Switzerland*

LESTER GUTTMAN *Solid State Sciences Division, Argonne National Laboratory, Argonne, Illinois, U.S.A.*

PETER HELLER *Physics Department, Brandeis University, Waltham, Massachusetts, U.S.A.*

R. I. JAFFEE *Battelle Memorial Institute, Columbus, Ohio, U.S.A.*

LAURENS JANSEN *Battelle Institute, Advanced Studies Center, Geneva, Switzerland*

G. JONA-LASINIO *Università degli Studi, Istituto di Fisica "Guglielmo Marconi", Roma, Italy*

B. D. JOSEPHSON *University of Cambridge, Cavendish Laboratory, Cambridge, England*

MARK KAC *The Rockefeller University, New York, New York, U.S.A.*

L. P. KADANOFF *Department of Physics, Brown University, Providence, Rhode Island, U.S.A.*

KYOZI KAWASAKI *Temple University, Department of Physics, Philadelphia, Pennsylvania, U.S.A.*

J. S. KOUVEL *Department of Physics, University of Illinois, Chicago, Illinois, U.S.A.*

J. D. LITSTER *Department of Physics, Massachusetts Institute of Technology, Cambridge, Massachusetts, U.S.A.*

WALTER MARSHALL *Atomic Energy Research Establishment, Harwell, Didcot, Berkshire, England*

P. C. MARTIN *Lyman Laboratory of Physics, Harvard University, Cambridge, Massachusetts, U.S.A.*

BARRY McCOY *Institute for Theoretical Physics, State University of New York at Stony Brook, Stony Brook, Long Island, New York, U.S.A.*

F. J. MILFORD *Battelle Memorial Institute, Columbus, Ohio, U.S.A.*

R. E. MILLS *University of Louisville, Department of Physics, Louisville, Kentucky, U.S.A.*

E. W. MONTROLL *Physics and Astronomy Department, University of Rochester, Rochester, New York, U.S.A.*

S. C. MOSS   Energy Conversion Devices, Inc., Troy, Michigan, U.S.A.
LARS ONSAGER   Chemistry Department, Yale University, New Haven, Connecticut, U.S.A.
A. QUATTROPANI   Ecole Polytechnique Fédérale, Laboratoire de Physique Théorique, Lausanne, Switzerland
O. K. RICE   Department of Chemistry, University of North Carolina, Chapel Hill, North Carolina, U.S.A.
T. M. RICE   Bell Telephone Laboratories, Inc., Murray Hill, New Jersey, U.S.A.
G. S. RUSHBROOKE   School of Physics, Department of Theoretical Physics, University of Newcastle-upon-Tyne, Newcastle-upon-Tyne, England
P. R. SIEVERT   Department of Physics, Battelle Memorial Institute, Columbus, Ohio, U.S.A.
H. E. STANLEY   Physics Department, Massachusetts Institute of Technology, Cambridge, Massachusetts, U.S.A.
HERBERT STENSCHKE   Battelle Memorial Institute, Advanced Studies Center, Columbus, Ohio, U.S.A.
M. J. STEPHEN   Physics Department, Massachusetts Institute of Technology, Cambridge, Massachusetts, U.S.A.
M. SUZUKI   Department of Physics, Cornell University, Ithaca, New York, U.S.A.
LEON VAN HOVE   CERN, Geneva, Switzerland
MATILDA VICENTINI-MISSONI   Università degli Studi, Istituto di Fisica "Guglielmo Marconi", Roma, Italy
JACQUES VILLAIN   Institut Max Von Laue-Paul Langevin, Grenoble, France
B. VOLOCHINE   Service de Physique du Solide et de Résonance Magnétique, Centre d'Etudes Nucléaires de Soclay, Gif-sur-Yvette, France
C. B. WALKER   Army Materials and Mechanics Research Center, Watertown, Massachusetts, U.S.A.
R. J. WARBURTON   Physics Department, Cornell University, Ithaca, New York, U.S.A.
BENJAMIN WIDOM   Chemistry Department, Cornell University, Ithaca, New York, U.S.A.
C. N. YANG   Institute for Theoretical Physics, State University of New York at Stony Brook, Stony Brook, Long Island, New York, U.S.A.

PREFACE

The fifth Battelle Colloquium on Materials Science was held in Geneva and Gstaad, Switzerland, September 7-12, 1970, on the topic of "Critical Phenomena". The colloquium started the second round of the four Battelle laboratories, the first colloquium on "Phase Stability in Metals and Alloys" having been held in Switzerland in May, 1966.

Critical phenomena embrace a wide variety of higher order phase transitions where the change of state occurs without a latent heat. The vapor-gas critical point discovered a century ago by Thomas Andrews is the classical example of a critical phenomenon. Over the past decade, research on critical phenomena has had a remarkable renascence, possibly stimulated by the expectation that advances made in the theoretical treatment might result in a general theory that could describe experimental behavior at the critical point in widely diverse fields.

The Organizing Committee of the Colloquium on Critical Phenomena stated their objectives as follows:

1. Elucidate recent progress in the experimental and theoretical studies of critical phenomena in alloys, magnets, and superconductors.
2. Examine the status of a general description of critical phenomena and of the knowledge of transport phenomena in the critical regime.
3. Suggest fruitful directions for future work.

The format used successfully in previous Battelle Colloquia was largely followed for Critical Phenomena. Thus, on the first day, introductory lectures were made by contributors who have made historically significant contributions to the topic, and which provided a historical perspective on the various facets of the subject from which modern research could be evaluated and projected. We were privileged to have as introductory lecturers Professors Lars Onsager, Chen Ning Yang, Mark Kac, and John Cahn. In the evening of the first day, another distinguished contributor, Dr. Leon Van Hove, presented an after-dinner address on the topic "The Changing Face of Physics", which is included in the front matter to these proceedings.

A departure from our customary format was employed in the second day sessions, which were devoted to "General Theory". To cope with a large number of theoretical papers and permit time for discussion, rapporteurs were used to introduce the papers and point out their relationship to each other and to the field itself. We were fortunate to have two excellent rapporteurs, Professor Ben Widom for the theoretical papers under the category "Thermodynamics and Scaling" and Professor Paul Martin for the theoretical papers on "Ising Systems and Field Theory". Professor Cyril Domb moderated the discussions and edited the resulting summary paper on the discussions, which are introduced by the rapporteur's remarks in summary form.

The balance of the colloquium was comprised of individual days devoted to three subdivisions of the general topic:

1. Development of spatial ordering
2. Magnets and superconductors
3. Transport phenomena

A typical day consisted of a morning session of experimental papers reporting new information on these topics. The first paper in each session was of a more general treatment of the topic, and subsequent papers covered current research of a more specific nature. After a free afternoon, the colloquium took up an agenda of questions and issues on the three topics led, respectively, by Professors Simon Moss, Richard Ferrell, and Leo Kadanoff. Summary papers of the three topical agenda discussions are included in the proceedings and may be taken to indicate the status of the various issues that have been resolved or are still in a state of controversy and some research directions which might resolve these issues.

In a final Agenda Discussion, Professor Elliott Montroll guided the Colloquium through an assessment of the overall field and an evaluation of the contributions made by the colloquium. The reader is commended to his summary paper for an overall assessment of

Critical Phenomena, circa 1970, approximately a century after the topic was brought forth as a special field of science.

It is the custom of the Colloquia to honor one of the distinguished participants. In the case of Critical Phenomena, this honor was paid to Professor Lars Onsager (quoting the inscription) "whose exact solution of the two-dimensional Ising model led the way to quantitative treatment of critical phenomena described in these proceedings." At the farewell banquet in Gstaad, which was held in his honor, Professor Onsager presented an autobiographical sketch, which is reproduced in the front matter, describing the circumstances attendant to his work on the Ising model.

The continued encouragement and financial support of the Battelle Institute Colloquia on Materials Science by Dr. Sherwood L. Fawcett, President of Battelle Memorial Institute, and Dr. Frederick J. Milford, Director of Research in the Physical Sciences of Battelle Institute, is gratefully acknowledged. We wish to thank Dr. Hugo Thiemann, Director of the Battelle-Geneva Laboratories, and members of his staff for facilities and support. Particularly, we acknowledge our debt and gratitude to Mlle. Anne-Marie Bonino, who was in charge of local arrangements and the secretariat, to Drs. Walter Bollmann, Hans Grimmer, and W. Kurtz for the technical arrangements, and to Mme. Jacqueline Juilliard and Mrs. R. I. Jaffee for the Ladies Program. Without the enthusiastic and effective support by these and many other people, the Colloquium could not have had its undoubted organizational success.

The Organizing Committee is grateful to Professors George Benedek and J. W. Cahn of MIT, J. W. Christian of Oxford, Michael Fisher of Cornell, Peter Heller of Brandeis, and Leo Kadanoff of Brown for their advice on the choice of subjects and participants for this colloquium.

The future will tell how successful we were in meeting the scientific objectives we set in organizing this colloquium. For ourselves, we believe that an honest attempt was made, for which we wish to thank the participants most sincerely.

Robert I. Jaffee
Edgar Ascher
Roger E. Mills

AUTOBIOGRAPHICAL COMMENTARY OF LARS ONSAGER

Bob Jaffee suggested that I give a bit of background of what I have been doing on these various problems that we have been discussing and how I got to doing them and so forth.

Actually, as Elliott (Montroll) mentioned, I got started, sort of cut my milk teeth, on problems in the theory of electrolytes back in Trondheim. And, at the time, I suspected some reciprocal relations, that seemed mighty hard to get around no matter how you did all the statistics and all the mechanics. It was not until I had spent nearly two years in Zürich (in the course of that period, I got acquainted with several persons worth knowing such as Peter Debye, his assistant Hückel, Scherrer, Weyl, Schroedinger, and Pauli) that I got a lead on these reciprocal relations. I was talking with Scherrer and mentioned the analogy with the principle of detailed balancing applied to some chemical reactions, which incidentally Professor Ritber had been studying back in Trondheim, and Scherrer said: "Yes, Lewis has paid much attention [that was Gilbert Newton Lewis] to this detailed balancing."

So I started to mull it over and turn questions around. Of course I had learned a few things; I learned I did not know everything and learned to respect somebody else's point of view. And, after all, it seemed reasonable that really the symmetry of past and future should be in the background of all of this and then all that was left, you see, was to get a handle on the problem; and as I had a little time on my hands, arriving in good time in the harbor on the channel to board a ship for the States, I thought of the fluctuations, and a few months later in the spring, I took it up and satisfied myself that it could be handled that way. So that was really done in the spring of 1928. But that did not bring me into much of the problems that have been interesting us—the topic of this conference; and not, I am sure, all your main interests, but certainly the one set of main interests that all here have in common.

The beginning of that was really that once the problems of electrolytes more or less settled—the first round anyway—how about the dielectrics that Debye had done? I was too lazy to go to the libraries and sat down and worked it out and lo and behold it came out quite different!

One of the things that I thought of was the importance of using this potential of average force. So I put that to use, and that gives a result quite different from the Lorentz-Lorenz. Well, this also then explained not so much why there are transition points but rather why in the dielectrics they don't crop up more often than they do. Of course, later, I realized that I'd slightly overcorrected Debye's computation, although this agreed beautifully with all the observations; but you have to bear in mind that although dielectric properties had been investigated very much at that time, when you look at liquids there is a natural selection, because once a system of dipoles gets ordered it is very likely to crystallize as well; thus all known ferroelectrics are solids. Nevertheless, that was in the end a contribution to ferroelectrics and ferromagnetics as well, with a considerable bearing on the transition point, at least on the locations of the transition points. And some years later in 1935, it was plainly high time to publish it.

By that time, Giauque and Stout had obtained some interesting results on the thermodynamic properties of ice; even a decade before Errera had obtained some *bewildering* results on the dielectric properties of ice. But a couple of years before, E. J. Murphy at Bell Telephone had indeed obtained some more reasonable results, and so it was much like the liquid in its dielectric properties, except for a much slower relaxation. It also turned out that ice showed a residual entropy, and as everybody jumped bandwagons in those days, several people thought about ortho-para transitions in analogy with hydrogen.

Of course, it was nothing of the sort. It was a matter of a variety of configurations, and Linus Pauling saw the light and even had a pretty good estimate for the number of configurations, which checked within the limits of error. So, I set myself to play with this and saw that Pauling's estimate might be a little on the low side but not much. A few years ago, John Nagle, working with me, made some careful computations and raised Pauling's estimate by about 1%. Well, I also played around with the dielectric properties; and then, while I was playing with that, it was getting toward 1938 or 1939, Busch in Zürich discovered that potassium di-hydrogen phosphate had some very interesting dielectric properties and goes ferroelectric at low temperatures; so I started expanding a little in that direction; this was doable and I was going to present that in Zürich in a conference in 1939 but then the war broke out.

I spent a lot of time trying to work out methods of improving the computations for these systems; and then a year later, spring 1940, I got a letter from Joe Meyer recommending a young fellow, indeed recommending him very highly. The young fellow had worked with Wannier before and spent a year with Joe and Maria (Meyer). On the strength of that recommendation, I used the strongest pressure I had at my disposal to wedge an opening and Elliott (Montroll) came to occupy it. I was very glad I had made that effort. And incidentally, so is everybody else in the Chemistry Department at Yale. I think everybody found Elliott very interesting to talk to. He brought the news of this development on the Ising model, that Wannier had brought home from Holland after working with Kramers. Well, this intrigued me considerably; by that time the methods they had been using were a little more advanced than the ones I'd been playing with, which got very near to using a transfer matrix. I did not do anything until I had a chance to talk to Wannier sometime between Christmas and New Year's, and he gave me a very vague idea how it was done. In the course of the spring, I started to reconstruct this dual transformation from imperfect information and, actually, very soon it turned out to be something much more general and I am not going to go into all the details because I told a lot of them on Monday ("The Ising Model in Two Dimensions", see p. 3). It was a sort of investigation where you got a good lead, and certainly you had to pursue that; and before you reached the end of that lead, up opened another, and this was, if anything, even more fascinating. Then one kept on going in several stages; one beautiful lead opening up after the other and every one much too good to abandon. Before I got to the end of it, it looked like a very good guess what the answer might be; it was something that showed up so regular that it was easy to extrapolate it all the way to an infinite crystal. It took a few months, though, to verify the guess, but it was doable; and that was how the Ising model was solved. And that, for those who are not aware of all the details, was a first model of anything, any kind of a system, that has what you call a continuous transition, that is one which occurs at a definite temperature but in such a manner that nothing changes by a finite amount. Every step is small. This was the first model of that sort that had ever been computed exactly, and the outcome demonstrated, once and for all, that the results of several more primitive theories had features that could not possibly be preserved in an exact computation. And for the rest, at the very least, it's taught us what *not* to do; what to stay away from. Long before that, though, I had been introduced to one other transition. In the course of my career, I spent one term at Johns Hopkins University; it was the first teaching position I had; then I spent five years as instructor at Brown, and since then I have been at Yale. But while at Brown, I was most intrigued by the solutions of metals in ammonia. These blue solutions burn bronze at high concentrations and have conductivity that varies from little better than an ordinary 1-1 electrolyte, say by a factor of 4, up to conductivities that compare with that of liquid mercury as you vary concentration. So here was just a straight, if gradual, transition from an electrolytic conductor to a metal. In those days, we did not know quite what to do with the theory of metals. I remember I thought it worthwhile to study the magnetic properties of these solutions and I tried to dream up a method, but I started on the wrong end, and the best I could dream up was not good enough. So, it remained for other folks to study those magnetic properties with quite interesting results, in fact, so interesting that nobody knows quite what to make of them yet. So this was an early empirical attack on the Mott transition; some of its major problems are still with us. Well, after I got out a paper on that Ising model, there was a young lady, Bruria Kaufman, and I had no idea what she might be like; first of all, she insisted on working with me. Well, I thought we might arrange a way to do that. Secondly, she insisted on working on the Ising model. Now, that was the kind of a task that I would never want to impose on a student. So I did my best to talk her out of it. But, no, she insisted. All right, then. I tried to dream up some problems that would be significant but reasonably safe. And for any ordinary good student, her assignment would have been a good problem. Well, she did some work on it, but she had a sort of a feeling that there was something better to be had; and after some months in the fall and a few months into the spring, her ideas crystallized into something a little more definite. And then I said now go ahead and see whether we can do something with that. So she went ahead and in a fairly short time it became very plain that something indeed could be done with that, and that was what we did from then on. As you know, that led to a whole lot of additional results; and it became possible to compute a lot of the detailed properties of the model, where before I had computed just the overall properties. And then, finally, it became apparent that we had a last problem, of the degree of order, and that turned out to need more invention.

Back in the days at Brown, I had a very good friend in the Mathematics Department; his name was Yasha Tamarkin. It is quite a number of years since he passed away, but Yasha introduced me to a couple of his friends who often came visiting. One was E. Hille with whom he worked a lot, then at Princeton, later at Yale, and the other was Norbert Wiener. Wiener came around to lecture on Fourier series and integrals. These were among the few lectures in mathematics I have attended and found most interesting. And, evidently, I learned a few other things besides—some I didn't learn—and, just the same, when I looked at these determinants of Bruria's, it looked like a difference equation. Then I tried to look at the infinite one; in a little while it dawned on me that this was just, in a difference equation, the same as the Milne equation in an integral equation; and so I tried the same technique (Wiener-Hopf) which solves the Milne equation. It solves the difference equation too. Then there was a little work on that which I described in more detail on Monday and we had a very neat result. Now, the reason that was not published the way it was computed was that, before I got around to it, I tried to find the general formula for the evaluation of determinants of the same type and lo and behold I found it. It was a general formula for the evaluation of Toeplitz matrices. The only thing I did not know was how to fill out the holes in the mathematics and show the epsilons and the deltas and all of that, and the limiting processes; I did not know just how it should be done and what mathematicians really knew about limiting processes in that ball park. Later, six years ago, at a meeting of the Mathematics Society, I talked to Hirschman, who had the best theorem to date. It turned out that if I had just gone ahead the way I had started and then used a theorem of Wiener's, that I had not known about, (fellows like Hirschman had known about the Wiener-Levy theorem long since) if I had just put that in, I would have had exactly the same conditions as Hirschman's.

So, with one thing or another, this is how some of these results in the theory of transitions were obtained. I had been playing with a few others, you know. Well, I had been playing with the theory of superfluids but not with the superconducting transition; so I think that gives you an idea roughly how, from various kinds of work, I got interested in the sort of problems that we have been talking about this week. And I must say that I have been greatly interested in learning what's being done in our days, because you can't possibly follow everything, particularly when you are active in other fields too.

I have enjoyed this meeting greatly. So, I want to thank you all for your contributions, and will you, Bob Jaffee, accept my thanks to Battelle, on behalf of the organization, for the most excellent arrangements.

Lars Onsager

Gstaad, Switzerland

AFTER-DINNER ADDRESS: THE CHANGING FACE OF PHYSICS

Leon Van Hove
CERN, Geneva

It is a great honor to address this distinguished gathering of scientists and guests at the close of a day of introductory lectures on critical phenomena, all the more so that the reasons for my being selected as after-dinner speaker, and therefore the choice of my theme, are not obvious. I have been a theoretical physicist long enough, however, to learn some of the tricks of our trade, for example, how to guess tentative explanations for obscure events, or how to pin down the theme of a lecture halfway through the lecturing period.

My theory for being here tonight shall be that I have had quite a vivid interest and a modest activity in several domains of physics, including some of the problems which occupy you this week. This is a dangerous theory, however, not only because it may be incorrect, which would not worry me too much, but because its systematic application would make life quite impossible in the world of today. How could a physicist with an interest in several fields still attend all the conferences which he would profit from? Thanks to the labor of the European Physical Society, we now know how incredibly numerous the physics meetings have become even on our side of the

Atlantic. The latest EPS list offers in Europe a choice of nine meetings for last week and eight for this week, not including the Battelle Symposium. All subjects are very specialized. Thus, you could have gone last week to Brussels for mass spectroscopy or to London for measurement techniques of spectroscopic data, while right now hadron spectroscopy is being talked about on Lake Balaton and electron spectroscopy in Uppsala. Quite a spectrum of spectroscopies for two brief weeks. And, as we all know, conferences are not enough for physics communication; summer schools are available in addition. According to the same EPS list, not less than twenty-five are being held this summer in Europe. Do we physicists not need a nonproliferation treaty of our own?

The fact of the matter is that the growth of physics research has been staggering in the last twenty years, thanks to ample public funding and sponsorship. Even specialization does not discourage high-level sponsorship, as illustrated by the International Congress on Thin Films announced for next month in Cannes under combined sponsorship of the French Ministry of the Interior, the French Ministry of Education, the French General Delegation for Scientific and Technical Research, and the International Union for Science, Technique and Applications of Vacuum.

Every branch of physics, big or small, had a more than fair chance to grow in the last two decades; most grew quite effectively, and a few led to truly important advances. Of course, none of the post-war achievements even approaches in fundamental significance the great discoveries of relativity and quantum physics which mark the first thirty years of our century. But such momentous steps are quite exceptional in the whole of science history, and it is neither funding nor official sponsorship which triggers them.

What intensive government support of research can achieve seems rather to be an acceleration of the normal growth of physics through experimental fact finding and improved instrumentation as well as through theoretical analysis and speculation. In addition, it makes the whole difference between building or not building certain giant instruments like large accelerators or reactors, large detectors, observation stations in outer space, and other expensive objects characteristic of big science.

I am now far enough in this address to know what I am talking about. My theme is the changing face of physics, how different it is now from some twenty years ago when, as a young mathematically minded theorist, I was increasingly fascinated by its fundamental problems and by the marvelous power of mathematics to analyze and predict its phenomena. The changes are profound indeed, not only by the new knowledge acquired but also in the style and atmosphere of research. In trying to summarize the latter, I am tempted to use two ugly words. Physics has gotten "chemicalized", and it has gotten "institutionalized". Let me explain what I mean, which is not quite as repulsive as the terminology I use.

It seems to me that physics now looks more like chemistry in the sense that, in percentage, a much larger fraction of the total research activity deals with complex systems, structures, and processes, as against a smaller fraction concerned with the fundamental laws of motion and interaction. This colloquium is a good example. Surely, we all believe that the fundamental laws of classical and quantum mechanics, of the electromagnetic interaction, and of statistical mechanics dominate the multivarious transition and critical phenomena you discuss this week; and I presume that none of you expects his work on such problems to lead to modification of these laws. You know the basic equations better than the phenomena. You are after the missing link between them, i.e., the intermediate concepts of mathematical or phenomenological nature which should allow a quantitative understanding and prediction of the phenomena on the basis of the fundamental equations. This is a difficult and fascinating task, ranging from strict mathematical treatment of simplified model systems to the intuitive hunt for phenomenologically relevant concepts and parameters. You are also after new and curious systems and processes, because the wealth and variety of behavior found when one arranges electrons and nuclei in different ways turns out to be absolutely amazing, and each new behavior has something to teach us.

Most of present physics is of this type, be it concerned with solids, liquids, neutral gases, plasmas, or atomic nuclei. Especially through the discovery of complex but remarkable systems and effects, many significant applications to technology have been found, again in analogy with good old chemistry. As a consequence, industry and government like physics more than in the past; and we are back to our earlier story of generous funding and prestigious sponsorship. This in turn brings us to the "institutionalization" of physics. The cause is big money; the consequence is elaborate management of research and mushrooming of scientific committees.

It would be a mistake to believe that funding and sponsorship of advanced research by the state is a recent development. Learning and research were generously patronized by the Ptolemaic dynasty which ruled Egypt after the conquests of Alexander the Great, relying heavily on brain-drain from Greece and elsewhere. Thus Strato, a distinguished natural philosopher and the second successor of

Aristotle as head of the Lyceum in Athens, was successfully attracted to Alexandria early in the third century B.C. by the first Ptolemy, not a bad act of scientific policy for a former general of Alexander. It is probably under the second Ptolemy that the famous Museum, or "Temple of the Muses", was founded with state money in Alexandria. In essence, it was a research university with a small amount of teaching. Its head, somehow the ancient analogon of our University Presidents or Laboratory Directors, was a high-priest. In addition to its celebrated Library, the Museum had lecture and study rooms, dissecting rooms, an observatory, a zoo, and a botanical garden. The investment, as we know, was a good one; among other things, it produced the Elements of Euclid and the first heliocentric system of astronomy due to Aristarchus of Samos (unfortunately superseded around 100 B.C. by Hipparchus's return to the geocentric system). Also the list of graduates is impressive; Archimedes is one of them.[1]

But I doubt that the Alexandrian Museum worried too much about efficient management of science, or that some of its senior scientists were spending half of their time in committee meetings. Why is our world of big science so fond of these things? Of course, one must accept that some of it is necessary if the flow of money is to be reasonably well used. After all, many decisions must be taken which no honest scientist, however competent he may feel, would like to take alone. What should you do when you have to make such a decision? You consult with other competent people in order to reduce the probability of mistake. If you have to do this five times a week, you set up a committee. Especially in Europe where science will only survive if it becomes international, the need to represent the interests and traditions of various countries is another factor in the multiplication of committees and the intricacies of management. Then, research at the frontier now requires very huge instruments. This leads to extreme centralization, basically a bad thing for creative research. Some of this is unavoidable; some of it is encouraged by the growing influence of an emerging class of scientific executives. Neither they nor the scientists have yet discovered how to reconcile managerial efficiency with scientific creativity.

Some of the dangers of the system are indeed already clear. Executive responsibility or membership of powerful committees confers to individual scientists an artificial increase of intellectual authority; they are rated more on the importance of decisions in which they participate than on their personal competence and achievements. Some of them grow so fond of administration and committee business that they keep in touch with science only through this business and imperceptibly lose the very foundation of their scientific judgment. While these evils can be effectively counteracted by turnover in executive tasks and committee membership, another danger is more subtle and more difficult to avoid. Some of the truly important decisions, let say on the best choice for a new large instrument, or on how and by whom an adventurous but costly experiment should be performed, require in an essential way the kind of purely intuitive, synthetic, and qualitative judgment which makes a good scientist move in the right direction without his being able to explain why. It is very hard to put such judgments in the analytic form which carries collective approval; and decision making by committees will unavoidably be poorer by the lack of them. Crystal gazing is part of the art of making wise scientific choices; and no amount of staring at documents, graphs, or budget figures can replace it. Hence, for some of the most important scientific decisions, informal, almost private discussion and consultation will be more reliable then official committee procedures. But how to circumvent the committees once they exist? And then, speaking biologically, they are so easily born and never die!

But back to the contents of present day physics and its chemical flavor. The bulk of it, as I mentioned earlier, consists of almost infinite variations on the themes of mechanics, statistics, and electromagnetic interactions. These studies are very rich, important, and worthwhile, just as chemistry is. I find it difficult to follow F. J. Dyson who, speaking at the dedication of a new physics laboratory in Princeton, recommended cosmic rays, biology, and astrophysics as orientations of research.[2] That some physicists move in these directions is excellent. But, as just mentioned, the more classical lines of physics research remain highly interesting and productive, and where will they be pursued in the spirit of pure science if not in the universities, where they must anyhow remain the backbone of physics teaching?

I have not said anything yet about the great open questions on the basic interactions. The fact that they are harder and that fewer people work on them do not make them less fundamental. If you bear with me a little longer, I shall devote my last remarks to them.

Experimental work on fundamental laws of physics proceeds along two main lines: push to higher precision in the study of known states of matter, or push matter into unknown states. In both directions, the hope is to find new effects revealing new features of the basic laws of motion and interaction. The high precision line has given one fundamentally new result since the Lamb shift of hydrogen triggered, in 1947, the renormalization technique for quantum electrodynamics; it is the 1964 discovery of CP violation in neutral kaon decay. For the rest, it impressively confirmed the validity of relativistic quantum mechanics and renormalized electrodynamics. The line of forcing matter into unusual states is most spectacularly represented by high energy physics. The main discoveries are the big new world of hadrons with its wealth, complexity, and curiously broken symmetries on the one hand, and on the other, the V-A coupling and the second neutrino for weak interactions.

The great disappointment is the absence of any visible progress toward a fundamental theory of strong and weak interactions. For the latter, we can at least recommend a crucial experiment, neutrino electron scattering at center of mass energies of a few hundred GeV; the difficulty is that it will probably be decades before we can perform it. For hadron physics, crucial experiments do not seem to exist; data gathering and model making are the order of the day. The pessimists may say that the models will not be better than the epicycles of Ptolemy the astronomer (second century A.D., not a member of the Ptolemaic dynasty mentioned above). But the optimists can retort that the remarkable mathematical regularities which are extracted from the data may turn out to be as significant as Kepler's discovery that the orbit of Mars is elliptic. As to the future Newtons of the strong and weak interactions, one hopes that they will come in due time but it might be naive to assume that they are already born.

It may be that the trouble lies in our inability to force matter in bulk into extreme conditions of density and energy and experiment with it there. What would we dream of in this direction? Even the heaviest of super-heavy nuclei are probably not very useful for fundamental laws of interaction. Now that superfluidity and superconductivity are understood in principle, bulk matter at very low temperature would not be our favorite either. Very high temperature at normal densities would become quite exciting if the average energy per volume is of GeV order or more, but this means a temperature of $T \gtrsim 10^{12}\,°K$. At lower temperatures, the really interesting densities would begin around 10^{16} g-cm^{-3}, when Fermi motion of baryons reaches the GeV range. Clearly, what we talk about is high-energy physics of bulk matter, and it looks perfectly utopian for laboratory conditions. In fact, the future here may lie in the heavens, through astrophysics, where we also can expect to learn something more about gravitation, the fourth fundamental interaction on which experimental research is just making a new and welcome start.

However this may be and while we have found no reason to doubt relativity and quantum mechanics, the mysteries around the basic interactions remain fascinatingly deep. Barring unlikely strokes of good luck, we shall still need much patient work at high-energy accelerators and much patient watching of the stars, detecting their photons, neutrinos, and gravitons, before we can hope to grasp the true picture. The experiments deal with such exotic phenomena, so far removed from normal conditions and so hard to measure, that the pace of progress can only be slow. Also this is a significant change in physics compared to earlier decades. Machines and detectors have grown to gigantic dimensions, but the scope of individual results is mostly very modest. We have no right to fool the taxpayer on this, because he pays our bill. And we should not fool ourselves nor our students, which may be harder.

But more important than the rate of progress is the undeniable fact that progress is being made, and we should constantly recall and reassert that elucidation of the fundamental laws remains the most essential task of physics. Our advanced society can certainly continue to afford the price of this pursuit, at least if, for the largest instruments, a sufficient selectivity is exercised and unnecessary duplication is avoided, so that in the long run a world-wide policy will become unavoidable. In a recent history of the Rochester Conferences, Marshak describes how an attempt was made in this direction as early as 1960 for high energy accelerators and how a cold war incident doomed it to failure.[3] He suggests that the time has come to take this matter up again. With CERN, high-energy physics has pioneered in European collaboration since almost twenty years ago. It is reasonable to expect that this field will be well suited to pioneer in world collaboration and that it will do so long before the end of the century. A very hopeful sign is the successful collaboration of CERN with the 70 GeV accelerator laboratory at Serpukhov near Moscow. There is a chance here for much more ambitious initiatives, and we should hope that it will be taken.

All this points to yet another change in the face of physics, a profoundly significant and entirely positive one. Twenty years ago, in the public mind, physics was unavoidably associated with nuclear weapons and the cold war. Now it gets more and more associated with world-wide international collaboration. Ironically, despite its many committees and complicated management, it is big science which is bringing this about by the very logic of its development. Isn't it obvious indeed that we should have one day the very best brains of all developed and that developing countries work together around the most advanced and costly research instruments which humanity will be able to afford and build to pursue its quest for fundamental knowledge?

One more question must be faced when we talk about the future. The best physics is done by young people, and we should therefore be concerned by Heisenberg's remark that most of the younger physicists show little interest for what he so aptly calls "die grossen Zusammenhänge", something like the unifying relationships.[4] This is certainly true, especially in comparison with the generation of scientific giants to which he belongs, and the unquestioned reign of specialization is unfortunate indeed. Not only does it foster intellectual shallowness, but it ignores the profound methodological unity of physics. Over and over again we see how techniques and concepts developed in one branch of physics find new and fruitful applications in other branches. The present trend to extreme specialization makes such cross-fertilizations ever more difficult to achieve, and one should try to counteract it.

Still, in another respect, the lack of proclaimed concern about the unifying relationships is also understandable and should perhaps not be the source of too much pessimism. Most young physicists simply do not work on the great problems of the basic interactions, and those who do learn soon that the road toward the ultimate goal is long and hard, the fare per mile very high, and the daily scientific rewards quite limited. When progress is slow, it may be better not to talk constantly about the glorious destination, all the more so that one travels without map and often falls victim to mirages. But here again, what really matters is the basic believe in progress; the basic conviction that in due time the fundamental laws can and will be elucidated. This conviction is shared by young and old alike, and now as in the past it provides the motivation and driving force to carry on.

REFERENCES

1. These data are taken from B. Farrington, "Science in Antiquity," 2d ed., chap. 8, Oxford University Press, 1969.
2. F. J. Dyson, "The Future of Physics," (lecture given at the dedication of Jadwin and Fine Halls, Princeton University, March 17, 1970), *Physics Today,* p. 23, Sept. 1970.
3. R. E. Marshak, The Rochester Conferences: The Rise of International Cooperation in High Energy Physics, Science and Public Affairs, *Bulletin of Atomic Scientists,* 26(6):92 (1970).
4. W. Heisenberg, Der Teil und das Ganze, *Gespräche im Umkreis der Atomphysik,* Piper Verlag, München, 1969.

CRITICAL PHENOMENA IN ALLOYS, MAGNETS, AND SUPERCONDUCTORS

Part One

INTRODUCTORY LECTURES

THE ISING MODEL IN TWO DIMENSIONS

Lars Onsager
Chemistry Department
Yale University
New Haven, Connecticut

The effort which led to the solution of the two-dimensional Ising Model started with more modest expectations.

In the Fall of 1940 Montroll brought the startling news that Wannier had located the critical point by way of a transformation. Wannier confirmed this when I saw him at a meeting during the Christmas week, and he gave me some sort of an idea what the result looked like. Still, my reconstruction during the Spring had to start almost from scratch.

For a simple try to associate order in a cold state with disorder in a hot one, it seemed reasonable to make the alternation of spins along a column correspond to an average product of two adjacent spins in a row. After a false start or two, the elements of a tentative transforming matrix were constructed by multiplying factors of the form

$$[\delta(\mu_j - \mu_{j+1}) + \mu'_j \delta(\mu_j + \mu_{j+1})]$$
$$= \tfrac{1}{4}(1 + \mu_j + \mu'_j - \mu_j \mu'_j)(1 + \mu_{j+1} + \mu'_j - \mu_{j+1}\mu'_j)$$

4 CRITICAL PHENOMENA

and normalizing the result thus

$$[(\mu)|L|(\mu')] = 2^{-(n+1)/2}\Pi_j[\delta(\mu_j - \mu_{j+1}) + \mu'_j\delta(\mu_j + \mu_{j+1})] \quad (1)$$

Applied to the transfer matrix which builds the quadratic crystal tier by tier, a similarity transformation by L produced the precise counterpart of Wannier's result for the screw construction. The matrix is orthogonal in the subspace of even functions; it annihilates odd ones, and the corresponding temperatures are given by the relation

$$\sinh\frac{2J}{kT}\sinh\frac{2J}{kT^*} = 1 \quad (2)$$

In addition, it was plain that the transformation interconverts the two factors which represent tier-wise and column-wise interaction. Moreover, it was noted that for a rectangular model with different interactions (J, J') along rows and columns, the transformation by L would still simply permute J and J' at the one temperature determined by the condition

$$\sinh\frac{2J}{kT}\sinh\frac{2J'}{kT} = 1 \quad (3)$$

and it seemed reasonable to suspect that this might be the critical condition for the rectangular model too. After a while it was indeed established that transformation by L always interconverts order and disorder. Thus, let an even eigenvector $\Psi(\mu)$ of the transfer matrix be expanded in products

$$\Psi[(\mu)] = 2^{1/2}a\left[1 + \sum C_{jk}\mu_j\mu_k + \sum C_{jklm}\mu_j\mu_k\mu_l\mu_m + \cdots\right] \quad (4a)$$

Then if

$$[L_1, +(\mu)] = \chi(\mu) \quad (4b)$$

the components of the latter are

$$\chi(+\cdots+) = \chi(-,\ldots-) = a$$
$$\chi[+\cdots+, -,\ldots -(m \text{ times}), +\cdots +] = ac_{j,j+m} \text{ etc.} \quad (4c)$$

Now if c_{jk} has a positive lower bound when sites j, k are far apart, then $+\psi$ must describe an ordered state, but χ must describe disorder because order is not compatible with a lower bound for the ratio

$$\chi(+ \cdots +, -, \ldots -, + \cdots +)$$

as the sequence of "wrong" $(-)$ spins is extended.

Meanwhile, for variety I studied the transfer matrix $\mathbb{W}(H, H)$ which describes the extension of a quadratic crystal along the diagonal. Even here transformation with the matrix (Eq. (1)) proved quite feasible and produced $\mathbb{W}'(H^*, H^*)$, i.e., the parameter was changed according to Eq. (2) and the matrix was transposed. It was just as easy to transform $\mathbb{W}, (H, H')$, which led to $\mathbb{W}'(H^*, H'^*)$. So far so good, but now it was possible to combine $\mathbb{W}$ with row or column interactions to produce the six-coordinated "triangular" lattice and the three-coordinated honeycomb graph. Transformation with L interconverted these. As these and a few more complicated cases were examined, it became evident that in every case the transformation would convert a graph to its topological dual. After that I satisfied myself that the transformation could be applied to any orientable graph, periodic or not, but without crossing vertex connections, with the strength of interaction along each vertex connection assigned independently of all the others. The result was a partition function for the dual graph, with each H replaced by the corresponding H^*. The restriction to orientable graphs was manifestly essential; the trick worked in two dimensions only!

Now to locate the transition points of the hexagonal models I needed one more tool, and that was where the star-triangle transformation came in. If we consider a vertex connected to just three neighbors—in general with different interaction constants—we can sum over the two values of its spin to obtain

$$2 \cosh(H_1 \mu_1 + H_2 \mu_2 + H_3 \mu_3)$$

This is an even function of the spins; the same is true of its logarithm, so that an expansion in spin products takes the form

$$\log[2 \cosh(H_1 \mu_1 + H_2 \mu_2 + H_3 \mu_3)]$$
$$= \log N(H_1, H_2, H_3) + H_{12} \mu_1 \mu_2 + H_{23} \mu_2 \mu_3 + H_{31} \mu_3 \mu_2 \quad (5)$$

Sum over alternate vertices in the honeycomb graph thus leads to the partition function of a triangular net with interaction constants (H_{12}, H_{23}, H_{31}). After that the dual transformation takes us back to the honeycomb graph with interaction constants $(H_{12}^*, H_{23}^*, H_{31}^*)$. The relations between these and (H_1, H_2, H_3) involve just the hyperbolic trigonometry of a totally improper triangle: The quantities of a hexagon whose corners are all right angles. The two triples of geodetics are mutually polar, and all the rules of spherical trigonometry have counterparts. For example, the ratios of hyperbolic sines are the same for all three pairs of opposite sides. At the critical points this ratio equals unity. For equal interactions between vertices of order $m = (3, 4, 6)$ the critical condition could be summarized in the form

$$2 \tan^{-1}(\tanh H) = gd2H = \frac{\pi}{m} \qquad (6a)$$

which was ultimately generalized to

$$\Sigma gd2H_j = \pi \qquad (6b)$$

The obvious is not always trivial.

Of course, the star-triangle transformation also relates the transfer matrix $W(H, H')$ which extends a rectangular crystal along the main diagonal to the two factors which appear when the crystal is built along a principal axis. At this stage of the development the latter were already described in the exponential form

$$V_1(H) = (2 \sinh 2H)^{n/2} \exp(H^* B)$$
$$V_2(H') = \exp(H'A) \qquad (7a)$$

with

$$A = \sum_j s_j s_{j+1}$$

$$B = \sum_j C_j$$

$$\{s_j, \psi[(\mu)]\} = \mu_j \Psi[(\mu)]$$

$$C_j \psi(\ldots \mu_{j-1}, \mu_j, \mu_{j+1} \ldots) = \psi[\ldots \mu_{j-1}, (-\mu_j), \mu_{j+1}] \qquad (7b)$$

The star-triangle transformation implies

$$W(H_1, H_2) V_1(H_3) = N(H_1, H_2, H_3) V_2(H_{12}) W(H_{13}, H_{23}) \quad (8a)$$

as well as

$$V_1(H_3) W(H_1, H_2) = N(H_1, H_2, H_3) W(H_{13}, H_{23}) V_2(H_{12}) \quad (8b)$$

If we differentiate these relations with regard to H_3^* and pass to the limit $H_3^* = 0$ (that is, $H_3 \to \infty$), we obtain the pair of differential equations

$$\sinh 2H_2 \cosh 2H_2 \frac{\partial W}{\partial H_1} + \cosh 2H_1 \sinh 2H_2 \frac{\partial W}{\partial H_2}$$

$$- n \cosh 2H_1 \cosh 2H_2 W = - BW + \sinh 2H_1 \sinh 2H_2 WA$$

$$= - WB + \sinh 2H_1 \sinh 2H_2 AW \quad (9)$$

whence W commutes with a linear combination of A and B

$$[W(H, H'), (B + kA)] = 0$$
$$k = \sinh 2H \sinh 2H' \quad (10)$$

It was interesting enough to discover that all correlations along the diagonal are determined by the single parameter k, which was already known to equal unity at the critical point. Beyond that, the discovery invited another effort: With the eigenwert problem of W reduced to the simpler one of $(B + kA)$, why not solve it for bias-cut strips 1, 2, 3, 4, 5 sites wide, which would involve no more than quartic equations?

Luck smiled. The first quartic (for $n = 4$) turned out to be a biquadratic. In retrospect this could have been foreseen inasmuch as a similarity transformation with $s_1 s_2 \ldots s_n C_2 C_4 \ldots C_n$ will just reverse the signs of both A and B for even n. The four characteristic numbers for even eigenfunctions of full dihedral symmetry were

$$\pm 8^{1/2} [1 + k^2 \pm (1 + k^4)^{1/2}]^{1/2}$$

$$= \pm 2 \left[1 - 2k \cos\left(\frac{\pi}{4}\right) + k^2 \right]^{1/2} \pm 2 \left\{ \left[1 - 2k \cos\left(\frac{3\pi}{4}\right) + k^2 \right]^{1/2} \right\} \quad (11)$$

and (for $H = H'$) the corresponding characteristic values of W

8 CRITICAL PHENOMENA

contained the factors

$$[1 + k \pm (1 - 2k \cos\omega + k^2)^{1/2}], \quad \omega = (2r - 1)\frac{\pi}{n} \qquad (12)$$

This suggested a system. Previous results for $n = 2, 3$ conformed, and the computations for $n = 5, 6$ were greatly facilitated by the correct advance guess. For $n = 7$ the system required one invariant constraint on the eigenvectors; for $n = 8$ three. These were soon found as attention was focused on configurations like (+++--+-) which have no reflexion symmetry. Eigenvectors of $(B + kA)$ and W belonging to diverse representations of the dihedral symmetry group were investigated through $n = 6$; all fell into a similar pattern. For odd eigenvectors the angles ω were even multiples of π/n instead of the odd multiples which appeared for even eigenvectors. While this sort of checking was in progress, Wannier was allowed to announce the result as a conjecture at a conference arranged by the New York Academy in February 1942, and he laid a four-to-one bet that it be right.

To prove the generality of the result took more invention, however. Finally, it dawned on me that if A and B were to be sums of operators

$$A = X_0^* + 2 \sum X_r^{*0} + X_n^*$$
$$B = X_0 + 2 \sum X_r + X_n \qquad (13)$$

with operators commuting except the pairs (X_r, X_r^*) then a result like Eq. (11) would be accounted for. Moreover, if so, then to find different linear combinations of the X_r it would suffice to compute successive commutators

$$[A, B], \ [(A, B), A]$$

etc. In succession appeared

$$[B, A] = 2 \sum (C_j s_j s_{j+1} + s_j s_{j+1} C_{j+1})$$
$$A_2 = \sum s_j C_{j+1} s_{j+2} \qquad (14)$$
$$A_3 = \sum s_j C_{j+1} C_{j+2} s_{j+3}$$

etc.

The starting operators fell into line as

$$B = -A_0$$

$$A = A_1$$

and the cycle closed at $A_{2n} = A_0$. With the conjecture (Eq. (13)) confirmed, the rest was mopping up. While the complete spectra of the transfer matrices were fairly evident from the initial explorations, a few years were to pass before the conjectured structure could be verified by a rigorous proof. Fortunately, it was not at all difficult to find the subspace which contains the principal eigenvector; this also had to contain the unique principal eigenvectors of V_1 and V_2. In organizing the material for publication, the eigenwert problem for $V_1 V_2$ (of Eq. (7)) seemed a little easier to explain than that for W, and as the work took shape, it proved the better route for planting flowers by the wayside. The reduction of elliptic integrals is usually a chore; on this occasion the elegance of the various relations greatly relieved the tedium.

The results are in the literature. Several other derivations have been found, and I see no need to repeat what is now quite widely known.

By the summer of 1945 I had a proof for the conjectured full spectrum of the transfer matrix $V_1 V_2$ (cf. Eq. (7)) and partial information about that of W. The approach was somewhat laborious: The combination

$$A_1 + A_{-1} = \prod (1 - C_j C_{j+1}) s_j s_{j+1} \qquad (15)$$

(of Eq. (14)) commutes with $A_0 = -B$. The eigenwert problem of the former defines a simple system of scalar spin waves (Fermi statistics, Bose symmetry). These are homogeneous functions of the spins μ, i.e., eigenvectors of B, at the same time eigenvectors of the operator described by Eq. (15) and in fact of all X_r. Other operators of the algebra generated by A and B can add or remove "Cooper pairs", whichever is possible. Before long, however, Bruria Kaufman had developed a much better strategy.

At Columbia University she first asked Willis E. Lamb to direct her work on order-disorder problems; but he was much too heavily engaged in an experimental effort, and I was asked to assume the responsibility. Unable to talk her out of the idea, I suggested that she explore the structure of W as well as the effect of joining crystal ends on a torus with a twist, etc. She made good progress; but as she got her bearings she was more intrigued by the ubiquitous trigonometric relations and decided to look for a possible connection with spinor theory. Why not? In fact, it seemed like very good sense, and so it was. The operators V_1, V_2 etc. were indeed identified as spin representations of complex-orthogonal transformations on a set of basic elements in the algebra. Now we only had to transform a set of 2n by 2n orthogonal matrices to canonical form, and the existence of the dual transformation was all the indication *she* needed to find the right line. By the summer of 1946 she had a beautifully compact computation of the partition function, bypassing all tedious detail.

By itself that was only a more elegant derivation of an old result; but the approach looked powerful enough to produce a few new ones. Very well, how about correlations? That, indeed, required almost everything that had been developed before. Now we had to specify explicitly the planes and the angles of the commuting rotations; this proved a bit tedious but the road was known. The trick was to transform spinors of various even order (2 for $s_1 s_2$, 4 for $s_1 s_3$ etc.) to the canonical frame of reference; then the Eq. (11) element would give us the desired expectation value. The successful performance of this task was another proof of her excellent intuition. The answers took the form of recurrent determinants, where the angles $\delta(\omega)$ of rotation involved in transforming the transfer matrix to diagonal form determined the elements according to the scheme

$$e^{i\delta} \equiv \sum c_m e^{im\omega}$$

$$D_{m+1} = \begin{vmatrix} c_0 & c_1 & \cdots & c_m \\ c_{-1} & c_0 & \cdots & c_{m-1} \\ \cdots & \cdots & \cdots & \cdots \\ c_{-m} & c_{1-m} & \cdots & c_0 \end{vmatrix} \qquad (16)$$

and $<s_j s_{j+m}>$ would involve the computation of D_m or in some cases more than one determinant of this type. For correlations along the diagonal the generating function is particularly simple

$$e^{i\delta} = \left(\frac{k - e^{i\omega}}{k - e^{-i\omega}} \right)^{1/2} \qquad (17)$$

$$<s_{11} s_{mm}> = D_{m-1}(e^{i\delta})$$

Now if we could compute the limit D_∞, we should know the degree of long-range order $\bar{s}$

$$D_\infty = (\bar{s})^2 \tag{18}$$

At the time it was only known that the limit of $(D_m)^{1/m}$ would be the geometric mean of the generating function. Otherwise it made sense that by Hadamard's estimate the value of the determinant could not exceed unity.

After a while, however, I realized that the associated difference equation was just the analog of a Milne integral equation, and I tried Wiener-Hopf technique. Above the Curie point $k < 1$ and the generating function may be written

$$e^{i\delta} = e^{i\omega} \left(\frac{1 - ke^{-i\omega}}{1 - ke^{i\omega}} \right)^{1/2}$$

Here, if to the first column I added multiples of the following, with coefficients taken from the power series of the cofactor

$$(1 - ke^{i\omega})^{1/2} = \sum a_m e^{-mi\omega}$$

I could make the first column vanish in the limit $m \to \infty$. Still utilizing essentially the Wiener-Hopf scheme, I could construct a resolvent for the difference equation below the Curie point. A little more manipulation of this sort produced a regular integral equation. An elliptic substitution led to a kernel of the type

$$K(u, v) = K_1(u + v) + K_2(u - v)$$

the eigenfunctions were of the form

$$e^{(2n+1)iu\pi/2K}$$

(K = complete elliptic integral) and the characteristic numbers were computed from the Fourier Series. The product turned out to be the ratio

$$\left(\frac{\vartheta_4}{\vartheta_3} \right)^{1/2} = \left[1 - \left(\frac{1}{k^2} \right) \right]^{1/4}$$

This was the basis for the first announcements of the result. After that, however, I tried to find a more general method for the evaluation of such Toeplitz determinants—and as soon as I tried rational functions of the form

$$\frac{\Pi(1 - \alpha_j e^{i\omega})}{\Pi(1 - \beta_k e^{-i\omega})}$$

(with all $|\alpha|, |\beta| < 1$) the general result stared me in the face. Only, before I knew what sort of conditions to impose on the generating function, we talked to Kakutani and Kakutani talked to Szegö and the mathematicians got there first.

ONE-DIMENSIONAL DELTA FUNCTION INTERACTION

C. N. Yang
Institute for Theoretical Physics
State University of New York
Stony Brook, New York

Since the question has been raised a few minutes ago about the early calculation of the spin-spin correlation in the Ising model, it might not be inappropriate for me to spend a few minutes to tell you about my involvement with the magnetization problem in the same model.

I became interested in the order-disorder problem when I was a graduate student in China during the war. Professor J. S. Wang, my master's thesis adviser, interested me in the order-disorder problem which was extensively studied in the mid-1930's, especially in Fowler's group in Cambridge, England. In the late 1940's, when I became a graduate student at Chicago, I made several attempts to understand the solution of Onsager for the Ising problem. I was probably not alone in bogging down in the complicated algebra.

Later on, in 1950, when I was in Princeton, I had a conversation with J. M. Luttinger during a bus ride from Palmer Square to the Institute for Advanced Study. He told me that Onsager's solution could be easily understood by the new method of Kaufmann and Onsager in which one introduces anti-commuting operators P and Q. That was the beginning of my understanding of Onsager's solution. In early 1951, it occurred to me that since the solution gave details of *all* the eigenfunctions, not just the eigenfunction with the largest eigenvalue, it should be possible to make use of such additional information. So, I launched into a computation of the magnetization. The computation turned out to be extremely involved. I remember that the effort was repeatedly stalled; and during several months in the Spring of 1951, I had given it up several times only to look at it again after a couple of weeks when irritation with the frustration had somehow subsided. I finally arrived, sometime in June, at the correct result which looked surprisingly simple. Since the calculation was rather involved and since some limiting procedure used had not been rigorously justified, I only became really convinced of the validity of the result when I compared its series expansion in powers of the parameter x up to x^{12} with known exact results and found complete agreement.

1. INTRODUCTION

A number of one-dimensional quantum mechanical problems and two-dimensional classical problems have in recent years been solved using a method which has come to be known as Bethe's hypothesis. These problems include:

1. spin-spin interaction in one dimension
2. quantum lattice gas in one dimension
3. two-dimensional ice problem
4. two-dimensional ferroelectric models
5. one-dimensional δ-function interaction

I shall try this morning to describe the essential steps that one goes through in solving such problems, using a one-dimensional boson system with δ-function interaction as an example. The essential steps in solving the other problems are similar in nature to those of this example. Each problem, however, involves specific variations of details, which have to be tackled individually.

2. FOUR STEPS IN THE SOLUTION OF THE ONE-DIMENSIONAL BOSON PROBLEM WITH δ-FUNCTION INTERACTION

The problem here is the one-dimensional Hamiltonian for N bosons

$$H = -\sum_i \frac{\partial^2}{\partial x_i^2} + 2c \sum_{i>j} \delta(x_i - x_j), \quad i,j = 1, 2, \ldots, N \quad (1)$$

To find the ground state wave function of this system, one assumes that in the region $x_1 < x_2 \cdots < x_N$ (for either a periodic box from $x = 0$ to $x = L$ or for infinite space) the wave function is of the form of a function of $N!$ terms, each of which is a simple exponential function of the x's with wave numbers which represent a permutation of N nonidentical k's: $k_1, k_2, \ldots, k_N$. In other words, one assumes

$$\psi = \sum_P \alpha_P \exp[i(k_{P1}x_1 + k_{P2}x_2 + \cdots + k_{PN}x_N)] \quad (2)$$

if

$$x_1 < x_2 \cdots < x_N \quad (3)$$

Here $P1, P2, \ldots, PN$ is a permutation P of $1, 2, \ldots, N$. The constants α_P are to be determined by matching this wave function with the wave function in other regions than in Eq. (3), which are obtained from Eq. (2) by the requirement of boson statistics.

Equation (2) is called Bethe's hypothesis. Notice that it's no more than the old reflection principle: In the N dimensional coordinate space $x_1 .. x_N$ the potential energies are confined to hyperplanes $x_i = x_j$. A plane wave repeatedly reflected from these hyperplanes gives rise to altogether $N!$ plane waves which are the individual terms in Eq. (2). It is to be emphasized, however, that it's rather accidental that this reflection principle (i.e., Bethe's hypothesis) gives the eigenfunction of the problem. If one had changed the Hamiltonian by giving the N particles different masses and/or different interaction strengths c_{ij} between different pairs of particles x_i and x_j, then Bethe's hypothesis would not hold for the ground state.

To satisfy the Schrodinger equation at the hyperplanes $x_i = x_j$ the constants α_P in Eq. (2) should obey

$$\frac{\alpha_{P'}}{\alpha_P} = -\exp[i\theta(k'-k)] \tag{4}$$

for every pair (P and P') of permutations for which

$$P'1\,P'2\ldots P'N$$
$$P1\,P2\ldots PN$$

are identical except for two neighboring columns, looking like

$$\ldots k'k\ldots$$
$$\ldots kk'\ldots$$

Notice that no two k's are identical. The function θ is defined by

$$\theta(x) = -2\tan^{-1}\frac{x}{c} \qquad |\theta(x)| < \pi \tag{5}$$

An important question is the consistency of Eq. (4). (There are $(N-1)(N!)$ equations for only $N!$ coefficients α_P.) The answer is that they are consistent. In fact, for any set of k's, all nonidentical, Eq. (4) gives a unique set of α's up to a normalization factor. (If one tries to use Bethe's hypothesis (Eq. (2)) for a case of unequal masses or unequal interaction strengths as mentioned above, one would find that the result (Eq. (4)) becomes inconsistent.)

The next step to be taken if we consider cyclic boundary conditions is to match the wave function (Eq. (2)) at the point

$$x_1, _2, \ldots, x_{N-1}, L$$

and the point

$$0, x_1, x_2, \ldots, x_{N-1}$$

Such matching leads to

$$\exp(ikL) = (-1)^{N-1}\exp\left[i\sum_{k'}\theta(k-k')\right] \tag{6}$$

for every k. There is no essential difficulty in carrying out this step in any of the problems mentioned in the introduction.

The third step is to take the logarithm of Eq. (6), yielding

$$kL = 2\pi I_k + \sum \theta(k - k') \tag{7}$$

where

$$I_k = \text{integer} \quad \text{if } N = \text{odd}$$

$$I_k = \text{integer} + \frac{1}{2} \text{ if } N = \text{even} \tag{8}$$

The numbers I_k are quantum numbers of the problem. To determine them for the ground state, one uses continuity arguments with respect to the variable c^{-1}. This is a characteristic step in these problems. A much more difficult problem arises if one asks about the quantum numbers I_k for the excited states. Here, much remains unknown for many of the problems.

The fourth step is to investigate the behavior of the solutions of Eq. (7) in the limit of infinite volume at a fixed density. In the repulsive boson case, one can rigorously prove for the ground state that the k's approach a distribution along the real axis with $L\rho(k)\,dk$ k's falling in the interval dk. This behavior allows one to write down the limit of (7) as an integral equation which, after some manipulation, leads to

$$1 = 2\pi\rho(k) - 2c \int_{-Q}^{Q} \frac{\rho(k')\,dk'}{c^2 + (k - k')^2} \tag{9}$$

The density of particles is

$$\frac{N}{L} = \int_{-Q}^{Q} \rho\,dk \tag{10}$$

and the energy per unit length of space is

$$\frac{E}{L} = \int_{-Q}^{Q} \rho k^2 \, dk \qquad (11)$$

While these four steps remain the essential ones for all of the problems, the execution of them for different problems presents difficulties of varying degrees, some of which are not yet solved.

3. ONE-DIMENSIONAL FERMION PROBLEM WITH δ-FUNCTION INTERACTION

For this problem, it turns out that it is algebraically advantageous to first deal with any symmetry of the wave functions for the Hamiltonian (Eq. (1)). One would then write down Bethe's hypothesis (Eq. (2)) for every order of the coordinates x_i. Equation (3) is, of course, only one of such possible orders. These different orders will be designated by the permutation Q. Thus, instead of the $N!$ coefficients α_P in (Eq. (2)), we now have $(N!)^2$ coefficients α_{PQ}. It is convenient to write these coefficients α_{PQ} as $(N!)$ column vectors α_P, each with $N!$ entries. In such a notation, Eq. (4) becomes replaced by $\alpha_{P'} = O\alpha_P$ where O is some operator.

In carrying out the next step, i.e., the imposition of cyclic boundary conditions, one arrives at, instead of Eq. (6), equations of the form

$$e^{ik_j L} \alpha_I = \Omega_j \alpha_I \qquad (12)$$

where I is the identity permutation and Ω_j is an operator. The solution of the eigenvalue problem (Eq. (12)) presented an essential difficulty which was resolved using Bethe's hypothesis (in a generalized form) a second time for a fermion problem with spin ½. For a fermion problem with spin J, it turns out that one has to use the generalized Bethe's hypothesis $2J$ times in addition to the original Bethe's hypothesis.

Once this is done, there is no difficulty in arriving at the ground state energy for the repulsive case $(c > 0)$.

To deal with the attractive case $(c < 0)$ for the ground state energy, a complication arises in the fourth step. Here, the k's move into the complex plane and execute rather involved paths. Fortunately, numerical calculation in special examples demonstrated that in the limit of infinite volume at a fixed density one could write down integral equations which are generalizations of Eq. (9). The complicated manner in which the k's move about in the complex plane indicates that the infinite volume limit and the limit $c \to 0-$ do not commute.

Recently, the ground state energy problem for a mixture of fermions and bosons was also solved. One merely studies the ground state energy for the Hamiltonian problem (Eq. (1)) with a somewhat complicated symmetry.

4. SOME UNSOLVED PROBLEMS

I should now like to mention three particular problems for the δ-function interaction model in one dimension which are very tantalizing but which are not yet solved.

1. The fermion problem at a finite temperature for both attractive and repulsive cases. The corresponding problem for the boson case with repulsive interaction was already solved. For the fermion problem, the main difficulty lies in steps 3 and 4.

2. Although the ground state wave function is known in great detail, it does not follow that one knows immediately the two particle-reduced density matrix, three particle-reduced density matrix, etc. This is a very important problem which, however, appears to be rather difficult. The point is that the wave function (Eq. (2)) is analytic and simple only in each of the $N!$ sections of phase space. Integration over redundant variables is therefore a very tedious process. Since one is especially interested in the long-range behavior of the off-diagonal elements of these reduced density matrices, this problem deserves further study.

3. Let us now concentrate on the boson case with repulsive interaction. The ground state energy is given by Eqs. (9) to (11). One could study also the excitations near the ground state (this was already done in 1963 by Lieb). Near the ground state the energy levels can be expressed in terms of certain "excitons." Now these excitons interact with each other, as one could demonstrate in either of two ways:

a. If the excitons do not interact, the thermodynamics of the system would be given by that of a system of noninteracting excitons with a fixed spectrum independent of the temperature. This is not in agreement with the known thermodynamic behavior of the problem.

b. One could compute the energy of two excitons in a box of length L to the order $(1/L)^0$. The energy is the same as the sum of the energies of the two excitons. But, to the order $1/L$ there is a correction term to this statement. The interpretation of this correction term is clearly that the excitons interact.

We have therefore a system with infinite degrees of freedom with certain "elementary particles," i.e., excitons which interact with each other. Further knowledge about the dynamics of these excitons would throw light on problems concerning systems of infinite degrees of freedom.

REFERENCES

S matrix
1. McGuire, J. B.: *J. Math. Phys.*, **5**:622 (1964).
2. Brezin, E., and J. Zinn-Justine: *Compt. Rend. Acad. Sci. Paris*, **B263**:670 (1966).
3. Berezin, F. A., and V. N. Sushko: *Zh. Eksperim. i Teor. Fiz.*, **48**:1293 (1965); *Soviet Phys. JETP*, **21**:865 (1965) (Eng. trans.).
4. Berezin, F. A.: *Moscow U. Vestnik*, **1**:21 (1964).
5. Yang, C. N.: *Phys. Rev.*, **168**:1920 (1968).

Bosons, $T = 0$
6. Lieb, E. H., and Liniger: *Phys. Rev.*, **130**:1605 (1963).
7. McGuire, J. B.: *J. Math. Phys.*, **5**:622 (1964).
8. Lieb, E. H.: *Phys. Rev.*, **130**:1616 (1963).

Fermions, $T = 0$
9. Yang, C. N.: *Phys. Rev. Lett.*, **19**:1312 (1967).
10. Sutherland, B.: *ibid.*, **20**:98 (1968).
11. Gaudin, M.: *Phys. Lett.*, **24A**:55 (1967); thesis, Faculté des Sciences d'Orsay, University of Paris, Nov. 1967.
12. Lai, C. K., and C. N. Yang: To be published.

Mixture of Fermions and Bosons, $T=0$
13. —— and ——: *Phys. Rev.*, in print.

Bosons, $T > 0$
14. Yang, C. N., and C. P. Yang: *J. Math. Phys.*, **10**:1115 (1969).

DISCUSSION *on Paper by C. N. Yang*

R. B. GRIFFITHS: What are the prospects for calculating correlation functions and Green's functions in this one-dimensional system, e.g., bosons with delta-function interactions? They would be of interest (for example) in question of superfluidity in one dimension.

C. N. YANG: The prospects did not seem good to me when I looked at this very important problem. The main difficulty lies in the integration over the redundant space coordinates. Since phase space is cut up into $N!$ regions, this integration is very messy.

B. WIDOM: You spoke of finding different solubilities of the Bosons in the Fermi fluid and of the Fermions in the Bose fluid. But does not finite solubility imply phase equilibrium? Yet you said that, as expected, there is no phase transition in your one-dimensional system.

C. N. YANG: I am sorry I used the wrong term. I should have said the energy of mixing instead of solubility.

G. FALK: Could you give some hints as to what the function $\Sigma(\kappa)$ looks like, especially as to its dependence on C?

C. N. YANG: When $C = +\infty$, the system is equivalent to a free Fermion system. So $\epsilon = -\mu + \kappa^2$. For $C \to 0+$ the limit is more complicated; to the lowest order the system approaches a free Boson system.

M. KAC: You mentioned the usefulness of continuity arguments in the limit $C \to 0$. How about the limit $C \to \infty$ in which exact results of A. Lenard are known?

C. N. YANG: The continuity argument is very important in the limit $C \to +\infty$ also. It provides in fact the essential argument for the boson case.

THE ROLE OF MODELS IN UNDERSTANDING PHASE TRANSITIONS

Mark Kac
The Rockefeller University
New York, New York

1. INTRODUCTION

I must begin with a double apology. First, because my topic is only weakly (very weakly, in fact) related to the subject of this conference (critical phenomena) and, second, because the title is even less descriptive of the content of the talk than most titles usually are. Assuming that I am forgiven on both counts let me begin.

Since the detailed nature of interactions in real physical systems is usually not known and since even the most far reaching simplifications still lead to enormous mathematical difficulties we must, in our attempts to understand phase transitions, rely upon a narrow class of models which are balanced (precariously!) between realism and solubility.

Since exact solubility is a rare event indeed, one latches on to exactly soluble models with vigor and tenacity not wholly justified by the illumination one might ultimately gain from them. In so doing one also runs the risk of attributing reality to accidental features of models.

What is "real" and what is accidental is not a question which has received enough attention and I should like to plead that it be taken up systematically and in earnest.

2. HOW MANY "FLAT PARTS"?

To dramatize the situation let me mention a recent remark by J. F. Nagle as extended by Hemmer and Stell.[1]

It has been known for some time that the equation of state of a one-dimensional gas with a hard core of size δ and an attraction of the form $\gamma\phi(\gamma r) \, (r > \delta)$

$$\phi > 0, \quad \int_0^\infty \phi(r)\,dr < \infty \tag{1}$$

becomes, in the limit $\gamma \to 0$, the van der Waals equation with the Maxwell construction yielding (for sufficiently low temperatures) *one* flat part of the isotherm.

By augmenting the hard core with a sufficiently strong soft core extending over the range $\delta < r < 2\delta$, Hemmer and Stell show that *any number* of flat parts can be obtained each with its own critical temperature.

For lattice gases they give a compelling and elegant argument that a second flat part can be expected if a sufficiently strong nearest neighbor repulsion is introduced in addition to the hard core and the attractive interaction. The argument is also so simple that I shall repeat it here.

Assume *infinite* repulsion for nearest neighbors (as well, of course, as the hard core, i.e., no site can be occupied by more than one particle). Then one might expect a first-order phase transition in two and more dimensions (in one dimension for a sufficiently large-range attraction the existence of such a phase transition was indeed proved by Dyson[2]). Since the density is less than ½ the flat parts of the isotherm will by necessity be confined to the interval $0 < \rho < 1/2$. If the nearest neighbor repulsion is finite but very large one expects (by continuity) only slight modifications of the critical point and the flat parts. But now the density can cover the full range $0 < \rho < 1$ and moreover by particle-hole duality if there is a flat part in $0 < \rho < 1/2$ there must also be a flat part in $1/2 < \rho < 1$. Thus for temperatures below the (modified) critical we should have two flat parts.

Are we dealing with spurious phenomena attributable to either dimensionality or to the lattice nature of the gas or are the new transitions ghostly reflections of hitherto unperceived reality?

One final remark. The discussion above makes it clear that the beautiful theorem of Yang and Lee on the roots of the grand partition function of the simple lattice gas is a happy accident. As soon as a sufficiently strong nearest neighbor repulsion is added, the distribution of roots becomes more complicated.

3. CORRELATION AT INFINITY: HIGH TEMPERATURES

For my next illustration I turn to the two-dimensional Ising model with nearest neighbor interactions.

This, of course, is sacred ground, but at the risk of sounding both sacrilegious and subversive I should like to suggest that exact solubility of the model is perhaps not as unmitigated a blessing as one has tended to assume.

I do not want to detract from either the beauty or the usefulness of exact results or from the power and ingenuity of the methods used in obtaining them. But it seems that the time has come to understand *qualitatively* and with one's "fingers" (to borrow a favorite expression of Professor Uhlenbeck) those features of the model which may have a bearing on physical reality. In particular, it would be of paramount importance to understand in this way the logarithmic singularity of the specific heat and, in particular, to what extent the nature of singularity is tied to the dimensionality of the model.

To be a little more specific let me discuss what appears to be a simpler problem namely the long-range behavior of the correlation function.

For temperatures above the Curie temperature T. T. Wu[3] has shown that

$$R(r) \sim \frac{e^{-\kappa r}}{\sqrt{r}}, r \to \infty \qquad (2)$$

where $R(r)$ is the correlation function between two spins placed either in the same row or the same column at distance r apart.

26 CRITICAL PHENOMENA

The result has the familiar Ornstein-Zernike appearance, the $\sqrt{r}$ being directly attributable to the model being two-dimensional. Yet Wu's calculation is of great complexity and ingenuity and is based on the representation of the correlation function as a Toeplitz determinant, a happy though no doubt fortuitous accident.

Can one convince oneself without relying upon the not wholly justified assumptions underlying the Ornstein-Zernike derivation that the form Eq. (2) is due to two-dimensionality of the model?

The answer is yes, and one can proceed as follows.

Consider a two-dimensional Ising model on a square ($N \times M$) lattice with the interaction energy

$$E = - \sum_{\substack{1 < k < k' < N \\ 1 < l < l' < M}} v(k, l; k', l') \mu_{k,l} \mu_{k',l'} \qquad (3)$$

where

$$v(k, l; k', l') = J \exp(-\sigma|k - k'|)[\delta_{l,l'} + \tfrac{1}{2}(\delta_{l',l+1} + \delta_{l',l-1})] \qquad (4)$$

This is the so-called Model A (Kac and Helfand[4]) and the correlation $R_M(r)$ defined as

$$R_M(r) = \lim_{N \to \infty} <\mu_{k,l} \mu_{k+r,l}> \qquad (5)$$

can be expressed in terms of the eigenvalues and eigenfunctions of the kernel

$K(x, y) =$

$$\exp\left[-\tfrac{1}{2} Q(x)\right] \prod_{k=1}^{M} \frac{\exp[-(y_k - x_k)^2/4 \sinh \sigma]}{2\sqrt{\pi \sinh \sigma}} \exp\left[-\tfrac{1}{2} Q(y)\right] \qquad (6)$$

where

$$Q(x) = \tfrac{1}{2} \tanh \tfrac{\sigma}{2} \sum_{1}^{M} x_k^2 - \sum_{k=1}^{M} \log \cosh \sqrt{\tfrac{\nu}{2}} (x_k + x_{k+1}) \qquad (7)$$

and where, as usual

$$\nu = \frac{J}{kT} \tag{8}$$

In fact (see, e.g., Kac,[5])

$$R_M(r) = \sum_{j=2}^{\infty} \left(\frac{\Lambda_j}{\Lambda_1}\right)^r \left(\int \tanh \sqrt{\frac{\nu}{2}} (x_l + x_{l+1}) \psi_1(x) \psi_j(x) dx\right)^2 \tag{9}$$

where the Λ_j's are the eigenvalues and the ψ_j's the corresponding (normalized) eigenfunctions of Eq. (6).

It goes without saying that

$$R(r) = \lim_{M \to \infty} R_M(r) \tag{10}$$

For high temperatures ($\nu \ll 1$) we use the approximation

$$Q(x) \sim \frac{1}{2} \tanh \frac{\sigma}{2} \sum_{k=1}^{M} x_k^2 - \frac{\nu}{4} \sum_{k=1}^{M} (x_k + x_{k+1})^2 \tag{11}$$

and the resulting (approximate) integral equation becomes easily soluble. The reason for this is that by an orthogonal change of variables

$$x_l = \sum_{s=1}^{M} a_{ls} \bar{x}_s$$

$$y_l = \sum_{s=1}^{M} a_{ls} \bar{y}_s \tag{12}$$

with

$$a_{ls} = \frac{1}{\sqrt{M}}, \; s = 1$$

$$a_{ls} = \frac{2}{\sqrt{M}} \sin \frac{\pi s(l-1)}{M}, \; s \text{ even}$$

$$a_{ls} = \frac{2}{\sqrt{M}} \cos \frac{\pi(s-1)(l-1)}{M}, \; s \text{ odd} (\neq 1) \tag{13}$$

28 CRITICAL PHENOMENA

the kernel becomes

$$\prod_{k=1}^{M} \exp\left(\pm \frac{1}{2} \omega_k^2 \bar{x}_k^2\right)\left[\exp \frac{(\bar{y}_k - \bar{x}_k)^2/4 \sinh\sigma}{2\sqrt{\pi} \sinh\sigma}\right] \exp\left(-\frac{1}{2} \omega_k^2 \bar{y}_k^2\right) \quad (14)$$

where

$$\omega_k^2 = \left(\frac{1}{2} \tanh\frac{\sigma}{2} - \nu\right) + \nu \sin^2 \frac{2\pi}{M}\left[\frac{k+1}{2}\right], \quad k = 0, 1, 2, \ldots, M-1 \quad (15)$$

and it is understood that for convenience M is assumed to be odd.

We now approximate the Λ_j's and the ψ_j's by the eigenvalues and normalized eigenfunctions of the approximate kernel and use the approximation

$$\tanh \sqrt{\frac{\nu}{2}}(x_l + x_{l+1}) \sim \sqrt{\frac{\nu}{2}}(x_l + x_{l+1}) \quad (16)$$

It is clearly advantageous to use the variables $\bar{x}_k$ and we finally obtain

$$R_M(r) \sim \frac{\nu}{2} \sum_{j=1}^{\infty} \left(\frac{\bar{\Lambda}_j}{\bar{\Lambda}_l}\right)^r \int [\Sigma(a_{l,r} + a_{l+1,r}) \bar{x}_r] \bar{\psi}_0 (\bar{\psi}_j d\vec{x})^2 \quad (17)$$

where the bars over Λ's and ψ's indicate that we are dealing with the eigenvalues and eigenfunctions of Eq. (15).

One obtains easily that

$$\psi_o = \prod_{k=1}^{M} \left(\frac{\sqrt[4]{\alpha_k}}{\sqrt[4]{\pi}}\right) \exp\left(-\frac{1}{2} \sum_{1}^{M} \alpha_k \bar{x}_k^2\right) \quad (18)$$

where

$$\alpha_k = \omega_{k-1} \sqrt{\frac{1}{\sinh\sigma} + \omega_{k-1}^2} \quad (19)$$

and ψ_j's which yield nonzero matrix elements in ( ) are already

$$\bar{\psi}_j = \sqrt[4]{\alpha_j}\,\bar{x}_j\,\bar{\psi}_1, j = 1, 2, \ldots, M \tag{20}$$

Thus

$$R_M(r) \sim \frac{\nu}{2} \sum_{j=1}^{\infty} \left(\frac{\bar{\Lambda}_j}{\bar{\Lambda}_l}\right)^r \frac{(a_{l,j} + a_{l+1,j})^2}{\alpha_j} \tag{21}$$

It is also quite easy to calculate the $\bar{\Lambda}_j$'s and one finds

$$\frac{\bar{\Lambda}_j}{\bar{\Lambda}_l} = \frac{1}{1 + 2\sinh\sigma(\alpha_j + \omega_j^2)} \tag{22}$$

Finally, for large M one can approximate the sum by an integral and one gets

$$R(r) \sim \frac{2\nu}{\pi} \int_0^{\frac{\pi}{2}} \frac{(1 + \cos^2\theta)}{\alpha(\theta)} \exp\{-r\log[1 + 2\sinh\sigma(\alpha(\theta) + \omega^2(\theta))]\} d\theta \tag{23}$$

where

$$\omega^2(\theta) = \left(\frac{1}{2}\tanh\frac{\sigma}{2} - \nu\right) + \nu\sin^2\theta \tag{24}$$

and

$$\alpha(\theta) = \omega(\theta)\sqrt{\frac{1}{\sinh\sigma} + \omega^2(\theta)} \tag{25}$$

It is now clear that the asymptotic behavior of $R(r)$ as $r \to \infty$ is of the form Eq. 2 and, moreover, the origin of this behavior is also clear. It is directly traceable to the density of eigenvalues $\bar{\Lambda}_2, \bar{\Lambda}_3, \ldots$ just below $\bar{\Lambda}_2$ and is clearly an effect of dimensionality.

4. CORRELATION AT INFINITY: LOW TEMPERATURES

More complicated and much more interesting is the situation for temperatures below the critical.

CRITICAL PHENOMENA

For the Ising model T. T. Wu loc. cit. finds that

$$\rho(\nu) - \rho(\infty) \sim \frac{e^{-\kappa r}}{r^2} \quad r \to \infty \qquad (26)$$

and the appearance of r^2 is striking indeed.

There is, however, evidence (of a nonrigorous nature) that r^2 may well be accidental.

The argument again uses "model A," i.e., the integral equation (Eq. 6). However, for very low temperatures one expects the eigenfunction to be centered near the minimum of

$$Q(\vec{x}) = \frac{1}{2} \tanh \frac{\sigma}{2} \Sigma x_k^2 - \sum_{k=1}^{M} \log \cosh \sqrt{\frac{\nu}{2}} (x_k + x_{k+1}) \qquad (27)$$

which is achieved at

$$x_1 = x_2 = \cdots x_M = \pm \frac{\eta}{\sqrt{2\nu}} \qquad (28)$$

where η is the nonnegative root of the equation

$$\eta \tanh \frac{\sigma}{2} = 2\nu \tanh \eta \qquad (29)$$

After a change of variables

$$x_k = x'_k \pm \frac{\eta}{\sqrt{2\nu}} \quad y_k = y'_k \pm \frac{\eta}{\sqrt{2\nu}}$$

we expand $Q(\vec{x})$ (and $Q(\vec{y})$) retaining only the quadratic term and then change variables again by the rotation

$$\bar{x}_l = \Sigma a_{ls} x'_s$$
$$\bar{y}_l = \Sigma a_{ls} y'_s$$

In this way we replace the operator (Eq. (6)) by *two* operators each of the form

$$\exp[MP(\nu)] \prod_{k=1}^{M} \exp\left(-\frac{1}{2} \omega_k'^1 \bar{x}_k^2\right) \left[\frac{\exp\left(-\frac{(\bar{y}_k - \bar{x}_k)^2}{4 \sin \sigma}\right)}{2\sqrt{\pi} \sinh \sigma} \right] \exp\left(-\frac{1}{2} \omega_k'^2 \bar{y}_k^2\right) \qquad (30)$$

where

$$P(\nu) = \frac{1}{4\nu}\left(\tanh\frac{\sigma}{2} \; \eta^2\right) - \log\cosh\eta \qquad (31)$$

and ω'^2_k are defined by the formula ( ) with ν replaced by

$$\nu' = \nu(1 - \tanh^2\eta) \qquad (32)$$

To be a little more precise, what all this means is that the eigenfunctions of the operator (Eq. (6)) are (for $v \gg 1$) approximately

$$\frac{1}{2}\psi_j\left(x_1 - \frac{\eta}{\sqrt{2\nu}}, \ldots, x_M - \frac{\eta}{\sqrt{2\nu}}\right) \pm \psi_j\left(x_1 + \frac{\eta}{\sqrt{2\nu}}, \ldots, x_M + \frac{\eta}{\sqrt{2\nu}}\right)$$
$$(33)$$

There is a corresponding doubling up of the spectrum (in close analogy with the standard Ising model) and an approximation to

$$\rho(r) - \rho(\infty)$$

can be calculated just as in Sec. 3. The result is that we still get $\sqrt{r}$ in the denominator and not r^2! It is true that my argument totally lacks in rigor (in fact I am not entirely comfortable with it myself) but it poses a dilemma and on balance I am inclined to think that T. T. Wu's r^2 is an accident of the Ising model with nearest neighbor interaction.

5. INTERLUDE

What is surely not accidental is the qualitative nature of the spectrum (both above and below the critical temperature) and the relation between the onset of the phase transition and the change in the "appearance" of the principal eigenfunction.

Above the critical temperature it has one maximum while below it has two.

If, therefore, there are (as everybody seems to believe) universal features of critical phenomena, there must be something universal about the way in which the principal eigenfunction changes its character as it passes through the critical point. It is not easy to see what it could be and I have no useful suggestions to offer.

6. SPECTRAL ANALYSIS AND ONE-DIMENSIONAL MODELS

Since I happen to believe that the most promising approach toward deeper understanding of phase transitions will be through spectral analysis of appropriate operations (e.g., transfer matrix, integral operators like Eq. (6)) let me conclude with a brief discussion of certain one-dimensional models to show once more how such an analysis can yield a far reaching qualitative (though nonrigorous) grasp of pretty subtle phenomena.

Consider a one-dimensional Ising model with interaction energy

$$E = - \sum_{1 < i < j < N} v(j - i) \mu_i \mu_j \qquad (34)$$

About v I shall assume that

1. $v \geq 0, \quad v(k) = v(-k)$

2. $\sum_{-\infty}^{\infty} v(k) < \infty$ (stability condition)

If in addition one assumes that

3. $\sum_{-\infty}^{\infty} |k| v(k) < \infty$

then it will come as no surprise to anyone (though the proof is not at all trivial) that the model will not exhibit a phase transition.

If on the other hand

4. $\sum_{-\infty}^{\infty} |k| v(k) = \infty$

the situation is much more interesting and intricate.

At first I conjectured that under condition (4) the model will exhibit long-range order for sufficiently low temperatures.

In this I was wrong, and Dyson[6] in a most ingenious way has shown that if

$$\sum_{-N}^{N} |k| v(k) \sim \log \log N$$

the model will *not* exhibit phase transition. On the other hand, by an even more ingenious argument Dyson[2] again has shown that if the divergence in condition 4 is sufficiently strong, long-range order will indeed be present for low enough temperatures.

The case of logarithmic divergence of condition 4 is left open although P. W. Andersen has given a heuristic argument in favor of long-range order.

Can we understand all this "with our fingers"? The answer is "yes"—if we use spectral analysis.

To get spectral analysis going, I shall assume that $v(k)$ is a linear combination of exponentials, that is

$$v(k) = \sum_{l=1}^{\infty} a_l \exp(-\sigma_l |k|) \qquad a_l \geq 0 \qquad (35)$$

and to simplify some calculations I shall also modify slightly the interaction energy and assume it to be

$$E = -\frac{1}{2} \sum_{i,j=1}^{N} v(i-j)(\mu_i + \mu_{i+1})(\mu_j + \mu_{j+1}) \qquad (36)$$

(with the understanding that the subscript $N + 1$ is to be replaced by condition 1).

Following the usual procedure with minor modifications caused by the modified form of the interaction energy, we arrive at the following result. The correlation

$$R(r) = \lim_{\substack{N \to \infty \\ (k \to \infty)}} < \mu_k \mu_{k+r} > \qquad (37)$$

for $r \geq 2$ is given by the formula

$$R(r) = \frac{1}{\lambda_1^2} \sum_{j=2}^{\infty} \left(\frac{\lambda_j}{\lambda_1}\right)^{r-1} [(K_+ - K_-)\phi_j, \phi_1]^2 \qquad (38)$$

CRITICAL PHENOMENA

where the λ_j and ϕ_j are the eigenvalues (ordered decreasingly) and normalized eigenfunctions of the kernel

$$K = K_+ + K_- \tag{39}$$

where

$$K_+ = \prod_k \exp[-s_k^2(x_k - b_k)^2]\left[\frac{\exp[(y_k - x_k)^2/2]}{\sqrt{2\pi}}\right]\exp[-s_k^2(y_k - b_k)^2] \tag{40a}$$

$$K_- = \prod_k \exp[-s_k^2(x_k + b_k)^2]\left[\frac{\exp[(y_k - x_k)^2/2]}{\sqrt{2\pi}}\right]\exp[-s_k^2(y_k + b_k)^2] \tag{40b}$$

and*

$$\left.\begin{aligned} s_k &= \sinh\frac{\sigma_k}{2} \\ b_k &= \frac{\sqrt{a_k}\, s_k \cosh(\sigma_k/2)}{s_k^2}\sqrt{\nu}\quad\left(\nu = \frac{1}{kT}\right) \end{aligned}\right\} \tag{41}$$

There appears immediately a difficulty caused by the fact that K_+ and K_- are infinitely dimensional. The maximum eigenvalues of each are of infinite degeneracy and so is without doubt (though I have no proof) the maximum eigenvalue of $K_+ + K_-$.

We are thus forced to interpret $R(r)$ in Eq. (38) as

$$\lim_{m\to\infty} R_m(r) \tag{42}$$

---

*It should perhaps be mentioned that the free energy ψ is given by the formula

$$-\frac{\psi}{kT} = \log 2 + \frac{1}{2}\sum \sigma_k + 2\nu \sum a_k \coth\frac{\sigma_k}{2} + \log\lambda_1$$

Convergence of $\sum a_k \coth \sigma_k/2$ is equivalent to stability condition 2 and to avoid trivial difficulties we may as well assume that $\sum \sigma_k$ converges too.

where $R_m(r)$ refers to a model in which $v(k)$ is replaced by $v_m(k)$ which is a sum of only m exponentials, that is

$$v_m(k) = \sum_{l=1}^{m} a_l e^{-\sigma_l |k|} \qquad (43)$$

This in turn calls for obvious modifications of the definitions of K_+ and K_-.

For large ν we treat K_+ and K_- as almost noninteracting and approximate the principal eigenfunction by

$$\frac{1}{\sqrt{2}} \{\phi_+ + \phi_-\} \qquad (44)$$

where

$$\phi_\pm(\vec{x}) = \prod_k \exp\left[-\frac{\alpha_k^2}{4}(x_k \mp b_k)^2 \frac{\sqrt{\alpha_k}}{\sqrt[4]{2\pi}}\right] \qquad (45)$$

and

$$\alpha_k = 2\sqrt[4]{s_k^2(1 + s_k^2)} \qquad (46)$$

Needless to say, ϕ_+ is (rigorously) the principal eigenfunction of K_+, and ϕ_- is the principal eigenfunction of K_-.

The question now arises as to the eigenfunction corresponding to the next to the highest eigenvalue (before we let $m \to \infty$).

The obvious candidate is the antisymmetric combination

$$\phi_2 \sim \frac{1}{\sqrt{2}} \{\phi_+ - \phi_-\} \qquad (47)$$

but if condition 3 is satisfied, one does better with

$$\phi_2 \sim \frac{1}{\sqrt{2}} x_m \{\phi_+ + \phi_-\} \qquad (48)$$

which yields the value zero (in the limit $m \to \infty$) for the matrix element

$$[(K_+ - K_-)\phi_2, \phi_1]$$

and hence (see Eq. (38)) no long-range order.

The same conclusion holds if the series in condition 4 diverges sufficiently slowly, and one is led to the conjecture that long-range order obtains (for sufficiently low temperatures) only if Eq. (47) is a "better" candidate for ϕ_2 than is Eq. (48).

In making my original conjecture, I was led astray by the thought that Eq. (47) is *always* better than Eq. (48).

This however is not all. If my model is in an external magnetic field, the energy is

$$E = - \sum_{1 \leq i < j \leq N} v(i-j)\mu_i \mu_j \sum_{i=1}^{N} i \qquad (49)$$

and setting

$$h = \frac{m}{kT} > 0 \qquad (50)$$

we find that the magnetization $\zeta(h)$ is given by the formula

$$\zeta(h) = \frac{1}{\lambda_{max}} \iint \phi_1(x;h)(e^h K_+ - e^{-h} K_-)\phi_1(y;h)\,dx\,dy \qquad (51)$$

where λ_{max} and ϕ_j are the maximum eigenvalue of the corresponding (normalized) eigenfunction of the kernel

$$e^h K_+ + e^{-h} K_- \qquad (52)$$

If

$$\lim_{h \to 0} \zeta(h) \neq 0 \qquad (53)$$

for sufficiently low temperatures one might again say that we have a phase transition. In fact, in the lattice gas terminology Eq. (53) is equivalent to having a flat part in the isotherm.

It is believed that Eq. (53) is equivalent to long-range order, but it should be clear that it need not be so because what happens to $\zeta(h)$ as $h \to 0$ depends only on what happens to the principal eigenfunction ϕ_1 in the limit, while long-range order is intimately connected with the degeneracy of λ_{max} and with the eigenfunction belonging to the next to the highest eigenvalue (for $h = 0$).

If the gaussian approximations can be relied upon, we should have (for sufficiently low temperature)

$$\phi_1(x; h) \sim \phi_+$$

and hence

$$\lim_{h \to 0} \zeta(h) \neq 0$$

if and only if condition 4 is satisfied.*

If we also have long-range order and if I am right that this means that

$$\phi_2 \sim \frac{1}{\sqrt{2}} (\phi_+ - \phi_-)$$

we are led to conjecturing that

$$\lim_{r \to \infty} R(r) = [\zeta(0 +)]^2 \tag{54}$$

I apologize for the crudeness of the arguments and for the total lack of rigor. But it would take more than stubbornness to deny that everything hangs together neatly and that we are dealing with some kind of an accident.

If I am right then (using lattice gas terminology), there are "gases" with flat parts but without a geometric separation of phases (as expressed by the linear combination character of the two-point distribution function) and there is even a distinct possibility (in borderline case of logarithmic divergence in condition 4 of two critical temperatures; one for the onset of long-range order and one for having nonzero magnetization at field zero (flat part of the isotherm). In the range between the two critical temperatures, we should have something like the "Mayer hat."

It is nice that partition functions can lead to such strange things, but the real world is probably a bit duller. Still there may be more to all this than statistical mechanicians have dreamt of.

Note added in proof. A recent example of Dyson disproves this statement.

REFERENCES

1. Hemmer, P. C., and G. Stell: *Phys. Rev. Lett.*, 24:1284 (1970).
2. Dyson, F. J.: *Comm. Math. Phys.*, 12:91 (1969).
3. Wu, T. T.: *Phys. Rev.*, 149:380 (1966).
4. Kac, M., and E. Helfand: *J. Math. Phys.*, 4:1078 (1963).
5. Kac, M.: "1966 Brandeis University Summer Institute in Theoretical Physics," vol. I, Gordon & Breach Science Publishers.
6. Dyson, F. J.: *Comm. Math. Phys.*, 12:212 (1969).

DISCUSSION on *Paper* by *M. Kac*

P. MARTIN: I would have thought that Stanley or McCoy would have commented on their suggestions of the same phenomena you have discussed in systems of more than one dimension. Stanley and Kaplan suggested that the two-dimensional Heisenberg model also has no long-range order but an infinite susceptibility. Likewise the McCoy-Wu model appears to have two T'_{cs}—one at what the susceptibility blows up and a second lower one at which the spontaneous magnetization sets in.

M. KAC: I did not want to imply that the phenomena discussed are restricted to one-dimensional models. All I claim is that *if* a model is describable in term of a transfer matrix or a kernel then what is decisive is the existence or nonexistence of a *gap* between the highest and next to the highest eigenvalue.

C. N. YANG: In relation to your first point about the zeros of the grand partition function, it is well known that repulsive forces (or equivalently antiferromagnetic forces) tend to make the zeros leave the unit circle. So I am not surprised by the remark you made.

There have been many recent generalizations of the unit-circle-theorem, I wonder whether Professor Suzuki could be induced to present a summary of these results in one of the evening sessions.*

H. THOMAS: With respect to the existence of spontaneous magnetization without long-range order: from the inequality

$$\frac{1}{N} \sum_k R(k) \geq \zeta^2$$

where $R(k) = \langle \mu_0 \mu_k \rangle$ is the correlation function and $\zeta = 1/N \langle \Sigma \mu_k \rangle$ is the magnetization, it follows that, for $\zeta \neq 0$, one must have $\Sigma_k R(k) \sim N$. Is this possible with $R(k) \to 0$ for $k \to \infty$?

*Cf. Suzuki, M: "Recent Development of Lee-Yang Circle Theorem and Analyticity of the Free Energy," this volume, p. 613.

M. KAC: One must distinguish between $R(k; h)$ (pair correlation in the presence of an external magnetic field) and $R(k) = R(k; 0)$, the pair correlation in the absence of an external magnetic field.

Now, unfortunately

$$\lim_{h \to 0} \lim_{k \to \infty} R(k; h)$$

is *not* necessarily equal to

$$\lim_{k \to \infty} \lim_{h \to 0} R(k; h) = \lim_{k \to \infty} R(k; 0)$$

and therefore there is no reason to believe that

$$\lim_{k \to \infty} R(k; 0) \geq \zeta^2(0+)$$

In fact, Griffiths proved the opposite inequality, that is

$$\zeta^2(0+) \geq \lim_{k \to \infty} R(k; 0)$$

G. BAKER: In your calculation of the correlations the model potential you used factors according to dimension. Is this feature essential to your results or is it merely a convenience?

M. KAC: No. The factorization is not essential and merely simplifies the kernels.

B. WIDOM: The same perturbation theory showed also in three dimensions that the Ornstein–Zernike holds below T_c as well as above.

M. KAC: I find the same on models like the one I have discussed.

C. GRUBER: Can you say anything about higher order correlation functions in the model where there is a spontaneous magnetization but no long-range order?

M. KAC: There is no difficulty in calculating higher order correlations but I have not done it. I am pretty sure though that if there is no long-range order the higher correlations will vanish too.

METASTABILITY, INSTABILITY, AND THE DYNAMICS OF UNMIXING IN BINARY CRITICAL SYSTEMS

John W. Cahn

Department of Metallurgy and Materials Science
Massachusetts Institute of Technology
Cambridge, Massachusetts

ABSTRACT

A single phase can be brought into the two-phase region of the phase diagram. If it is outside the spinodal, the phase's metastability is governed by nucleation. Close to the coexistence curve the metastable phase can exist for geological times; but for systems with a large diffusivity, nucleation theory predicts a limit to the accessible region of the phase diagram that seems consistent with experiment.

For systems with small diffusivities, the entire two-phase region is accessible. Within the spinodal unmixing dynamics can be explained in terms of a diffusion equation with a negative diffusion coefficient.

For solids there are at least two phase diagrams. The equilibrium one is attainable only if each phase forms its own lattice and imposes no mechanical constraint on the other. For the commonly observed coherent phase diagram, the two phases form a distorted single crystal which can be in metastable equilibrium. The two critical points are sometimes separated by hundreds of degrees, and critical point opalescence is only associated with the metastable one.

1. INTRODUCTION

The Van der Waals theory for the equation of state in the gas-liquid critical region gives two possible pressure-density isotherms below the critical temperature; one describes equilibrium with a two-phase region where the isotherm is horizontal; the other describes a single phase and where this is not the equilibrium situation the isotherm loops up and down (Fig. 1) forming two regions of positive compressibility in which the single phase is thought to be metastable, separated by a central region of negative compressibility in which the single phase is unstable. The boundary between the metastable and unstable region is a locus of points of infinite compressibility that was termed the spinodal by Van der Waals[1] (Fig. 2). Similar loops have since been suggested for the two-component liquid-liquid or solid-solid unmixing (using the isotherm of chemical potential of one component μ vs. its density at constant pressure, or the molar free energy F vs. mole fraction of one component C, related to μ by $[1/(1-c)]\, \partial \mu/\partial c = \partial^2 F/\partial c^2$).

The qualitative concept that a single phase can be brought into a metastable and unstable condition by cooling below the critical point has been very useful in metallurgy and material science. I will try to review the main ideas and their consequences. It will be convenient

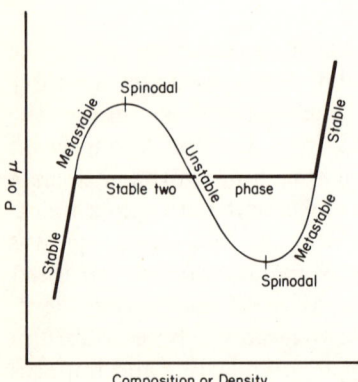

Fig. 1 *The Van der Waals loop showing the equilibrium (heavy) and nonequilibrium (light) isotherm with stable, metastable, and unstable portions. For a one component gas-liquid transition the axes could be pressure, fugacity, or chemical potential vs. density; for two components, activity or chemical potential of one components vs. its composition. The spinodals are at the extrema of this curve.*

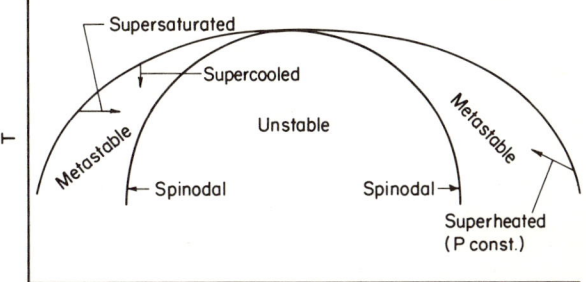

Fig. 2 *Schematic phase diagram showing the two-phase equilibrium and the metastable and unstable regions for a single phase. The common experimental paths leading to metastability are indicated (superheated refers to one component only) at constant P.*

to treat the unstable or spinodal region and the metastable or "nucleation and growth" region separately. In addition, because of the elastic problems, solids will require a special discussion.

The theoretical foundations were laid down by Gibbs[2] who inquired into the thermodynamic nature of metastability by considering two continuous paths leading to the formation of the second phase. The first of these is by the gradual change of a large region of the sample. He found certain necessary conditions for (meta)stability to such changes: positive compressibility and heat capacity, and increasing chemical potential with increasing density. The second of these continuous changes is the continuous expansion of a small droplet of the new phase starting from a vanishingly small size. For fluids a positive surface tension for the surface between the new phase and the old is a necessary condition for metastability to such a change. Positive surface tension has been found generally for phase changes that also obey the first condition since the first condition implies a finite difference between the new and old phase. Conversely, when the first condition fails, transition phases infinitesimally close to the original phase become possible. The surface between these phases would have a vanishingly small surface tension.

The spinodal corresponds to the instability with respect to continuous change over large regions. Outside the spinodal the system is metastable to all infinitesimal changes but becomes unstable to some large enough changes. Finding what changes are sufficient to render the system unstable and assessing how frequently they occur is the subject of nucleation theory.

2. THE METASTABLE REGION

There are two points I wish to make about the metastable region. First, the metastable state is easily accessible for experimental study even quite near the critical point. Second, the droplet theory is quite adequate to explain the experimentally accessible domain of metastability.

Except for the critical density or composition, it is relatively easy to bring a single phase a small distance into the two-phase region without obtaining the second phase that is necessary for full equilibrium. This metastable condition can be called supersaturated if the density or composition exceeds the equilibrium for that temperature (temperature and pressure in a binary); supercooled if the temperature is below the equilibrium for condensation for that pressure of gas or equilibrium precipitation for that composition, a binary; and superheated if the temperature is above the equilibrium for boiling for that pressure of a liquid. Experiments indicate that in clean carefully prepared systems that are not too far into the two-phase region this metastable condition can last indefinitely. Rigorous statistical mechanics indicates that an infinite system will never be metastable or that a large but finite system will not last indefinitely in the metastable condition; but, as I shall show later, we have here one of these few times in physics where the infinity, that is a convenient tool for theory, is so unrealistic for experiment that it should be modified.

Away from the critical point, experience with the metastable state is legion. It is such an everyday occurrence that it hardly needs documenting. Cloud chambers and more recently bubble chambers are worth noting because the former provided us with one of the earliest tests of nucleation theory. The two combined complete the demonstration of two metastable regions in a liquid-gas transition. The only question that remains is how close to the critical point does observable metastability exist. The Van der Waals theory predicts that it should exist for densities or compositions arbitrarily close to the critical point, but that it should not exist for the critical composition or density.

We have recently completed a study of nucleation in the two-component C_7F_{14}–C_7H_{14} liquid-liquid immiscibility system.[3,4] Figure 3 gives the equilibrium phase diagram for this system. The critical temperature is at 46.1°C. We measured for a variety of compositions how far we were able to undercool into the two-phase region (below the equilibrium temperature for that composition). The limit, which I shall call the cloud point, was a sudden increase in

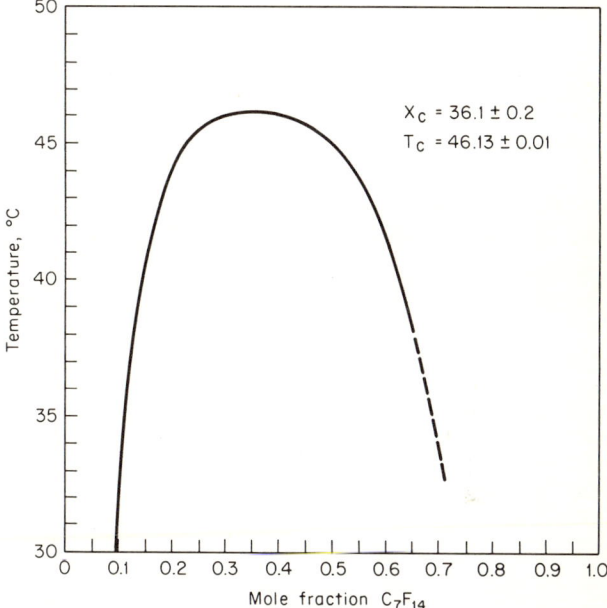

Fig. 3 *Phase diagram for the system perfluoromethylcyclohexane (C_7F_{14})-methylcyclohexane C_7H_{14}). After Heady.*[3]

opalescence followed by rapid separation into two much clearer layers. Figure 4 gives the undercooling as a function of the difference between the composition and the critical composition. The undercooling was insensitive to cooling rates which varied from 10^{-4} to 10^{-1} °C/min.

The equilibrium temperature was measured either by slowly heating a two-phase mixture until one of the phases disappeared or by cooling a single-phase liquid slowly in the presence of a surface that catalysed the nucleation of the second phase. The two methods gave consistent results. The undercooling was measured for the same sample with the nucleation catalyst out of contact. Undercooling was observed for all noncritical compositions. Samples that were close to the critical composition exhibited much critical opalescence while metastable, but no unmixing until the unmistakable cloud point. The reduction in opalescence after unmixing is simply due to the fact that the equilibrium compositions are further from the critical point than the single metastable phase. The experiments indicate that while the range of undercooling diminishes as the critical composition is reached, the phenomenon of metastability definitely occurs in the critical region.

46 CRITICAL PHENOMENA

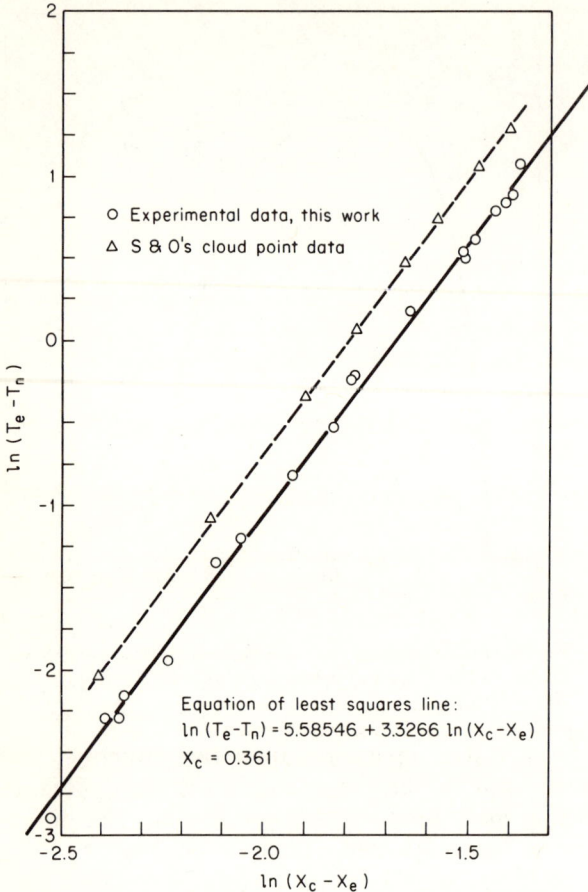

Fig. 4 *Undercooling below the two-phase boundary as a function of a deviation from critical composition C_7H_{14}-C_7F_{14}. After Heady.*[3]

The Gibbs theory of nucleation[2,5] inquired about the minimum amount of work required to render a metastable phase unstable by considering how large a droplet of the new phase must be for its growth to continue. Consider a single-component metastable phase at a pressure P_o. Because the old phase is metastable, if we bring it in contact with a new phase at the same pressure, the latter would grow at the expense of the former. There exists a pressure P_n which can be determined from thermodynamics that when applied to the new phase alone will just restrain it from growing. When the new phase is at P_n, it will be in chemical equilibrium with the old phase at P_o. The chemical potentials of its components species are then equal in the two phases. This equating of μ defines P_n and incidentally for the multicomponent systems a unique composition of the nucleus phase. Since the surface surrounding a droplet of the new phase exerts an additional pressure $2\sigma/r$ where σ is the surface tension and r the radius of the droplet, we may define a critical radius $r*$ by

$$\Delta P = P_n - P_o = \frac{2\sigma}{r*} \tag{1}$$

Any droplet smaller than $r*$ is expected to shrink, because the pressure $P_o + 2\sigma/r > P_n$ and the new phase is under too great a pressure; while those droplets that exceed $r*$ grow since $P_n > P_o + 2\sigma/r$ and the pressure exerted by P_o and the surface tension cannot restrain the new phase (Fig. 5). For a superheated liquid P_n is very closely approximated by the equilibrium vapor pressure for that temperature; where the new phase is incompressible $P_n - P_o$ is very closely approximated by ΔF_v, the change in Helmholz free energy per unit volume of the new phase formed.

The work of forming a drop of critical radius $r*$ of the new phase was given by Gibbs to be the difference between the work needed to create the surface and the work done by the new phase to expand its volume.

$$W = 4\pi r*^2 \sigma - \frac{4\pi}{3} r*^3 \Delta P \tag{2}$$

or by using Eq. (1) we may successively eliminate $r*$

$$W = \frac{16\pi \sigma^3}{3(\Delta P)^2} \tag{3a}$$

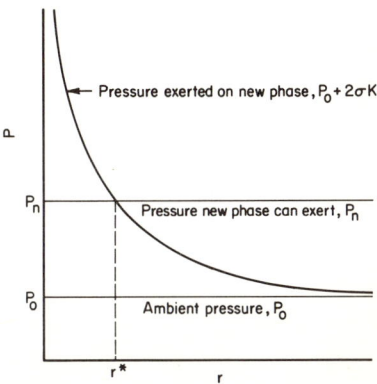

Fig. 5 *The critical size of a nucleating droplet is determined by the balancing of pressure due to surface curvature against the thermodynamic pressure difference that will prevent the more stable phase from growing. The equilibrium at $r*$ is unstable. Smaller droplets will shrink; larger ones will grow.*

or ΔP

$$W = \frac{4\pi}{3} r*^2 \sigma \qquad (3b)$$

Since ΔP rises slowly from zero when the phase first crosses into the metastable region, both $r*$ and W begin by being infinite. Hence there is a very large resistance to the initial formation of a droplet of the new phase and, unless there is a surface upon which the new phase can form, the old phase has a region of metastability except where Gibbs' necessary conditions for metastability are violated.

A brief example of how resistant ordinary materials are to phase change can be illustrated with liquid water at $110°C$ and $P_o = 1$ atm. Because the vapor pressure P_n at $110°C$ is 1.6 atm, $\Delta P = 0.6$ atm or 6×10^5 dynes/cm. The surface tension is 60 dynes/cm. From Eq. (1), $r* = 2\mu$ and $W = 10^{-5}$ ergs. At $383°K$, $kT = 5 \times 10^{-14}$ ergs, $(W/kT) = 10^{+8}$. Exp (W/kT) is larger than an astronomical number. If we filled the volume of the universe with water at $110°C$ and 1 atm (difficult to do in the presence of gravity) and waited for the age of the universe, we might expect no fluctuations that are larger than $W/kT = 10^3$. Hence it is safe to say that water would never boil at $110°C$ without nucleation catalysts.

Actually an infinite system of superheated water would appear quite metastable even for times as long as 10^{10} years. In that time nucleation events will have produced bubbles separated by $10^{(3 \cdot 10^7)}$ km. Even if they grew with the velocity of sound these bubbles would have grown to only 10^{18} km and vapor would occupy a negligible fraction of the volume.

Nucleation is thought to occur by thermal fluctuations in systems with high diffusivities whenever $W/kT = 70 \pm 20$. This number is crudely determined by asking what is the probability of having such fluctuations occur in a convenient time $10^{-3}-10^{+3}$ seconds in a convenient size system $10^{20}-10^{24}$ molecules with an attempt frequency of $10^{12} \sec^{-1}$ molecules^{-1}.

In a system with adequate diffusional mobility in the absence of heterogeneous nucleation sites, the practical limit to metastability occurs at undercoolings that according to Eq. (3a) correspond to

$$(\Delta P)_{\lim} = \left(\frac{16\pi\sigma^3}{210kT}\right)^{1/2} \qquad (4a)$$

when

$$W = 70kT$$

with a limiting critical radius

$$r^*_{\text{lim}} = \left(\frac{210kT}{4\pi\sigma}\right)^{1/2} \qquad 4(b)$$

Degassed water can be subjected to large negative pressures in agreement with both the Van der Waals and nucleation theory. Degassed water at 1 atm can be heated to about 300°C, where finally the combination of a very large ΔP and a reduced σ brings W into the range where a bubble of critical size can be created by thermal fluctuation and the sample explodes.[6] The "stick" vacuum of a McCloud gage means that mercury supports a negative pressure and that even the small gas bubble at the tip of the column is too small to expand until the mercury reaches a sufficiently large negative pressure. Water saturated with dissolved gases at about 140–270 atm, depending on the gas, will not cavitate at 1 atm.[7] It does cavitate at the container interface at slightly greater pressures. Equation (4) indicates that away from the interface water should be metastable to dissolved gas supersaturated to about 1400 atm.

At more extreme supersaturations the droplet theory of nucleation has several shortcomings:

1. In many cases r^*_{lim} (given by Eq. (4b)) is too small to be meaningful. It is of atomic size when σ is large (say 1000 ergs/cm²) and temperature is low ($T = 100°K$) as is encountered in metal condensation. Near the critical point σ is low leading to a very large r^*_{lim}, but the coherence length or the interface thickness is also large and all of these quantities scale with similar exponents. For the $C_7H_{14}-C_7F_{14}$ system we estimate that r^*_{lim} is only about twice the interface thickness.

2. The droplet theory predicts a limit to the metastable range quite independent of the spinodal. When this limit defines a domain entirely within the metastable domain bounded by the spinodal, the nucleation limit becomes the practical limit. This is the case for the $C_7H_{14}-C_7F_{14}$ system where the cloud limit occurs along a line about 3 percent of the miscibility gap. For such a system the accessible domain is only a small fraction of the domain bounded by the spinodal.

When the nucleation domain extends beyond the spinodal, it would seem that the spinodal becomes the practical limit. However, a nucleation theory which does not make the droplet assumption for fluids[8] shows that the nucleation domain always lies well within the metastable region if diffusivities are high. This theory also shows that the droplet theory should be valid at low supersaturations ($r*$ greater than interface thickness) while at higher supersaturations the nucleus becomes a fluctuation large in extent and small in amplitude whose W goes to zero continuously as the spinodal is approached from the metastable side (Figs. 6 and 7).

Assuming now that limiting supersaturations are small and that the droplet theory is valid near the critical point we may derive some relations for $\Delta P, W$ and $r*$ for the cloud limit.

For small supersaturations $(C - C_{eq})$, the quantity ΔP is given by

$$\Delta P = \left[\frac{\partial^2 F}{\partial C^2} \frac{(C'_{eq} - C_{eq})^2}{V_m} \right] \frac{C - C_{eq}}{C'_{eq} - C_{eq}} \quad (5)$$

where V_m is the molar volume; C, C'_{eq}, and C_{eq} are the sample composition and two equilibrium compositions respectively; $C'_{eq} - C_{eq}$ is the range of immiscibility; and $C - C_{eq}$ is the supersaturation. Table 1 gives the scaling exponents for these quantities.

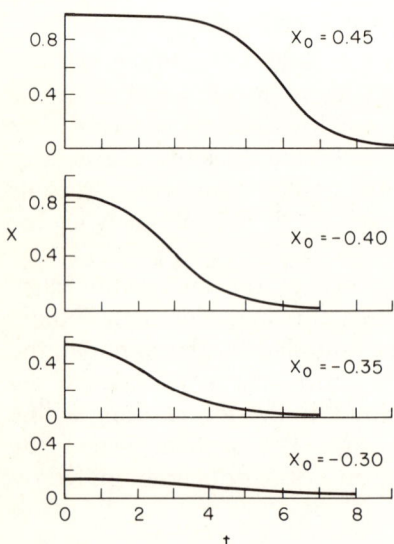

Fig. 6 *Theoretical composition profiles of critical nuclei at supersaturations, various matrix compositions approaching the spinodal.[8] Note that the nucleation event begins to resemble a diffuse composition fluctuation. X and t are reduced composition and radius respectively.*

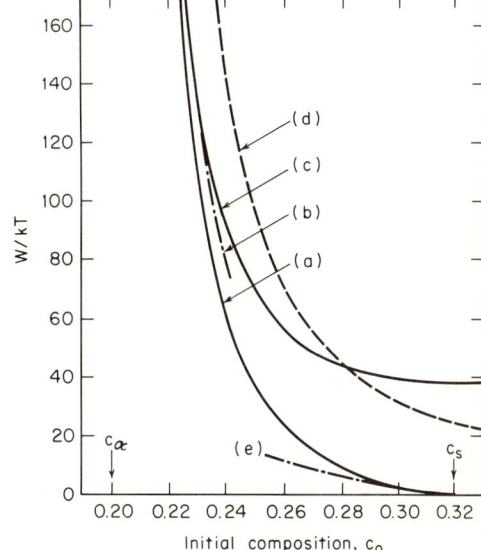

Fig. 7 *(a) Calculated W for the fluctuations of Fig. 6 compared with those (b–d) obtained from various forms of the droplet theory. The droplet theory overestimates the difficulty of forming the new phase.*[8]

TABLE 1. Scaling Laws in Nucleation

Quantity	Exponent	Magnitude	Comment
$(C'_{eq} - C_{eq})$	β	0.3	Fluids, coherent isotropic
		0.5	Incoherent
$\dfrac{\partial^2 F}{\partial c^2}$	γ'	1.3	At $C = C_{eq}$
		1.0	Incoherent
σ	χ	1.3	Coherent, coherent isotropic
	β	0.5	Incoherent, aligned
	0	0	Incoherent, not aligned
ΔP_{lim}	$3\chi/2$	1.9	Eq. (4a)
r^*_{lim}	$-\chi/2$	-0.6	Eq. (4b)
W_{lim}	0	0	
$\dfrac{(C - C_{eq})}{(C'_{eq} - C_{eq})}$	$3\chi/2 - \gamma' - 2\beta$	0	Fluids
	$-\gamma' - \beta/2$	-1.25	Incoherent aligned

The limiting supersaturation as a fraction of the immiscibility range using Eq. (4a) and (5) is

$$\frac{C - C_{eq}}{C'_{eq} - C_{eq}} = \frac{V_m (16\pi\sigma^3/210kT)^{1/2}}{\partial^2 F/\partial C^2 (C'_{eq} - C_{eq})^2} \qquad (6)$$

With empirical exponents this ratio diverges slowly (Table 1). However, within the experimentally accessible range it is still small. Converting Eq. (6) into variable of Fig. 4 is accomplished by noting that $C'_{eq} - C_{eq} = 2(C_c - C_{eq})$ and that undercooling U is given by

$$U \alpha \Delta T^{1/\beta - 1 + 3X/2 - \gamma' - \beta}$$

The slope of Fig. 4 thus should be $1/\beta - 1 + 3X/2 - \gamma' - \beta$. The actual slope is somewhat greater.

Experimentally the cloud point is very sudden. It should not be completely sharp yet there is good theoretical justification in the droplet model for the suddenness. Consider a change in conditions that affect σ or P, e.g., a change in composition C of the sample (Eq. 5). From Eq. (3) we obtain

$$dW = W(3 d\ln\sigma - 2 d\ln\Delta P) \qquad (7)$$

We also have

$$dW = - kT d\ln\dot{N} \qquad (8)$$

where $\dot{N}$ is the nucleation rate. Neglecting changes in σ we have at the cloud point $(W = 70kT)$

$$d\ln\dot{N} = 140 d\ln\Delta P \qquad (9)$$

For a system like $C_7F_{14}-C_7H_{14}$ this becomes at the cloud point

$$d\ln\dot{N} \sim 8 \times 10^3 d\left(\frac{C - C_{eq}}{C'_{eq} - C_{eq}}\right) \qquad (10)$$

This number is sufficiently large to give the impression of a sharp transition. In liquid-solid nucleation $\dot{N}$ has been measured with sufficient accuracy over a small range of undercoolings to verify the theory.

A part of the metastable region is thus experimentally accessible for convenient times and system size. It is possible to do reversible experimental measurements on such system subject only to the constraint that the new phase not appear. In the absence of catalysts the system imposes these constraints on itself until the cloud limit. Strictly speaking it isn't equilibrium. Given enough time the stable phase will form. From a statistical mechanical point of view, the metastable part corresponds to an almost vanishingly small part of the total partition function Z. The system spends such a long time in this small part of the phase because there are very few continuous paths that link the metastable part to the remainder of phase space.

Consider a metastable phase that is far from the cloud point, e.g., water at 110°C at 1 atm. We evaluate its partition function as an expansion, starting only with configurations that have no droplets of what might be considered the new phase and then adding configurations that correspond to system having droplets of the new phase of progressively increasing size dispersed in the system. Such a cluster expansion will appear to converge very rapidly. For water at 110°C configurations containing vapor cluster in excess of 10Å, but below 2μ, will contribute nothing to Z. A lab-sized system might contain a 100kT bubble once a year. A 100kT bubble has a radius of only 10Å.

If we stop the summation at this apparent convergence, we would have a rigorous calculation of the metastable state.

It is also clear that there would be no detectable singularity in the free energy curve on crossing into the metastable phase. One might even venture that the metastable free energy is extrapolatable until the cloud point since even a 60kT fluctuation is sufficiently rare to contribute little to a partition function, although it is a frequent enough occurrence to limit the lifetime of the metastable state. This is why most approximate theories have no difficulty in yielding metastable states which should please anyone caught in this finite world where metastability is demonstrably real and long lasting.

THE SPINODAL REGION

In the previous section we saw that in a binary system of highly mobile components the spinodal is not accessible by taking the system through the metastable region. However, in solids and viscous liquids, such as silicate glasses, that have critical points it is sometimes possible to quench sufficiently rapidly to stop phase separation in the metastable region. We are then in a position to observe the mechanism of phase separation when the initial phase is unstable.

Our understanding of what happens[9-11] is couched entirely in terms of an equation in which the interdiffusional flux of A is related to the gradient of the difference in chemical potentials

$$-j = M\nabla(\mu_A - \mu_B) \tag{11}$$

in which the proportionality constant M must be positive for thermodynamic reasons. For a system with small composition gradients

$$\nabla(\mu_A - \mu_B) = \left(\frac{\partial^2 F}{\partial C^2}\right)\nabla C - 2K\nabla^3 C \tag{12}$$

where C is the concentration of A and K is the gradient energy coefficient (assumed positive)[1,2]

$$-j = M\left(\frac{\partial^2 F}{\partial C^2}\right)\nabla C - 2MK\nabla^3 C \tag{13}$$

If we neglect the last term and define the diffusion coefficient by the ratio $-J/\nabla c$, we see that inside the spinodal D is negative. Without the last term Eq. (13) would predict decomposition into two phases on an infinitely fine scale, which would be inconsistent with thermodynamics in that final mixture would have an infinite surface area.

By taking the divergence of Eq. (13) we obtain after linearization

$$\frac{\partial c}{\partial t} = M\frac{\partial^2 F}{\partial c^2}\nabla^2 c - 2MK\nabla^4 c \tag{14}$$

which has the solution

$$c = \exp[i\beta \cdot r + R(\beta)t] \tag{15}$$

where

$$R(\beta) = -\beta^2 M\left(\frac{\partial^2 F}{\partial c^2} + 2K\beta^2\right) \tag{16}$$

UNMIXING IN BINARY CRITICAL SYSTEMS

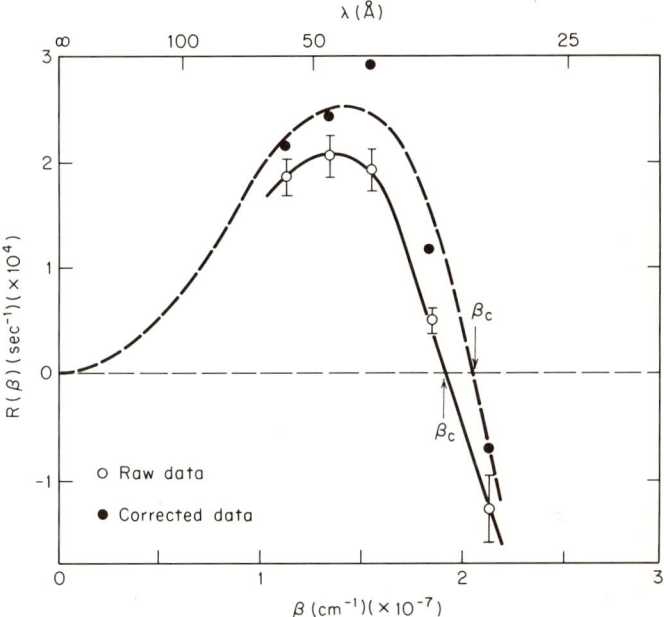

Fig. 8 *The amplification factor as a function of wave number β. Data from Al-Zn.*[15]

This is a stationary plane wave growing or shrinking in amplitude according to whether $R(\beta)$ is positive or negative. $R(\beta)$ is plotted in Fig. 8.

Outside the spinodal $R(\beta)$ is always negative and the system is stable to all small amplitude composition fluctuations. Inside the spinodal the system remains stable to short wavelength composition fluctuations* but is unstable to long wavelengths. The critical wavelength $2\pi/\beta_c$ depends on the instability $(-\partial^2 F/\partial c^2)$

$$2K\beta_c^2 = -\left(\frac{\partial^2 F}{\partial c^2}\right) \qquad (17)$$

The fastest growing waves have a wave number

$$\beta_m = \left(\frac{\beta_c}{\sqrt{2}}\right)$$

---

*This incidentally provides a possible means of calculating the free energy of a system in the unstable region by considering only configurations which are homogeneous on a large scale.

Since $(-\partial^2 F/\partial c^2)$ is expected to be linear in undercooling below the spinodal $(\Delta T_s) \beta_c$ and β_m should vary as $(\Delta T_s)^{1/2}$.

The time constant for unmixing is given by

$$\left(\frac{1}{R(\beta_m)}\right) = 2KM\beta_m^4$$

The diffusion Eq. (14) and its solutions (Eqs. (15) and (16)) have been tested both for stable[13,14] and unstable[15,16] systems by observing how small angle x-ray scattering evolves with time. The empirical coefficients obtained this way indicate that K is not very sensitive to temperature and is of order $N_v kT_c \lambda^2$ where λ is 2–5Å for metals and ca 20Å for silicates. Simple theoretical estimates of λ indicate that it should be of order of a weighted mean distance for intermolecular forces.[17]

With these magnitudes spinodal decomposition results typically in a two-phase mixture with a 10–200Å size scale. The time constant is roughly $10^{12} |D|$ sec. In ordinary liquids $D = 10^{-5}$ cm/sec and only microseconds are involved. In the glasses studied $|D| < 10^{-13}$. For solids just below their melting points $D \sim 10^{-8}$ and a solid in which the time scale is convenient should have its critical point at $T < 0.7 T_m$.

Since β_c can be measured everywhere in the unstable region and K estimated by extrapolation from high temperature measurements, Eq. (17) affords a means of approximately measuring $\partial^2 F/\partial c^2$ in the unstable region and locating the spinodal from below. Because the quench necessarily prevents the "equilibrium" fluctuations from forming, the unstable free energies and the spinodal determined in this way may be different from that obtained by extrapolation from above.[18]

What does a two-phase structure look like after spinodal decomposition?[19,20] For the asymmetric composition where one phase occupies a small volume, the structure would be isolated particles not randomly dispersed but showing a strong inclination to be uniformly dispersed with a distance of $2\pi/\beta_m$ appearing frequently. For the middle of the miscibility gap the predicted structure is shown in Fig. 9a. Compare this with micrographs (Fig. 9b and c) of a glass heat-treated above and below the spinodal.

Thus the simple diffusion equation seems to explain all the facts about the unmixing dynamics within the spinodal for glasses in terms of empirical quantities such as the curvature of unstable portions of the free energy curve, the gradient energy coefficient and the diffusional mobility.

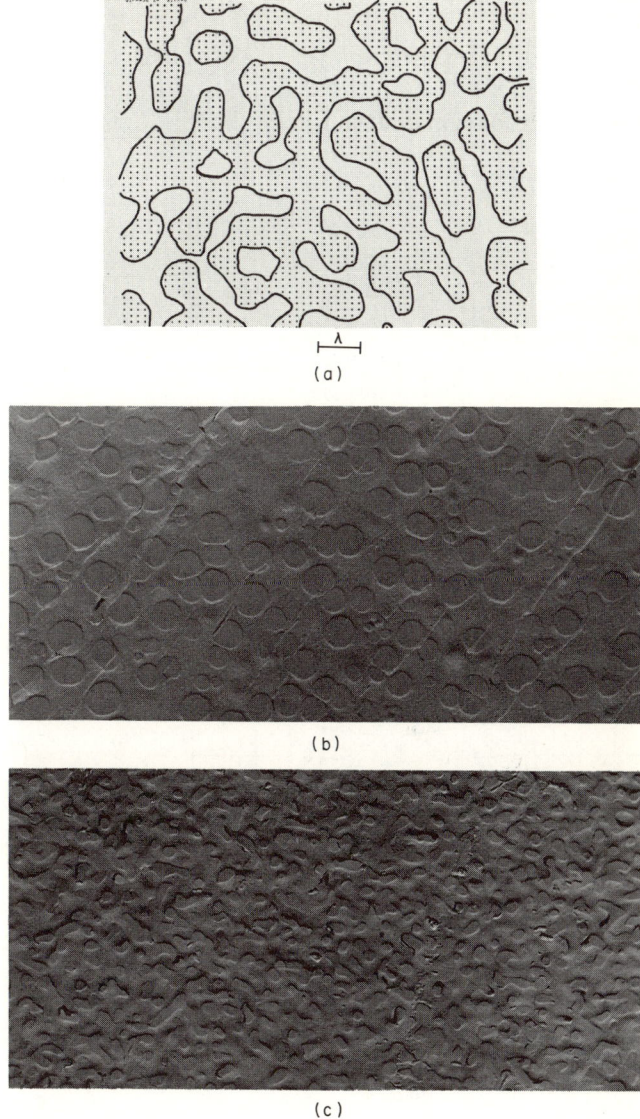

Fig. 9 *(a) Two-phase morphology calculated from spinodal decomposition. (b) Vycor glass two phases formed in the metastable region by nucleation and growth (715°C, 24 hrs). (c) The same vycor glass transformed in the spinodal region (650°C, 100 hrs).*

BINARY SOLIDS

In solids at least two types of two-phase states are observed. In the ideal incoherent state the two phases are mechanically separate but able to exchange matter freely. Each is under the same hydrostatic

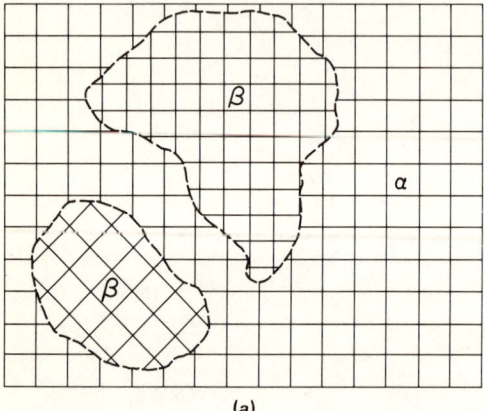

(a)

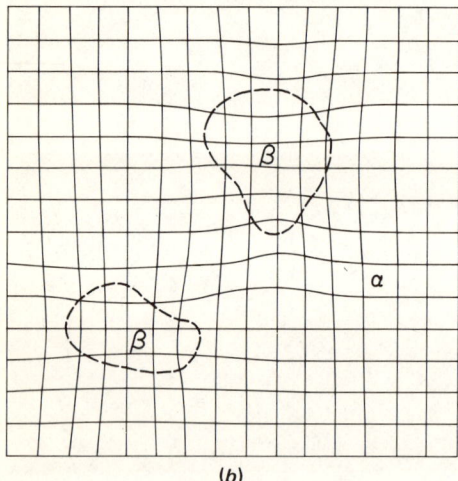

(b)

Fig. 10 *(a) Equilibrium two-phase structure in which each phase has its own lattice parameter and exerts no stress on the other. If the phases are aligned, σ will vary as the lattice parameter difference; if they are not aligned, σ will not vanish at the critical point. (b) Coherent two-phase structure in which the two phases with different lattice parameters form a single crystal. Each phase is under stress due to the coherency.*

stress. These are the conditions necessary to insure full equilibrium, but a good approximation to this state is a polycrystalline solid consisting of large grains of the two phases, well annealed to insure that the macroscopic internal stresses are uniform and hydrostatic. All of our good thermodynamic data for single phase materials are applicable to the phases that participate in such equilibria and most of the careful phase diagrams are for this state.

The other important two-phase state that I wish to consider is the fully coherent state. Here the two phases have either the same or closely similar crystal structures and are so arranged that the two phases are part of the same single crystal with all lattice planes continuous through both phases (Fig. 10). When the lattice parameters of the two phases differ, this structure is a large distorted single crystal. To avoid problems due to interfacial-area dependent terms, consider the two phases dispersed on a scale fine compared to the crystal size yet coarse enough that each particle may be considered as a bulk phase exposed to a macroscopic stress due to coherency and that therefore the interfacial-area dependent terms may be neglected. Under these coarse-dispersion conditions, it is clear that the coherent two-phase state always represents a metastable or unstable state which would go spontaneously to the incoherent state if a mechanism for coherency loss is available. (For a very fine dispersion the coherent state may be stable with respect to loss of coherency which raises the interfacial free energy, but it never represents equilibrium.)

If we consider that the free energy of each phase is empirically given in the hydrostatically stressed state, we may calculate both the incoherent and coherent equilibria.[21] The former is a quite straightforward minimization of free energy or equating of chemical potentials. The latter requires a calculation of the additional free energy due to the elastic effects of coherency. This free energy is a function of the mismatch in lattice parameters which depends on the composition assigned to each of the two phases, the volume fractions of the two phases and their geometrical arrangement and the elastic moduli. Coherent equilibrium is obtained by adding the elastic energy to the stress-free energy and then minimizing the total with respect to composition and arrangement of the two phases but maintaining the lattice coherent. Solutions to this problem are surprisingly simple.

1. If the two phases have the same elastic constant and are elastically isotropic, and if their stress-free lattices differ by a pure dilatation, then the elastic energy is independent of the geometrical arrangement, degree of dispersion, as long as it approaches neither

the atomic nor the crystal size, and depends only on the stress-free mismatch strain δ and the volume of the phases V and V' and $E/(1 - \nu)$ where E is Young's modulus and ν is Poisson's ratio. The elastic-free energy due to coherency of the crystal is

$$F_{el} = \frac{E\delta^2}{1 - \nu} \frac{VV'}{V + V'}$$

For cubic systems δ is $(a' - a)/a$ where a' and a are the stress-free lattice parameters of the two phases. Where a is linear in composition, this result can be generalized to consider the coherency stresses of a slowly varying but otherwise arbitrary composition distribution

$$F_{el} = \frac{E\eta^2}{1 - \nu} \int (c - \bar{c})^2 dV$$

where $\eta \equiv d\ln a/dc$ and $\bar{c}$ is the mean composition. It is thus clear that the coherency energy plays a role similar to the stress-free volume energy F and can be combined with it.

$$F_{coh.} = \int \left[\left(\frac{F}{V_m} \right) + \frac{E\eta^2}{1 - \nu} (c - \bar{c})^2 \right] dV$$

Coherent equilibria are for minima of F_{coh} subject to

$$\int (c - \bar{c}) dV = 0$$

As long as $\partial^2 F/\partial c^2 + 2E\eta^2 V_m/1 - \nu > 0$, the only solution is $c = \bar{c}$. But this condition includes a region below the incoherent critical point where $\partial^2 F/\partial c^2 < 0$. At temperatures where for a range of compositions $\partial^2 F/\partial c^2 + 2E\eta^2 V_m/(1 - \nu) < 0$ the coherent equilibrium is two-phase but spans a range that lies within the incoherent equilibrium. The coherent critical point is the highest temperature for which $\partial^2 F/\partial c^2 + 2E\eta^2 V_m/(1 - \nu) = 0$ for some composition, and lies well below the incoherent critical point. These results are shown schematically in Fig. 11.

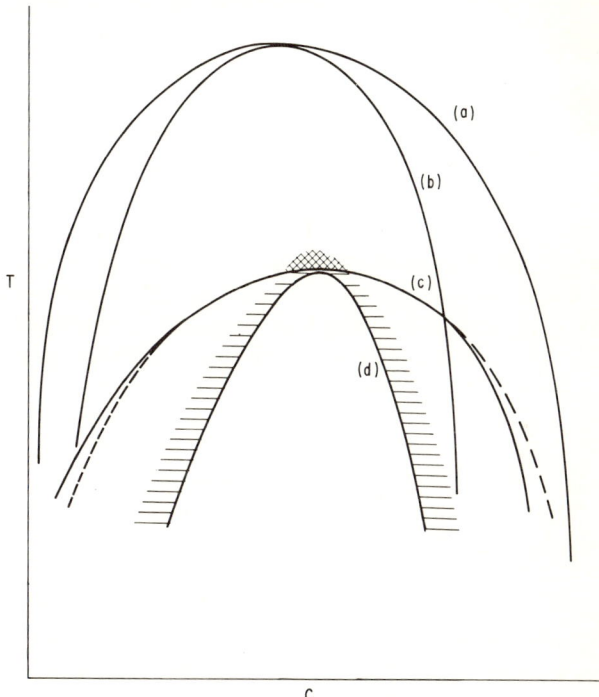

Fig. 11 *The equilibrium (a) and the coherent (c) phase diagram have separated critical points.[21] The critical phenomena occur only with the metastable coherent critical point. The coherent spinodal (d) is the one that determines instability. The cross-hatched region is where critical opalescence would be observed.*

2. If the above conditions apply except that the phases are elastically anisotropic, the elastic energy remains independent of dispersion size but not dependent on shape, orientation, and relative position. For a given δ and volumes the minimum energy is for parallel stacks of plates of the two phases lying in elastically soft planes. For cubic these are either (100) planes if $C_{44} > (C_{11} - C_{12})/2$ or (111) planes if $C_{44} < (C_{11} - C_{12})/2$. For the more general case the elastic energy depends only on the direction and amplitude of Fourier components of the composition. The coherent phase diagram is now not only depressed from the incoherent one, but at the coherent critical point it is only possible to form plates of the two phases on the soft planes. Only as the temperatures are lowered below the coherent critical point do more complex shapes for the second phase become possible since these contain Fourier components of other orientations.

3. With no assumptions at all except coherency, it is possible to prove that the coherent two-phase region cannot lie outside the equilibrium incoherent diagram since it then becomes possible to construct a perpetual motion machine.

Coherency loss requires the creation of high-energy defects (dislocations) at the interface. As long as the particle size or the fluctuation are either small in extent or small in amplitude this defect is energetically unfavored. If the elastic energy is also too large, the fluctuation or particles do not form to any large scale.

There is no acceptable evidence that in a two-component alloy system there is any singular behavior at the incoherent critical point. For the isotropic solid one expects the usual critical point behavior at the coherent critical point if one suppresses the formation of the incoherent phases. For the anisotropic solid the directional effects should spread critical behavior over a range of temperatures and obscure singularities.

The incoherent interface free energy should vary with the same exponent as the lattice parameter difference if the two phases are epitaxially aligned. Hence the exponent is 0.5. (See Table 1.) The value for the exponent γ' should be classical as well. The supersaturation required for nucleation thus diverges and incoherent nucleation should never occur near the critical point, except on heterogeneities. Heterogeneous nucleation alone seems to be observed.

Where the two phases are not aligned, the interfacial free energy does not vanish at the critical point but approaches that of a grain boundary. Heterogeneous nucleation of incoherent particles does tend to give particles that are aligned.

The directionality has a profound effect on spinodal decomposition and gives rise to simple directional periodic structures (Fig. 12). As was mentioned for the isotropic spinodal, $\partial^2 f/\partial c^2$ could be measured by measuring β_c on the assumption that the gradient energy coefficient is slowly varying. By measuring β_c as a function of orientation it is now possible to use the easily measured elastic anisotropy as a measure of both K and $\partial^2 f/\partial c^2$.

The magnitude[22] of the critical point depression in metals varies from 400°C for Au-Ni where η is about 15 percent to less than 10°C for Al-Zn where it is slightly more than 1 percent. In the Au-Ni system one can hold a system at any composition and temperature within a few hundred degrees below T_c without encountering any coherent unmixing or spinodal decomposition. Only incoherent precipitates appear, growing from grain boundaries and the free surface. Even though diffusion is clearly fast enough for fluctuations,

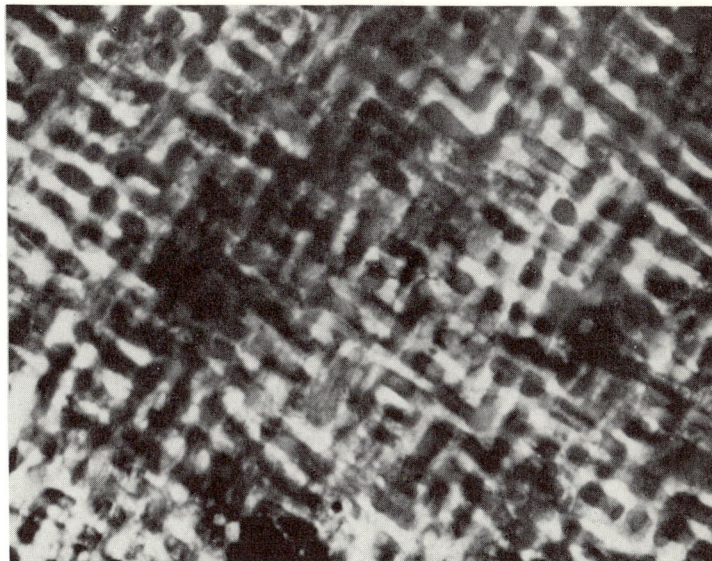

Fig. 12 *Transmission micrograph of a solid-state spinodal reaction in a Cu-Ni-Fe alloy.*[23]

the phase is metastable everywhere until the incoherent precipitate front comes by. Coherent precipitates form at very low temperatures, but the alloys can be homogenized by heating above the coherent critical point even though the system is inside the incoherent two-phase region.

ACKNOWLEDGMENT

The author is grateful to the National Science Foundation for support of this work and to Dr. Ronald B. Heady for much valuable and stimulating discussion and permission to quote results from his thesis prior to publication.

REFERENCES

1. Van der Waals, J. D., and P. Kohnstamm: "Lehrbuch der Thermodynamik," 1st ed., p. 133, Maas and van Suchtelen, Amsterdam, 1908.
2. Gibbs, J. W.: "Scientific Papers," pp. 105 and 252, Dover, 1961.
3. Heady, R. B.: doctoral dissertation, Massachusetts Institute of Technology, 1969.
4. Sundquist, B. E., and R. A. Oriani: *J. Chem. Phys.*, 36:2604 (1962).
5. Landau, L. D., and E. M. Lifshitz: "Statistical Physics," p. 474, Pergamon Press, 1958.
6. Apfel, R. E.: *J. Chem. Phys.*, 54:62 (1971); Biggs, L. J.: *J. Appl. Phys.*, 26:1001 (1955).

7. Hemminsen, E. A.: *Science,* 167:1493 (1970).
8. Cahn, J. W., and J. E. Hilliard: *J. Chem. Phys.,* 31:688 (1959); Sarkies, K. W., and N. E. Frankel: *J. Chem. Phys.,* 54:433 (1971).
9. Hillert, M.: *Acta Met.,* 9:525 (1961).
10. Cahn, J. W.: *Ibid.,* 9:795 (1961).
11. Hilliard, J. E.: "Phase Transformations," chap. 12, ASM, 1970.
12. When K is negative, see Cook, H. E., D. de Fontaine, and J. E. Hilliard: *Acta Met.,* 17:765 (1969).
13. Cook, H. E., and J. E. Hilliard: *J. Appl. Phys.,* 40:2191 (1969).
14. Philofsky, E. M., and J. E. Hilliard: *Ibid,* 40:2198 (1969).
15. Rundman, K. B., and J. E. Hilliard: *Acta Met.,* 15:1025 (1967).
16. Neilsen, G. F.: *Phys. Chem. Glasses,* 10:54 (1969).
17. Cahn, J. W., and J. E. Hilliard: *J. Chem. Phys.,* 28:258 (1958); Kikuchi, R., and J. W. Cahn: *J. Phys. Chem. Solids,* 23:137 (1962).
18. De Fontaine, D.: Private communication.
19. Cahn, J. W.: *J. Chem. Phys.,* 42:93 (1965).
20. Cahn, J. W., and R. J. Charles: *Phys. Chem. Glasses,* 6:181 (1965).
21. Cahn, J. W.: *Acta Met.,* 10:907 (1962).
22. Hilliard, J. E.: To be published.
23. Nicholson, R. B., and P. J. Tufton: *Z. Angew. Physik,* 21:59 (1966).

DISCUSSION *on Paper by J. W. Cahn*

H. THOMAS: As has been mentioned by Dr. Cahn, at the gas-liquid transition, nucleation by droplet formation sets in at a much smaller supersaturation than that corresponding to the spinodal line (to be defined by vanishing diffusion constant). Is this behavior universal, or are systems known which behave in the opposite way, i.e., in which even at the spinodal line the rate of droplet formation is completely negligible?

J. CAHN: Yes, under three different conditions:

  1. At the critical density or composition the system is brought on cooling directly from the single phase region into the spinodal region.

  2. Where the two-phase boundary is at such a low temperature that the diffusion distance is negligible in the time needed to cool the system to the spinodal. Glasses and solids fall in this group.

  3. In solids in the region between the coherent and incoherent phase diagram the chemical spinodal can often be reached without appreciable "droplet" formation of the incoherent phase. In this region the coherent precipitate will not form because it would raise the elastic free energy, and the incoherent precipitate will not form because the incoherent interfacial free energy σ_i is too large. Since σ_i scales as β, the work of forming the incoherent nucleus diverges at the incoherent critical point.

G. BAKER: Actual calculations[1] have been made of the apparent location of the spinodal curve by series and Pelé methods for the

Ising model. They found that, in the M,T plane, the spinodal has the same critical index as the magnetization or order-parameters but with a different amplitude. The ratio of the amplitudes is approximately 1.20 for the b.c.c. lattice versus $\sqrt{3}$ for mean field theory. This spinodal curve seems likely to represent a line of analytic singularities. The measurements inside this curve may possibly be dependent on the path used to reach them, in contrast to the thermodynamic region. I wonder how the distance penetrated into the "metastable" region compares with the calculated location of the spinodal curve for the Ising model.

J. W. CAHN: In the system C_7H_{14}—C_7F_{14}, the cloud point occurs along a curve which is between 3 and 5 percent of the distance between the coexistence curve and the critical composition. The percentage seems to decrease with temperature. Thus we do not come close to approaching the spinodal, in agreement with predictions from nucleation theory for this system; this is also a necessary condition for the validity.

It is difficult to extrapolate our thermodynamic data from a single phase region and obtain a reliable estimate for the spinodal, but we agree that $\sqrt{3}$ is too large; however, 1.20 seems too small.

REFERENCES

1. Gaunt, D. S., and G. A. Baker: *Phys. Rev.*, B1:1184 (1970).

Part Two
GENERAL THEORY

A. Thermodynamics and Scaling

Rapporteur's Introduction:
THERMODYNAMICS AND SCALING

B. Widom
Chemistry Department
Cornell University
Ithaca, New York

Three main themes are represented by the papers of this session.

The first is scaling, further subdivided into thermodynamic scaling, which is the subject of the papers of Josephson and Cooper, and correlation function scaling, which is the subject of Domb's paper.

The second theme is that of lambda lines, i.e., the effect on a critical point of secondary variables, and this, in the context of the lambda transition in liquid helium, is the subject of Rice's paper.

The third theme is that of mean field (Landau-type) descriptions of first- and higher-order phase transitions in ferromagnets and ferroelectrics, represented by the papers of Stenschke and Falk. The purpose of this introduction is to describe the major issues that are raised in these papers.

The thermodynamic scaling idea is that the equation of state near a critical point takes on a simple form, asymptotically, as the critical point is approached. In the language of a liquid-vapor critical point the idea may be expressed as follows.

There are two ways of being displaced from a critical point, by the temperature deviating from the critical temperature T_c and by the density deviating from the critical density ρ_c. If these two deviations are measured on comparable scales, e.g., by expressing $\rho - \rho_c$ alternatively as the temperature $T_c - T(\rho)$ where $T(\rho)$ is the phase transition temperature at the density ρ, then $|T - T_c|$ and $T_c - T(\rho)$ may be anticipated to have comparable effects on the thermodynamic properties of the system. The "comparability" manifests itself in the free energy being, in effect, a homogeneous function of these two variables, and so a power of either one times a function of the ratio of the two. In this way the equation of state takes the form of one thermodynamic variable, properly scaled, being a function of only one other variable, also properly scaled. In the notation used by Josephson in his paper, for example, the magnetic equation of state is

$$\frac{M}{m(T)} = F\left[\frac{H}{h(T)}\right]$$

where M = magnetization, H = magnetic field, $m(T) \sim |T_c - T|^\beta$, and $h(T) \sim |T_c - T|^\Delta$.

The scaling function F is now a function of only one argument, there there may be one function F for $T > T_c$ and another for $T < T_c$.

In the early days of the scaling idea it was recognized that "normal" critical behavior required well-behaved scaling functions F. Josephson now discusses the consequences of assuming the scaling function to be singular at the coexistence curve, i.e., at zero value of its argument. It is then necessary to take explicit account of the asymptotic nature of scaling by first taking F to be f_t (here $\tau = T - T_c$) and letting $\tau \to 0$ *after* differentiation and *after* the argument of f_t is set equal to 0. If then $f_t(0)$ or $f'_t(0)$ vanish or diverge as $\tau \to 0$, the critical point exponent relations differ from "normal"; yet in a true sense the equation of state may still be said to be of scaling form. That is the major point of Josephson's paper.

Cooper, reporting on the results of his collaboration with Green and Levelt-Sengers, raises the question of which are the right variables in terms of which to write a scaled equation of state (the "right" variables are those in which the symmetry and analyticity implicit in scaling actually hold in real substances), and the further question of what is the nature of the corrections to scaling outside the asymptotic critical region. One of the results of this analysis is that the diameter of the coexistence curve is found to reflect the singularity in the specific heat, so that the law of rectilinear diameters fails. This has already been found to be a property of the penetrable sphere model (Widom and Rowlinson) and is now seen to be expected on general thermodynamic and scaling grounds. In liquid-vapor equilibrium the whole question of symmetry in density and of analyticity of the chemical potential seems destined soon to become one of the most important topics in theory and experiment in the subject of critical phenomena in fluids.

Just as the equation of state has a scaling form near the critical point, so also does the pair correlation function $G(R)$. Near the critical point and for sufficiently large R the correlation function becomes proportional to a power of the distance R times a function of κR alone, where κ^{-1} is the correlation length, diverging at the critical point.

This $G(R)$, as well as all the thermodynamic functions including the logarithm of the partition function and the compressibility (susceptibility), can be approximated by retaining only certain dominant classes of terms in high-temperature expansions—the self-avoiding walk approximation of Domb. The coefficients in the nth terms of such expansions in this approximation are expressible as quantities related to self-avoiding walks of n steps. Such a walk may be imagined to have a limiting "shape" which scales with the mean-square end-to-end distance. This scaling in the context of Domb's approximation then implies a scaling of $G(R)$. An apparent deviation from strict scaling of the walk distribution for small r is likewise related to an apparent breakdown of "strong" scaling of $G(R)$.

A quite different class of problems is concerned with the effect of secondary variables on the critical point, with particular emphasis here on the example of the effect of pressure on the lambda point of liquid helium. Rice in his paper inquires into the consistency with the thermodynamic data of the idea that the interactions that are responsible for the nonsingular part of the free energy (the "lattice" contributions), and the interactions that are responsible for the singular part (the "spin" contribution), are only weakly coupled. In this weak coupling limit the isobaric-isothermal partition function may be imagined factored into a lattice part Z_l and a spin part Z_s, the latter factor being singular when kT is some invariant constant multiple of a characteristic interaction enthalpy J that is itself a slowly varying function of the pressure. Rice points out that an earlier attempt to evaluate the lattice contribution from the low-temperature phonon spectrum did not seem to allow a decomposition $Z = Z_l Z_s$ consistent with experimental data. He shows now, however, that if the nonsingular contributions are obtained instead by a more or less objective interpolation across the lambda line then the decomposition *is* consistent with the data. Thus, the weak coupling idea provides considerable insight into the nature of the lambda phenomenon.

Stenschke and Falk both consider mean-field equations of state like that in the earlier Devonshire theory of ferroelectrics, though the application Stenschke has in mind is to ferromagnets. He shows that if the parameter α measuring the magnitude of the magnetostrictive effect is small ($< 2/3$) then the transition is second order, and there is no metastability, so that below T_c there are no metastable paramagnetic states, above T_c there are no metastable ferromagnetic states, and the free energy and energy associated with the stable states are both continuous at T_c. By contrast, when $\alpha > 2/3$ there is a range of T below T_c where metastable paramagnetic states exist and a range of T above T_c where metastable ferromagnetic states exist. In this case the free energy is continuous at $T = T_c$, but not its derivative, so the transition at T_c is first order. The existence of metastable states in association with a first-order transition may then, as emphasized by Stenschke, be the fundamental characteristic of such a transition.

Falk shows that when the coefficients of the p^2 and p^4 terms in his free energy are both linear functions of the temperature (p is the polarization), then there is a critical field E_c above which there is no transition and at which the polarization, electric susceptibility, and specific heat are characterized by rather unusual values of critical point indices. For example, $p - p_c \sim |T - T_c|^{1/3}$ at $E = E_c$. With $E = E_c$ considered the analog of the critical isobar in a fluid, this result for $p - p_c$ is merely a reflection of the classical $\delta = 3$. With the view that $p - p_c$ is the order parameter, however, the very same result may be interpreted as meaning not $\delta = 3$ but $\beta = 1/3$. Taking consistently the latter point of view, the whole set of otherwise classical exponents becomes now the set of nonclassical exponents $\alpha = 2/3$, $\beta = 1/3$, $\gamma = 2/3$, $\delta = 3$. It is interesting that even with this interpretation the two scaling relations $\alpha + 2\beta + \gamma = 2$ and $\delta = 1 + \gamma/\beta$ are still satisfied, just as with the mean-field set $\alpha = 0$, $\beta = 1/2$, $\gamma = 1$, $\delta = 3$.

The papers in this session have touched on some most interesting problems, and they point to developments in the near future that may be of considerable importance.

ASYMMETRY OF CRITICAL EXPONENTS AND THE ASYMPTOTIC SCALING FOR CRITICAL PHENOMENA

B. D. Josephson
University of Cambridge
Cavendish Laboratory
Cambridge, England

ABSTRACT

It is shown that a system can exist whose equation of state satisfies the scaling laws asymptotically as the critical point is approached, and yet does not satisfy all of the relations for critical exponents generally considered to follow from the scaling laws (such as $\gamma = \gamma'$). The 3D Ising model may be an example of such a system.

In this paper it will be shown that under certain conditions the scaling laws[1-4] may be valid at a critical point while some of the standard relations between critical exponents are not valid. For simplicity we shall confine our attention to magnetic systems undergoing a ferromagnetic to paramagnetic phase transition at a temperature T_c. The scaling-law hypothesis states that the equation of state near T_c may be put in the form

$$\frac{M}{m(T)} = \begin{cases} F_+[H/h(T)] & (T > T_c) \\ F_-[H/h(T)] & (T < T_c) \end{cases} \quad (1)$$

where $m(T) = m_0|\tau|^\beta$, $h(T) = h_0|\tau|^\Delta$, with m_0, h_0, β, and Δ being constants while τ denotes $(T/T_c) - 1$ and F_+ and F_- are two temperature-independent functions (M and H denote the magnetization and the magnetic field respectively. Equation (1) has the implication that all isotherms on one side of T_c may be deformed into each other by adjusting the scales of the M and H axes.

If Eq. (1) is exact, it can be shown that all critical exponents can be expressed in terms of the two parameters β and Δ. Consider, for example, the zero-field susceptibility χ_0. According to Eq. (1) this is given by $(m_0/h_0)(F^+)'(0)|\tau|^{\beta-\Delta}$ above T_c and $(m_0/h_0)(F^-)'(0)|\tau|^{\beta-\Delta}$ below T_c (the prime denotes differentiation). As a result we obtain the relation $\gamma = \gamma' = \Delta - \beta$. Relations such as this lead to a number of relations between the critical exponents for thermodynamic quantities.

Unfortunately, Eq. (1) can be correct only in an asymptotic sense. If it were exact in some finite region near the critical point, analyticity would require it to be exact everywhere, and this would clearly be unrealistic. Let us put this asymptotic property into mathematical language.

We first define a set of functions f_τ, dependent on the parameter τ. We do this by requiring the following equation to be satisfied exactly

$$\frac{M}{m(T)} = f_\tau\left\{\frac{H}{h(T)}\right\} \quad (2)$$

Let us now say that the scaling law is satisfied asymptotically if nontrivial functions F_+, F_- exist such that

$$f_\tau(x) \to \begin{cases} F_+(x) & \text{as } \tau \to 0^+ \\ F_-(x) & \text{as } \tau \to 0^- \end{cases} \quad (3)$$

We obtain deviations from the critical exponent relations by assuming that some kind of singular behavior of $f_\tau(x)$ is present as one approaches the critical point along the critical isochore from below T_c. This assumption is not too implausible, since some types of singular behavior have been found in this region by Kadanoff[5] and by Wu[6] in the case of the 2D Ising model and by Stanley and Kaplan[7] and Dyson[8] for the Heisenberg ferromagnet.

The first possibility that we consider is that $(f_\tau)'(0)$ may tend to infinity as $\tau \to 0^-$ (this is consistent with asymptotic scaling). To be precise, let us define a critical exponent σ such that $(f_\tau)'(0) \sim$ const. $\times |\tau|^{-\sigma}$ as $\tau \to 0^-$. Then from Eq. (2) we find

$$\chi_0(T) = \frac{m_0}{h_0} f_\tau'(0) |\tau|^{\beta-\Delta}$$

$$\sim \text{const.} \times |\tau|^{\beta-\Delta-\sigma} \quad \text{as } \tau \to 0^-$$

Consequently $\gamma' = \Delta - \beta + \sigma$, which is at variance with the usual prediction $\gamma' = \Delta - \beta$.

By a similar line of argument one can show that the critical exponent for the zero-field magnetization is not necessarily the same as the exponent β which appears in the definition $m(T) = m_0 |\tau|^\beta$. This situation arises if $f_\tau(0)$ tends to a zero or infinite limit as $\tau \to 0^-$.

The most obvious candidate for applicability of the ideas outlined here is the 3D Ising model, where there is some evidence that γ and γ' may be distinct quantities.

ACKNOWLEDGMENT

I should like to thank the Physics Department of the Massachusetts Institute of Technology for their hospitality during the period in which this paper was written.

REFERENCES

1. Widom, B.: *J. Chem. Phys.*, **43**:3898 (1965).
2. Domb, C., and D. L. Hunter: *Proc. Phys. Soc.*, **86**:1147 (1965).
3. Kadanoff, L. P.: *Physics*, **2**:263 (1966).
4. Patashinskii, A. Z., and V. L. Pokrovskii: *Soviet Phys. JETP*, **19**:677 (1964).
5. Kadanoff, L. P.: *Il Nuovo Cimento*, **44**:B276 (1966).
6. Wu, T. T.: *Phys. Rev.*, **149**:380 (1966).
7. Stanley, H. E., and T. A. Kaplan: *Phys. Rev. Lett.*, **17**:913 (1966).
8. Dyson, F. J.: Unpublished manuscript.

DISCUSSION on *Paper* by *B. D. Josephson*

R. B. GRIFFITHS: There is something which I do not understand. Scaling normally means taking a function of two variables and reducing it to a function of one variable. However, in Dr. Josephson's description the second variable comes in as a subscript and this seems to go back to two variables again.

B. D. JOSEPHSON: We do not expect scaling to hold except asymptotically as $\tau \to 0$, and in that limit there is only one variable.

R. B. GRIFFITHS: How does this go beyond previous scaling where it was well understood to be asymptotic?

B. D. JOSEPHSON: This goes beyond earlier formulation by allowing the possibility of the limiting scaling function being singular at the coexistence curve and this allows asymmetry in the critical exponents above and below T_c.

C. DOMB: The spherical model also has a peculiarity at the coexistence curve and this leads the susceptibility and its derivatives to be infinite on the low temperature side.

J. S. KOUVEL: Similarly for spin waves in the Heisenberg model.

GENERALIZED SCALING AND THE PARAMETRIC EQUATION OF STATE IN THE CRITICAL REGION

Martin J. Cooper

Statistical Physics Section
National Bureau of Standards
Washington, D.C.

ABSTRACT

A description of the critical region is developed using a local parametric characterization of the thermodynamic potential surface. The physical consequences of the formalism are presented and its predictions compared with some of the experimental data. Results indicate an extended description of the thermodynamic properties over a much enlarged region about the critical point.

1. INTRODUCTION

Considerable understanding of the physical properties in the critical region of second-order phase transitions has resulted from the

Widom-Kadanoff-Domb scaling hypothesis.[1] Such limiting behavior has been found near the critical points of various mathematical models[1] and also appears to be an accurate qualitative description for real systems.[2]

However, in any analysis of the bulk thermodynamic data there remain two rather "curious" points. First, there is the problem of selecting those thermodynamic coordinates which satisfy the basic symmetry required by simple scaling. For some systems, this choice is easily based upon an intrinsic or known symmetry in certain of the physical variables, whereas for many others (such as fluids), the actual assignment is less certain. Secondly, there is a problem in estimating the actual range over which scaling of the thermodynamic properties remains valid. While theoretical arguments would suggest that scaling is an asymptotic representation valid only close to the actual critical point, empirical observation indicates that in real systems scaling appears to hold over a rather large region.[2]

The present article summarizes an attempt by the author, working with Melville S. Green and J. M. H. Levelt-Sengers, to provide answers to these two points. Beginning from the fundamental thermodynamic potential surface given explicitly in terms of intensive variables, a description of the thermodynamic properties over an extended region about the critical point is developed. This is accomplished by mapping the physical fields onto a new coordinate set defined by a "tangent" plane of the potential surface and a local parameterization of that plane by algebraic powers in only one of the parametric variables. To the lowest order, the formulation recovers Schofield's[3] parametric representation of the Widom-Kadanoff homogeneous equation of state. A new critical index appears in the structure and is related to the usual critical exponents using an invariance property first noted by M. S. Green. Elimination of the parametric variables and inversion of the coordinate transformation leads to expressions for the critical isochore, critical isotherm, and coexistence curve which contain modifications beyond simple scaling. Deviations from a completely symmetric structure are in general expected and the consequences of such asymmetry are discussed. Finally, the predictions of this formalism are compared with some of the experimental data on real fluids.

2. FORMULATION

The conditions for thermodynamic equilibrium between two coexisting phases of a single substance are conveniently phrased in terms of the intensive physical variables which Griffiths and Wheeler[4] refer to as "fields." These thermodynamic coordinates $\{\Phi_i | i = 0, 1, 2\}$ have the simple property of being equal in each of the two coexisting phases, $\Phi_i^I = \Phi_i^{II}$, and form a fundamental thermodynamic equation when any one is regarded as an explicit function of the other two, $\Phi_0 = \Phi_0(\Phi_1, \Phi_2)$. The conjugate physical quantities, the reduced extensive "densities," are defined as the partial derivatives of the dependent field with respect to the independent ones

$$\rho_j = -\frac{\partial \Phi_0}{\partial \Phi_j} \qquad j = 1, 2 \qquad (1)$$

and assume different values in each of the two phases $\rho_j^I \neq \rho_j^{II}$. In a fluid system, e.g., one might choose the three fields as the (negative) pressure $-p$, the chemical potential μ, and the temperature T respectively, with ρ_1 the number density and ρ_2 the entropy density per unit volume.

While the selection of any one particular physical field as the dependent variable is somewhat arbitrary, we require only that the form $\Phi_0(\Phi_1, \Phi_2)$ be regarded as an explicit equation for the thermodynamic potential surface. It is believed that this surface is everywhere smooth except possibly for a single finite line along which a discontinuity in slope occurs. Geometrically, this just means a simple fold in the thermodynamic surface extending up to a critical point $(\Phi_0^c, \Phi_1^c, \Phi_2^c)$ which corresponds to the phase-boundary along which the two different phases coexist (their difference going continuously to zero as the critical point is approached). The plane "tangent" to this surface at the critical point is given by the equation

$$\Phi_0 - \Phi_0^c = \left(\frac{\partial \Phi_0}{\partial \Phi_1}\right)^c (\Phi_1 - \Phi_1^c) + \left(\frac{\partial \Phi_0}{\partial \Phi_2}\right)^c (\Phi_2 - \Phi_2^c) \qquad (2)$$

(Here the term "critical point tangent plane" is used in the loose mathematical sense to mean that plane defined by Eq. (2) which just touches the surface $\Phi_0(\Phi_1, \Phi_2)$ at its critical point.) The separation between the fundamental surface $\Phi_0(\Phi_1, \Phi_2)$ and this plane defines a relative thermodynamic potential

$$\Pi(\Delta\Phi_1, \Delta\Phi_2) = \Delta\Phi_0(\Delta\Phi_1, \Delta\Phi_2) + \rho_1^c \Delta\Phi_1 + \rho_2^c \Delta\Phi_2 \qquad (3)$$

where each $\Delta\Phi_i = \Phi_i - \Phi_i^c$. It is from this quantity that Schofield[3] developed his parametric representation for the equation of state.

To describe the properties near the critical point, it is convenient to introduce a transformation from the physical coordinates $\{\Phi_i\}$ to a new set $\{\Psi_i | i = 0, 1, 2\}$ characterizing the thermodynamic surface in this region. One such set of local coordinates may be constructed from thermodynamic surface and its critical point plane as follows:

1. Take two noncolinear vectors which lie on the plane and are directed away from the critical point to define two axes Ψ_1 and Ψ_2;
2. Let the coordinate Ψ_0 be the difference (distance) between the thermodynamic surface and the plane, i.e., the $\Pi(\Delta\Phi_1, \Delta\Phi_2)$ function, as directed out from the plane along some suitable direction, say the Φ_0 axis.

This triplet $\{\Psi_i\}$ completely satisfies a new nonorthogonal basis centered at the critical point and related to the original physical coordinates by a linear transformation

$$\Delta\Phi_i = \sum_{j=0}^{2} A_{ij} \Psi_j \qquad (4)$$

The matrix elements A_{ij} are determined by the construction and the geometric properties of the thermodynamic surface: $A_{00} = 1$, $A_{10} = A_{20} = 0$; $A_{0j} = A_{1j} + A_{2j}$ for $j = 1, 2$; $A_{01} = \rho_1^c$, $A_{02} = \rho_2^c$.

Each point on the critical point plane is located by two numbers designating its position relative to the Ψ_1 and Ψ_2 axes, respectively. It is also possible to describe any point on this plane by two other parameters r and θ, where r is a measure of its "distance from the actual critical point (origin)" and θ the displacement along a contour of constant r. This scheme has the useful features of Eq. (1) characterizing the points by their geometric distance from the critical point, and of Eq. (2) providing an easy description of lines or contours leading to the critical point. We shall *assume* that this parameterization of the tangent plane is given by the general form

$$\Psi_i = r^{\epsilon_i} V_i(r, \theta) \qquad \epsilon_i \text{ positive} \qquad (i = 1, 2) \qquad (5a, b)$$

where the $V_i(r, \theta)$ are simple, smooth analytic functions in the real variables r and θ. The third element of this local coordinate basis Ψ_0 is also defined in terms of a distance representing a displacement from the actual critical point—the difference between the thermodynamic surface and the critical point plane. This suggests that the Π-variable also depends upon the distance from the critical point and is an explicit function of the planar parameters (r, θ). For this we *assume*

$$\Psi_0 = r^{\epsilon_0} V_0(r, \theta) \tag{5c}$$

where V_0 is a smooth analytic function, ϵ_0 positive.

Of the three algebraic powers or exponents $\{\epsilon_i\}$ appearing in the local coordinate scheme, the two associated with those vectors which lie in the plane may be related through an invariance argument given by M. S. Green. Since any physical property must remain unaltered by the choice (direction) of the two vectors Ψ_1 and Ψ_2 it follows that any change in Ψ_1 for a fixed Ψ_2 must result in a compensating shift in the coordinate transformation (and conversely).[5] Because of the algebraic structure of these two vectors, this is possible if and only if the difference between the two forms is itself analytic in r. This requires that

$$|\epsilon_1 - \epsilon_2| = n \tag{6}$$

where n is some integer 0, 1, 2, ...

Within this formalism, each one of the physical quantities $\{\Phi_i\}$ has a similar algebraic structure in the parametric variables. After inverting the transformation (Eq. (4)) and defining new analytic functions $U_i(r, \theta)$, the resulting expressions are found to be of the form

$$\Delta\Phi_0 = r^{\epsilon_0} V_0(r, \theta) + r^{\epsilon_1} U_0(r, \theta) \tag{7a}$$

$$\Delta\Phi_1 = r^{\epsilon_1} U_1(r, \theta) \tag{7b}$$

$$\Delta\Phi_2 = r^{\epsilon_2} U_2(r, \theta) \tag{7c}$$

The corresponding densities are easily determined from their definition (Eq. (1) and the convenient properties of Jacobians) to find that deviations from the critical value ρ_i^c have the general form

$$\Delta\rho_i = \rho_i - \rho_i^c = r^{\epsilon_0 - \epsilon_i} M_i(r, \theta) \tag{8}$$

As the critical point is approached from different directions, the behavior of the various quantities is described by (combinations of) algebraic powers in a single parameter r, characteristic of the "distance away from the actual critical point." This is like the homogeneity hypothesis of Widom[1] and leads to a "scaling description" with a set of power-law relations between the physical variables. Moreover, by a proper assignment of labels and a suitable choice of functions, these expressions reduce to the parametric scheme proposed by Schofield.[3]

3. PHYSICAL CONSEQUENCES

It is necessary only to make appropriate choices for the three physical fields $\{\Phi_i\}$ and then eliminate the parametric variables in order to obtain a complete description of the thermodynamics in the critical region of any system. (The discussion here is limited to fluids, although it is easily adapted to other systems (magnets) or an alternate choice of coordinates.) Thermodynamics alone, however, cannot dictate any preferred assignment of the independent variables, such as $p(\mu, T)$ or $T(p, \mu)$. The various ensemble representations of statistical thermodynamics offer many alternate but not necessarily simpler choices as, for example, the Grand Potential which expresses p/T as an explicit function of the variables μ/T and $1/T$. We shall choose $p = p(\mu, T)$ and examine the consequences of setting

$$\Delta p = \Delta \Phi_0 \ , \ \Delta \mu = \Delta \Phi_1 \ , \ \Delta T = \Delta \Phi_2 \text{ with } \Delta \rho = \Delta \rho_1 \qquad (9)$$

The critical isotherm is characterized by the parametric equation $V_2(r, \theta) = 0$, whose solutions $\theta = \theta_0(r)$ describe the constant temperature contour $T = T_c$, $\Delta T = 0$. Along this line, elimination of the parametric variables yields an expression for the chemical potential in terms of the density

$$\Delta \mu = (\Delta \rho)^{\epsilon_1/(\epsilon_0 - \epsilon_1)} \{h_0 + h_1^{\pm} |\Delta \rho|^{1/(\epsilon_0 - \epsilon_1)} + \cdots \} \qquad (10)$$

where the $h_0, h_1^{\pm}, \ldots$ are just numerical constants. By a similar procedure along the critical isochore $\rho = \rho_c$, in the single-phase region $(T \geq T_c)$, the chemical potential is given in terms of the temperature

$$\Delta \mu = (\Delta T)^{\epsilon_1/\epsilon_2} \{\mu_0 + \mu_1 (\Delta T)^{1/\epsilon_2} + \cdots \} \qquad (11)$$

Below the transition temperature $T < T_c$, there is a smooth curve on the thermodynamic surface along which two physically different densities ρ_L and ρ_G coexist. This implies there are two algebraic branches

$$\Delta \rho^+ = \rho_L - \rho_c \quad \text{and} \quad \Delta \rho^- = \rho_G - \rho_c \qquad (12)$$

which form simultaneous nontrivial solutions to the conditions for thermodynamic stability of the two-phase equilibrium

$$\Delta p^+(\rho_L, T_L) = \Delta p^-(\rho_G, T_G) \qquad (13a)$$

and

$$\Delta \mu^+(\rho_L, T_L) = \Delta \mu^-(\rho_G, T_G) \qquad (13b)$$

for

$$\Delta T^+(\rho_L) = \Delta T^-(\rho_G) = -t < 0 \qquad (14)$$

Along this curve, the parametric quantities on either branch $(r^\pm, \theta^\pm)$ may be expanded in arbitrary algebraic powers of t and substituted into the p, μ, T equilibrium relations.[6] The requirement that the resulting equations be algebraically consistent for all values of t then establishes relations between the otherwise arbitrary powers. After some manipulation it is found that along the coexistence curve the density

$$\Delta \rho^\pm = t^{(\epsilon_0 - \epsilon_1)/\epsilon_2} \{ m_0 + m_1^\pm t^{1/\epsilon_2} + \cdots \} \qquad (15)$$

If the leading terms in these expressions (Eqs. (10), (11), and (15)) are identified with the usual power law description of thermodynamic anomalies close to the critical point, the three exponents ϵ_0, ϵ_1, ϵ_2 can be related to the familiar critical indices

$$\frac{\epsilon_1}{\epsilon_0} = \frac{\delta}{(\delta + 1)} \qquad (16)$$

$$\frac{\epsilon_2}{\epsilon_0} = \frac{1}{\beta(\delta + 1)} \qquad (17)$$

The additional terms in the expansions represent corrections to such asymptotic behavior and permit an extended description of the thermodynamic properties over an enlarged region about the critical point. (See discussion in last section.)

Another very important result of this formalism is that the parametric functions provide a means of accommodating deviations from the simple symmetry required in the usual scaling treatment.[7] This is especially important for fluids (gases) where the antisymmetry of the chemical potential is only approximate and the coexistence curve is very asymmetric.[2] In the usual scaling of fluids, the assumed antisymmetry of $\Delta\mu$ in the $\Delta\rho$ variable implies that only the pressure, and not the chemical potential, may contribute to the specific heat anomaly; also the coexistence curve must be symmetric about $\Delta\rho$.[2] These limitations are overcome within this extended description as both μ and p have been taken to be fundamental thermodynamic fields. A short calculation shows that both the pressure and the chemical potential have highly singular second-temperature derivatives at T_c with the sum of their individual contributions resulting in an α' or α type ($T \lessgtr T_c$ respectively) divergence in the specific heat. There is also the possibility of $\alpha \neq \alpha'$. Within this formalism, the diameter of the phase-boundary is given by

$$\frac{\rho_L + \rho_G}{2\rho_c} = 1 + \rho_m t^{(\epsilon_0 - \epsilon_1 + 1)/\epsilon_2} \qquad \rho_m = \frac{m_1^+ - m_1^-}{2\rho_c} \qquad (18)$$

When the relations among the various exponents, including $|\epsilon_1 - \epsilon_2| = n$, are used, the diameter is found to be characterized by the critical index $(\epsilon_0 - \epsilon_1 + 1)/\epsilon_2 = 1 - \alpha'$. This could also be obtained directly from the specific heat calculation, independent of the previous invariance argument. A similar result has recently been suggested for several model fluids.[8]

4. EXPERIMENTS AND CONCLUSIONS

To test the hypothesis that this extended parametric representation describes the thermodynamic properties in an enlarged region about the critical point, we have attempted to fit some of the available experimental data to the predicted expansions. Unless otherwise stated, the data used here is that cited by the NBS group in their systematic analysis of scaling in the critical region.[2]

1. The liquid and vapor sides of the coexistence curve of CO_2 and O_2 were fit to separate three-term expressions of the form suggested by Eq. (15), allowing for different exponents on the two branches. It appears that for both fluids, $\beta_L = \beta_G = \beta(= 0.35)$ and $(\epsilon_2)_L = (\epsilon_2)_G$

with $1/\epsilon_2 \simeq 0.6 - 0.7$. Only the first two terms in the expansion were required over a wide range in density (at least 40 percent from ρ_c). This value of $1/\epsilon_2$ compares favorably with that obtained using the established numerical values for the critical indices $1/\epsilon_2 = 1 - \alpha'(= 0.04) - \beta(=0.35) = 0.61$.

2. The $t^{1-\alpha'}$ form predicted for the phase-boundary diameter has the curious feature of curving into the critical point with zero slope. We suspect that because of the extremely small numerical value of $\alpha'(\alpha' \lesssim 0.05)$, any observable curvature occurs in a region quite close to the critical point (experimentally almost inaccessible), making its detection quite difficult.

3. To determine the behavior along the critical isotherm, we have graphically integrated the expression $d\mu = dp/\rho$ using critical temperature data of CO_2, and following a suggestion by H. Meyer,[9] have constructed $\mu(\rho_L) \pm \mu(\rho_G)$. The difference yields the lowest order term of simple scaling $\Delta\rho|\Delta\rho|^{\delta-1}$, $\delta = 4.5$. The sum, which represents corrections to scaling, varies like $\Delta\rho$ to the power $p = 6.2 \pm 0.2$ (coefficient 1.3) up to about $\Delta\rho = 0.8\rho_c$. Similar results have also been reported by H. Meyer from an independent study of his He^3 data. With the relation between the various exponents, $1/\epsilon_2 = \beta(p - \delta)$, this yields a value for $1/\epsilon_2$ of about 0.6 also.

It is most interesting to note the similarity between the estimated value for $1/\epsilon_2 \sim 0.6$ and variously reported experimental values of the critical index ν of 0.6–0.7. This latter index characterizes the correlation scale in terms of a temperature distance $\xi \sim \Delta T|^{-\nu}$, while the parametric relation for the thermodynamic distance from the critical point is $r \sim |\Delta T|^{1/\epsilon_2}$. Curiously, neither index, $1/\epsilon_2$, nor ν is required to describe the bulk thermodynamic properties close to the critical point.

While the parametric formulation presented here is not entirely unique, it does appear to provide an adequate description of the thermodynamic properties in an enlarged region about the critical point including any asymmetric behavior such as that found in fluids. In the rather special case where $1/\epsilon_2$ is arbitrarily set equal to β, the formalism reproduces many of the results of a previously proposed expanded formulation of thermodynamic scaling obtained from a scaled free-energy functional form.[10]

ACKNOWLEDGMENTS

The author wishes to thank Drs. M. Green and Levelt-Sengers for their contributions to the work presented here.

REFERENCES

1. Widom, B.: *J. Chem. Phys.*, **43**:3898 (1965); Kadanoff, L. P.: *Physics*, **2**:263 (1966); Domb, C., and D. L. Hunter: *Proc. Phys. Soc. (London)*, **86**:1147 (1965). See also Griffiths, R. B.: *J. Chem. Phys.*, **43**:1958 (1965).
2. Vicentini-Missoni, M., R. Joseph, M. S. Green, and J. M. H. Levelt-Sengers: *Phys. Rev.*, **B1**:2312 (1970); also *NBS J. Res.*, **73A**:563 (1969).
3. Schofield, P.: *Phys. Rev. Lett.*, **22**:606 (1969).
4. Griffiths, R. B., and J. C. Wheeler: *Phys. Rev.*, **A2**:1047 (1970).
5. Halmas, P. R.: "Finite-dimensional Vector Spaces," 2d ed., D. Van Nostrand, Princeton, 1958.
6. This is essentially a generalization of the procedure used by van Laar and Baehr. See references in J. M. H. Levelt-Sengers AIChE Meeting, Washington, 1969.
7. The use of a parametric scheme to accommodate the asymmetry of real fluids has also been suggested by M. E. Fisher at the Irvine Conference on Critical Phenomena, Jan. 1970.
8. Rowlinson, J., and B. Widom: *J. Chem. Phys.*, **52**:1670 (1970). Also see Hemmer, P., and G. Stell: *Phys. Rev. Lett.*, **24**:1284 (1970).
9. Meyer, H.: Private communication; Wallace, B., and H. Meyer: *Phys. Rev.*, **A2**:1610 (1970).
10. Cooper, M. J.: "Expanded Formulation of Thermodynamic Scaling in the Critical Region," *NBS J. Res.*, **75A**:103 (1971); also M. J. Cooper: *Int'l. Conf. Magnetism 1970 J. de Physique*, in press.

DISCUSSION on Paper by M. J. Cooper

C. N. YANG: Why is $\epsilon_1 - \epsilon_2 = n$ (n: integer)?

L. P. KADANOFF: If you allow V_1 and V_2 to be independent of r, then you can always make the transformation replacing r by r to some power and the transformation just changes ϵ_1, ϵ_2.

The condition $\epsilon_1 - \epsilon_2$ equals to an integer seems from that point of view to be a condition which has no meaning. However, if you demand also that V_1 and V_2 are analytic in r about $r = 0$, then the absolute values of ϵ_1 and ϵ_2 become significant. In particular, if you require that the expandibility condition remains invariant under linear transformations like $\Psi_1 \to \Psi^*_1 = a\Psi_1 + b\Psi_2$, then this demand requires $\epsilon_1 - \epsilon_2$ equals an integer.

G. A. BAKER: Gaunt and Baker[1] and Baker and Rushbrooke[2] have examined the corrections to scaling numerically and theoretically for the Ising and Heisenberg model in mean field theory and find

$$M = \tau^{1/\delta}(1 - c\tau^{1-1/\delta} + \cdots)$$

How does this result compare with your corrections?

C. DOMB: Further to Dr. Baker's question, may I go into more details about correction terms for the Ising model. If certain plausible assumptions are made about the analytic form of the susceptibility and derivatives for the Ising model

$$\chi_0 = t^{-\gamma}\phi(t) + \Psi(t)$$

where $\Psi(t)$ and $\phi(t)$ are analytic at $t = 0$, it is possible to derive correction terms. One finds a whole series of corrections terms

$$\Delta\mu = \Delta\rho^\delta \Big[Q_0 + |\Delta\rho|^{1/\beta} + Q_2|\Delta\rho|^{2/\beta} + \cdots$$
$$+ b_1|\Delta\rho|^{\gamma/\beta} + b_2|\Delta\rho|^{(\gamma+1)/\beta} + \cdots$$
$$+ c_1|\Delta\rho|^{2\gamma/\beta} + c_2|\Delta\rho|^{(2\gamma+1)/\beta} + \cdots \Big]$$

I think the first series corresponds to Dr. Cooper's term but I suspect that the condition of $\epsilon_1 - \epsilon_2$ being an integer gives rise to a power series instead of a single correction term. (That is, I think, equivalent to what Dr. Kadanoff said, i.e., the fact that one assumes analyticity leads to a power series.)

However, there will not be just one correction term, but a series of correction terms as above (Baker and Gaunt consider only one such series). Therefore, I think that the situation can be much more complicated; but this paper represents a possible first exploration of correction terms.

R. B. GRIFFITHS: I have not been able to sort out these correction terms and I wonder if Baker's, Domb's correction terms, and Cooper's might be put in similar notations. How do they compare on the critical isotherm?

M. J. COOPER: Putting our equations into the magnetic language $H(M,T)$, I find that the behavior along the critical isotherm is given by

$$H = M^{\epsilon_0/\epsilon_1}\{h_0 + \cdots + h_r M^{r/\epsilon_1} + \cdots\}$$
$$+ M^{1+1/\epsilon_1}\{g_0 + \cdots + g_s M^{s/\epsilon_1} + \cdots\}$$

with r, s integer.

By an appropriate identification of terms ($\epsilon_0 = \beta\delta$, $\epsilon_1 = \beta$), this expression can be cast into a form quite like that found by Domb for the two- and three-dimensional Ising models.

REFERENCES

1. Gaunt, D. S., and G. A. Baker: *Phys. Rev.,* B3:1, 1184 (1970).
2. Baker, G. A., and G. S. Rushbrooke: *Ibid.,* to be published.

SELF-AVOIDING WALKS AND SCALING

C. Domb
Physics Department
King's College
University of London
London, England

ABSTRACT

The geometrical properties of self-avoiding walks can give insight into the nature of the critical behavior of the Ising model. It is shown that the property of tending to a limiting shape for a self-avoiding walk is equivalent to the standard scaling form of the spin pair correlation function for the Ising model. The failure of "strong scaling" in three dimensions can be associated with the anomalous behavior of a self-avoiding walk near the origin suggested by McKenzie and Bloomfield.

1. INTRODUCTION

A self-avoiding walk on a lattice is a random walk subject to the condition that no lattice site may be visited more than once in the walk. Although the mathematical problem of calculating the

TABLE 1. Passage from a Random to a Self-avoiding Walk and Ising Analogs

	Normal random walks	Restricted walks (order r)	Self-avoiding walks Two dimensions	Self-avoiding walks Three dimensions	Ising analog
Number of walks of n steps	$c_{n0} = q^n$	$c_{nr} = A_{lr}\lambda_{lr}^n + \cdots A_{Nr}\lambda_{Nr}^n$	$c_n \sim A\mu_n^{n\,1/3}$	$A\mu_n^{n\,1/6}$	Magnetic susceptibility
Number at origin after n steps	$u_{n0} \sim B_0 q^n/n^{d/2}$	$u_{nr} \sim B_{lr}\lambda_{lr}^n/n^{d/2}$	$u_n \sim B\mu_n^{n}n^{-3/2}$	$B\mu_n^{n}n^{-3/4}$	Specific heat (zero field)
Fraction at origin after n steps	$u_{n0}/c_{n0} \sim B_0/n^{d/2}$	$u_{nr}/c_{nr} \sim B_{lr}/A_{lr}n^{d/2}$	$u_n/c_n \sim n^{-11/6}$	$n^{-23/12}$	
Mean square length	$\langle R_{n0}^2 \rangle = n$	$\langle R_{nr}^2 \rangle \sim C_{0r}n$	$\langle R_n^2 \rangle \sim C_n^{3/2}$	$Cn^{6/5}$	Mean square correlation range
Probability distribution of x coordinate of end point	$F_{0n}(x) \sim \dfrac{1}{\sqrt{2\pi n}}\exp\left(-\dfrac{x^2}{2n}\right)$	$F_{rn}(x) \sim \dfrac{1}{\sigma_{rn}\sqrt{2\pi}}\exp\left(-\dfrac{x^2}{2\sigma_{rn}^2}\right)$	$F_n(x) \sim \exp\left(-\dfrac{x^4}{\sigma_n}\right)$	$F_n(x) \sim \exp\left(-\dfrac{x}{\sigma_n}\right)^{5/2}$	Spin-pair correlation function

properties of self-avoiding walks is formidable, and there are no exact analytical solutions, there have been many numerical investigations from which a pattern of behavior has been conjectured. This behavior differs significantly from that of a normal random walk on the lattice, or a restricted walk of finite order r (defined as a random walk in which all polygonal closures of order $\leq r$ are forbidden). The differences are summarized in Table 1 (taken from Ref. 1).

There is a close correspondence between geometrical properties of self-avoiding walks and physical properties of the Ising model, as indicated in the last column of Table 1. The thermodynamic property of the Ising model can be expanded in a high temperature series whose nth coefficient is obtained by enumerating certain configurations of $l(\leq n)$ lines on the crystal lattice. Among these configurations the most significant are those corresponding to the self-avoiding walk.

For example, the logarithm of the partition function for the Ising model can be expanded as a series of $w(= \tanh \beta J)$ where J is the energy of interaction of a pair of parallel spins

$$\ln Z = \sum_n a_n w^n \qquad (1)$$

and the coefficients a_n can be expressed as

$$a_n = \sum_{l\ n} p_{lx} g_{lx} \qquad (2)$$

Here the configurations which enter are multiply connected graphs of l lines, the suffix x being used to differentiate the types of graph. p_{lx} is the number of such configurations which can be constructed from l bonds of the lattice (called the "lattice constant" or number of weak embeddings), and the g_{lx} are simple weighting numbers of order unity. The most significant graph is the self-avoiding polygon of n sides p_n.

For the magnetic susceptibility in zero field we have an analogous expansion

$$\frac{kT\chi_0}{m^2} = \sum_n d_n w^n \qquad (3)$$

where m is the magnetic moment of a single spin. The coefficients d_n

can be expressed as

$$d_n = \sum c_{lx} k_{lx} \qquad (4)$$

where the c_{lx} refer to multiply connected graphs which have been broken between two points (for further details see Ref. 2). The most significant graph is now the simple chain c_n.

If we ignore all terms other than those corresponding to self-avoiding walks we obtain the "self-avoiding walk approximation," for example,

$$\ln Z = \sum p_n w^n \qquad (5)$$

$$\frac{kT\chi_0}{m^2} = \sum c_n w^n \qquad (6)$$

Numerical studies have established the asymptotic behavior recorded in columns 3, 4 of Table 1 ($u_n = np_n$). Hence if we substitute in Eqs. (5) and (6) we find that the critical temperature is given by

$$w_c = \frac{1}{\mu} \qquad (7)$$

where μ is a constant of the lattice called the "connective constant" or "self-avoiding walk limit." The critical exponents, using standard terminology,[3] are given by

$$\gamma = \tfrac{4}{3} \quad \alpha = \tfrac{1}{2} \text{ in 2 dimensions}$$
$$\gamma = \tfrac{7}{6} \quad \alpha = \tfrac{1}{4} \text{ in 2 dimensions} \qquad (8)$$

The value of w_c is in error by about 10 percent in two dimensions and 2 percent in three dimensions; the correct values of the exponents are

$$\gamma = \tfrac{7}{4} \quad \alpha = 0 \text{ in 2 dimensions}$$
$$\gamma = \tfrac{5}{4} \quad \alpha = \tfrac{1}{8} \text{ in 3 dimensions} \qquad (9)$$

We thus see that the approximation is far better in three dimensions than in two dimensions, and, in fact, it improves as the dimension increases when a self-avoiding walk approaches a normal random walk.

It is possible to add correction terms to Eqs. (5) and (6) in a systematic manner[2] toward Eqs. (1) and (3). However, we shall not pursue this matter here, but will instead concern ourselves with the form of the pair correlation function in the self-avoiding walk approximation. We shall find that this furnishes considerable insight into the scaling hypothesis and the failure of strong scaling reported recently.[4,5]

The spin-pair correlation function between two sites R distance apart is defined by

$$\Gamma(\underline{R}) = \langle \sigma_0 \sigma_{\underline{R}} \rangle \tag{10}$$

and can be expanded in a parallel manner to Eq. (3)

$$\Gamma(\underline{R}) = \sum d_n(\underline{R}) w^n \tag{11}$$

Here $d_n(\underline{R})$ are the subgroup of configurations of d_n whose ends are distant $\underline{R}$ apart. The self-avoiding walk approximation to $\Gamma(\underline{R})$ is

$$\Gamma(\underline{R}) = \sum c_n(\underline{R}) w^n \tag{12}$$

where $c_n(\underline{R})$ are self-avoiding walks whose ends are $\underline{R}$ distance apart. Equation (12) will be the basis of our discussion in the rest of the paper.

2. SHAPE OF A SELF-AVOIDING WALK

If we wish to test whether a particular type of walk tends to a limiting shape as n increases, we must first reduce all the walks to scale, and a convenient "standard length" for this purpose is the rms length

$$\sigma_n = \sqrt{\langle R^2 \rangle} \simeq A n^\nu \tag{13}$$

We therefore write $\underline{R}/\sigma_n = \underline{u}$, and to ensure correct normalization in

d dimensions we construct the function

$$f_n(\underline{u}) = \frac{c_n(\underline{R})}{c_n}(\sigma_n)^d \tag{14}$$

and investigate whether it tends to a limiting form $f(\underline{u})$ as $n \to \infty$.

Let us first consider a normal random walk on a lattice[6] for which σ_n is equal to $\sqrt{n}$ in all dimensions. The resulting distribution is Gaussian. In one dimension we have

$$\frac{c_n(\underline{R})}{c_n} \sim \frac{1}{\sqrt{2\pi n}} \exp\left(-\frac{R^2}{2n}\right) \tag{15}$$

so that we find

$$f(u) = \frac{1}{\sqrt{2\pi}} \exp\left(-\frac{u^2}{2}\right) \tag{16}$$

In two dimensions

$$\frac{c_n(\underline{R})}{c_n} = \frac{1}{\pi n} \exp\left(-\frac{R^2}{n}\right) \tag{17}$$

so that

$$f(\underline{u}) = \frac{1}{\pi} \exp(-u^2) \tag{18}$$

In three dimensions

$$\frac{c_n(\underline{R})}{c_n} = \left(\frac{3}{2\pi n}\right)^{3/2} \exp\left(-\frac{3R^2}{2n}\right) \tag{19}$$

so that

$$f(\underline{u}) = \left(\frac{3}{2\pi}\right)^{3/2} \exp\left(-\frac{3}{2}u^2\right) \tag{20}$$

It is easy to establish the general form in d dimensions

$$f(\underline{u}) = \left(\frac{d}{2\pi}\right)^{d/2} \exp\left(-\frac{d}{2} u^2\right) \qquad (21)$$

For a self-avoiding walk Domb, Gillis, and Wilmers[7] examined statistics of exact enumerations of $c_n(\underline{R})$ and made the following conjectures:

1. $f_n(\underline{u})$ tends to a limiting function $f(\underline{u})$ as $n \to \infty$.
2. The limiting function has spherical (or circular) symmetry, and can therefore be written as $f(u)$.
3. The shape of $f(u)$ is no longer Gaussian and the decay can be represented by $\exp - u^k$ with k equal to 4 for the simple quadratic lattice, and 5/2 for the simple cubic lattice.

Subsequent investigations by D. S. McKenzie[8] suggested that the shape function $f(u)$ is independent of lattice structure as for standard random walks.

Although Domb et al.[7] obtained an expression for the function $f(u)$, they concentrated on the region away from the origin, and their estimates of behavior near $u = 0$ were inaccurate. A subsequent investigation by McKenzie and Bloomfield[9] examined the form of $f(u)$ much more carefully, and came to the conclusion that near $u = 0$ in three dimensions a different scaling length to σ_n was required if the functions $f_n(u)$ were all to tend to the same limit.

We illustrate these points in more detail by direct plots of the function $f_n(u)$ from Eq. (14) in Fig. 1 for the S.Q. lattice ($n = 16$–18) and in Fig. 2 for the S.C. lattice ($n = 12, 13$). We first notice that already for these values of n the points fall well into shape, and deviations arising from lattice structure, anisotropy, and different n are quite small. We also note the very significant difference between $f(u)$ for self-avoiding walks and normal random walks. This difference is most marked at $u = 0$ where $f(u)$ is zero for self-avoiding walks. It is much less noticeable in the length distributions of a self-avoiding walk since the function $f(u)$ is then multiplied by u^{d-1} (see Figs. 4 and 5 of Ref. 7).

Let us now investigate more carefully the behavior near $u = 0$. For any value of u for which $f(u)$ is non-zero we know from Eq. (14) that

$$c_n(R) \sim c_n \sigma_n^{-d} \qquad (22)$$

for large n. In two dimensions ($\nu = 3/4$) this means

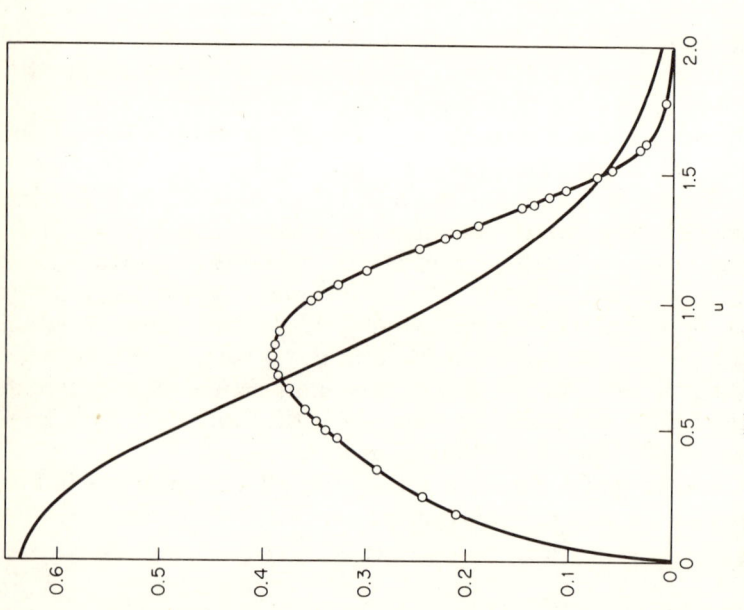

Fig. 1 Plot of $(c_n^{(R)}/c_n)\sigma_n^d$ for self-avoiding walks on the simple quadratic lattice. Data correspond to $n = 16-18$ and the smooth curve represents limiting behavior. The second curve represents the same function for a normal random walk.

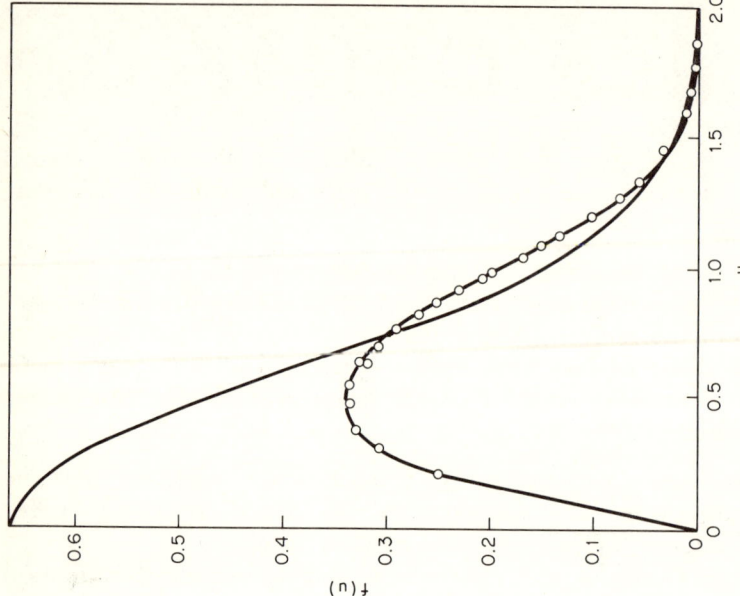

Fig. 2 Plot of $(c_n^{(R)}/c_n)\sigma_n^d$ for self-avoiding walks on the simple cube lattice. Data correspond to $n = 12, 13$ and the smooth curve represents limiting behavior. The second curve represents the same function for a normal random walk.

$$\frac{c_n(R)}{c_n} \sim n^{-3/2} \qquad (23)$$

and in three dimensions ($\nu = 3/5$)

$$\frac{c_n(R)}{c_n} \sim n^{-9/5} \qquad (24)$$

However, at the origin $c_n(R)/c_n$ is given by u_n (in fact the self-avoiding condition ensures that $c_n(0)$ is always zero; but u_n is equal to $c_{n-1}(1)$ at the nearest neighbor distance). From Table 1 it is clear that u_n/c_n differs significantly from the values in Eqs. (23) and (24), the difference being well above any likely numerical error in the estimation of indices.

Hence we conclude that

$$\frac{c_n(1)}{c_n} \sigma_n^d \simeq \frac{1}{n^\epsilon} \qquad (25)$$

where $\epsilon = 1/3$ in two dimensions and $7/60$ in three dimensions. This means that $f(0)$ is zero, but we note that any positive value of ϵ would have ensured this same result. If we represent the behavior of $f(u)$ near $u = 0$ by u^g we can then describe as "strong scaling"[5] the situation in which g has exactly the value required to satisfy Eq. (25), that is,

$$g = \frac{\epsilon}{\nu} \qquad (26)$$

If Eq. (26) is not satisfied we describe the situation as "weak scaling." We have already mentioned numerical evidence[9] that in three dimensions strong scaling does not hold. In this case a second length τ_n enters to measure the region near the origin in which strong scaling fails; we must have $\tau_n/\sigma_n \to 0$ as $n \to \infty$ to preserve weak scaling.

3. SCALING FORM OF SPIN-PAIR CORRELATION FUNCTION

We now use the results of the last section to calculate the spin-pair correlation function in the self-avoiding walk approximation. From

Eqs. (12) and (14) we have

$$\Gamma(\underline{R}) = \sum_n c_n (An^\nu)^{-d} f\left(\frac{R}{An^\nu}\right) w^n$$

$$= BA^d \sum_n n^{\gamma-\nu d-1} \exp[n \ln(\mu w)] f\left(\frac{R}{An^\nu}\right) \text{ (taking } c_n \simeq Bn^{\gamma-1}\text{)} \tag{27}$$

Now $\mu = 1/w_c$, and for T near to w_c write

$$\ln(\mu w) = \ln \frac{w}{w_c} = -t \simeq -a\left(\frac{T}{T_c} - 1\right) \tag{28}$$

where

$$a = \left(\mu - \frac{1}{\mu}\right)\tanh^{-1} \frac{1}{\mu} \tag{29}$$

is ~ 1 for large μ. Also for the small t we can replace the sum by an integral in a new variable $v = nt$

$$\Gamma(\underline{R}) = BA^d \int_0^\infty \left(\frac{v}{t}\right)^{\gamma-\nu d-1} t^{-1} \exp\left[-vf\left(\frac{Rt^\nu}{Av^\nu}\right)\right] dv = t^{\nu d-\gamma} \Phi\left(\frac{Rt^\nu}{A}\right) \tag{30}$$

where

$$\Phi(x) = BA^d \int_0^\infty v^{\gamma-\nu d-1} \exp - vf(xv^{-\nu}) dv \tag{31}$$

If we compare Eq. (30) with the standard form[3]

$$\Gamma(\underline{R}) \sim R^{2-d-\eta} \Psi(\kappa R) \tag{32}$$

we see that a complete identification is possible if we take

$$\kappa = \frac{t^\nu}{A} \tag{33}$$

and

$$\Phi(x) = x^{(\gamma-d\nu)/\nu}\Psi(x) \tag{34}$$

We then have the relation

$$2 - \eta = \frac{\gamma}{\nu} \tag{35}$$

which gives the numerical values for η of 2/9 in two dimensions (correct value 1/4) and 1/18 in three dimensions (very close to exact numerical estimates). We also have, using a new variable of integration $V = vx^{-1/\nu}$,

$$\Psi(x) = BA^d \int_0^\infty V^{\gamma-1-d\nu} \exp(-Vx^{1/\nu}) f(V^{-\nu}) dV \tag{36}$$

To proceed further we must use a suitable analytic representation of $f(u)$, and one which simulates the behavior adequately for small and large u is[9]

$$f(u) = Cu^g \exp(-u^k) \tag{37}$$

This leads to an integral for $\Psi(x)$ of the form

$$\int_0^\infty V^{-h} \exp[-Vx^{1/\nu} - V^{-(k-1)}] dV \tag{38}$$

where

$$h = (d + g)\nu - \gamma + 1 \tag{39}$$

and we have used Fisher's relation

$$\nu k + 1 = k \tag{40}$$

We note that the exponential factors at $V = 0$ and $V = \infty$ ensure the convergence of the integral in Eq. (38).

4. DETAILED CALCULATION OF THE SPIN-PAIR CORRELATION FUNCTION

The integral in Eq. (38) can be evaluated analytically for $k = 2$ (normal random walk, $\nu = 1/2$), and special values of h. Thus we find that when $h = 1$

$$\int_0^\infty V^{-1} \exp[-Vx^2 - V^{-1}]dV = 2K_0(2x) \qquad (41)$$

and this corresponds to a normal random walk in two dimensions. For d dimensions we require the integral corresponding to $h = d/2$, and when d is even we can obtain this from Eq. (41) by transformation and differentiation. These exact results can serve as a useful check on calculations for arbitrary k and h.

More generally for large x we can use the saddle point method to obtain an asymptotic expansion for Eq. (38). The saddle point V^* is given by

$$x^{1/\nu} - (k-1)V^{*-k} = 0$$

$$V^* = (k-1)^{1/k} x^{-1/(k-1)} \qquad (42)$$

Hence the dominant contribution to the integral is

$$Dx^{[h-1/2(k+1)]/(k-1)} \exp(-cx) \qquad (43)$$

where

$$c = \nu^{-\nu}(1-\nu)^{-(1-\nu)} \qquad (44)$$

and

$$D = (2\pi)^{1/2} k^{-1/2} (k-1)^{(1/2-h)/k} \qquad (45)$$

It was pointed out by Moore[9] that we can use the fact that the Ornstein-Zernicke treatment is valid *away* from the critical point to determine the value of g. We then have for the index of the power multiplying the exponential (using Eqs. (34) and (39))

$$\frac{\gamma - d\nu}{\nu} + \frac{1}{k-1}[(d+g)\nu - \gamma + 1] - \frac{(k+1)}{2(k-1)} \qquad (46)$$

and this must reduce to the Ornstein-Zernicke value

$$\frac{1}{2}(1 - d) \tag{47}$$

which is obtained by giving the exponents in Eq. (46) their classical values. Hence from Eqs. (46), (47), and (40) we find that

$$g = \frac{1 - d/2 + d\nu - \gamma}{1 - \nu} \tag{48}$$

This value is appropriate for large x and is unlikely to be consistent with the value corresponding to small x from Eq. (26). We should really choose a second value g^* to deal with behavior for small x.

If we proceed to derive an asymptotic series multiplying Eq. (43) it is easy to show that the form is

$$1 + b_1/x + b_2/x^2 + \cdots \tag{49}$$

and that the coefficients $b_1, b_2, \ldots$ are similar expressions to Eq. (45).

For small x we can obtain an alternative expansion of the integral Eq. (38) in powers of $x^{1/\nu}$. When $x = 0$ it is easy to see that the integral reduces to

$$\frac{1}{k-1} \Gamma\left(\frac{h-1}{k-1}\right) \tag{50}$$

Differentiating successively with respect to $x^{1/\nu}$ we derive the expansion

$$\frac{1}{k-1} \left[\Gamma\left(\frac{h-1}{k-1}\right) - x^{1/\nu} \Gamma\left(\frac{h-2}{k-1}\right) + x^{2/\nu} \Gamma\left(\frac{h-3}{k-1}\right) \cdots \right] \tag{51}$$

When h is an exact integer there are logarithmic terms, as for example in the expansion of Eq. (41). Expansion (51) serves as a useful guide since no analytic information is available in this region for the Ising model.

We finally discuss briefly the consequences of a breakdown in scaling at the origin as mentioned in §2. We now expect a second scaling length $r_n = A^* n^{\nu^*}$, and an exponent ϵ^* which varies with distance satisfying

$$\frac{c_n(R)}{c_n} \sigma_n^d \simeq \frac{R^{g*}}{n^{\epsilon*}} \tag{52}$$

when R varies between zero and τ_n. When $R \sim \tau_n$ we expect ϵ^* to attain the true scaling value ϵ_1, and when $R \to 0$, $\epsilon^* \to \epsilon_0$, the value corresponding to Eq. (25). We must now add to Eq. (38) a term which takes into account the deviation from strong scaling as follows

$$\sum_n n^{\gamma-1-d\nu} R^{g*} \exp[-nt(n^{-\epsilon*} - n^{-\epsilon_1})] \tag{53}$$

This can be converted into an integral as before

$$\frac{R^{g*}}{t^{\gamma-1-d\nu}} \int_0^\infty v^{\gamma-1-d} \left[\left(\frac{v}{t}\right)^{-\epsilon*} - \left(\frac{v}{t}\right)^{-\epsilon_1}\right] \exp(-vt)\, dv \tag{54}$$

but we must remember that ϵ^* is to be treated as a function of R/τ_n or $Rt^{\nu*}/A^*v^{\nu*}$. The above treatment and formulas bring to mind those of Stell.[5] For a known variation of ϵ^* we could evaluate Eq (54) but in the absence of precise information we do not pursue the matter further.

ACKNOWLEDGMENT

The author is indebted to Dr. D. S. McKenzie for helpful discussion and for communicating his results before publication. This research has been supported (in part) by the U.S. Department of the Army through its European Research Office.

REFERENCES

1. Domb, C.: in K. E. Schuler, ed., "Advances in Chemical Physics," vol. 15, p. 229, John Wiley & Sons, Inc., New York, 1969.
2. ———: *J. Phys. Chem.*, **3**:256 (1970).
3. Fisher, M. E.: *Rep. Prog. Phys.*, **30**:615 (1967).
4. Ferer, M., M. A. Moore, and M. Wortis: *Phys. Rev. Lett.*, **22**:1382 (1969).
5. G. Stell: *Ibid.*, **24**:1343 (1970).
6. Montroll, E. W., and G. H. Weiss: *J. Math. Phys.*, **6**:167 (1965).
7. Domb, C., J. Gillis, and G. Wilmers: *Proc. Phys. Soc. (London)*, **85**:625 (1965).
8. McKenzie, D. S.: Private communication.

9. —— and V. A. Bloomfield: "End-to-End length Distribution of a Self-Avoiding Walk on a Regular Crystal Lattice," to be published.

DISCUSSION *on Paper by C. Domb*

G. B. BENEDEK: Is it possible to use this approximation to determine the function in the critical equation of state?

C. DOMB: The critical function involves determining the higher derivatives of the susceptibility with respect to magnetic field and this involves the interference of many chains. Whereas the approximation I described is meaningful for a single chain or polygon it gets more complex and less adequate for many chains. One must then pass to the droplet picture as introduced by Fisher but it is essential to take account of the exclusion property of droplets which Fisher ignores. It may be possible to combine the two into a coherent picture to obtain the critical function but this requires a good deal more thought.

M. KAC: If you assume, e.g., that your approximate formula for $\log Z$ in terms of the numbers u_n of nonself-intersecting random walks is *exact* is there a model which is consistent with this assumption?

Presumably the interactions would be temperature dependent but perhaps for some temperatures the corrections are not serious.

C. DOMB: This would mean that all higher order diagrams must vanish. I will look into this possibility.

L. GUTTMAN: Can you give more details on the deviation of your pair correlation function from the Ornstein-Zernike form?

C. DOMB: If the shape function is Gaussian one obtains the Ornstein-Zernike form. In fact, the shape differs from Ornstein-Zernike in two important features:
 1. The decay is more rapid than Gaussian ($\exp(-u^4)$ in two dimensions, $\exp(-u^{5/2})$ in three dimensions). This leads to a modification of Ornstein-Zernike to: $R^{2-d-\eta} \Psi(\kappa R)$.
 2. The origin is a "peculiar" point at which the shape function is zero and scaling is no longer essential at this point. It is still possible for scaling to hold but this would be a coincidence (as perhaps in two dimensions).

SECONDARY VARIABLES IN CRITICAL PHENOMENA, WITH APPLICATION TO λ TRANSITION IN LIQUID HELIUM

O. K. Rice and Do-Ren Chang

Department of Chemistry
University of North Carolina
Chapel Hill, North Carolina

ABSTRACT

Any critical point is described in terms of a pair of variables directly connected with the order-disorder phenomenon, e.g., pressure and molal volume in the liquid-vapor transition, magnetic field and magnetization in the magnetic transition, certain hidden variables in the λ transition of liquid helium, etc. Secondary variables can also influence the critical point; pressure and molal volume are secondary variables in the magnetic and helium transitions. If a locus of critical points or λ points in the $P - V$ plane (where P and V are secondary variables) is a locus along which one of the specific heats, C_v or C_p (with the subscripts again referring to secondary variables), tends to infinity, infinities occur in other thermodynamic functions. If the lattice or body forces, which depend directly on V or P, and the order-disorder forces (depending, typically, in the magnetic case, on spin orientation) are mutually independent, the partition function can be factored. Thus, if C_p tends to infinity, we write for the isothermal-isobaric partition function, $Z = Z_l Z_s$, where Z_l is a lattice factor and Z_s a "spin," or order-disorder, factor. Then changes of entropy, and molal volume, along the λ line can then be expressed in terms of lattice quantities and other changes along the λ line, essentially independent of spin quantities; an expression for C_v can also be obtained, which involves lattice quantities.

These ideas have been applied to liquid helium. If it be assumed that Z_l depends only on the phonons, it is not possible to get a completely satisfactory resolution of Z into factors. However, allowing for the possibility that other factors influence Z_l, a satisfactory resolution can be obtained, and the various equations are satisfied. The lattice contributions to the specific heat, the coefficient of expansion, and the compressibility, at the λ line, are found and discussed.

1. GENERAL THEORY

In the liquid-vapor transition the primary variables associated with the order-disorder phenomenon are the pressure P and the molal volume V. In a magnetic transition the corresponding primary variables are magnetic field and the magnetization of the material. The λ transition of a magnetic material is well known to be the critical point of a magnetic transition, the rest of the coexistence curve not appearing if the magnetic field is held at zero. The λ point of liquid helium is also a critical point, but the primary order-disorder variables are hidden, or at least cannot be manipulated by an experimenter.

In each of the latter two cases, and in many others, the pressure or the density affect the critical phenomena. Thus in liquid helium the λ temperature T_λ is a function of pressure, and in the $P - T$ plane there is a λ line. Then P and V are *secondary* variables. Other secondary variables are possible; here, however, we shall consider only P and V and apply the equations to liquid helium. We shall be especially interested in the specific heat when the secondary variables V or P are held constant; these will be denoted as C_v and C_p, respectively. In the case of liquid helium, since the primary variables cannot be experimentally controlled, there is no ambiguity in this notation. Likewise in a magnetic transition at fixed zero magnetic field no ambiguity would arise.

The thermodynamics of this situation has been fairly thoroughly discussed.[1-3] If C_v tends to become infinite a thermodynamic instability results, but this is not the situation if C_p tends to become infinite. There have been some suggestions that the instability or the infinity will be avoided due to the interaction between the motions of the individual atoms and the magnetic or other ordering.[4] However, it does seem reasonable to suppose that C_p approaches infinity in helium,[3] and the instability when C_v tends to infinity does in certain instances seem to have been observed experimentally.[5]

Garland and Renard[5] suggested that, as an approximation, the lattice properties of the order-disorder phenomenon could be considered separately, with the canonical partition function Q written as $Q_l Q_s$, where Q_l is the lattice part and Q_s is the "spin" part, where here we use the term "spin" as a semantic tool, in analogy to the Ising lattice. Q_s was supposed to depend upon J/kT where J is an energy factor (an energy associated with the relative direction of adjacent spins in the Ising lattice). They assumed J to depend only on V, so that C_v would tend to become infinite at a certain value of J/T.

In the case of helium, since it is C_p that tends to infinity, it is more natural to use[6] the isothermal-isobaric partition function

$$Z = \sum_L e^{-H_L/kT} \tag{1}$$

where H_L is the enthalpy of the Lth state of an assembly held at constant pressure, and where we write

$$Z = Z_l Z_s \tag{2}$$

Here Z_s depends only on J/kT where J is an enthalpy quantity that depends only on P. Then, for a certain value of J/T, the value of C_p becomes infinite. Since the Gibbs free energy is given by

$$G = -kT \ln Z \tag{3}$$

the thermodynamic functions can be found by appropriate differentiation, and they will all be composed of lattice and spin parts.

The following equations are readily derived by differentiation, designating the lattice part simply by the appropriate symbol with

108 CRITICAL PHENOMENA

the subscript l. Z'_s is the first, Z''_s the second derivative of Z_s with respect to J/kT

$$H = H_l - \frac{Z'_s}{Z_s} J \tag{4}$$

$$C_p = C_{p,l} - \frac{H_s^2}{kT^2} + \frac{Z''_s}{Z_s} \frac{J^2}{kT^2} \tag{5}$$

$$V = V_l - \frac{Z'_s}{Z_s} \frac{dJ}{dP}$$

$$= V_l + \frac{H_s}{J} \frac{dJ}{dP} \tag{6}$$

$$\left(\frac{\partial V}{\partial T}\right)_P = \left(\frac{\partial V}{\partial T}\right)_{P,l} + \frac{C_{p,s}}{J} \frac{dJ}{dP} \tag{7}$$

$$\left(\frac{\partial V}{\partial P}\right)_T = \left(\frac{\partial V}{\partial P}\right)_{T,l} + \frac{H_s}{J} \frac{d^2 J}{dP^2} - C_{p,s} \frac{T}{J^2} \left(\frac{dJ}{dP}\right)^2$$

$$= \left(\frac{\partial V}{\partial P}\right)_{T,0} - C_{p,s} \frac{T}{J^2} \left(\frac{dJ}{dP}\right)^2 \tag{8}$$

where $(\partial V/\partial P)_{T,0}$ is defined by the equation.

It is also possible to show in several different ways that

$$J^{-1} \frac{dJ}{dP} = \frac{T_\lambda^{-1} dT_\lambda}{dP} \tag{9}$$

Perhaps the best way to derive this equation is to note that, since Z_s depends on J/kT, the value of J/T must remain constant along the λ line. We have then

$$d\frac{J}{T} = -\frac{J}{T^2} dT_\lambda + T^{-1} \frac{dJ}{dP} dP_\lambda = 0$$

from which Eq. (9) follows immediately.

SECONDARY VARIABLES IN CRITICAL PHENOMENA 109

Since it is seen by Eqs. (4) and (5) that $C_{p,s}$ is a function of J/T, and since J/T is constant along the λ line, it follows from the integral

$$\int_0^{T_\lambda} \frac{C_{p,s}}{T} dT \bigg|_{P \text{ const}} = \int_0^{T_\lambda/J} \frac{C_{p,s} J}{T} d\frac{T}{J} \bigg|_{P \text{ const}}$$

that S_s is constant along the λ line. Thus we may write, since S_λ changes only on account of the lattice contribution

$$\frac{T_\lambda dS_\lambda}{dT} = T_\lambda \left(\frac{\partial S}{\partial T}\right)_{P,l,\lambda} + T_\lambda \left(\frac{\partial S}{\partial P}\right)_{T,l,\lambda} \frac{dP_\lambda}{dT}$$

$$= C_{p,l,\lambda} - T_\lambda \left(\frac{\partial V}{\partial T}\right)_{P,l,\lambda} \frac{dP_\lambda}{dT} \qquad (10)$$

(In earlier work the possibility was considered that the constancy of S_s might not hold, but it is now seen that this is inconsistent.)

A somewhat similar expression may be obtained for dV_λ/dT. We start by considering a locus in the $P - V$ plane such that C_p is constant. Along such a locus

$$\left(\frac{\partial V}{\partial T}\right)_{C_p} = \left(\frac{\partial V}{\partial T}\right)_P + \left(\frac{\partial V}{\partial P}\right)_T \left(\frac{\partial P}{\partial T}\right)_{C_p}$$

$$= \left(\frac{\partial V}{\partial T}\right)_{P,l} + \frac{C_{p,s}}{J} \frac{dJ}{dP} + \left[\left(\frac{\partial V}{\partial P}\right)_{T,0} - \frac{T}{J^2}\left(\frac{dJ}{dP}\right)^2 C_{p,s}\right]\left(\frac{\partial P}{\partial T}\right)_{C_p}$$

from Eqs. (7) and (8). Since $(\partial V/\partial T)_{C_p}$ approaches dV_λ/dT as C_p approaches infinity, the factor multiplying $C_{p,s}$ must approach zero. In Appendix I we show that

$$C_{p,s}\left[\left(\frac{\partial T}{\partial P}\right)_{C_p} - \frac{T}{J}\frac{dJ}{dP}\right]$$

(the bracket of which vanishes, giving at first an indeterminate form) approaches zero, so that, since $(\partial P/\partial T)_{C_p}$ itself remains finite

$$\frac{dV_\lambda}{dT} = \left(\frac{\partial V}{\partial T}\right)_{P,l,\lambda} + \left(\frac{\partial V}{\partial P}\right)_{T,0,\lambda} \frac{dP_\lambda}{dT} \tag{11}$$

The interest in Eqs. (10) and (11) lies in the fact that they express properties of the λ line in terms of the lattice quantities. Analogous equations for the case where $C_v \to \infty$ were obtained by Renard and Garland.[8]

From the thermodynamic relation between C_p and C_v

$$C_v = C_p + T\left(\frac{\partial V}{\partial T}\right)_P^2 \left(\frac{\partial P}{\partial V}\right)_T \tag{12}$$

using Eqs. (7)–(9) we obtain the following expression for C_v

$$C_v = C_{p,l} - \frac{a + (T_\lambda/T)b + a^2/4C_{p,s}}{1 - (T_\lambda/T)b/C_{p,s}} \tag{13}$$

where

$$a = 2T_\lambda \left(\frac{\partial V}{\partial T}\right)_{P,l} \frac{dP_\lambda}{dT} \tag{14}$$

and

$$b = T_\lambda \left(\frac{\partial V}{\partial P}\right)_{T,0} \left(\frac{dP_\lambda}{dT}\right)^2 \tag{15}$$

Equation (13) is equivalent to the expression obtained previously,[7] except that it contains no approximations. On the λ line Eq. (13) becomes

$$C_{v,\lambda} = C_{p,l,\lambda} - a_\lambda - b_\lambda \tag{16}$$

If we reduce Eq. (16) by the use of Eq. (11) then use Eq. (10) in the result we find

$$C_{v,\lambda} = \frac{T_\lambda dS_\lambda}{dT} - T_\lambda \frac{dV_\lambda}{dT} \frac{dP_\lambda}{dT} \tag{17}$$

SECONDARY VARIABLES IN CRITICAL PHENOMENA

Equation (17) was derived on general thermodynamic grounds by Buckingham and Fairbank.[3]

2. APPLICATION TO THE λ TRANSITION OF LIQUID HELIUM

A comparison of Eqs. (16) and (17) has been made for liquid helium[7] (without the full realization at the time that Eq. (17) could be derived from Eq. (16)). In order to do this, dV_λ/dT and dP_λ/dT were taken from the data of Kierstead,[9] values of $T_\lambda \, dS_\lambda/dT$ were taken from Buckingham and Fairbank,[3] and $(\partial V/\partial T)_{P,l}$ and $(\partial V/\partial P)_{T,0}$ were evaluated on the assumption that the lattice properties were to be deduced from the properties of the phonon spectrum. Since the compressibility is insensitive to temperature $(\partial V/\partial P)_{T,l}$ was found from the value of $(\partial V/\partial P)_T$ at low temperatures,[10] and for the other term in $(\partial V/\partial P)_{T,0}$ the value of H_s was obtained from data of Kramers, Wasscher, and Gorter,[11] combined with those of Lounasmaa and Kojo.[12] $(\partial V/\partial T)_{P,l}$ was evaluated from the formula for the coefficient of expansion obtained from the phonon spectrum[15]

$$\alpha_l = \frac{1}{V}\left(\frac{\partial V}{\partial T}\right)_{P,l} = \frac{16\pi^5 k^4 T^3}{15h^3 c^3}\left(\frac{1}{c}\frac{dc}{dP} + \frac{1}{3}\kappa_T\right) \tag{18}$$

where c is the velocity of sound and κ_T the compressibility, these being given their values far from the λ point. In similar fashion $C_{p,l}$ was evaluated from the phonon formula

$$C_{p,l} = \frac{16\pi^4 k^4 T^3}{15h^3 c^3 \rho} \tag{19}$$

where ρ is the density, and found to be negligible.

Fair agreement was found between Eqs. (16) and (17), but there were definite discrepancies. However, Eqs. (10) and (11) are much more sensitive, and by their use the difficulties are brought out much more graphically. The data and calculations for selected λ temperatures are presented in Table 1.

It is seen from the table that, although the calculated value of dV_λ/dT agrees with the observed value about as well as the two values of $C_{v,\lambda}$ agree,[7] there is a definite discrepancy, and there is a very serious discrepancy as concerns $T_\lambda \, dS_\lambda/dT$.

TABLE 1. Summary of Calculations on Liquid Helium Based on Eqs. (18) and (19)

Calculation of $(\partial V/\partial T)_{P,l,\lambda}$ from Eq. (18)				
T_λ	c	$c^{-1}dc/dP$	κ_T	$(\partial V/\partial T)_{P,l,\lambda}$
°K	m-sec^{-1}	atm^{-1}	atm^{-1}	cm^3-mole^{-1}-deg^{-1}
2.172	238	0.0353	0.0118	0.304
2.000	323	0.0144	0.0060	0.035
1.762	384	0.0102	0.00425	0.008

Calculation of $(\partial V/\partial P)_{T,0,\lambda}$				
T_λ	$-(\partial V/\partial P)_{T,l,\lambda}$	$H_{s,\lambda}$	$-T_\lambda^{-1}d^2T_\lambda/dP^2$	$-(\partial V/\partial P)_{T,0,\lambda}$
°K	cal-atm^{-2}-mole^{-1}	cal-mole^{-1}	atm^{-2} × 10^4	cal-atm^{-2}-mole^{-1}
2.172	0.00782	2.68	2.85	0.00858
2.000	0.00348	2.10	1.21	0.00373
1.762	0.00228	1.58	2.55	0.00268

Comparisons of dV_λ/dT and $T_\lambda dS_\lambda/dT$ and data along the λ line							
T_λ	ρ_λ	P_λ	$C_{p,l}$	$T_\lambda dS/dT_\lambda$		dV_λ/dT	
°K	g-cm^{-1}	atm	cal-mole^{-1}-deg^{-1}	cal-mole^{-1}-deg^{-1}		cm^3-mole^{-1}-deg^{-1}	
				obs.*	calc. Eq. (10)	obs.	calc. Eq. (11)
2.172	0.1461	0.050	0.2	5.6	2.0	43.8	39.7
2.000	0.1670	14.69	0.06	1.74	0.18	11.0	11.3
1.762	0.1804	29.81	0.02	1.40	0.04	5.2	6.1

*See Appendix II.

There are a number of possible reasons for discrepancies to appear. In the first place, the separation of the partition function into two independent factors may not be exact. It is also possible that J may depend on some other variable than pressure. It seems likely that even in this case Eq. (10) or a very near equivalent would hold (with $dS_{s,\lambda}/dT = 0$), but this cannot be proved if the isothermal-isobaric partition function is divided according to Eq. (2). It is also to be noted that Eq. (18) contains quantities which have singularities at the λ point, but these singularities were ignored.

We feel, however, that the most important difficulty lies with the way in which the division between lattice and spin quantities has been effected. It was, in fact, suggested by Buckingham and Fairbank[3] that the specific heat should be divided into *three* parts, the lattice part, a portion similar to that expected for a

Bragg-Williams type of transition with no infinity but a sharp drop in the specific heat at the λ point, and finally a logarithmically singular portion which they supposed might be symmetrical around the λ point. Since then it has become well known that the three-dimensional Ising lattice, in which only the order-disorder part is considered, has a singularity which is not symmetrical around the λ point. It appears unnecessary to assume a division into three parts, but it does seem likely that the rotons contribute something to the lattice portion as well as to the order-disorder part. The rotons are of course excitations, and these could build up without a singularity in the specific heat. The singularity may well be associated with interactions *between* the rotons.

In the light of these considerations, it seems reasonable to investigate whether the thermodynamic data for liquid helium can be consistently represented using a somewhat different division between lattice and spin parts. If one examines the plot of a thermodynamic quantity taken as a function of temperature at constant pressure one can interpolate the nonsingular part across the λ region, and thus obtain an estimate of the lattice part of the particular thermodynamic quantity. In actually evaluating the lattice part, however, it may be convenient to modify this procedure somewhat.

The specific heat, near the λ point at the vapor pressure (which is essentially C_p at a very low pressure) was expressed by Buckingham and Fairbank[3]

$$C_p = 4.55 - 3.00 \log_{10}(T_\lambda - T) \text{ joule } g^{-1} \text{ deg}^{-1} \quad T < T_\lambda$$
$$C_p = -0.65 - 3.00 \log_{10}(T - T_\lambda) \text{ joule } g^{-1} \text{ deg}^{-1} \quad T > T_\lambda \quad (20)$$

In order to extend the temperature range, Buckingham and Fairbank inserted certain exponential factors and added the Debye specific heat (which is negligible except at low temperatures). These expressions include both the lattice and spin parts, and our present assumption is that the Debye term does not completely account for the lattice part. By inspection we conclude that at the λ point $C_{p,l}$ is equal to about 1.5 joule-g^{-1}-deg^{-1}, so we propose to modify the expressions of Buckingham and Fairbank to get the representations of $C_{p,s}$ which follow

$$C_{p,s} = [92.4 - 90.9 \log_{10}(T_\lambda - T)] e^{-7.4/T} \text{ joules } g^{-1} \text{ deg}^{-1}$$
$$T < T_\lambda \quad (21)$$
$$C_{p,s} = [-11.77 - 16.42 \log_{10}(T - T_\lambda)] e^{-3.7/T} \text{ joules } g^{-1} \text{ deg}^{-1}$$
$$T > T_\lambda$$

These are equivalent to $C_p - 1.5$ of Eq. (20) near the λ point. Now as we have seen above, $C_{p,s}$ should be a function only of J/T or T_λ/T. We write $T = xT_\lambda$ and substitute into Eq. (21). We realize, then, that all terms and factors which do not contain x should be independent of T_λ. Since Eq. (21) was set up for the normal λ point, $T_\lambda = 2.172$, we may evaluate $C_{p,s}$ numerically for any T_λ by substituting $T_\lambda = 2.172$. We thus obtain

$$C_{p,s} = [61.8 - 90.9 \log_{10}(1-x)]e^{-3.4/x} \text{ joules g}^{-1} \text{ deg}^{-1} \qquad T < T_\lambda$$

$$C_{p,s} = [-17.30 - 16.42 \log_{10}(x-1)]e^{-1.7/x} \text{ joules g}^{-1} \text{ deg}^{-1}$$

$$T > T_\lambda \qquad (22)$$

These equations may not hold over the complete range of temperatures, but should hold over a fairly extended range about the λ point. $C_{p,l,\lambda}$ was chosen to best fit the data at higher pressures and in all cases $C_{p,l}$ was assumed to vary as $T^{5.27}$. The reason for this is that we found that $(\partial V/\partial T)_{P,l}$ could be well represented by a function $AT^{5.27}$ and it seemed a reasonable guess to make them parallel each other. In Fig. 1 we show the curves obtained for the specific heat and compare them with the experimental data, which are seen to be well represented.

In the case of $(\partial V/\partial T)_P$ the situation is somewhat complicated because there are positive and negative contributions. $(\partial V/\partial T)_P$ is positive at low temperature, but eventually becomes negative at $x \simeq 1/2$, and again positive slightly above T_λ. So $(\partial V/\partial T)_{P,l}$ was evaluated at the points[10] on the saturated vapor curve where $(\partial V/\partial T)_P = 0$ by using Eq. (7), since the last term in that equation could be calculated. It was found that the form $AT^{5.27}$ fit these two points, and this was used to evaluate $(\partial V/\partial T)_{P,l}$ at other temperatures. Adjusting the value of A, the same formula was used at other pressures. The values of A were adjusted to make $(\partial V/\partial T)_P = 0$ at the proper temperature above the λ point.[10b,13] The temperature at which $(\partial V/\partial T)_P = 0$ below the λ point[10a,14] could then be calculated. At 14.69 atm. the value thus calculated is 0.68° and the observed value is 0.66°, while at 29.81 atm. the calculated value is 0.59° and the observed value is less than 0.6°. The expressions used resulted in the values of $(\partial V/\partial T)_{P,l,\lambda}$ shown in Table 2. The calculated values of the expansion coefficient, by Eqs. (7) and (22), are compared with the experimental data in Fig. 2.

Having $(\partial V/\partial T)_{P,l}$, we may find $(\partial V/\partial P)_{T,0}$ from Eq. (11) [and $(\partial V/\partial P)_{T,l,\lambda}$ may then be found by estimating $H_{s,\lambda}$ using Eq. (22) (see Table 2)]. It is then possible to calculate $(\partial V/\partial P)_T$ from Eq. (8)

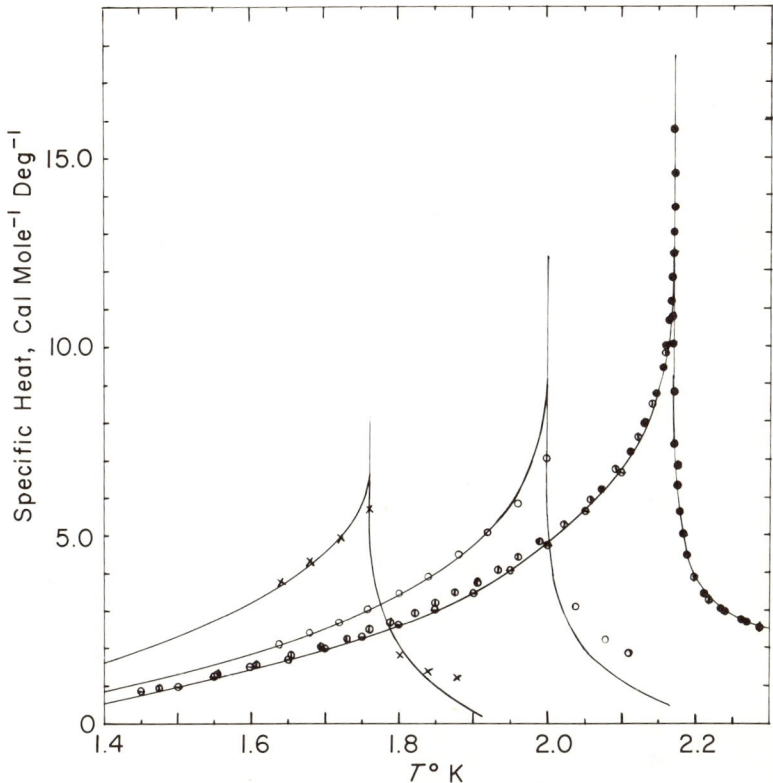

Fig. 1 Specific heats. I. C_p at normal saturated vapor pressure: $T_\lambda = 2.172°K$, $P_\lambda = 0.05$ atm. Solid line: Present calculations. ●Buckingham and Fairbank,[3] ⓪: Lounasmaa and Kojo,[12] ⊖: Kramers, Wasscher and Gorter, Ref. 11. II. C_v at $P = 14.69$ atm., $T_\lambda = 2.00°K$. Solid line: present calculations. ○: Lounasmaa, Ref. 16 at $\rho = 0.1650$, $T_\lambda = 2.02°K$, $P_\lambda = 13.0$ atm. for direct comparison, points should be shifted about $0.02°$ to left. III. C_v at $P = 29.8$ atm, $T_\lambda = 1.762°K$. Solid line present calculations. x: Lounasmaa,[16] supercooled liquid ($T < T_\lambda$).

and C_v from Eq. (13). The latter is shown in Fig. 1 for the higher pressures and compared with experimental data and $(\partial V/\partial P)_T$ is compared with data in Figs. 3 and 4. $(\partial P/\partial T)_V$ is calculated from $-(\partial V/\partial T)_P/(\partial V/\partial P)_T$ and shown against the data in Fig. 5. Equation (11) is valid, of course, only in the immediate neighborhood of the λ line, where the lines of constant C_p parallel the λ line. Its use outside this region serves to give only a rough idea of the magnitude of $(\partial V/\partial P)_{T,0}$. This is satisfactory for C_v because just in the region where this procedure breaks down the terms in b start to predominate in the fraction in Eq. (13) and the error cancels. The

TABLE 2. Values of Thermodynamic Quantities for Liquid Helium Based on Present Calculations

P_λ atm	$C_{p,l,\lambda}$ cal-mole^{-1}-deg^{-1}	$(\partial V/\partial T)_{P,l,\lambda}$ cm^3-mole^{-1}-deg^{-1}	$H_{s,\lambda}$ from Eq. (22) cal-mole^{-1}	$-(\partial V/\partial P)_{T,l,\lambda}$ cal-atm^{-2}-mole^{-1}	$T_\lambda\, dS_\lambda/dT$ from Eq. (10) cal-mole^{-1}-deg^{-1}
0.050	1.43	0.70	2.1	0.00877	5.5
14.69	0.333	0.44	1.9	0.00326	1.88
29.81	0.256	0.49	1.7	0.00162	1.41

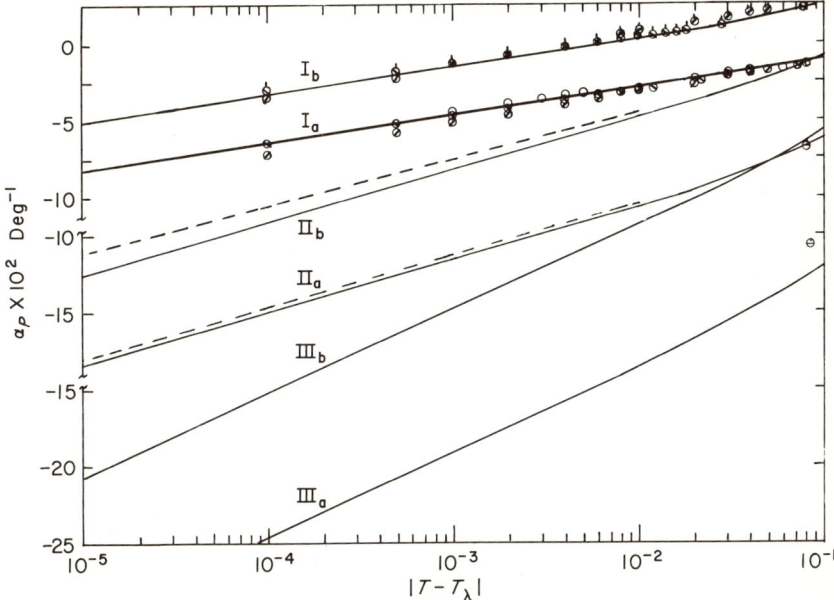

Fig. 2 *Thermal expansion coefficients. Cases I, II, and III as in Fig. 1. Solid lines: present calculations, a refers to $T < T_\lambda$; b and upward bar attached to experimental points refer to $T > T_\lambda$. I.* ○: *Atkins and Edwards,*[15] *and Phys. Rev., 97:1429 (1955).* ⊘: *Kerr and Taylor: Ann. Phys., 26:292 (1964).* ⊘: *Chase, Maxwell, and Millett: Physica, 27:1129 (1961). II. Dashed line: estimated from Ref. 10b. III.* ⊖: *Mills and Grilly: R. O. Davis (ed.), Proc. 8th Int. Conf. on Low Temps., 1963.*

values of $(\partial V/\partial P)_T$ and $(\partial P/\partial T)_V$ can be properly compared with experimental data only close to the λ point.

We now have the data on hand to make another test of Eq. (10). In Table 2 we collect the values of the various quantities along the λ line and show the new calculated values of $T_\lambda dS_\lambda/dT$ from Eq. (10). These may be compared with the observed values shown in Table 1. It will be noted that the agreement is now reasonably good, probably within experimental error. This is, perhaps, not surprising, since the values of the lattice quantities were chosen to give good agreement with other experimental data. We may conclude, in any case, that the scheme of separation of lattice and spin variables which we have outlined is capable of giving a consistent analysis of the experimental data.

The comparison of Tables 1 and 2 bears out the general idea that the roton excitation contributes to the lattice quantities. The values

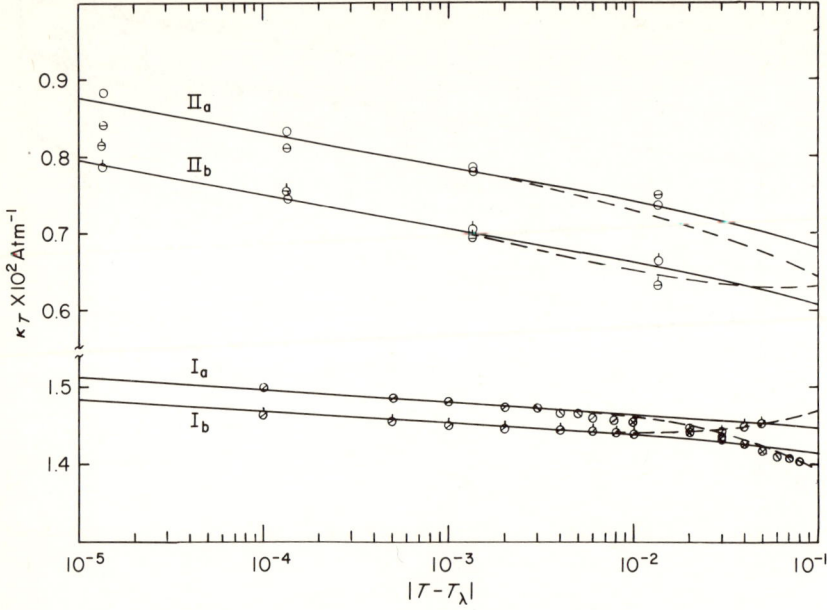

Fig. 3 *Compressibility. Cases Ia, Ib, IIa, IIb, (IIIa, and IIIb in Fig. 4), significance of solid lines and upward bar on experimental points, as in Fig. 2. Dashed lines from velocity of first sound. I.* ○: *Atkins and Edwards.* ⊘: *Chase, Maxwell, and Millett. For references see Fig. 2. II. Data points calculated from empirical formulas,* ○: *Grilly: Phys. Rev., 149:97 (1966).* ⊖: *Elwell and Meyer.*[10b]

$C_{p,l,\lambda}$ shown in Table 2 are considerably larger than those shown in Table 1, calculated from the phonon spectrum. The values of $(\partial V/\partial T)_{P,l,\lambda}$ in Table 2 are also greater than those shown in Table 1, and it will be noted that they are positive, while the contributions of individual rotons to $(\partial V/\partial T)_P$ are negative. That the latter statement is true is indicated by the fact that $(\partial V/\partial T)_P$ becomes negative between 1.1° and 1.2°K, where interaction between the rotons is negligible. It was noted by Atkins[15] and Edwards that the coefficient of expansion could be divided into a positive phonon part and a negative roton part, the former being calculated from Eq. (18) and the latter from the properties of the individual rotons (including the energy of excitation and its variation with density). This suggests that the positive contribution of the rotons to the lattice part of the coefficient of expansion is, at least in part, an indirect one. Indeed, Eq. (18) must break down in the neighborhood of the λ point since the terms in the parentheses diverge.

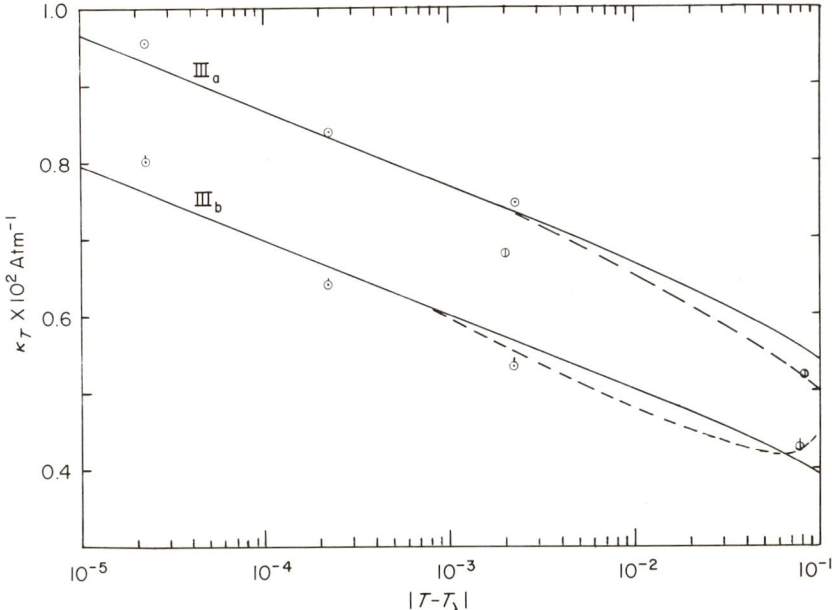

Fig. 4 *Compressibility, continuation of Fig. 3. III. ⊙: Kierstead,[13] ⊕: Mills and Grilly (see Fig. 2).*

The temperature behavior of $(\partial V/\partial T)_{P,l}$ is of interest. In Fig. 6 we have plotted our representation of $(\partial V/\partial T)_{P,l}$, namely $AT^{5.27}$, against $x = T/T_\lambda$ and compared it with the value calculated from Eq. (18). It is seen that at $x = 0.5$, close to the temperature at which the observed $(\partial V/\partial T)_P$ becomes negative, $AT^{5.27}$ is lower than the value obtained from Eq. (18), reflecting to some extent the negative effect of the rotons. At lower values of x, however, around 0.2 or 0.3, the two curves should coincide, for at this point only phonon effects persist. It thus is seen that $AT^{5.27}$ cannot be a perfect representation of $(\partial V/\partial T)_{P,l}$ and, indeed, considering the varied factors which contribute to the latter quantity, this could not be expected. Nor can the representation, either $AT^{5.27}$ or that of the spin part as derived from Eq. (22), be perfect above the λ point, since discrepancies begin to be seen when comparison is made with experimental data. We may hope, however, that our representation is good near the λ line.

Of considerable interest is the fact that at the higher pressures $-(\partial V/\partial P)_{T,l,\lambda}$ is smaller than the value shown in Table 1, which is

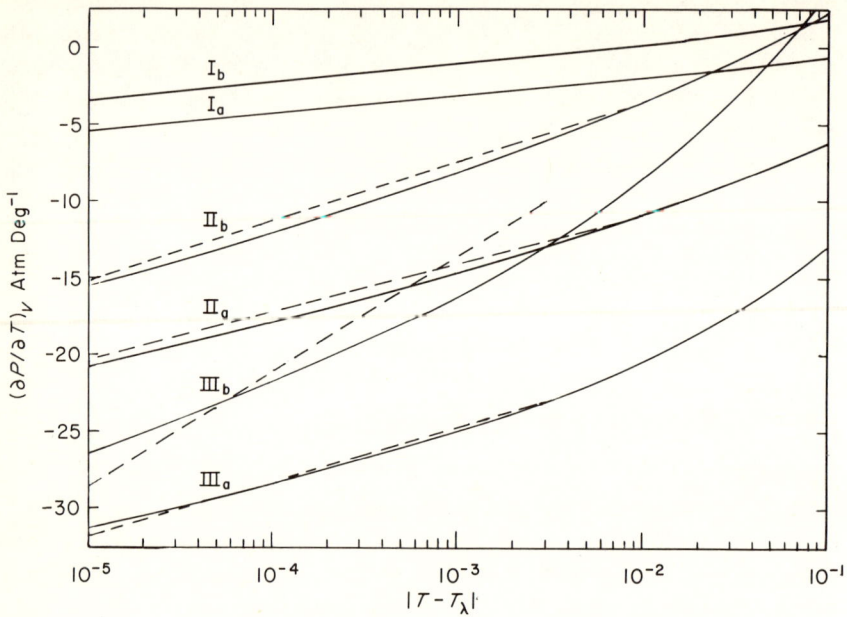

Fig. 5 *Pressure coefficient. Cases Ia, Ib, IIa, IIb, IIIa, IIIb, significance of solid lines, as in Fig. 2. II. Dashed lines: Elwell and Meyer[10b] III. Dashed lines: Kierstead.[13]*

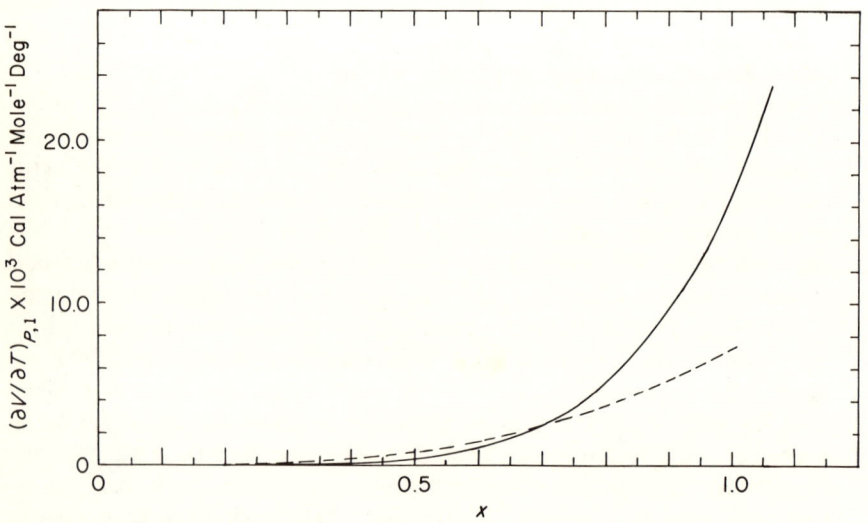

Fig. 6 *Temperature dependence of $(\partial V/\partial T)_{P,1}$. Solid line: present calculation. Dashed line: From Eq. (18), using values at $0°K$.*

SECONDARY VARIABLES IN CRITICAL PHENOMENA

based on the value of $-(\partial V/\partial P)_T$ at $0°K$. One would not expect the lattice compressibility to decrease with increasing temperature, but this appears to be a secondary effect, which occurs because of the negative coefficient of expansion. At the saturated vapor pressure the contraction between $0°K$ and the λ point is about 0.7 percent, but at 25 atm. it is 2.0 percent, which would seem to be sufficient to have an appreciable effect on the lattice compressibility, affecting it in more or less the same way as compression by an external force.

ACKNOWLEDGMENT

This work has been supported by the Army Research Office, Durham, and the National Science Foundation.

APPENDIX I

Since $(\partial T/\partial P)_{C_p} = -(\partial C_p/\partial P)_T/(\partial C_p/\partial T)_p$ the evaluation of the indeterminate form $\chi = C_{p,s}[(\partial T/\partial P)_{C_p} - (T/J)\,dJ/dP]$ will depend on the behavior of $C_p \approx C_{p,s}$ in the neighborhood of the λ line. We therefore explore forms of $C_{p,s}$ of the kind which seems to be empirically observed.

For example, let us suppose that for $T < T_\lambda$

$$C_{p,s} = -A \ln(T_\lambda - T)$$

where A is a function of P. Then

$$\left(\frac{\partial T}{\partial P}\right)_{C_p} = \frac{d \ln A}{dP}(T_\lambda - T)\ln(T_\lambda - T) + \frac{dT_\lambda}{dP}$$

and

$$\left[\frac{\partial}{\partial T}\left(\frac{\partial T}{\partial P}\right)_{C_p}\right]_P = -\frac{d \ln A}{dP}[\ln(T_\lambda - T) + 1]$$

whereas

$$\left(\frac{\partial C_{p,s}^{-1}}{\partial T}\right)_P = -A^{-1}[\ln(T_\lambda - T)]^{-2}(T_\lambda - T)^{-1}$$

122 CRITICAL PHENOMENA

By comparison of the last two expressions it will be seen that the limiting value of χ is zero. This holds also for $T > T_\lambda$, if $C_{p,s} = -A \ln(T - T_\lambda)$.

We may also consider, for $T < T_\lambda$, a more general form

$$C_{p,s} = A(T_\lambda - T)^\alpha = Ae^{\alpha \ln(T_\lambda - T)}$$

where both A and α are functions of P, and $-1 < \alpha < 0$. In this case

$$\left(\frac{\partial T}{\partial P}\right)_{C_p} = \frac{d \ln A}{dP} \frac{T_\lambda - T}{\alpha} + \frac{d \ln \alpha}{dP}(T_\lambda - T)\ln(T_\lambda - T) + \frac{dT_\lambda}{dP}$$

$$\left[\frac{\partial}{\partial T}\left(\frac{\partial T}{\partial P}\right)_{C_p}\right]_P = -\frac{1}{\alpha}\frac{d \ln A}{dP} - \frac{d \ln \alpha}{dP}[\ln(T_\lambda - T) + 1)]$$

and

$$\left(\frac{\partial C_{p,s}^{-1}}{\partial T}\right)_P = \frac{\alpha/A}{(T_\lambda - T)^{\alpha+1}}$$

so that again χ is seen to vanish at the λ line, and again a similar result holds for $T > T_\lambda$.

APPENDIX II

The observed values of $T_\lambda \, dS_\lambda/dT$ given in Table 1 were obtained by plotting the values of S_λ tabulated by Lounasmaa.[16] In this way we actually estimated the value of $T_\lambda \, dS_\lambda/dT$ for $T_\lambda = 2.172$ to be 6.4 cal-mole^{-1}-deg^{-1}. However application of the thermodynamic equation[3]

$$\frac{T_\lambda \, dS_\lambda}{dT} = C_p - T_\lambda \left(\frac{\partial V}{\partial T}\right)_P \frac{dP_\lambda}{dT}$$

using Eq. (20) for C_p in the neighborhood of the λ point shows that $T_\lambda \, dS_\lambda/dT = 5.6$ cal-mole^{-1}-deg^{-1} gives better values of $(\partial V/\partial T)_P$.

REFERENCES

1. Rice, O. K.: *J. Chem. Phys.*, 22:1535 (1954).
2. Pippard, A. B.: *Phil. Mag.*, 1:473 (1956).
3. Buckingham, M. J., and W. M. Fairbank: in C. J. Gorter (ed.), "Progress of Low Temperature Physics," vol. 3, p. 80, John Wiley & Sons, Inc., New York, 1961.
4. Fisher, M. E.: *Phys. Rev.*, 176:257 (1968); C. A. Baker, Jr., and J. W. Essam: *Phys. Rev. Lett.*, 24:447 (1970); H. Wagner: *ibid.*, 25:31 (1970).
5. Garland, C. W., and R. Renard: *J. Chem. Phys.*, 44:1120 (1966) and following articles.
6. Rice, O. K.: *Phys. Rev.*, 153:275 (1967).
7. ———: *J. Am. Chem. Soc.*, 91:7682 (1969).
8. Renard, R., and C. W. Garland: *J. Chem. Phys.*, 45:763 (1966).
9. Kierstead, H.: *Phys. Rev.*, 162:153 (1967).
10. (a) Baghosian, C., and H. Meyer: *Phys. Rev.*, 152:200 (1966); (b) Elwell, D. L., and H. Meyer: *ibid.*, 164:245 (1967).
11. Kramers, H. C., J. D. Wasscher, and C. J. Gorter: *Physica*, 18:329 (1952).
12. Lounasmaa, O. V., and E. Kojo: *Ann. Acad. Sci. Fennicae*, A6, No. 36 (1959).
13. Kierstead, H.: *Phys. Rev.*, 153:258 (1967).
14. Mills, R. L., and S. G. Sydoriak: *Ann. Phys.*, 34:276 (1965).
15. Atkins, K. R.: "Liquid Helium," pp. 63, 65, Cambridge University Press, Cambridge, Mass., 1959.
16. Lounasmaa, O. V.: *Cryogenics*, 1:212 (1961).

DISCUSSION *on Paper Presented by O. K. Rice*

P. C. CLAPP: I cannot understand how the weak coupling approximation can be valid near the critical point. Can anyone clarify this?

G. A. BAKER: In your work, you make a form of the weak coupling approximation, that is, you can treat the behavior of the lattice system and the spin system as separate functions in the partition function. However, when you assume that the exchange energy is a function of the pressure alone (rather than the interatomic spacing) then this assumption corresponds to a particular method of taking the spin lattice coupling into account (see Wagner and Swift, and Baker and Essam, preprints). In particular I would expect in this case that you would find C_v to be renormalized so that it remains finite and a second-order transition is obtained rather than a first-order one, which one sees in the usual weak coupling approximation where the exchange energy depends only on the mean interatomic spacing.

O. K. RICE: The data do not go so close to the λ line, but that the assumption of weak coupling seems reasonable. The Pippard relations should hold near the λ line. They depend only on thermodynamics and certain general properties of the transition, as described in slightly different ways by Pippard and myself.[1,2]

Note Added in Proof: The remarks of Baker cannot refer specifically to the treatment of the experimental data on liquid helium where C_v is always finite and the transition is not first order. There are some first-order transitions where P and V are secondary variables, and these necessarily occur in regions where C_v tends, or tries, to exceed its stability limit,[3] even though it may not actually tend to become infinite.

REFERENCES

1. Rice, O. K.: *J. Chem. Phys.*, 22:1535 (1954).
2. Pippard, A. B.: *Phil. Mag.* 1:473 (1956).
3. Wheeler, J. C., and R. B. Griffiths: *Phys. Rev.*, 170:249 (1968).

THERMODYNAMICS OF A MAGNETIC MODEL EXHIBITING A FIRST-ORDER PHASE TRANSITION

*Herbert Stenschke**
Advanced Studies Center
Battelle Memorial Institute
Columbus, Ohio

ABSTRACT

The thermodynamic states of a particular magnetic model are obtained in terms of the temperature, external magnetic field, and magnetization. The magnetization plays the role of the order parameter in the usual Landau theory of phase transitions. Absolute and local minima of the free energy as function of the order parameter correspond, respectively, to absolutely stable and metastable states of the system.

While understanding of second-order phase transition exhibited, starting in 1944, a number of quantum jumps, the thermodynamics of first-order phase transitions is still vegetating. The major obstacle

*Now at Institut für Theoretische Physik, Freie Universität, 1 Berlin 33, Arnimstrasse 3, W. Germany.

to a comparable progress is the fact that for systems undergoing first-order transitions, metastable states obscure the actual phase transition. Many imponderabilities determine the lifetime of these metastable states and, thus, affect the reproducibility of experimental investigations. From the theorists point of view, metastable states complicate the problem likewise. The usual procedure of calculating the partition function of a system with Hamiltonian $\mathcal{H}$

$$Z = Tr \exp\left(-\frac{\mathcal{H}}{kT}\right) = \sum_{\text{all configurations}} \exp\left(-\frac{E}{kT}\right) \quad (1)$$

gives only information about absolutely stable states. From the partition function Z one cannot obtain any property of the system in a thermodynamic state with a finite lifetime. Being interested in both absolutely stable and metastable states, one may, in principle at least, proceed as follows. Instead of calculating the partition function Z (Eq. (1)) one carries out the statistical mechanics in two steps. First step consists of calculating a restricted partition function considering only configurations with given magnetization m (in particular, we deal with a magnetic system in an external magnetic field H and in a bath of temperature T)

$$Z_m(T,H) = \sum_{\substack{\text{all configurations} \\ \text{with magnetization} = m}} \exp\left(-\frac{E(H)}{kT}\right) \quad (2)$$

From $Z_m(T,H)$ one obtains a restricted free energy

$$F(T,H;m) = -kT \ln Z_m(T,H) \quad (3)$$

In equilibrium, the magnetization m is, of course, a function of T and H. In our procedure, equilibrium is established in a second step. This step consists of minimizing the restricted free energy, Eq. (3), as function of the magnetization. This means, in equilibrium, the following conditions hold

$$\frac{\partial F(T,H;m)}{\partial m} = 0 \quad (4a)$$

$$\frac{\partial^2 F(T,H;m)}{\partial m^2} > 0 \quad (4b)$$

Absolute and local minima of $F(T,H;m)$ as function of m then correspond, respectively, to absolutely stable and metastable states of the system.

That the usual formulation of statistical mechanics gives only information about the lowest stable state is seen as follows. First, we observe from definitions Eqs. (1) and (2)

$$Z = \int Z_m(T,H)\,dm \tag{5}$$

and consequently

$$F = -kT \ln \int \exp\left(-\frac{F(T,H;m)}{kT}\right) dm \tag{6}$$

We now evaluate Eq. (6) by expanding $F(T,H;m)$ about its minima as defined in Eqs. (4a) and (4b). Ignoring the fluctuations about the minima, we obtain

$$F = -kT \ln \sum_{i=0}^{n} \exp\left[-\frac{F(T,H;m_i)}{kT}\right] \approx -kT \ln \exp\left[-\frac{F(T,H;m_0)}{kT}\right]$$

$$\cdot \left\{ 1 + \sum_{i=1}^{n} \exp\left[-\frac{F(T,H;m_i) - F(T,H;m_0)}{kT}\right] \right\} \tag{7}$$

where m_i labels the $n+1$ minima with m_0 the lowest. Each $F(T,H;m_i)$ is proportional to the particle number N and thus for real physical systems, $N \to \infty$

$$F_{N\to\infty} \simeq F(T,H;m_0) \tag{8}$$

In the following we will demonstrate the procedure described above using as an example a Heisenberg model with equal exchange interaction between all pairs of particles. Thus, we have

$$\mathcal{H} = g\mu \sum_{i=1}^{N} H\sigma_{zi} - \frac{2J}{N} \sum_{i<j=1}^{N} \sigma_i \cdot \sigma_j \qquad (9)$$

where H is the applied magnetic field. The energies and degeneracies have been calculated previously,[1] and one obtains the partition function for given total spin M

$$Z_M = \sum_{L=0}^{L=(N/2-M)} \frac{N!(N + 2M + 1)}{M!(N - M + 1)!}$$

$$\cdot \exp\left\{-\frac{2J}{kT}\left[\frac{2M(N - M + 1)}{N} - \frac{(N-1)}{2}\right] + \frac{g\mu H}{kT}\left(\frac{N}{2} - M - L\right)\right\} \qquad (10)$$

If one sums over L, for large N, the partition function can be written as

$$Z_M = (N - 1)! \cdot \exp\left[-\frac{NF(T,H;m)}{kT}\right] \qquad (11)$$

where $F(T,H;m)$ is the free energy per particle as a function of the magnetization per particle $m = (1/N)(N/2 - M)$

$$F(T,H;m) = 4Jm^2$$
$$+ kT\left[\left(\frac{1}{2} - m\right)\ln\left(\frac{1}{2} - m\right) + \left(\frac{1}{2} + m\right)\ln\left(\frac{1}{2} + m\right)\right] - g\mu Hm \qquad (12)$$

At this point, we generalize the model by allowing the exchange energy to be a function of the magnetization, that is,

$$J = J_0(1 + \alpha m^2) \qquad (13)$$

Such a relation is conceivably derivable from a magnetostrictive effect, i.e., a particular magnetic state will affect the lattice constant, which in turn will affect the exchange constant J. The properties of our model, for different values of α, will be discussed below.

Equilibrium states (stable and metastable) are those states for which the magnetization minimizes the free energy; that is to say, the conditions, Eqs. (4a) and (4b), relate the thermodynamical value of the magnetization to temperature and magnetic field, $m = m(T,H)$. That means $m(T,H)$ is the solution of the equation

$$2m = \tanh \frac{1}{kT} \left[2m \left(2J + m \frac{\partial J}{\partial m} \right) - \frac{1}{2} g\mu H \right] \qquad (14)$$

Equation (14) is of the form of the equation for the magnetization for the spin ½ Weiss mean field theory.[2] Note that in general Eq. (14) will have a number of solutions, corresponding to the stable and metastable states.

The magnetization per particle m plays in our theory the role of a Landau order parameter,[3] having the following properties:

1. m is an internal variable in the sense that its equilibrium value $m(T,H)$ is determined by the condition of equilibrium, Eqs. (4a) and (4b).

2. For $H = 0$, an instability of the system as a function of the magnetization m occurs for a critical temperature $T_{SC} = 2J_0/k$, such that

$$\left. \frac{\partial^2 F(T,0;m)}{\partial m^2} \right|_{\substack{T=T_{SC} \\ m=0}} = 0 \qquad (15)$$

The consequence of this instability is a magnetic phase transition.

The order of the phase transition can be determined by using as criteria Gibbs' conditions of critical phases,[4] according to which a phase transition is of second order if

$$\left. \frac{\partial^3 F(T,0;m)}{\partial m^2} \right|_{\substack{T=T_{SC} \\ m=0}} = 0 \quad \text{and} \quad \left. \frac{\partial^4 F(T,0;m)}{\partial m^4} \right|_{\substack{T=T_{SC} \\ m=0}} > 0 \qquad (16)$$

These conditions are fulfilled if $\alpha < 2/3$. Near $T_C = T_{SC}$, the free energy then can be approximated by the first terms of its Taylor series

$$F(T,0;m) = -kT \ln 2 + 2k(T - T_C)m^2 + 2kT_C \left(\frac{2}{3} - \alpha \right) m^4 \qquad (17)$$

This is, of course, an example of a Landau expression[3] of the free energy of a system which undergoes a second-order phase transition.

For $\alpha > 2/3$, Gibbs' conditions are not satisfied; the phase transition is thus of first order. In this case, we obtain (Fig. 1) for four values of the temperature functionally different forms of the free energy, just as in Devonshire's theory of first-order ferroelectric transitions.[5] For temperatures higher than T_{SH}, the maximal superheating temperature, the free energy as a function of m has the form of curve I, only the paramagnetic phase ($m = 0$) exists, i.e., there is one absolutely stable state and no metastable states. Curve 2 ($T < T_{SH}$) has its absolute minimum at $m = 0$ but also has a local minimum at a finite value of the magnetization. This means physically that the paramagnetic phase is absolutely stable. The ferromagnetic phase (a metastable state) also exists at this temperature but has a finite lifetime.[7] In curve 3, the situation is the reverse of that of curve 2; the ferromagnetic phase is absolutely stable and the paramagnetic phase is metastable. For temperatures below T_{SC}, the minimal supercooling temperature, only the ferromagnetic phase exists as shown by curve 4.

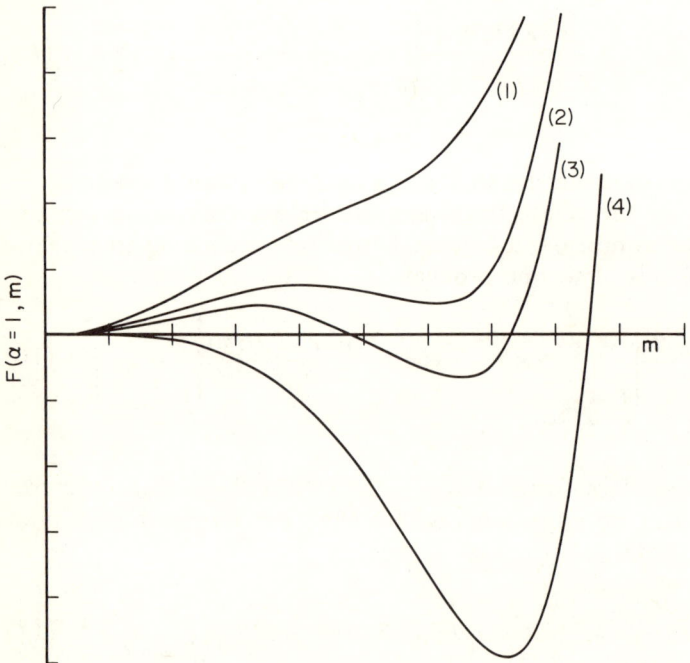

Fig. 1 *The free energy as function of the magnetization for four typical values of temperature. The parameter α has been chosen to be 1.*

In Fig. 2, the free energy of the ferromagnetic phase (broken line) is compared with that of the paramagnetic phase (solid line). The dotted curve corresponds to the maxima of the free energy which exist between the two minima. Different temperature ranges are labelled according to the type of the free energy as a function of m as classified in Fig. 1. We note that for a system which exhibits a first-order phase transition, the paramagnetic phase and the ferromagnetic phase overlap. The region where they exist simultaneously is called a region of metastability. For a second-order phase transition, on the other hand, the region of metastability shrinks to zero; paramagnetic phase and ferromagnetic phase, in this case, join each other. Finally, we observe in Fig. 2 that the phase transition at T_C is accompanied by a jump of the derivative of the free energy with respect to temperature. This corresponds to the appearance of latent heat, the commonly thought of characteristic of a first-order phase transition. However, it is clear from the previous discussion that the existence of a region of metastability is a more basic characteristic of a first-order phase transition than latent heat.

Figure 3 shows the inverse susceptibility as a function of the temperature. In the paramagnetic region, one observes the usual Curie-Weiss behavior with T_{SC} as critical temperature

$$\chi_{\text{para}}^{-1} \sim (T - T_{SC}) \tag{18}$$

The ferromagnetic susceptibility diverges at the maximal superheating temperature T_{SH} which is the stability limit of the ferromagnetic phase. As long as $T_{SH} \neq T_{SC}$, it turns out

$$\chi_{\text{ferro}}^{-1} \sim m(T_{SH})(T_{SH} - T)^{1/2}$$

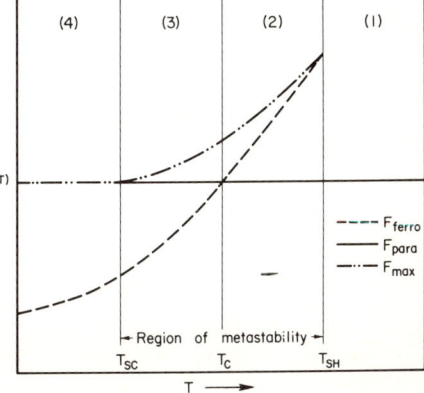

Fig. 2 *The free energy of the ferromagnetic phase F_{ferro} and of the paramagnetic phase F_{para} as function of the temperature. F_{max} corresponds to the free energy maxima between the ferromagnetic and paramagnetic free energy minima.*

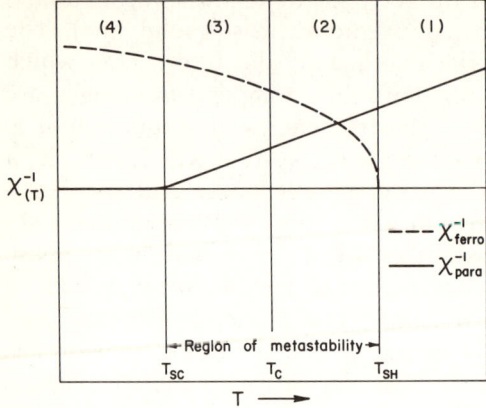

Fig. 3 *The inverse susceptibility as function of temperature.*

in the neighborhood of T_{SH}. Hence, $\gamma' = \gamma/2$ as long as the phase transition is of first order. This asymmetry reminds one of the experimentally observed asymmetry of the half width of light scattered from SF_6. Saxman and Benedek[9] intended to investigate SF_6 at its critical point. Since the critical density was not completely achieved, the phase transition of the substance was of first rather than of second order. Therefore, the high and low temperature data lie on two curves which intersect instead of aiming to the same critical temperature as it should be in case of a second-order phase transition. The two curves also have considerably different slopes.

ACKNOWLEDGMENT

I wish to thank Dr. J. I. Kaplan for collaboration on parts of this work.

REFERENCES

1. Kaplan, J. I.: *Naval Res. Lab. Rept.* 4777, 1956; Kittel, C., and H. Shore: *Phys. Rev.,* **138A**:1165 (1965); Peterson, G. A.: private communication.
2. Weiss, J.: *J. Phys. Appl.,* **6**:661 (1907).
3. Landau, L. D., and E. M. Lifshitz: "Statistical Physics," chap. 14, Pergamon Press, Inc., New York, 1958.
4. "The Scientific Papers of J. W. Gibbs," vol. 1, p. 131, Dover Publications, Inc., New York, 1961; Falk, G.: "Theoretische Physik II," chap. B, Springer-Verlag, Inc., Berlin, 1968.

5. Devonshire's phenomenological theory[6] describes a ferroelectric material (for example, BaTiO$_3$) by a thermodynamic potential of the following form

$$f(T,P) = f_0(T) + A(T - T_1)P^2 + B(T - T_2)P^4 + CP^6$$

where P is the polarization; $A,B,C > 0$; and $T_2 > T_1$.
6. Fatuzzo, E., and W. J. Merz: "Ferroelectricity," North-Holland Publishing Company, Amsterdam, 1967.
7. The lifetime of a metastable state is proportional to $\exp\{-\Delta F(T)/kT\}$, where $\Delta F(T)$ denotes the barrier of the free energy (Fig. 1) which separates the metastable state from the absolutely stable state.[8]
8. Langer, J. S.: *Ann. Phys.*, **54**:258 (1969).
9. Heller, P.: *Rep. Prog. Phys.*, **30**(II):804 (1968).

DISCUSSION on Paper by H. Stenschke

R. B. GRIFFITHS: I think it is worthwhile to point out that there are severe limitations in using molecular field type models of the sort considered by Stenschke in trying to understand metastability. It is known[1] that $F(T, H;m)$ in the thermodynamic limit for a system with short-range forces *cannot* exhibit a double minimum in m, and thus it is impossible to identify the metastable state with a local minimum in the free energy in the manner suggested by mean-field models (which correspond to forces of infinite range). Indeed, for systems with short-range forces (ordinary fluids, for example), it seems to me it is necessary to consider the dynamics of the system in order to understand metastability, and the situation is considerably more complicated than mean field theory would suggest.

REFERENCES

1. Griffiths, R. B.: *Phys. Res.*, **152**:240 (1966).

CRITICAL STATES OF FERROELECTRICS?

G. Falk

Institut für Mathematische Physik
Universität Karlsruhe
Karlsruhe, West Germany

The following considerations are based on the assumption that the first-order phase transitions shown by ferroelectrics are described by means of a thermodynamic potential being essentially identical with the expression used by Devonshire.[1] A ferroelectric material is then in the simplest approximation characterized by five constants two of which, T_1 and T_2, are usually introduced as temperatures and three (A, B, C) as coefficients in the thermodynamic potential

$$g(T,P) = g_0(T) + A(T - T_1)P^2 + B(T - T_2)P^4 + CP^6 \qquad (1)$$

P is the polarization.

In the presence of an external electric field E the thermodynamic potential assumes the form

$$G(T,P,E) = g(T,P) \mp EP - \frac{E^2}{8\pi} \qquad (2)$$

where the plus sign in front of the second term occurs if E and P have opposite directions. From Eq. (2) it is easily inferred[1] that for not too large fields E the transition remains first order and that with increasing E the transition temperature will be shifted to higher values while at the same time the discontinuities of the extensive variables decrease steadily. Hence there will be a certain field $E = E_c$ for which the transition will be continuous indicating the existence of a critical state.

In order to investigate the stability of the thermodynamical system described by Eq. (2) it is convenient to introduce instead of the constants T_1 and T_2 the critical values T_c, P_c, and E_c of the system taking these as its physical characteristics. If one then applies

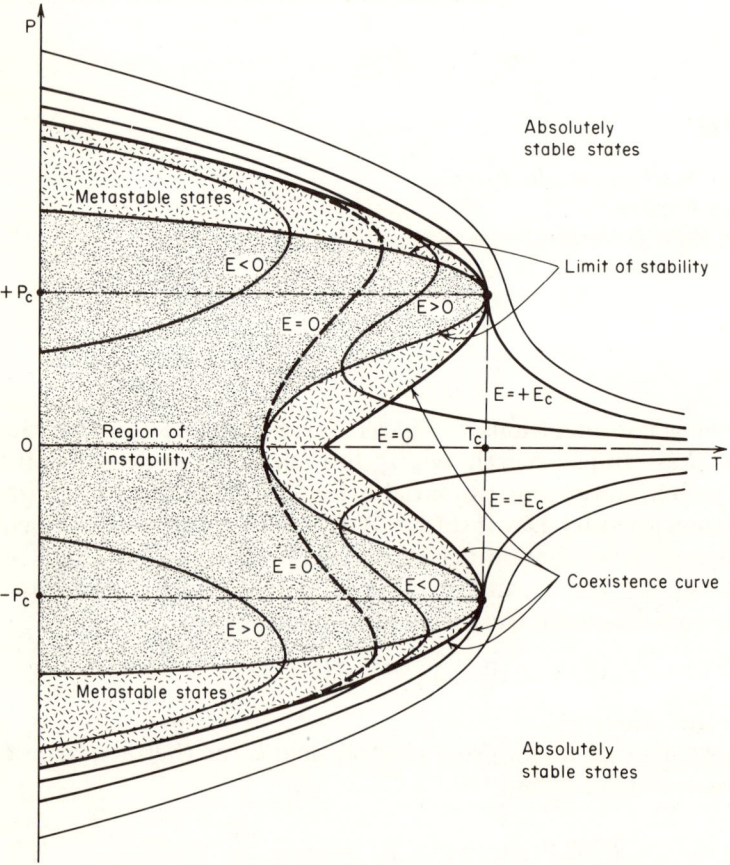

Fig. 1 *Stability chart of a ferroelectric.*

the stability condition

$$\frac{\partial^2 G(T,P,E)}{\partial P^2} < 0 \qquad (3)$$

one obtains the following results.[2]

For $T = T_c$, $P = P_c$, $E = E_c$, the first three derivatives of G vanish while the fourth is positive yielding a critical state as expected. Keeping $E = E_c$ = const. one finds for the critical dependences of the polarization, the susceptibility, and the specific heat

$$P - P_c \propto (T - T_c)^{1/3}$$
$$\chi \propto (T - T_c)^{-2/3} \qquad (4)$$
$$C \propto (T - T_c)^{-2/3}$$

and for the critical isotherm

$$E - E_c \propto (P - P_c)^3 \qquad (5)$$

The values (4) and (5) of the critical exponents are, of course, influenced by the linear temperature dependence of the coefficients assumed in Eq. (1) but their mutual relations are nevertheless not quite as usual. We draw particular attention to the fact that under the critical condition $E = E_c$ (and also for other values of E, except $E = 0$) there is no interval of constancy for the order parameter in equilibrium as the order parameter is different from zero below and above T_c. For values $E > E_c$ there is no phase transition at all whereas for $E < E_c$ a first-order transition results leading to the well-known double hysteresis. The details are immediately obtained from the stability chart of the system.

REFERENCES

1. Devonshire, A. F.: *Adv. in Phys.*, **3**:85 (1954).
2. Dukek, G., and G. Falk: *Z. Physik*, **240**:93 (1970).

DISCUSSION on *Paper by G. Falk*

H. THOMAS: Do not critical point exponents that you find depend in an essential way on the assumption that the coefficient of the

P^2 and the P^4 terms in the free energy are linear in the temperature?

G. FALK: Yes, the linearity in T is essential.

Part Two

GENERAL THEORY

B. Ising Systems and Field Theory

Rapporteur's Introduction:
ISING SYSTEMS AND FIELD THEORY

P. C. Martin
Lyman Laboratory of Physics
Harvard University
Cambridge, Massachusetts

Although the division of the material in this theoretical session into two parts may suggest otherwise, it seems to me that the material I shall discuss in this session, like the material Ben Widom summarized in the previous one, is basically concerned with the scaling properties of the second-order phase transition. The Ising model serves as a test, and field theory as a method for study, of these notions. Since different tests, examples, and tools have no logical linear ordering, the thread running through this session tends to be a tortuous and tattered one.

I shall therefore try to put forth one theorist's view of the nature of the problem of critical phenomena, and the present status of our knowledge, and leave it to you to assess the bearing of the papers in this session to these questions. Of course, several of the papers have an interest of their own, apart from their connection with scaling.

That interest will be apparent to the reader and requires no special accompanying prose from me.

What are the basic questions we must answer with regard to critical phenomena? At the risk of repetition let me summarize what seem to be the main three.

1. THE DEGREE OF HOMOGENEITY OR SCALING

1.1. Thermodynamics

To what extent is it true that the singular behavior of the correlation functions near the critical point is homogeneous? In what sense is it true, if we speak in magnetic terms, that we can deduce

$$\frac{dM}{dH} = \frac{1}{g(\epsilon)} \Phi\left(\frac{|H|}{h(\epsilon)}\right) \qquad (1)$$

where M and H are the magnetization and field, $\epsilon = (T - T_c)/T_c$ the reduced temperature, and g and $h \to 0$, as $\epsilon \to 0$? Under what conditions do $g(\epsilon)$ and $h(\epsilon)$ behave as $g(\epsilon) \sim (\epsilon)^\gamma$ and $h(\epsilon) \sim (\epsilon)^\Delta$ for positive ϵ and as $(-\epsilon)^{\gamma'}$ and $(-\epsilon)^{\Delta'}$ for negative ϵ?

What are the smoothness properties of g, h, and Φ? In particular are they sufficient to insure that the critical exponents γ and Δ above T_c are identical with the indices γ' and Δ' along the coexistence curve? There are no serious experimental or theoretical counter examples—at least none in which the claimed disagreement is greater than 5 percent—but there are some suggestions notably in accurate specific heat measurements in helium that there may be some difficulties. Discrepancies could be pushed aside if they violated rigorous inequalities of the type discussed by Griffiths, Rushbrooke, Buckingham, Gunton, and here by Baker. Unfortunately, the discrepancies always seem to have the proper sign for consistency.

1.2 Static Correlations

To what extent is it true that $S(q, H, \epsilon)$, the Fourier transform of the correlation function $G(r, H, \epsilon)$ of the order parameter (which is measured in elastic scattering by neutrons and X-rays), depends entirely at large distances near T_c on a single inverse length $K(\epsilon, |H|)$? That is to say, to what extent can its asymptotic dependence on q be cast in the form

$$S(q, K) = \frac{1}{\rho(K)} s\left(\frac{q}{K}\right) \cong \frac{1}{g(\epsilon)} \Phi\left(\frac{|H|}{h(\epsilon)}\right) s\left(\frac{q}{K}\right)? \qquad (2)$$

Does $\rho(K) \sim K^{\gamma/\nu}$; or equivalently, does $g(\epsilon) \sim \epsilon^{\gamma}$ and $K \sim \epsilon^{\nu}$ and what are the smoothness properties of ρ? Are they sufficient to guarantee that $\nu = \nu'$?

1.3. Dynamic Correlations

For fixed q/K, do the time-dependent correlations have a frequency spectrum $\chi(\omega)$ for which

$$\chi(\omega, K) \sim \frac{1}{\rho(K)} \chi\left(\frac{\omega}{\Gamma(K)}\right) s\left(\frac{q}{K}\right)? \qquad (3)$$

(With appropriate normalizations and approximation the zero frequency fluctuations reduce to the structure function $S(q, K)$.) Can one understand the relationship between the degree of kinematic slowing down as described by $\Gamma(K) \sim K^{\gamma*/\nu} \sim \epsilon^{\gamma*}$ and the increase in susceptibility as described by $S \sim (\epsilon)^{-\gamma}$? In other words, under what circumstances does the transport coefficient, whose temperature singularly is $(\epsilon)^{-\tau} \sim (\epsilon)^{-\gamma*+\gamma} \sim K^{-\tau/\nu}$, represent a diverging collision time (as it seems to in fluids, critical mixtures, and Heisenberg magnets) and under what conditions does it represent a vanishing collision time (as Suzuki has demonstrated for the stochastic Ising model)?

2. UNIVERSALITY

If the weakest form of the scaling does apply, what determines the asymptotic form of the functions discussed above? Do the nature of the interaction or its range, the crystal structure, etc., play a role in determining the asymptotic form of the scaling functions or do they depend only on the dimension and the degree of symmetry of the underlying interaction? The latter possibility is called universality. How is universality affected by nonergodic behavior. For example, do systems with random impurities behave in a different way, as the work of Rushbrooke and McCoy suggests?

If stronger forms of scaling also hold, are the stronger forms of scaling also universal? And, of course, we have the basic unsolved question: If there are exponents and they have a degree of universality, how can they be calculated or their values understood in terms of the symmetries and dimensionality of the system? As Mills has stressed, no real progress has been accomplished in this direction by any of the microscopic theories.

3. RELATIONS INVOLVING SPATIAL DIMENSION AND STRONG SCALING

To what extent is the inverse coherence length K the only length that contributes to the thermodynamic properties at least in two and three dimensions?

Alternatively, how strong are the reasons for believing that the asymptotic formula for the instantaneous correlation function holds not only when q and K are much smaller than the inverse microscopic lengths a^{-1} in the problem (weak scaling) but even when $Ka \to 0$ but qa remains finite (strong scaling)? Strong scaling is the basis for the proposed relation between a correlation length, a thermodynamic parameter, and the dimension of the system (for example, $2 - \alpha = \nu d$). There are no large discrepancies with this conclusion in two and three dimensions for forces of finite range, but particularly in three, there are suggestions of small discrepancies. Some of those appear in Stanley's paper in this section. Furthermore, as stressed here by Jona-Lasinio, the microscopic arguments in favor of weak scaling of the correlation function are far more convincing than those for strong scaling. Nonetheless, experiments on the properties of superfluid behavior in three dimensions, and the Onsager solution of the Ising model in two dimensions, appear to support strong scaling, and no truly definitive denial has been presented.

Where does that leave us? I think it leaves experimenters with little to do. To rule out the strongest scaling assumption, experiments would have to convincingly demonstrate 5 percent deviations. That seems a rather unlikely and thankless task. It leaves theorists with the same problems they faced several years ago. Perhaps, the microscopic approaches of the Russians, Jona-Lasinio, and Stell will shed some better light on the limitations of the scaling ideas; perhaps studies of analytic properties in temperature and field like that of Glasser will lead to new insights. Unless they do, future conferences on critical phenomena will be more frustrating than those of the past five years.

1. INTRODUCTION

Most current theories of ferromagnetic phase transitions[1] parameterize the singularity in the behavior near T_c of macroscopic quantities like the specific heat, spontaneous magnetization, and zero field susceptibility in terms of a power law

$$|T - T_c|^{\text{critical exponent}} \tag{1}$$

In addition the spin-spin correlation function $\langle \sigma_r \sigma_{r'} \rangle$ is usually assumed to behave when $|r| \to \infty$ as

$$[\text{spontaneous magnetization}]^2 + \text{const} \, |r - r'|^{-a} \exp\left(\frac{-|r-r'|}{\xi}\right) \tag{2}$$

$$\text{if} \quad T \neq T_c$$

and as

$$|r - r'|^{-\text{critical exponent}} \quad \text{if} \quad T = T_c \tag{3}$$

Such parameterizations are usually justified on three gounds: They are simple, they can be made to agree with many existing experiments, and they are the forms of the results of exact calculations on the two-dimensional Ising lattice studied by Onsager. However, recent experiments on specific heats indicate that, if T is close enough to T_c, divergences which appeared to be logarithmic or of the form of Eq. (1) smooth out and behave continuously (and even differentiably) at T_c.[2] In these experiments there exists a temperature scale Δ such that if

$$\Delta \ll \frac{|T - T_c|}{T_c} \ll 1 \tag{4}$$

then the Eq. (1) form describes the singularity in the specific heat. However, if

$$\frac{|T - T_c|}{T_c} \ll \Delta \tag{5}$$

the exponent in Eq. (1) changes from a value near zero to a value greater than 1. Indeed, in the Eq. (5) range there is no assurance that there is any singularity of the Eq. (1) form at all.

RANDOM IMPURITIES IN THE TWO-DIMENSIONAL ISING MODEL

Barry M. McCoy
Institute for Theoretical Physics
State University of New York
Stony Brook, New York

ABSTRACT

We present the results of a study of the critical behavior of a two-dimensional Ising model with frozen-in random impurities. Only the spontaneous magnetization is found to be of the critical exponent form. The specific heat has an infinitely differentiable essential singularity at T_c. The zero field susceptibility is infinite for a range of temperatures from T_c to some T_1 with $T_c < T_1$. When T is close enough to T_c, the correlation function $\langle \sigma_{oo} \sigma_{om} \rangle$ approaches its $m \to \infty$ limit at least as slowly as some inverse power of m while if $T = T_c$ it vanishes slower than any power law permits.

The critical exponent forms Eqs. (1)-(3) are, ultimately, a mathematical abstraction of the results of Onsager's two-dimensional Ising model. Accordingly, there are many differences between real systems and Ising systems to which one can look for an explanation of this experimental breakdown of the critical exponent form. For example, one may argue that finite size effects are becoming important or that the experimenters have failed to allow the sample to reach thermal equilibrium. These effects can, of course, be studied by using larger samples or by waiting longer and are not of fundamental importance for equilibrium statistical mechanics. There are, however, two other possible explanations which are more intriguing. One is the influence of long-range dipole-dipole forces. The other is the presence of a nonzero density of random impurities. In this paper we will examine the consequences of adding random impurities to Onsager's two-dimensional Ising model and will see that these impurities destroy the critical exponent behavior of all quantities except the spontaneous magnetization.

2. THE MODEL

There are at least two different kinds of impurities that are of interest: mobile and frozen-in.

Mobile impurities can migrate through the lattice in times short when compared to the time of thermalization. These impurities can distribute themselves according to some sort of thermal equilibrium and it is appropriate to average the *partition function* over all impurity distribution. They have been studied in the context of the Ising model by Syozi and others[3] where they have been found to lead to "renormalization" of the critical exponent in the sense of Fisher.[4]

The kind of impurities considered here will be the frozen-in variety. In this situation the impurity distribution is fixed but unknown. These impurities can only migrate from site to site on a time scale very much longer than the time required for the rest of the system to thermalize. In general these impurities will destroy the symmetry which the underlying crystal lattice has under discreet translation by a lattice vector [not only globally but also locally]. It is the destruction of this translational symmetry group which is ultimately responsible for the deviation from critical exponent behavior that we find.

To incorporate frozen-in impurities into the two-dimensional Ising model we modify the two-dimensional lattice of Onsager (which is specified by the strength of the horizontal interaction E_1 and the strength of the vertical interaction E_2) as follows: All bonds E_1 are left unchanged; all vertical bonds $E_2(j)$ between row j and row $j+1$ are constrained to be equal; $E_2(j)$ is allowed to vary randomly from row to row in such a way that for $j \neq j'$ $E_2(j)$ and $E_2(j')$ are independent random variables characterized by the probability distribution function $P(E_2)$.

In contrast to the treatment of mobile impurities it is not possible to study these impurities by averaging the partition function. The physically interesting object is the free energy and, if it depends on the detailed distribution of bonds in the lattice, it is useless to suppose that knowledge of $P(E_2)$ is sufficient to specify the thermodynamics of the system. But fortunately we are interested in the lattice in the thermodynamic limit of infinite size. In this limit we can show that the free energy per site of each lattice of the collection specified by $P(E_2)$ approaches, with probability one, the same value. In other words, the free energy of the large but finite system has a probability distribution, determined by $P(E_2)$, which is so sharply peaked about one value that it becomes a delta function in the thermodynamic limit. Therefore, with probability one, the Curie temperature and the spontaneous magnetization of any two lattices from the collection are the same. However not all quantities of interest have such sharply peaked probability distributions. For example, the spin-spin correlation functions do depend on the detailed arrangement of the interaction energies. For such quantities one needs more than just an average value, one wants the entire probability distribution function.

3. RESULTS

The details of the mathematical treatment of this model have been published elsewhere[5] and do not lend themselves to short summary. Therefore, I will skip all derivations and present the results.

3.1. Critical Temperature and Specific Heat

For any $P(E_2)$ we determine the critical temperature T_c as the unique solution of the equation

$$\int_{-\infty}^{\infty} dE_2 P(E_2) \ln|z_2| = \ln \left| \frac{1 - z_1}{1 + z_1} \right| \qquad (6)$$

where

$$z_i = \tanh(E_i/k_B T) \tag{7}$$

At this value of T we have been able to show that for any $P(E_2)$ the specific heat $C_{H=0}$ fails to be analytic. However, at T_c $C_{H=0}$ does not diverge but rather is an infinitely differentiable function of T and hence possesses no singularity of the Eq. (1) form. Ideally, we would like to show that $C_{H=0}$ fails to be analytic *only* at the T_c determined by Eq. (6). This we have not been able to do in general. The farthest we have been able to go is to show that if T is low (high) enough so that if all bonds were decreased (increased) to the minimum (maximum) value allowed by $P(E_2)$ that the resulting Onsager lattice would still be below (above) T_c then $C_{H=0}$ will be an analytic function of T. Furthermore, if we specialize to the distribution

$$P(E_2) = \begin{cases} N^{-1}\lambda_0^{-N}\lambda^{N-1}\dfrac{d\lambda}{dE_2} & \text{if } 0 \leq \lambda \leq \lambda_0 \\ 0 & \text{otherwise} \end{cases} \tag{8}$$

with

$$\lambda = z_2^2 \tag{9}$$

and if N is large and

$$T - T_c = C_1 N^{-2}\delta \tag{10}$$

where δ is of order 1 and C_1 is a constant, then to leading order in N

$$C_{H=0} = C_2 \left\{ \int_0^\infty d\phi \left[\frac{\partial^2}{\partial \delta^2} \ln K_\delta(\phi) - (\phi+1)^{-1} \right] + \ln N^2 \right\} + C_3 \tag{11}$$

where $K_\delta(\phi)$ is the modified bessel function of the third kind of order δ. This function does have the hoped for property that for real values of δ it is analytic except at $\delta = 0$.

3.2. Average Boundary Spin-Spin Correlation Functions

Because our model does not possess translational symmetry in the vertical direction the correlation function $\langle \sigma_{lm} \sigma_{l'm'} \rangle$ (l is the row index and m is the column index) is a function of the three variables l, l', and $m - m'$ and not of the two variables $l - l'$ and $m - m'$. However, in many measurements (such as neutron scattering), it is not so much the correlation $\langle \sigma_{lm} \sigma_{l'm'} \rangle$ that is of interest but rather its average over all l, l', m, m' with $l - l'$ and $m - m'$ held fixed. Such an average will be the same as the average of the correlation function over all sets $\{E_2(j)\}$.

For the boundary row (called 1) we have studied the correlation function $\langle \sigma_{1o} \sigma_{1m} \rangle$ for the situation where a magnetic field $\mathcal{H}$ interacts only with the boundary row. Denote the average over all sets $\{E_2(j)\}$ by $\langle \cdots \rangle_{E_2}$. We have found for the Eq. (8) distribution the following results which are valid to leading order in N when N is large:

1. There exists a range of temperature $-T_{(1)} < T_c < T_{(1)}$ in which the average zero field boundary susceptibility $\langle \chi_1(0) \rangle_{E_2}$ is infinite. There exists a still larger temperature range around T_c in which, even though $\langle \chi_1(0) \rangle_{E_2}$ exists, $\langle M_1(\mathcal{H}) \rangle_{E_2}$ is not an analytic function of $\mathcal{H}$ at $\mathcal{H} = 0$.[6]

2. When $T = T_c$ and $\mathcal{H} \ll N^{-1}$

$$\langle M_1(\mathcal{H}) \rangle_{E_2} \sim - C_4 \operatorname{sgn}(\mathcal{H}) N^{-1} \left[\ln\left(\frac{\mathcal{H} N}{k_B T}\right) \right]^{-1} \quad (12)$$

3. If $\mathcal{H} = 0$, $T - T_c = 0(N)^{-2})$ and $m \gg N^2$, then $\langle\!\langle \sigma_{1o} \sigma_{1m} \rangle\!\rangle_{E_2}$ approaches its limiting value $\langle M_1^2(0) \rangle_{E_2}$ as an inverse power of m. The value of the power depends linearly on $T - T_c$.

4. If $T = T_c$, $\mathcal{H} = 0$, and $m \gg N^2$, then

$$\langle\!\langle \sigma_{1o} \sigma_{1m} \rangle\!\rangle_{E_2} \sim C_5 N^{-2} \left[\ln\left(\frac{m}{N^2}\right) \right]^{-1} \quad (13)$$

5. As $T \to T_c -$

$$\langle M_1(0) \rangle_{E_2} \sim C_6 (T_c - T) \quad (14)$$

6. $M_1(0)$ has the probability distribution

$$\bar{P}(M_1) = 2C_7 N \exp[-(C_7 N M_1)^2] \frac{(C_7 N M_1)^{2|\delta|-1}}{\Gamma(|\delta|)} \quad \text{if } T < T_c$$

$$= \delta(M_1) \quad \text{if } T > T_c \quad (15)$$

3.3. Bulk Spin-Spin Correlation Functions

We expect the behavior of the correlation function in the bulk of two spins in the same row to be qualitatively the same as the boundary correlations. Such is the case with Onsager's lattice. However, bulk calculations are much harder to perform than boundary ones, in part because the general case requires a study of the asymptotic behavior for large dimensions of block Toeplitz determinants of 2×2 matrices. So far we have only been able to study averages over sets $\{E_2(j)\}$ with the symmetry property

$$E_2(j) = E_2(-j - 1) \qquad [-\infty < j < \infty] \qquad (16)$$

In this case the block Toeplitz determinants reduce to a product of two scalar Toeplitz determinants and Szegö's theorem can be used. In this case we can show that if $P(E_2)$ is given by Eq. (8) then

$$M = \langle \lim_{m \to \infty} \langle \sigma_{oo} \sigma_{om} \rangle^{1/2} \rangle_{E_2} \qquad (17)$$

behaves as

$$M \sim C_8 (T_c - T) \qquad (18)$$

as $T \to T_c -$. This behavior is the same as that exhibited in Eq. (14) for the boundary magnetization. This should be contrasted with the fact that in Onsager's lattice the bulk magnetization vanishes as $(T_c - T)^{1/8}$ while the boundary magnetization vanishes as $(T_c - T)^{1/2}$. Furthermore, the fact that for the Eq. (16) special case M vanishes at the same T_c at which $C_{H=0}$ is singular and at which the boundary magnetization vanishes for an unrestricted set $\{E_2(j)\}$ persuades us to believe that for the case of a general $\{E_2(j)\}$ M will still vanish at the T_c defined by Eq. (6). If we accept this, then we may use Griffith's inequalities to show that for $T \geq T_c$ the average values of the boundary quantities displayed in B 1-5 are lower bounds on the corresponding bulk quantities. In particular $\chi(0)$ is infinite in at least a range of temperature $T_c < T_{(1)}$ and for $T = T_c$ the $\mathcal{H} \to 0$ behavior of $M(\mathcal{H})$ is not of the critical exponent form $\mathcal{H}^{1/\delta}$. Furthermore, the correlation function $\langle \sigma_{oo} \sigma_{1m} \rangle$ will not have the critical exponent form of Eqs. (2) and (3).

4. DISCUSSION

All of the results obtained from the Eq. (8) narrow distribution indicate that significant deviations from Onsager's lattice occur only when

$$T - T_c = 0(N^{-2}) \qquad (19)$$

The correlation length ξ of Onsager's lattice diverges as $|T - T_c|^{-1}$. Therefore we may rephrase Eq. (19) by associating with the impurities a length scale of the order of magnitude N^2. Over lengths much smaller than N^2 the lattice appears to be translationally invariant. If the correlation length is less than this impurity length, an individual spin will (in some sense) behave as if it were in a translationally invariant environment and the macroscopic properties of the lattice will be well described by their values in an Onsager lattice with the same value of E_1 and T_c. However, if in this pure lattice the correlation length is greater than the impurity length scale, there is no sense in which the lattice looks translationally invariant.

A second observation concerns the distribution of magnetization throughout the lattice. To be more precise, we may consider

$$M^2(l) = \lim_{m \to \infty} \langle \sigma_{lo} \sigma_{lm} \rangle \qquad (20)$$

as a measure of the magnetization in the lth row. (The possibility of such a measure is a consequence of the horizontal translational symmetry. In a general lattice with all bonds random the limit in Eq. (20) will not exist.) The probability that $M(l)$ will have a given value for a random lattice in the collection is expected to be qualitatively the same as the corresponding boundary probability (Eq. (15)). This distribution has the feature that for T sufficiently close to T_c – it is infinite at $M = 0$ but has a long tail that extends out to values of M comparable to those that would obtain in an Onsager lattice of the same T_c. When $T > T_c$ each $M(l)$ will vanish. However, when $T < T_c$ even though the arithmetic mean of $M(l)$ will be different from zero due to the long tail there will exist large strips in the lattice where $M(l)$ is much smaller than the arithmetic mean (but not *identically* zero) due to the large peak at $M = 0$. At $T \to T_c$ the strips in which $M(l)$ is comparable to the arithmetic mean are expected to become more and more separated by strips where $M(l)$ is extremely small.

This observation on the nonuniformity of the local magnetization can in principle be made more precise and the relation of the size of these strips to the impurity length scale can be investigated if we could study $\langle \sigma_{oo} \sigma_{lo} \rangle$ and $\lim_{m \to \infty} \langle \sigma_{oo} \sigma_{om} \sigma_{lo} \sigma_{lm} \rangle$, that is, the correlation functions that are perpendicular to the impurity strips. However, for these correlation functions the knowledge of the boundary correlations is no help at all and at present their study is a major unsolved problem in the theory of this model.

The results of these model calculations make more concrete the somewhat vague notion that the random impurities can "smear out" phase transitions, and they demonstrate that the behavior of real systems near T_c may be considerably more complex than that indicated by the "critical exponent" forms. The experimental understanding of the mechanism whereby the "critical exponent" form can be broken if T is close enough to T_c represents an important area for future study of magnetic critical phenomena.

ACKNOWLEDGMENT

This work was supported in part by United States Atomic Energy Commission Contract No. AT(30-1)-3668B.

REFERENCES

1. An extensive bibliography to this literature is given by L. Kadanoff, W. Götz, D. Hamblen, R. Hecht, E. A. S. Lewis, V. V. Palciauskas, M. Rayl, F. Swift, D. Aspenes, and J. Kane: *Rev. Mod. Phys.*, **39**:395 (1967).
2. EuS was studied by B. J. C. van der Hoeven, D. T. Teaney, and V. L. Moruzzi: *Phys. Rev. Lett.*, **20**:719 (1968); Ni was studied by P. Handler, D. E. Mapother, and M. Rayl: *Ibid.*, **19**:356 (1967); $RbMnF_3$ (an antiferromagnet) was studied by D. T. Teaney, V. L. Moruzzi, and B. E. Argyle: *J. Appl. Phys.*, **37**:1122 (1966); and dysprosium aluminum garnet (an antiferromagnet) was studied by B. E. Keen, P. P. Landau, and W. P. Wolf: *Ibid.*, **38**:967 (1967).
3. Syozi, I.: *Prog. Theoret. Phys. (Kyoto)*, **34**:189 (1965); Syozi, I., and S. Miyazima: *Ibid.*, **36**:1083 (1966); Essam, J. W., and H. Garelick: *Proc. Phys. Soc. (London)*, **92**:136 (1967).
4. Fisher, M.: *Phys. Rev.*, **176**:257 (1968).
5. McCoy, B. M., and T. T. Wu: *Phys. Rev.*, **176**:631 (1968) and **188**:982 (1969). Also McCoy, B. M.: *Ibid.*, **188**:1014 (1969) and *Ibid.*, **2**:2795 (1970).
6. A related lack of analyticity of the bulk magnetization at $H = 0$ has been discovered in another two-dimensional Ising model with frozen-in impurities by R. Giffiths: *Phys. Rev. Lett.*, **23**:17 (1969).
7. Griffiths, R.: *J. Math. Phys.*, **8**:478 (1967).

DISCUSSION on *Paper* by *B. M. McCoy*

G. A. BAKER: Are your results valid for a single representative choice of random exchange energies or are they valid for the average over an ensemble of such systems?

B. M. McCOY: The properties of this model are characteristic *not* just of one particular collection of bonds but are characteristic of all collections with the exception of a set of measure zero.

C. DOMB: What happens if E_2 assumes some positive value E with probability p and 0 with probability $q = 1 - p$?

B. M. McCOY: When the probability distribution is

$$(1 - p)\delta(E_2 - E^°_2) + p\delta(E_2)$$

then $T_c = 0$, so none of the critical properties exist.

RANDOMLY DILUTE ISING AND HEISENBERG SYSTEMS

G. S. Rushbrooke

Department of Theoretical Physics
University of Newcastle-upon-Tyne
Newcastle-upon-Tyne, England

ABSTRACT

The paper discusses Ising and Heisenberg ferromagnetics in which there are random, fixed, nonmagnetic impurities. The (paramagnetic) susceptibility is expanded either in powers of the concentration p of magnetic systems or in powers of a suitable high-temperature variable. The underlying theory is sketched, as is its connection with other problems in percolation theory. Prima facie conclusions regarding the critical temperature $T_c(p)$ and susceptibility exponent $\gamma(p)$ are presented graphically. Comparison is made with a series approach to the Syozi model. This predicts $T_c(p)$ accurately, but does not adequately predict the renormalization of γ.

1. INTRODUCTION

This paper is primarily concerned with the Ising and Heisenberg ferromagnetic problems defined with respect to a random lattice $L(p)$ in which each site has independently the probability p of "existing" as far as the spin Hamiltonian is concerned. Physically, a fraction $1 - p$ of the lattice sites are occupied by nonmagnetic impurities, which are both randomly distributed and fixed in position (irrespective of energetic, or thermodynamic, considerations). Explicitly, the Hamiltonian is

$$\mathcal{H} = -2J \sum_{\langle ij \rangle} [\alpha(S_i^x S_j^x + S_i^y S_j^y) + S_i^z S_j^z] - g\beta H \sum_i S_i^z \quad (1)$$

with $\alpha = 1$ for the Heisenberg model $H(s)$, and $\alpha = 0$ for the Ising model $I(s)$. The symbols have their customary meanings $\langle ij \rangle$ running over each pair of neighboring sites. But in Eq. (1) i and j are confined to sites occupied by magnetic systems.

Griffiths and Lebowitz[1] have shown that, at least for $I(1/2)$, spontaneous magnetization does occur for sufficiently large p (and at sufficiently low temperatures), but, of course, cannot occur for $p < p_c^s$, where p_c^s is the critical probability for the site "percolation" problem,[2-4] since for $p < p_c^s$ there are no "infinite" clusters of magnetic systems. Our concern here however is not with the spontaneous magnetization but with the paramagnetic (high-temperature) susceptibility χ, and $T_c(p)$ denotes the temperature at which this diverges. $T_c(1)$ is the Curie point for the pure substance.

We expand χ in the two forms

$$\frac{\chi kT}{g^2 \beta^2} = \frac{1}{3} s(s+1) N p [1 + a_1(T) p + \cdots + a_n(T) p^n + \cdots] \quad (2)$$

and

$$\frac{\chi kT}{g^2 \beta^2} = \frac{1}{3} s(s+1) N p [1 + a_1(p) x + \cdots + a_n(p) x^n + \cdots] \quad (3)$$

where $x = J/kT$. For $I(1/2)$ we use in Eq. (3) not x but the natural high-temperature variable $v = \tanh(J/2kT)$. Such expansions were derived some years ago by several authors, starting with Behringer;[5-10] recently they have been extended. Before discussing conclusions drawn from them we shall first outline the underlying theory.

2. GENERAL THEORY

The theory is due to many authors.[11-16] If g is any graph completely specified by its edges (regarded as labeled) and $A(g)$ any function whose value is determined by g, we can write

$$A(g_1) = \sum_{g_2 \subseteq g_1} c(g_2) \qquad (4)$$

where

$$c(g_1) = \sum_{g_2 \subseteq g_1} (-1)^{l_1 - l_2} A(g_2) \qquad (5)$$

Here g_2 is any subset of edges from g_1 and l_i denotes the number of edges in g_i. Equations (4) and (5) constitute Möbius inversion formulas equivalent to the principle of inclusion and exclusion.[17] Moreover, from Eq. (5), if A is an *additive* function, such that for any disconnected graph $g = g' + g''$, $A(g) = A(g') + A(g'')$, then for any disconnected graph $c(g) = 0$. Equation (5) then becomes

$$c(g_1) = \sum_{g_3} (-1)^{l_1 - l_3} A(g_3) \qquad (6)$$

where g_3 is a connected subset of edges of g_1 such that all edges of g_1 not included in g_3 pass through vertices of g_3.

When the A's and thus the c's are specified only by the topology of the graph concerned, it is appropriate to draw up a list of connected graphs which may possibly occur in Eqs. (4) or (6). We denote by (m, l, τ) a connected graph having m vertices and l edges, where τ distinguishes between alternative topological types, and conveniently list these so that m and l occur in "dictionary" order. Then Eq. (4) becomes

$$A(g) = \sum_{i \geq 1} t^g_{(i)} c(i) \qquad (7)$$

where i is short for (m, l, τ) and $t^g_{(i)}$ is the number of times i, as an unlabeled (free) graph, can be located on g. We can now remove the restriction to graphs having no free vertices. If $i = 0$ denotes a free vertex and $c(0) = A(0)$, then Eqs. (6) and (7) become

$$A(G) = \sum_{i \geq 0} t^g_{(i)} c(i) \qquad (8)$$

and

$$c(i) = \sum_{j \geq 0} (-1)^{l_i - l_j} s^i_{(j)} A(j) \qquad (9)$$

where $s^i_{(j)}$ is the number of times j can be located on i such that all unused edges of i pass through vertices of j.

If now in place of G we have a random graph $G(p)$, in the sense that each site of G has an independent probability p of existing, then averaging A over all realizations of $G(p)$ weighted according to their probabilities of occurrence, yields

$$A[G(p)] = \sum_{i \geq 0} p^{m_i} t^G_{(i)} c(i) \qquad (10)$$

with the $c(i)$ still given by Eqs. (8) or (9). And it is straightforward to show, using Eq. (9), that Eq. (10) rearranges to

$$A[G(p)] = \sum_{i \geq 0} p^{m_i} (1-p)^s t^G_{[i]_s} A(i) \qquad (11)$$

where $t^G_{[i]_s}$ is the number of strong embeddings of i on G having s neighbors (not in i) on G. To count strong embeddings[13] of i on G, we identify vertices of i with vertices of G requiring that all edges of G between these vertices are matched by edges of i.

For a finite graph G, Eq. (11) is a tautology for $p = 1$. For an "infinite" lattice L, when A and t are proportional to N, the number of lattice sites, we shall assume that Eq. (10) is valid provided the right-hand side converges, and that it sums to Eq. (11) for $p \leq p_c^s$.

There have been several applications of this formalism. In particular, (1) taking $A(i) = 1$, which counts the number of clusters in $L(p)$, Sykes and Essam[18] in very beautiful work located p_c^s for the plane triangular lattice; (2) taking $A(i) = m_i, m_i \leq n$, $= 0$ otherwise, de Gennes et al.[3] used Eq. (11) to discuss the basic percolation problem; (3) taking $A(i) = m_i^2$, several authors, in particular Sykes and Essam,[19] have estimated p_c^s from the radius of convergence of Eq. (10). This series diverges when the average number of systems which can be "reached" from a given magnetic system diverges. Here, however, we are primarily concerned with the choice (4) $A(i) = \chi(i)$, where χ is the susceptibility of the cluster i according to either Ising or Heisenberg prescription. Equation (10) then yields the series Eq. (2). The series Eq. (3) is a rearrangement of this based on the high-temperature expansion of each $\chi(i)$. The coefficients $a_n(p)$ are fully known, since $c(i)$ starts[12,15] with a term in x^{l_i}. Essentially, as Eq. (10) shows, Eq. (2) sums J-interaction graphs according to the number of their vertices ($n + 1$ for α_n), and Eq. (3) according to the number of their edges.

3. SUSCEPTIBILITY SERIES

We shall confine our attention to the face-centered cubic lattice. In this case, for $H(1/2)$, Eq. (2) is now known through α_5 and Eq. (3) through a_9. For $I(1/2)$ there has been less work on Eq. (2), but Eq. (3) is known through a_8. The series Eq. (3) are long enough to analyze by conventional ratio and Padé approximant techniques. The series Eq. (2), however, are too short for this, and we must have recourse to such approximations as $p_c^{(n)}(T) = \alpha_n(T)^{-1/n}$ or $p_c^{(n)}(T) = \alpha_{n-1}(T)/\alpha_n(T)$ without being able to perform a very satisfactory final extrapolation. Here $p_c^{(n)}(T)$ is the nth estimate of $p_c(T)$, the inverse function to $T_c(p)$. Alternatively, we can invert Eq. (2) to obtain a (truncated) series in p for $1/\chi$, and determine $p_c(T)$ from $1/\chi = 0$. These methods lead[6,20] to $T_c(p) = 0$ at $p = p_c^s$, since, at $T = 0$, $\chi kT/g^2\beta^2$ equals $ms(ms + 1)/3$ for $H(s)$ and equals $m^2 s^2, m \geq 2$, for $I(s)$: which essentially reduces application (4) to application (3) as $T \to 0$. But the conclusion is not a rigorous one.[9]

Figure 1 shows perhaps the most plausible inferences regarding $T_c(p)$ for the $I(1/2)$ and $H(1/2)$ models. The upper parts of these curves, above $p \sim 0.6$, are obtained from the series Eq. (3) and the lower parts from the series Eq. (2). We would emphasize, however, that the lower part of the Heisenberg curve is really very uncertain, whereas the upper part is apparently well determined. By way of illustration, Fig. 2 shows $p_c^{(5)}(T)$, or $T_c^{(5)}(p)$, from $\alpha_4(T)/\alpha_5(T)$ and from the last approximation to $1/\chi = 0$. Earlier approximations yield similar curves and for low p satisfactory extrapolation from them is barely possible. This is true also of approximations based on $\alpha_n(T)^{-1/n}$, successive curves interweaving for small p.[10] There are fewer difficulties in the Ising case. On the other hand, as Fig. 3 shows, above $p \sim 0.6$ different Padé approximants to the expansion Eq. (3) for $H(1/2)$ yield essentially the same curve for $T_c(p)$.

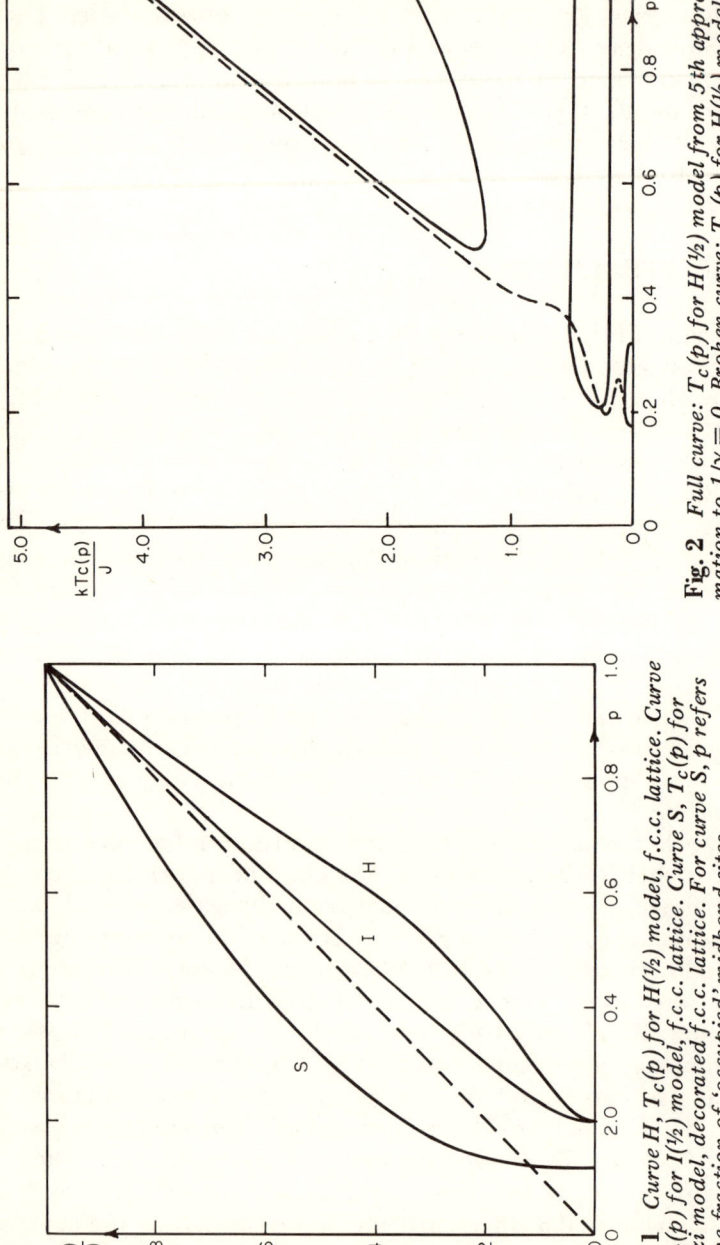

Fig. 1 Curve H, $T_c(p)$ for $H(\frac{1}{2})$ model, f.c.c. lattice. Curve I, $T_c(p)$ for $I(\frac{1}{2})$ model, f.c.c. lattice. Curve S, $T_c(p)$ for Syozi model, decorated f.c.c. lattice. For curve S, p refers to the fraction of 'occupied' midbond sites.

Fig. 2 Full curve: $T_c(p)$ for $H(\frac{1}{2})$ model from 5th approximation to $1/\chi = 0$. Broken curve: $T_c(p)$ for $H(\frac{1}{2})$ model from $\alpha_4(T)/\alpha_5(T)$. Both these approximations are based on Eq. (2)

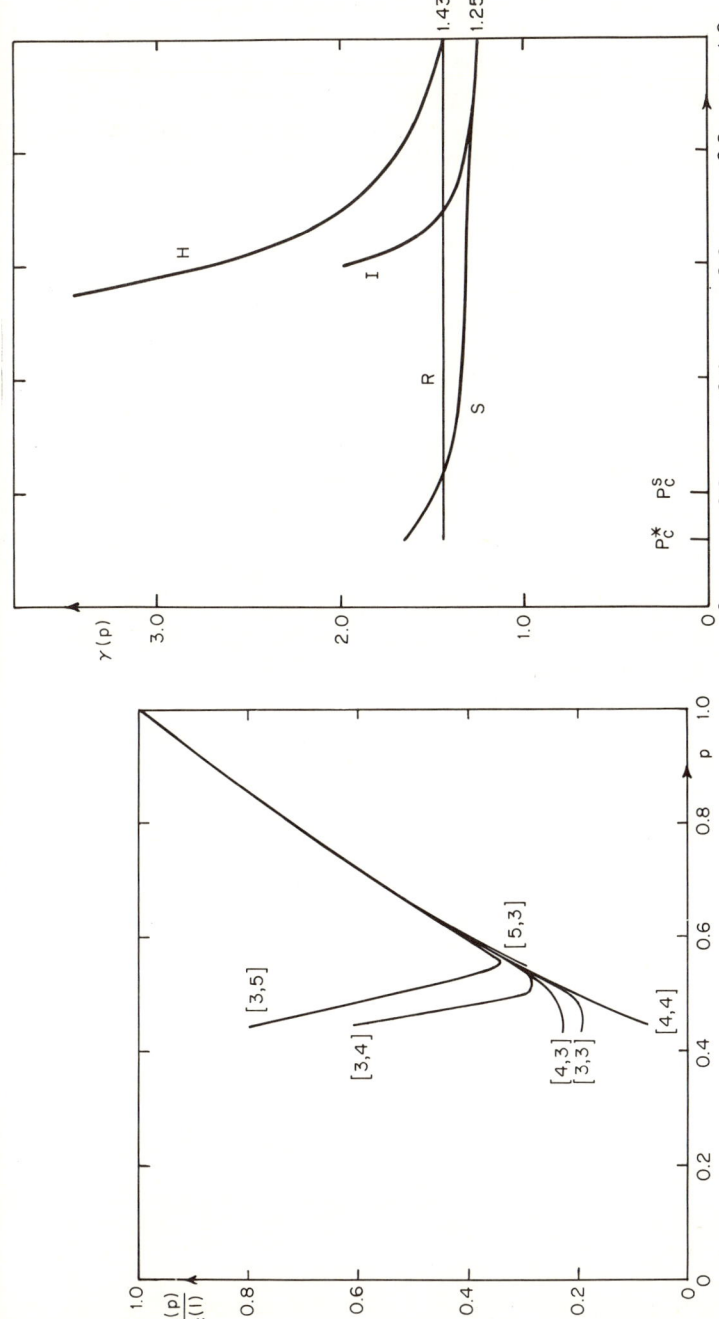

Fig. 3 $T_c(p)$ from different Padé approximants to the logarithmic derivative of Eq. (3), $H(\frac{1}{2})$ model. $[D,N]$ has denominator of degree D and numerator of degree N.

Fig. 4 Curve H, $\gamma(p)$ for $H(\frac{1}{2})$ model, f.c.c. lattice. Curve I, $\gamma(p)$ for $I(\frac{1}{2})$ model, f.c.c. lattice. Curve S, $\gamma(p)$ from $[8,8]$ Padé approximant to logarithmic derivative of series (3) for Syozi model, decorated f.c.c. lattice. Curve R, true, 're-normalized,' γ for Syozi model. p_c^* refers to the Syozi model: p_c^s to the site percolation problem.

Confining attention to the series Eq. (3), we may use the Padé and ratio methods to examine the susceptibility critical index γ. The curves labeled I and H in Fig. 4 show the results, as functions of p, for the two models. A curve similar to that given here for the Ising model was published recently by Oguchi[21] (based on a $a_1 \ldots a_5$ for the simple cubic lattice). The results are surprising, particularly when viewed in the light of the renormalization theory developed by Fisher[22] on the basis of Essam and Garelick's solution[23] of the Syozi model. Of course this theory is strictly not applicable to the present models, since our impurities are not in thermodynamic equilibrium with the rest of the system. If it were, however, $\gamma(p)$ should remain equal to $\gamma(1)$ for $H(1/2)$, since the pure Heisenberg model probably does not have an infinite specific heat,[24] and jump from 1.25 at $p = 1$ to 1.43 for $p_c^s < p < 1$ for the Ising model. We have found a curve very like that for $H(1/2)$ for the classical Heisenberg model $H(\infty)$.

In the hope of shedding light on this, we have derived Eq. (3) for the Syozi model corresponding to the decorated f.c.c. lattice. This is a very open lattice, with mean coordination-number only 3.43 even when $p = 1$. But the coefficients $a_1 \ldots a_8$ for the pure Ising model[25] (and the corresponding energy coefficients[26]) enable us to obtain Eq. (3) through the term in v^{27}, and so form the [8,8] Padé approximant to the logarithmic derivative of χ. The results for $T_c(p)$ and γ are shown in the curves labeled S in Figs. 1 and 4.

Curve S of Fig. 1 reproduces Essam and Garelick's exact result with remarkable accuracy, within the thickness of the curve, right down to $T_c(p) = 0$. More precisely, for $0.2 \leq p \leq 1$ the error in $T_c(p)$ never exceeds 2 parts in 1000; and $T_c(p) = 0$ at $p = 0.116$, whereas the exact value is 0.115. On the other hand, as Fig. 4 shows, the series results for γ certainly do not exhibit full renormalization (which would imply R in Fig. 4). But curves I and H of Fig. 4 remain rather anomalous. Oguchi ascribes curve I to diminishing (topological) dimensionality of the "infinite" cluster responsible for ferromagnetism: but this begs an unsolved mathematical question. Another possibility is that the series Eq. (3), for the random models (as for the Syozi model for $p < 0.2$), increasingly underestimate $T_c(p)$ as p decreases: though it is not clear why they should do so. Meanwhile the matter deserves further investigation, both theoretically and experimentally.

ACKNOWLEDGMENT

The work described has been done in collaboration with Dr. J. Eve, close to p_c^*. This is complementary to Fisher and Scesney's finding that renormalization will not be revealed experimentally (for practicably realizable temperature ranges) except very close to p_c^*.

For the randomly dilute Ising and Heisenberg models one ought to examine whether the anomalous behavior of γ is due to the closing in of nonphysical singularities in χ, in the same way as Dr. Stanley has suggested that $\gamma = 1.43$ for $H(1/2)$ is produced by nonphysical singularities (whose existence Dr. Baker and I mentioned in 1967). But I should like to comment more on this later.

To suggest that γ increases owing to decreasing dimensionality of infinite cluster begs, as I have said, an unsolved mathematical problem: What is the topological dimensionality of the infinite cluster?

R. B. GRIFFITHS: In view of what McCoy has told us about his model, I wonder if the experts on series could comment on the possibility of calculating series for the low-temperature susceptibility for the model Rushbrooke has considered in order to see if the low-temperature susceptibility diverges at a temperature below that at which the high-temperature susceptibility diverges.

C. DOMB: It is very difficult to construct any low-temperature series expansions for a random model. Normal low-temperature expansions are excitations from an ordered lowest energy state. The lowest energy state of a random model is very complex and it is difficult to think of constructing a lengthy series of excitations. One is more likely to succeed by coming down from infinite field on the low-temperature side as Baker and Rushbrooke have done for the Heisenberg model.

Dr. K. Pirnie, Dr. R. L. Stephenson, and, notably, Dr. R. A. Muse. We wish to thank Dr. D. S. Gaunt, of King's College, London, for suggesting to us the parallel attack in the Syozi model which has certainly shown that series are capable of investigating $T_c(p)$, even if they do not always do full justice to γ.

REFERENCES

1. Griffiths, R. B., and J. L. Lebowitz: *J. Math. Phys.*, **9**:1284 (1968).
2. Frisch, H. L., and J. M. Hammersley: *J. Soc. Ind. Appl. Math.*, **11**:894 (1963).
3. de Gennes, P. G., P. Lafore, and J. P. Millot: *J. Phys. Chem. Solids*, **11**:105 (1959).
4. Kasteleyn, P. W., and C. M. Fortuin: *J. Phys. Soc. Japan*, **26**(Suppl.): 11 (1969).
5. Behringer, R. E.: *J. Chem. Phys.*, **26**:1504 (1957).
6. Rushbrooke, G. S., and D. J. Morgan: *Mol. Phys.*, **4**:1 (1961).
7. Morgan, D. J., and G. S. Rushbrooke: *Ibid.*, **4**:291 (1961).
8. Elliott, R. J., and B. R. Heap: *Proc. Roy. Soc. (London)*, **A265**:264 (1962).
9. Morgan, D. J., and G. S. Rushbrooke: *Mol. Phys.*, **6**:477 (1963).
10. Heap, B. R.: *Proc. Phys. Soc.*, **82**:252 (1963).
11. Abe, R.: *Prog. Theor. Phys.*, **31**:412 (1964).
12. Rushbrooke, G. S.: *J. Math. Phys.*, **5**:1106 (1964).
13. Sykes, M. F., J. W. Essam, B. R. Heap, and B. J. Hiley: *Ibid.*, **7**:1557 (1966).
14. Essam, J. W., and M. F. Sykes: *Ibid.*, **7**:1573 (1966).
15. Jasnow, D., and M. Wortis: *Ibid.*, **8**:507 (1967).
16. Essam, J. W.: *Ibid.*, **8**:741 (1967).
17. Rota, G.-C.: *Z. Wahrscheinlichkeitstheorie*, **2**:340 (1964).
18. Sykes, M. F., and J. W. Essam: *J. Math. Phys.*, **5**:1117 (1964).
19. ——— and ———: *Phys. Rev.*, **133**:A310 (1964).
20. Elliott, R. J., B. R. Heap, D. J. Morgan, and G. S. Rushbrooke: *Phys. Rev. Lett.*, **5**:366 (1960).
21. Oguchi, T.: *J. Phys. Soc. Japan*, **26**:580 (1969).
22. Fisher, M. E.: *Phys. Rev.*, **176**:257 (1968).
23. Essam, J. W., and H. Garelick: *Proc. Phys. Soc.*, **92**:136 (1967).
24. Baker, G. A., Jr., H. E. Gilbert, J. Eve, and G. S. Rushbrooke: *Phys. Rev.*, **164**:800 (1967).
25. Domb, C., and M. F. Sykes: *J. Math. Phys.*, **2**:63 (1961).
26. Sykes, M. F., J. L. Martin, and D. L. Hunter: *Proc. Phys. Soc.*, **91**:671 (1967).

DISCUSSION on *Paper* by *G. S. Rushbrooke*

C. DOMB: Martin made a distinction between the two models (of McCoy and Rushbrooke), that the strong singularity of the first model makes it different from the second.

M. SUZUKI: The dependence of critical exponent $\gamma(p)$ to the concentration p may be explained qualitatively by observing that the "effective" dimensionality of such a random system decreases as the parameter p decreases. This was originally pointed out by Oguchi.

G. S. RUSHBROOKE: For the Syozi model it is clear that even long series do not yield a "renormalized" value of γ except for p very

ANALYTICAL PROPERTIES OF A CLASS OF PLANAR FERROELECTRIC MODELS

M. L. Glasser
Advanced Studies Center
Battelle Memorial Institute
Columbus, Ohio

ABSTRACT

This report concerns the analytic properties of a number of solvable planar ferroelectric models which incorporate the so-called ice condition. These include the Slater KDP and Rhys F models. The free energies of these models are discussed as well as their complex temperature Riemann structures with particular emphasis on those features relating to critical behavior.

1. INTRODUCTION

For many years, due to its unique role as an exactly solvable physically realistic system undergoing a phase transition, the two-dimensional Onsager-Ising model has been a touchstone for statistical mechanics. In particular, in agreement with the ideas put forth by Yang and Lee, its second-order phase transition is associated with a mathematical singularity of the free energy on the real axis on (the physical sheet of) the complex temperature plane. Figure 1 shows the physical sheet of the Riemann surface for Onsager's expression for the free energy. The heavy lines denote logarithmic branch cuts and the dashed lines denote those of square-root type. The relevant singularity for the ferromagnetic case is the branch point at T_c. Knowledge of the Riemann structure is also necessary to determine the limits of validity of series expansion as has been emphasized by Majumdar[1] for the Ising model.

Over the past few years several "new" models, which, however, have been advanced and studied numerically in the past, have been solved exactly. These include the two-dimensional Slater KDP and Rhys F model, proposed to describe hydrogen bonded ferro- and antiferroelectrics, whose exact energies were found by Lieb.[2] The purpose of this report is to summarize the mathematical aspects of the solutions to these and several related models. To unify the presentation we consider first a generalized model which effectively interpolates between the KDP and F models and whose free energy

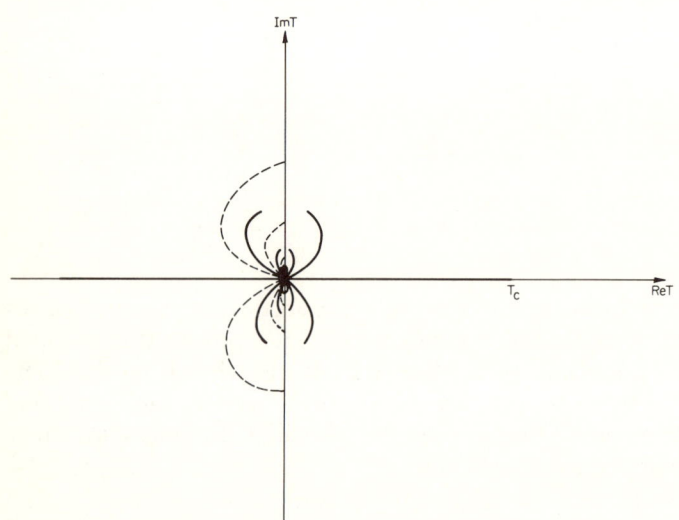

Fig. 1 *Physical sheet of the Riemann surface for the free energy of the ferromagnetic Onsager-Ising model. The heavy lines denote logarithmic branch cuts, the dashed line denotes first-order branch cuts. The "spider" has an infinity of pairs of legs converging to $T = 0$.*

PROPERTIES OF PLANAR FERROELECTRIC MODELS

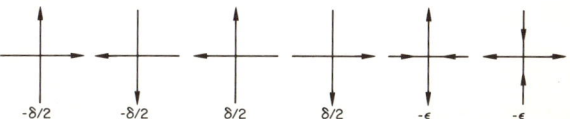

Fig. 2 *Vertex assignments for the generalized model.*

can be obtained by a simple extension of Lieb's method. This model is described in Fig. 2, where Yang and Southerland's[3] notation has been adopted. The six vertices shown are to be arranged on an $N \times N$ square lattice so that two arrows enter and leave each lattice point. Each configuration is given a statistical weight consisting of the product of the activities $\eta = e^{\beta \delta}$, η^{-1}, or $\zeta = e^{2\beta\epsilon}$ (where $\beta = 1/kT$) corresponding to each vertex and the canonical partition function is obtained by summing over all allowed configurations. The relevance of this model to ferroelectrics is seen by assigning each vertex an electric moment pointing in the quadrant bounded by the outgoing arrows (none to the ϵ vertices). The free energies are given by the largest eigenvalue of a transfer matrix, which Lieb calculated by relating the transfer matrix to the Hamiltonian for the anisotropic Heisenberg chain.[4] In the thermodynamic limit the high-temperature free energy becomes

$$-\beta F = \frac{1}{4}\beta\delta + \frac{1}{4\mu}\int_0^\infty dx \operatorname{sech}\left(\frac{\pi x}{2\mu}\right)\ln R(x) \qquad (1)$$

where

$$R(x) = \frac{[(2\Delta - \eta)^2 + 1](\cosh x + \Delta) - 2(2\Delta - \eta)(1 + \Delta \cosh x)}{(\eta + \eta^{-1})(\cosh x + \Delta) - 2(1 + \Delta \cosh x)} \qquad (2)$$

The quantity $\Delta = -\cos\mu = 1/2(\eta + \eta^{-1} - \zeta)$ is the Heisenberg chain asymmetry parameter and the high-temperature region corresponds to $|\Delta| < 1$. $R(x)$ is strictly positive (for real parameters) so, since $d\mu/d\Delta = (1 - \Delta^2)^{-1/2}$, critical behavior can occur only for $\Delta = -1 (\delta < 2\epsilon)$ or $\Delta = 1 (\delta > 2\epsilon)$. As shown in Fig. 3, this divides the $\epsilon - \delta$ plane into regions I and II. (Due to the obvious symmetry $\delta \to -\delta$, we consider only $\delta > 0$.) For purposes of simplicity we

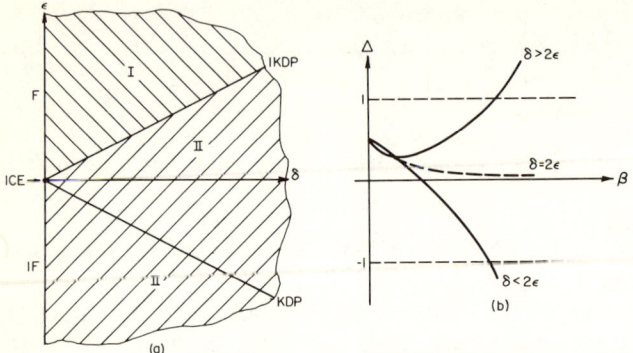

Fig. 3 *(a) Parameter plane for the generalized model. Each special case corresponds to a ray making an angle φ with the δ-axis. The ice model $\delta = \epsilon = 0$ is the unique $T = \infty$ limit in all cases. (b) Behavior of the anisotropy parameter vs. inverse temperature.*

eliminate negative vertex energies by adding ϵ to each vertex in region I and $\delta/2$ to each vertex in region II.

In region II, for $\Delta = 1$ we find $F = 0$. But, since βF is a nonnegative monotonic quantity, we must have $F = 0$ for $T < T_c$ defined by $\cosh[\delta/(kT_c)] - 1/2 \, e^{2\epsilon/kT_c} = 1$. In this region, which includes the KDP model, we find as $T \to T_c^+ (\Delta \to 1^-, \mu \to \pi^-)$ that

$$\beta F \simeq \frac{(\mu - \pi)^2}{4(1 - \eta)}$$

$$\text{Internal Energy} = U \simeq \frac{\delta(\eta^2 - 1)/2\eta - \epsilon\zeta}{2^{3/2}(1 - \eta)} \quad (3)$$

so the system undergoes a first-order phase transition to a ferroelectrically aligned state at which the specific heat diverges as $(T - T_c)^{-1/2}$. For the case $\eta = 1$ (IF model) we have

$$\beta F \simeq \frac{1}{2}(\pi - \mu)$$

$$U \simeq \frac{1}{2}|\epsilon|(\pi - \mu) \quad (4)$$

and the system undergoes a second-order phase transition at $T_c = 0$.

In region I, where $\Delta = -1$ corresponds to $\mu = 0$, the free energy can be expanded formally in powers of $\mu^2 = \{\arccos(-\Delta)\}^2$; however

PROPERTIES OF PLANAR FERROELECTRIC MODELS   169

it can be shown that this series has a vanishing radius of convergence. Since μ^2 is an analytic function of the temperature, F has bounded temperature derivatives of all orders at T_c defined by $(1/2)e^{2\epsilon/kT_c} - \cosh(\delta/kT_c) = 1$, yet it is not analytic. Thus in region I, which includes the F model, the system undergoes an infinite-order phase transition to an antiferroelectrically ordered state. The low-temperature expression for the free energy in this region ($\Delta < -1$) is

$$F = -\epsilon$$

$$+ \frac{1}{4\beta} \ln \left\{ \frac{1}{2}\eta \left[\frac{\Delta([2\Delta - \eta]^2 + 1) - 2[2\Delta - \eta]}{\beta\delta - 1} \right] \left(\frac{1 + (1-a^2)^{1/2}}{1 + (1-b^2)^{1/2}} \right) \right\}$$

$$+ \frac{1}{2\beta} \sum_{n=1}^{\infty} n^{-1} \operatorname{sech} n\lambda \left\{ \left(\frac{1 - (1-b^2)^{1/2}}{b} \right)^n - \left(\frac{1 - (1-a^2)^{1/2}}{a} \right)^n \right\} \quad (5)$$

where

$$\Delta = -\cosh\lambda$$

$$a = \frac{[(2\Delta - \eta)^2 + 1] - 2\Delta(2\Delta - \eta)}{2(2\Delta - \eta) - \Delta[(2\Delta - \eta)^2 + 1]}$$

$$b = \frac{\cosh\beta\delta - \Delta}{1 - \Delta\cosh\beta\delta}$$

In this case F shows no singular behavior down to $T = 0$.

The IKDP model, which separates region I from region II, undergoes no phase transition at all; the free energy is an analytic function of temperature in a region including $0 < T < \infty$. However, as will be seen, even this case has a nontrivial singularity structure.

Each special case of the general model corresponds to a ray in Fig. 3a. The behavior of the transition temperature as one proceeds along a circle about the origin in the $\delta - \epsilon$ plane is shown in Fig. 4. It is interesting that the F and KDP models have the same transition temperature $T_c = \epsilon/k \ln 2$.

The gross features of the complex temperature Riemann structures for these models arise as follows. Because the free energy depends

170 CRITICAL PHENOMENA

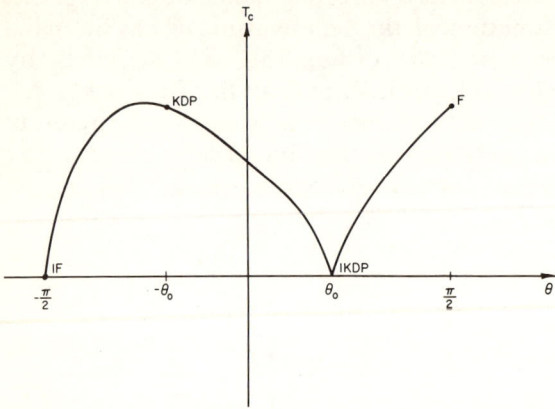

Fig. 4 *Behavior of the transition temperature in the ϵ–δ plane.*

on temperature through $\mu = \arccos(-\Delta)$, each strip of the μ plane (parallel to the imaginary axis) of width 2π corresponds to an entire Δ plane; these are joined by branch cuts $(1, \infty)$, $(-\infty, -1)$ and are connected as indicated in Fig. 5. In addition, there may be other branch cuts and additional sheets arising from branch points in the explicit μ dependence. Next, each Δ plane corresponds to a β plane and, because going from Δ to β involves taking a logarithm, each branch cut on a Δ sheet corresponds to a periodic array of parallel cuts on the corresponding β sheet. Finally, in going from β to T, by taking the reciprocal, the parallel branch cuts on the β plane become sets of circles on the corresponding T plane. We shall give the detailed

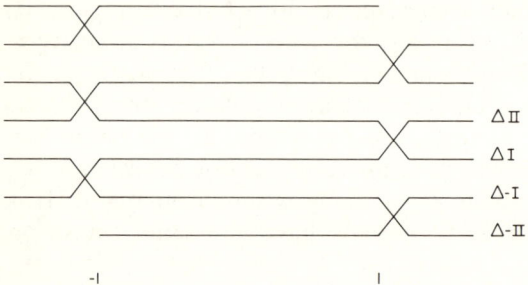

Fig. 5 *Connectivity of Δ sheets mapping into the μ-plane.*

PROPERTIES OF PLANAR FERROELECTRIC MODELS

structures only for the simplest model typifying each type of behavior.

2. KDP MODEL

The free energy in this case can be expressed in closed form[5]

$$-\beta F = \begin{cases} \ln\left[\dfrac{2\mu}{\pi}\cot\left(\dfrac{\pi^2}{2\mu}\right)\csc\mu\right] & T > T_c \\ 0 & T < T_c \end{cases} \qquad = \dfrac{\epsilon}{k\ln 2} \qquad (6)$$

where

$$\cos\mu = -\dfrac{1}{2}e^{\beta\epsilon}$$

The analytic structure for the low-temperature free energy is trivial. The high-temperature free energy has logarithmic branch cuts at

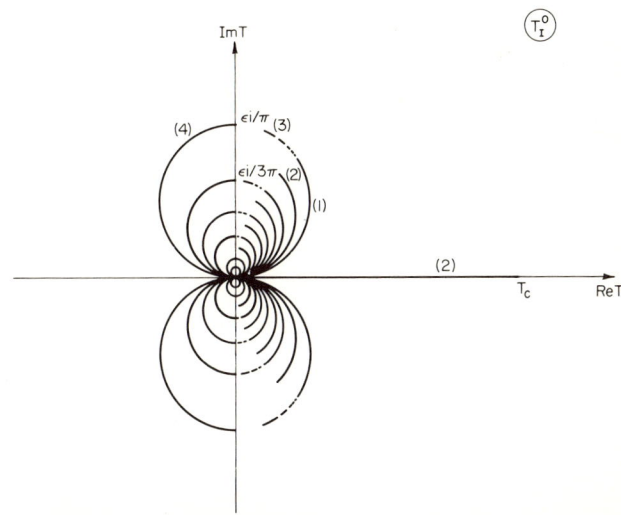

Fig. 6 *Physical sheet for the free energy of the Slater KDP-model. The heavy lines denote logarithmic branch cuts.*

$\mu = \pi/n$, πn, $n = \pm 2, \pm 3, \ldots$, which we associate with cuts $\pm [\pi/2|n|, \pi/2(|n| + 1)]$ and cuts starting at $\mu = n\pi$ and going to ∞ parallel to the imaginary μ axis. Thus the point $\mu = 0$ is an accumulation point of branch points and is an essential singularity. Tracing this structure through to the complex temperature plane, as indicated above, we find for the physical sheet the structure shown in Fig. 6. We thus see that the first-order phase transition is associated with a logarithmic branch point at $T = T_c$ corresponding to $\mu = \pi$.

3. F MODEL

Lieb's expression for the free energy of the F model can be transformed to

$$-\beta(F - \epsilon) = \int_0^\infty \frac{dx}{x} \sinh \mu x \left(1 - \frac{\tanh \mu x}{\tanh \pi x}\right) \qquad \cos \mu = \frac{1}{2} e^{2\beta\epsilon} - 1 \tag{7}$$

which is valid for $0 \leq \text{Re}\,\mu < \pi$. This can be continued analytically

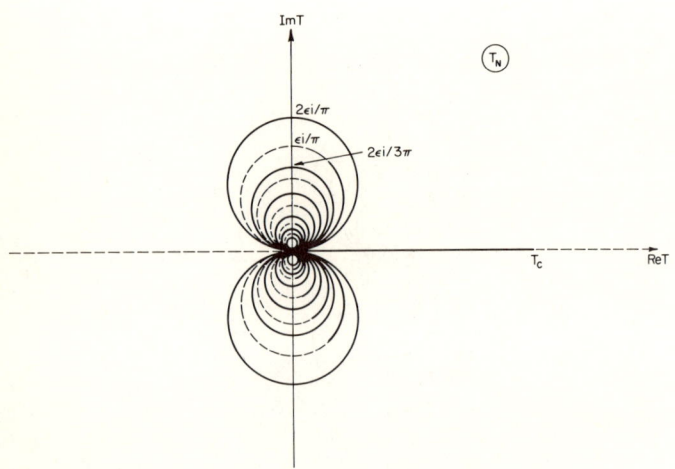

Fig. 7 *Riemann structure for the free energy of the Rhys F-model. The heavy lines denote logarithmic branch cuts. The dashed lines, absent on the physical sheet, denote natural boundaries.*

PROPERTIES OF PLANAR FERROELECTRIC MODELS   173

throughout the entire μ plane with the exception of a natural boundary extending along the entire negative μ axis. The complex temperature plane is shown in Fig. 7. On the physical sheet we have only the circular branch cuts and a cut extending from $T = 0$ to $T = T_c$. However, F is well behaved at T_c so this causes no trouble. On the first unphysical sheet, however, the natural boundaries shown as dashed lines occur resulting in the infinite order behavior at T_c. It is interesting to note that the behavior at $T = 0$ is also nonanalytic.

4. IKDP MODEL

The IKDP model is the only other case for which the free energy can be expressed in elementary terms. Here we have

$$\beta F = \ln \left| \frac{\pi}{2\mu} \tan \mu \tan \left(\frac{\pi^2}{2\mu} \right) \right| \qquad (8)$$

where

$$\cos \mu = -\frac{1}{2} e^{-\beta \epsilon}$$

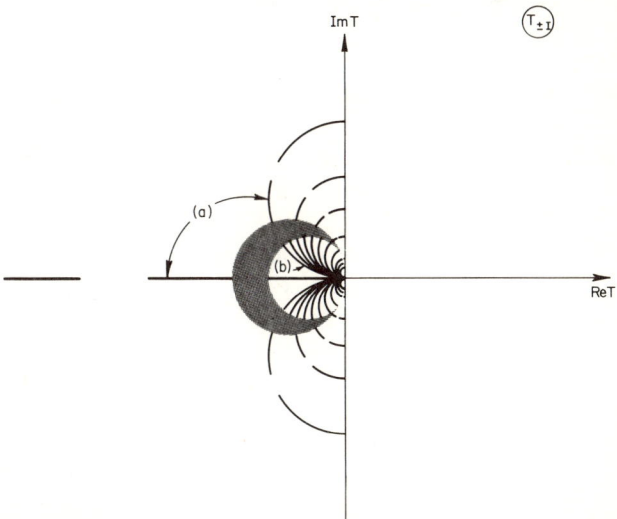

Fig. 8 *Physical sheet for the free energy of the IKDP model. The shading denotes continuation of the sequence of small branch cuts converging to the boundary circle of the inner region of cuts.*

As for the KDP model, there is an infinite sequence of small branch cuts along the real μ axis converging to $\mu = 0$. In addition we have equally spaced cuts, again along the real μ axis, from $\pm(2n - 1)\pi/2$ to $\pm n\pi$, $n = 2, 3, \ldots$. This leads to the Riemann structure whose physical sheet is shown in Fig. 8. The shading denotes a region of sequences of branch cuts converging to the perimeter of a circle containing the cuts labeled (b). In this case the half plane $ReT > 0$ is completely free of singularities, although this ceases to be the case on other sheets.

ACKNOWLEDGMENTS

Portions of this work were done in collaboration with Drs. E. H. Lieb and D. B. Abraham.

REFERENCES

1. Majumdar, C. K.: *Phys. Rev.*, **145**:158 (1966).
2. Lieb, E. H.: "Lectures in Theoretical Physics," vol. XIB, University of Colorado Press, Boulder, Colo., 1969.
3. Yang, C. P.: *Phys. Rev. Lett.*, **19**:586 (1967).
4. Yang, C. N., and C. P. Yang: *Phys. Rev.*, **150**:327 (1966).
5. Glasser, M. L.: *Ibid.*, **184**:539 (1969).

DISCUSSION on *Paper* by *M. L. Glasser*

M. L. GLASSER: I wish to emphasize the following interesting points which emerge from this calculation:
  1. The singularities of the F-type models which are responsible for the infinite order phase transition occur on an unphysical sheet of the Riemann surface.
  2. The IF model has a second-order phase transition at $T = 0°K$.
  3. All possible six vertex models have natural boundaries in their Riemann structures with the sole exception of KDP and IKDP whose free energies can be expressed in elementary terms.
  4. Unlike the phase transition for the 2D Ising model, which can be called dynamic, the phase transition in these are kinematic in that they are due only to the fact that the free energy depends on μ. This is due, in some way, to the constraint imposed by the ice condition.

M. SUZUKI: The peculiar singularity of ferroelectrics seems to be mathematically connected with two-dimensional distribution of zeros of the partition function in the complex fugacity plane (i.e., complex field z plane: $z = e^{-mH/kT}$). Fisher and I have proved that the Lee-Yang circle theorem is valid in ferroelectrics at all temperatures below T_c. On the other hand, Katsura has shown numerically that the distribution of zeros is two-dimensional above T_c.

R. B. GRIFFITHS: Would someone here like to comment on the physical origin of the peculiar (as compared to "normal" Ising model behavior) analytic properties of these two-dimensional ferroelectric models? Do they arise from the ice condition or from the two-dimensional nature of the models?

C. N. YANG: I believe they arise from the ice condition. If one tries to convert the two-dimensional ferroelectric problem with the ice condition into a lattice gas problem, one obtains infinitely long chain molecules, rather than ordinary short molecules. This is probably the cause of the "peculiar behavior."

C. DOMB: I should like to mention that it is possible to use series expansion for these ferroelectric models in two and three dimensions as was done by Nagle before Lieb's exact solutions. For some models there are no excited states at low temperatures and hence one can see that the partition function is given by the lowest energy state over a finite range.

CONTRIBUTIONS OF BROKEN-SYMMETRY MODES IN SECOND-ORDER PHASE TRANSITIONS

Roger E. Mills

Department of Physics
University of Louisville
Louisville, Kentucky

ABSTRACT

A review is made of some recent applications of field theory to the study of critical phenomena. A closed and exact set of equations, in which the central object is the two-particle correlation function, is derived as a basis for comparison. The effect of the symmetry change at the phase transition is considered explicitly for the case of ferromagnetism in a system of itinerant fermions. The Goldstone modes are shown to be both a necessary consequence of the symmetry breaking, and an essential factor in completing the self-consistency in the equations. The latter feature has been considered explicitly in only one of the previous types of calculation. The possibility of achieving a solution to the full set of equations is discussed.

1. INTRODUCTION

The thermodynamic characterization of certain phase transitions in magnets, fluids, and alloys as being of (or nearly of) the second order does not adequately describe the striking similarity at the transition point of corresponding properties of the various materials. Landau[1] has shown that the change in symmetry at such transitions implies that near the critical point the properties can be described in terms of similar monomial functions. These functions have the form of a leading coefficient, dependent on the material, times a simple power or root of the deviation of the intensive variables from their values at the critical point. Many careful and sophisticated experiments have shown that the similarity exists which is predicted by the Landau theory for these critical exponents, but that the observed values differ from the Landau values. Reviews of these experiments have been given by Heller[2] and by Kadanoff et al.[3]

The Landau exponents agree exactly with those of several earlier approximate theories. Onsager's exact solution of the two-dimensional Ising model and subsequent work on this model by Yang have shown that here the exponents are clearly different from those of Landau (and of experiment). Unfortunately the mathematical tools needed for the exact solution of the Ising problem are such that their use gives little insight into ways of improving upon the earlier approximations in cases where exact solutions are unobtainable.

Many different efforts have been made to extend the accuracy of approximate calculations and to achieve better understanding of various models through series expansions and use of Padé approximants. A comprehensive review of this work by Fisher[4] shows that much detailed information has been gained on a considerable variety of systems. It still remains to determine what, if any, common features of critical phenomena will generally produce a departure of the exponents from the Landau values which nevertheless leaves the exponents fairly well clustered together.

It has been shown by Brout[5] and others that a self-consistent or mean field approach can be used to describe many different transitions. It has been the general experience that such an approach leads in the Hartree-Fock or random phase approximations to the Landau results again. It has been shown by Tahir-Kheli[6] that in magnetic problems ad hoc modifications of the mean field could be made which gave a magnetization exponent in agreement with experiment. However a more general consideration of such modified mean field theories[7] has shown that the probability distribution for the excitations in such theories leads to a restrictive relation between the specific heat and magnetization exponents which is not in agreement with either experimental results or with results of series expansions.

A more general theory of phase transitions or fermion systems has been given by Mattuck and Johansson.[8] Subsequent work by Johansson[9] and independent work by Morandi[10,11] has stressed the importance of the appearance of a mode of the Goldstone type[12-15] at the onset of the transition. Meissner[16] obtained similar results for a phonon system. It has been suggested by Umezawa and collaborators[17-19] that the Goldstone mode is not an incidental consequence of the symmetry change at the transition, but that it is involved in a significant way in a nonlinear rearrangement, through the dymanics, of the basis for the realization of the symmetry.

A very general description of spontaneous ordering in systems of many bosons or fermions has been given by DeDominicis and Martin.[20] In order to make a more explicit study of the symmetry aspect of the problem and of the role of the Goldstone mode, more restrictive equations describing ferromagnetism in a system of itinerant fermions are outlined in the next section. A closed, exact set of equations is derived, and their adequacy for the description of critical phenomena is discussed. These equations are also used to provide a common basis for comparison of several different efforts at providing descriptions of critical phenomena beyond the mean field approximation.

2. FERROMAGNETISM IN A SYSTEM OF ITINERANT FERMIONS

The occurrence of ferromagnetism in a system of fermions of spin ½ has been discussed extensively by Rajagopal et al.[21] Their attention was directed to conditions necessary for the existence of the ordered states for several types of interaction. Their treatment was not limited to the Hartree-Fock approximation, but it did not make full use of the equations which will be outlined below. Although these equations have been available at various places in the literature, the possibility of using them for a self-consistent calculation of critical phenomena does not seem to have been discussed elsewhere. It is the purpose of this paper to raise this possibility primarily for the advantage of the perspective it gives on several sets of work which consider one or another aspect of the problem. In order to conserve space, the derivation of Rajagopal et al.[21] or of Baym and Kadanoff[22] will be used as far as is practicable.

The Hamiltonian of the fermions is given by

$$\mathcal{H} = \sum_\sigma \int d^3 1 \psi_\sigma^+(1) \left(\frac{-\nabla_1^2}{2m} + V(r_1) \right) \psi_\sigma(1)$$

$$+ \frac{1}{2} \sum_{\sigma_1, \sigma_2} \int d^3 1 d^3 2 dt_2 \psi_{\sigma_1}^+(1) \psi_{\sigma_2}^+(2) \mathcal{V}(1-2) \psi_{\sigma_2}(2) \psi_{\sigma_1}(1) \quad (1)$$

in terms of the creation and annihilation operators $\psi_{\sigma_1}^+(1)$, $\psi_{\sigma_1}(1)$ associated with particles of spin σ_1, position r_1, and time t_1. The operators have the usual anticommutation algebra. $V(r_1)$ is a background potential, while $\mathcal{V}(1-2)$ is the interaction potential between the fermions. $\mathcal{V}(1-2)$ is assumed here to be short-ranged and instantaneous. Time dependence of operators is taken in the Heisenberg representation. The expectation values $\langle O \rangle$ of an operator O can be considered here as being in the grand canonical ensemble.

The magnetization operator is given by

$$M = \int d^3 1 \rho_z(1) = \int d^3 1 [\psi_+^+(1) \psi_+(1) - \psi_-^+(1) \psi_-(1)] \quad (2)$$

The physics of the many-fermion medium is examined by studying the propagation in it of externally produced excitations. The propagation characteristics are expressed through one- and two-particle Green's functions

$$G_1(1\sigma_1, 2\sigma_2) \equiv -i \langle T\{\psi_{\sigma_1}(1) \psi_{\sigma_2}^+(2)\} \rangle \quad (3)$$

$$G_2(1\sigma_1 2\sigma_2, 3\sigma_3 4\sigma_4) \equiv (-i)^2 \langle T\{\psi_{\sigma_1}(1) \psi_{\sigma_2}(2) \psi_{\sigma_3}^+(3) \psi_{\sigma_4}^+(4)\} \rangle \quad (4)$$

where T is the Wick time ordering operator. The sources of the excitations are classical fields $U(1\sigma_1, 2\sigma_2)$ which enter through a perturbation

$$\mathcal{H}' \equiv \sum_{\sigma_1 \sigma_2} \int d^3 1 d^3 2 \, U(1\sigma_1, 2\sigma_2) \psi_{\sigma_1}^+(1) \psi_{\sigma_2}(2) \tag{5}$$

These fields U can be interpreted in terms of nonlocal fluctuations in the internal magnetic field or the chemical potential. The effects of the sources are traced by means of first-order response functions, or functional derivatives.[22] One such is the two-particle correlation function

$$L(1\sigma_1 2\sigma_2, 3\sigma_3 4\sigma_4) = G_2(1\sigma_1 2\sigma_2, 3\sigma_3 4\sigma_4) - G_1(1\sigma_1, 3\sigma_3) G_1(2\sigma_2, 4\sigma_4)$$

$$= \left. \frac{\delta G_1(1\sigma_1, 3\sigma_3)}{\delta U(4\sigma_4, 2\sigma_2)} \right|_{U=0} = \left. \frac{\delta G_1(2\sigma_2, 4\sigma_4)}{\delta U(3\sigma_3, 1\sigma_1)} \right|_{U=0} \tag{6}$$

The magnetization $\langle M \rangle$ can be expressed in terms of G_1, while the average energy $\langle \mathcal{H} \rangle$ can be expressed in terms of G_1 and G_2 or L. In addition the susceptibilities and the specific heat are closely related to L.[21] Thus this correlation function is quite important in expressing the physical content of the problem.

The functions G_1 and G_2 or L are also related through the equation of motion of G_1.[21]

$$\left\{ \left[i \frac{\partial}{\partial t_1} + \frac{\nabla_1^2}{2m} - V(1) - i\mathcal{U}(1-\bar{4}) G_1(\bar{4}, \bar{4}^+) \right] \delta_\sigma(1-\bar{3}) - U(1-\bar{3}) \right\}$$

$$\times G_1(\bar{3}, 2) + i\mathcal{U}(1-\bar{3}) L(1\bar{3}, 2\bar{3}^+) = \delta_\sigma^{(4)}(1-2) \tag{7}$$

The bar over repeated numerals implies an integration over that coordinate. The spin subset on the delta functions is to recall the now implicit spin labels.

A function $G_1^{-1}(1\sigma_1, 2\sigma_2)$ is inverse to G_1 in the sense that

$$G_1^{-1}(1, \bar{3}) G_1(\bar{3}, 2) = G_1(1, \bar{3}) G_1^{-1}(\bar{3}, 2) = \delta_\sigma^{(4)}(1-2) \tag{8}$$

The limit of G_1^{-1} for vanishing interaction $\mathcal{U}$ is denoted $G_{1,0}^{-1}$. These quantities are used to obtain a self-energy function $\Sigma(1\sigma_1, 2\sigma_2)$ which expresses the manner in which the interaction $\mathcal{O}$ affects the propagation of an excitation in the medium

$$\Sigma(1,2) = G_{1,0}^{-1}(1,2) - G_1^{-1}(1,2) \tag{9}$$

The variation of this quantity due to the presence of a second excitation gives a measure of the interaction between the excitations in the medium, $I(1\sigma_1, 2\sigma_2, 3\sigma_3, 4\sigma_4)$

$$I(12, 34) \equiv \frac{\delta \Sigma(1,3)}{\delta G_1(4,2)} = \frac{\delta \Sigma(2,4)}{\delta G_1(3,1)} \tag{10}$$

An equation of motion for the correlation function L follows upon functional differentiation of Eq. (8) and use of I[22]

$$L(12, 34) = -G_1(1,4)G_1(2,3) + G_1(1,\bar{5})G_1(\bar{6},3)I(\bar{5}\bar{8},\bar{6}\bar{7})L(\bar{7}2, \bar{8}4) \tag{11}$$

Iterated scattering of two excitations can be expressed using a reducible two-particle interaction or vertex $\Gamma(1\sigma_1, 2\sigma_2, 3\sigma_3 4\sigma_4)$. Γ is related to I by an integral equation

$$\Gamma(12, 34) = I(12, 34) + I(1\bar{5}, 3\bar{6})G_1(\bar{7},\bar{5})G_1(\bar{6},\bar{8})\Gamma(\bar{8}2, \bar{7}4) \tag{12}$$

It can be verified directly that solutions of Eq. (12) give solutions of Eq. (11) in the form

$L(12, 34)$

$$= -G_1(1,4)G_1(2,3) - G_1(1,\bar{5})G_1(\bar{6},3)\Gamma(\bar{5}\bar{8},\bar{6}\bar{7})G_1(2,\bar{8})G_1(\bar{7},4) \tag{13}$$

It may be that the homogeneous solutions of Eqs. (11) or (12) are nonvanishing. Such solutions represent bound states of two excitations in the medium, while the regular part of the solution to the full equations involves only the scattering of the excitations. If the bound states exist, their normalization must be determined by an independent condition.

If Γ or I are set to zero, one obtains just the Hartree-Fock approximation in Eq. (7). In order to improve on this approximation (which gives again the Landau exponents), it is necessary to give careful attention to Γ or I. It is not sufficient, for example, to approximate I by the bare interaction, for this leads to inconsistent results in Eqs. (7) to (13).

Since the striking feature of the phase transition is the change in symmetry upon ordering or disordering, the condition this places on the equations above is now investigated. Recall that the function G_1 can be expressed as a 2 x 2 matrix in the spin indices, and as such expanded in terms of the identity τ_0 and the Pauli matrices τ_1, τ_2, τ_3.[21] If a source field $U_{++} = -U_{--} = H_z$ is turned on, only the τ_0 and τ_3 matrices have nonzero components $G_1^{(0)}$ and $G_1^{(3)}$, respectively. The latter then leads to nonzero values of $\langle M \rangle$. Similar expansions can be made of the matrix forms of Σ, G_1^{-1}, $G_{1,0}^{-1}$. The third component of the latter is $H_z \delta(1-2)$. The effect of an infinitesimal rotation through an angle $\Delta\theta$ about the x axis is produced by an operator of the form $R_x(\Delta\theta) = \tau_0 - i\Delta\theta\tau_3$. The change in G_1^{-1} under this operation then is

$$\Delta G_1^{-1} = R_x G_1^{-1} R_x^{-1} - G_1^{-1} = -i\Delta\theta[H_z \delta(1-2) - \Sigma^{(3)}(1,2)]\tau_2 \tag{14}$$

Thus G_1^{-1} and by extension G_1 is invariant under arbitrary rotations about the x axis only if $\Sigma^{(3)}$ is zero when H_2 vanishes. But from Eq. (9)

$$\Delta G_1^{-1}(1,2)\Big|_{H_z=0} = -\Delta\Sigma(1,2)\Big|_{H_z=0}$$

$$= \frac{\delta\Sigma(1,2)}{\delta G(\bar{4},\bar{3})} G_1(\bar{4},\bar{5}) G_1(\bar{6},\bar{3}) \Delta G_1^{-1}(\bar{5},\bar{6}) \tag{15}$$

Further manipulation gives

$$[G_1^{-1}(1,\bar{3}) G_1^{-1}(\bar{4},2) - I(1\bar{3},2\bar{4})]\Delta G(\bar{3},\bar{4}) = 0 \tag{16}$$

Comparison with Eq. (11) shows that Eq. (16) is an eigenfunction equation associated with the homogeneous equation for L. Thus the existence of bound-state terms (or collective excitation terms, depending on their spectrum) in L or G_2 is directly associated with the nonvanishing of the $\Sigma^{(3)}$ coefficient which implies nonzero $\langle M \rangle$ and the loss of invariance under rotation. These terms also act in Eq. (7) as anomalous source terms.

The basic conservation laws of energy, momentum, and mass are satisfied in any approximation to L or I which meets well-defined conditions.[22] The spin indices bring in a new relation. The quantity

$$\langle j_0(1 \pm) \rangle \equiv \langle \psi_-^+(1) \rangle = -iG_1(1+, 1^+ -) \tag{17}$$

184 CRITICAL PHENOMENA

is a measure of the probability of a spin-flip event at 1. A related current is expressed as

$$\langle j(1\pm)\rangle \equiv \frac{1}{2im}[(\nabla_1 - \nabla_{1'})\langle\psi_-^+(1')\psi_+(1)\rangle]_{1'\to 1^+}$$

$$= -\frac{1}{2m}[(\nabla_1 - \nabla_{1'})G_1(1+,1'-)]_{1'\to 1^+} \quad (18)$$

For general U, it follows that

$$\frac{\partial}{\partial t_1}\langle j_0(1\pm)\rangle + \nabla_1 \cdot \langle j(1\pm)\rangle =$$

$$[U(1+,\bar{3})G_1(\bar{3},1^+-) - G_1(1+,\bar{3})U(\bar{3},1^+-)] \quad (19)$$

Thus with Eq. (6)

$$\frac{\partial}{\partial t_1}L(1+2-,1^+-2^++)$$

$$+ \nabla_1 \cdot \left[\frac{1}{2im}(\nabla_1 - \nabla_{1'})L(1+2-,1'-2^++)\right]_{1'\to 1^+} = -\langle \rho_z(1)\rangle \delta(1-2) \quad (20)$$

where ρ_z, from Eq. (2), is the local magnetization. Then in the limit of vanishing U Eqs. (19) and (20) become

$$\frac{\partial}{\partial t_1}\langle j_0(1\pm)\rangle + \nabla_1 \cdot \langle j(1\pm)\rangle = 0 \quad (21)$$

and

$$\langle T\left\{\left[\frac{\partial j_0(1\pm)}{\partial t_1} + \nabla_1 \cdot j(1\pm)\right]j_0(2\mp)\right\}\rangle = -\langle \rho_z(1)\rangle \delta(1-2) \quad (22)$$

Thus the spin-flip current is conserved regardless of the state of the magnetization, but the correlation function does not satisfy such a relation for equal times in the ordered state.

If the system is uniform, $\langle \rho_z(1)\rangle$ is independent of position, and one can use the Fourier transform function $\Lambda_i(k,\omega)$

$$\Lambda_i(\mathbf{k},\omega) = \int d^4 1 \exp\{i[\omega(t_1 - t_2) - \mathbf{k}\cdot(\mathbf{r}_1 - \mathbf{r}_2)]\}\langle T\{j_i(1\pm)j_0(2\pm)\}\rangle$$

(23)

to obtain

$$\omega\Lambda_0(\mathbf{k}_1\omega) - \mathbf{k}\cdot\boldsymbol{\Lambda}(\mathbf{k},\omega) = \langle\rho_z\rangle \qquad (24)$$

This equation is closely related to the Ward's identity derived by Morandi.[11] In the present form, the arguments of Lange concerning the Goldstone mode apply more directly.[14] One concludes that when $\langle\rho_z\rangle$ does not vanish, i.e., when the continuous rotation R_x is not an invariant operation, the functions Λ_i must have singular parts when ω and $\mathbf{k}$ vanish if Eq. (24) is to be satisfied. These terms come from the solutions to the Bethe-Salpeter type equations (15) or (16). The normalization of these functions is specified by requiring that they satisfy Eq. (24). One might anticipate that such terms would be large in L near T_c, but unless simple scattering of the excitations is negligible there, both types of terms must be present in L to achieve self-consistency. The two types of terms appear additively in L, and so in $\langle\mathcal{H}\rangle$ and other expectation values.

Ideally, one could begin with a suitable approximation to L and solve Eq. (7) for G_1. The functions Γ and I then follow from Eqs. (13) and (12), and the results can be used to derive a second form for L from Eqs. (11) and (16), including any bound states or collective excitations. If the original form and the second form agree, one can go on to calculate $\langle\mathcal{H}\rangle$, etc. If not, iteration to convergent results can be attempted.

This brief and superficial sketch of an ideal calculation is not meant to disguise the difficulty of effecting it either analytically or numerically. The calculation by Kadanoff and Martin[23] of properties of superconductors illustrates the difficulty, for there where the calculation is already extensive, the vertex I is approximated by $\mathcal{O}$. Thus the calculation cannot be self-consistent in L, Γ, or I.

The Goldstone mode appears in many second-order phase transitions where a continuous symmetry is broken, so the equations above are indicative of a more general problem than the restriction to magnetism might otherwise indicate.

3. SOME OTHER FIELD-THEORETIC APPROACHES

The set of equations discussed in the preceding section suggested that a study of the correlation function L might give results conveniently in common with both theory and experiments. However this is not the only possible starting point, for one might also proceed from the vertex functions Γ or I. A study of critical phenomena of bose fluids was carried out on this basis by Patashinskii and Pokrovskii.[24] Order of magnitude assumptions appropriate to the critical region, requirements of consistency between G_1 and the vertex, and application of dimensionality arguments gave the result that near T_c the self-energy function Σ is dominated by a term proportional to $|k|^{3/2}$, in the momentum picture. Such a term gives a logarithmic specific heat in agreement with experiment, but other quantities such as the compressibility do not agree. Abe[25] applied this same technique to the Ising model, and found parallel points of agreement and disagreement. In both cases, the vertex and self-energy functions were restricted to monomials, and no attempt was made to compare scattering with collective terms.

More recently, Polyakov[26] and Migdal[27] have used methods from relativistic field theory to attack this problem. Polyakov treated Ising systems and Migdal the bose fluid, but the form of their work was similar. The Green's functions were continued in the complex momentum plane to obtain functions to which techniques exploiting the unitarity of the S matrix could be used to calculate the absorbtive parts of G_1 and I. This approach has reproduced many previous results relating to scaling laws. However, Feibelman[28] has pointed out with regard to Migdal's work, and the same criticism is applicable to Polyakov's, that neglect of the inhomogeneous terms removes from the calculation information essential to complete the physical character of the solutions. Moreover in obtaining the solutions both Migdal and Polyakov assume that the functions separate into powers of $|k|$ and homogeneous functions. Thus their derivation of the scaling laws is to be expected, but it is not on a more fundamental basis than other derivations.

A study of the renormalization group of nonlinear equations of motion can give information concerning the form of the solutions, apart from subtraction constants. DiCastro and Jona-Lasinio[29] have applied this technique to critical phenomena. A report on this work will be given elsewhere in these proceedings.

The work by Umezawa and his collaborators[17-19] has concentrated on the renormalization of operators when a system undergoes a change of symmetry. Ordinarily, one considers the "dressing" of a model field ψ, because of nonlinear effects in the equations of motion, to get a physical field ϕ. But if the symmetry is changed, there may also be a Goldstone field B to be considered. Then, according to Umezawa, one has the mapping $\psi \to \psi[\phi, B]$. It is assumed that a substitution of this new form for ψ into the original Hamiltonian gives a new form in which both ϕ and B appear as free particles. Corresponding changes occur also in the forms of other observables. In particular, the original conserved current $j[\psi]$ takes a new form including additive contributions involving B. The new form of the current operator still obeys the original conservation law, leading Umezawa to speak of the dynamical rearrangement of symmetry.

Calculations following this program have been carried out in a pair approximation and give results for phase transitions into lattices[30] and into superconducting states.[31,32] No special attention has been given to critical phenomena, but the attention given to the Goldstone field as an active, indeed essential, part of the dynamical problem is similar in many respects to the attention given in the exact equations of the previous section. It may only be in the pair approximation that the fields ϕ and B can be considered as free, for one would expect that in general quasi-particle excitations and collective excitations would decay into each other and achieve stability only on the average. It remains to be determined whether or not the concepts of the dynamical rearrangement of symmetry program will be more fruitful than the unitarity-analyticity program in establishing complete, self-consistent approximate solutions to the equations for L, Γ, or I. Such an approximate solution is necessary to give a quantitative measure of the contribution through L to the thermodynamic functions of regular scattering of excitations and of the collective excitations arising in connection with the broken symmetry.

ACKNOWLEDGMENT

A portion of this work was performed as part of the laboratory-supported basic research of the Solid State Physics Division, Battelle Memorial Institute, Columbus.

REFERENCES

1. Landau, L., and E. M. Lifschitz: "Statistical Physics," Pergamon Press, London, 1958.
2. Heller, P.: *Rep. Prog. Phys.*, **30**:731 (1967).
3. Kadanoff, L. P., W. Gotze, D. Hamblen, R. Hecht, E. A. S. Lewis, V. V. Palciauskas, M. Rayl, J. Swift, D. Aspnes, and J. Kane: *Rev. Mod. Phys.*, **39**:395 (1967).
4. Fisher, M. E.: *Rep. Prog. Phys.*, **30**:615 (1967).

5. Brout, R.: "Phase Transitions," W. A. Benjamin, New York, 1965.
6. Tahir-Kheli, R. A.: *Phys. Rev.*, **132**:689 (1963).
7. Mills, R. E.: *Phys. Lett.*, **29A**:184 (1969).
8. Mattuck, R. D., and B. Johansson: *Adv. Phys.*, **17**:509 (1968).
9. Johansson, B.: *J. Phys.*, **C3**:50 (1969).
10. Morandi, G.: *Ibid.*, **A2**:487 (1969).
11. ——: *Nuo. Cimento,* **66B**:77 (1970).
12. Goldstone, J.: *Ibid.*, **19**:154 (1961).
13. Anderson, P. W.: "Concepts in Solids," W. A. Benjamin, New York, 1963.
14. Lange, R. V.: *Phys. Rev. Lett.*, **14**:3 (1965).
15. Wagner, H.: *Z. Physik,* **195**:273 (1966).
16. Meissner, G.: *Ibid.*, **205**:249 (1967).
17. Umezawa, H.: *Nuo. Cimento,* **38**:1415 (1965); *ibid.*, **40**:450 (1965).
18. ——: *Acta Phys. Hung.*, **19**:9 (1965).
19. Leplae, L., R. N. Sen, and H. Umezawa: *Nuo. Cimento,* **49**:1 (1967).
20. DeDominicis, C., and P. C. Martin: *J. Math. Phys.*, **5**:14, 31 (1964).
21. Rajagopal, A. K., H. Brooks, and N. R. Ranganathan: *Nuo. Cimento Suppl.*, **5**:807 (1967).
22. Baym, G., and L. P. Kadanoff: *Phys. Rev.*, **124**:287 (1961).
23. Kadanoff, L. P., and P. C. Martin: *Ibid.*, **124**:670 (1961).
24. Patashinskii, A. Z., and V. L. Pokrovskii: *Zh. Eksperim. Teor. Fiz.*, **46**:994 (1964) (*Sov. Phys. JETP,* **19**:677 (1964)).
25. Abe, R.: *Progr. Theoret. Phys.*, **33**:600 (1965); *ibid.*, **38**:568 (1967).
26. Polyakov, A. M.: *Zh. Eksperim. Teor. Fiz.*, **55**:1026 (1968); *ibid.*, **57**:271 (1969) (*Sov. Phys. JETP,* **28**:533 (1969); *ibid.*, **30**:151 (1970)).
27. Migdal, A. A.: *Zh. Eksperim. Teor. Fiz.*, **55**:1964 (1968) (*Sov. Phys. JETP,* **28**:1036 (1969)).
28. Feibelman, P.: *Phys. Rev. Lett.*, **22**:1091 (1969).
29. DiCastro, C., and G. Jona-Lasinio: *Phys. Lett.*, **29A**:322 (1969).
30. Coniglio, A., and M. Marinaro: *Nuo. Cimento,* **48B**:262 (1967).
31. Leplae, L., and H. Umezawa: *Ibid.*, **44B**:410 (1966).
32. —— and ——: *J. Math. Phys.*, **10**:2038 (1969).

RENORMALIZATION GROUP AND THEORY OF PHASE TRANSITIONS

G. Jona-Lasinio
Istituto di Fisica dell'Università Roma
Istituto Nazionale di Fisica Nucleare—Sezione di Roma
Rome, Italy

ABSTRACT

In this paper we make an attempt to describe those aspects of the critical point which seem to depend on general structural properties of classical and quantum statistical mechanics. The main tool of the analysis is the multiplicative Renormalization Group which represents the most general scaling property of many body theory. We recover in this framework some results of the usual scaling theory of phase transitions and we indicate explicitly the mathematical conditions under which they are expected to hold. The critical exponents and the coherence distance are explicitly given in terms of appropriate vertex functions.

1. INTRODUCTORY REMARKS

The scaling theory of the critical point[1] with its remarkable successes represents today a considerable challenge for both classical and quantum statistical mechanics. In fact, while it has been possible to calculate the critical behavior of thermodynamic quantities within the framework of special models, it is not clear in general which are the dynamical features leading to an approximate or rigorous validity of the scaling hypothesis. At the phenomenological level scaling theory introduces various kinds of technical assumptions.[2]

The first hypothesis is that there is a kind of contraction of variables: the dynamical behavior of the system near the critical point can be described in terms of a single parameter, the coherence distance ξ.

The second hypothesis states that certain correlation functions are homogeneous functions of their space or momentum argument and the coherence length ξ. One requires that this hypothesis applies only when the space variable is large compared to the range of the forces. On the other hand, no precise criterion really exists specifying the meaning of the expression "near the critical point." It is required however that no restriction exists on the ratio r/ξ where r is the space variable. When this ratio grows from zero to infinity we cover regions where very different physical phenomena are involved and it is therefore clear that in the present form the second assumption is actually a collection of separate hypotheses concerning considerably different physical situations.

In the following we shall make an attempt to determine those aspects of the critical point which seem to depend mainly on general structural properties of classical and quantum statistical mechanics. The approach has the advantage of showing clearly that the various results obtained from the usual scaling theory stand on very different footings as far as dynamical implications at the microscopic level are concerned.

The main tool of our analysis will be the multiplicative Renormalization Group (R.G.). The relevance of the R.G. for the theory of phase transitions comes from the fact that it represents the most general scaling property satisfied by many body theory. It is then natural to argue that the more restricted scaling laws used in the description of the critical point should somehow be connected with special realizations of the R.G.[3] Of course, the R.G., being an exact property of theory, holds at all temperatures and provides a restriction on microscopic quantities such as the correlation functions. Generally speaking these restrictions are rather weak and it is only under very special conditions that the R.G. provides an effective tool to determine relevant structural properties. It appears that the critical point is just one of those lucky cases. In fact the phenomenological study of phase transitions indicates that the critical region is an asymptotic region in the sense that various physical quantities exhibit a singular behavior: this means that we can concentrate on leading terms only and obtain a considerable simplification of the mathematical problem.

In Sec. 2 we shall indicate how to formulate the R.G. invariance for the case of classical systems. This will have mainly a pedagogical interest.

In Sec. 3 we shall develop some heuristic considerations showing the connection between the R.G. and scaling theory.

In Sec. 4 we shall try to connect scaling and the R.G. in a more systematic way.

We cannot close these introductory remarks without mentioning the work of Migdal and Polyakov.[4,5] These authors develop a strong coupling theory of phase transitions and a comparison between their work and ours will be reported elsewhere.

2. THE R.G. IN CLASSICAL STATISTICAL MECHANICS

The partition function of any classical system can be given a form resembling that of the S matrix of an euclidean field theory with the bare propagator proportional to the interaction potential. For definiteness in the following we shall refer to the Ising model. In this case the partition function can be written

$$Z = e^{\Delta} \exp\left[\sum_r \ln \cosh(\phi_r)\right]\Bigg|_{\phi=0} \quad (1)$$

where

$$\Delta = \frac{\beta}{2} \sum_{r,r'} V_{r,r'} \frac{\partial}{\partial \phi_r} \frac{\partial}{\partial \phi_{r'}} - \beta \sum_r h_r \frac{\partial}{\partial \phi_r}$$

$V_{r,r'}$ is the interaction potential and h_r is the external magnetic field. In field theoretic language, Z has the structure of the vacuum expectation value of the S matrix generated by the nonpolynomial

Lagrangian density

$$L_r = \ln\cosh(\phi_r) \qquad (2)$$

A formula like Eq. (1) is known as Hori formula. We introduce an auxiliary parameter g by modifying Eq. (2)

$$L_r(g) = \ln\cosh(g\phi_r) \qquad (2a)$$

g can be interpreted as an interaction strength. It is trivial to see that the following scaling invariance holds

$$Z(g, V, h) = Z(gL^{-1/2}, VL, hL^{1/2}) \qquad (3)$$

Our aim is to exploit this property to obtain information on the correlation functions generated by Z. Equation (3) by itself has no obvious dynamical content. As a first step toward dynamics we analyze the structure of the order parameter correlation function defined by

$$G(r_1, r_2) = \left.\frac{\partial^2 \ln Z}{\partial(\beta h_{r_1})\partial(\beta h_{r_2})}\right|_{h=0} \qquad (4)$$

A diagrammatic analysis (βV plays the role of a bare line) shows that the Fourier transform of G satisfies the equation

$$G(K) = \pi(K) + \beta V(K)\pi(K)G(K) \qquad (5)$$

where $\pi(K)$ in the terminology of diagrams is "one particle" irreducible. This means that it contains no diagram which can be divided into two parts just by cutting one line. From Eq. (3) it also follows

$$G(K, gL^{-1/2}, VL) = L^{-1}G(K, g, V)$$

$$\pi(K, gL^{-1/2}, VL) = L^{-1}\pi(K, g, V) \qquad (6)$$

The steps to the construction of the renormalization group equations now follow the usual scheme[6] by defining a dressed propagator via the equation

$$D(K) = \beta V(K) + \beta V(K)\pi(K)D(K) \qquad (7)$$

D satisfies the scaling condition

$$D(K, gL^{-1/2}, VL) = \overline{L}D(K, g, V) \qquad (8)$$

One then defines the function d

$$d(K) = D(K)[\beta V(K)]^{-1} = [1 - \beta V(K)\pi(K)]^{-1} = G(K)\pi^{-1}(K) \qquad (9)$$

and shows that multiplication by the scaling factor L is equivalent to a redefinition of the coupling g and a subtraction in the self-energy term $\beta V\pi$. This enables us to normalize d for appropriate values of its arguments. In the present case we are interested in the dependence of d on k and β and we shall have a normalization condition of the form $d(K = \lambda, \beta = \beta^0) = 1$. The next step consists in expressing d in terms of dimensionless variables

$$x = \frac{K^2}{\lambda^2} \qquad y = \frac{T - T_c}{\lambda^2} \qquad y_i = \frac{\mu_i^2}{\lambda^2} \qquad (10)$$

the variable $T - T_c$ rather than T has been chosen for convenience and the μ_i represent all the dimensional quantities of the theory, e.g., the range of the forces or other parameters characterizing the potential. The R.G. functional equation can finally be written

$$d(x, y, \{y_i\}, g) = d(t, y^0, \{y_i\}, g)\, d\left(\frac{x}{t}, \frac{y}{t}, \left\{\frac{y_i}{t}\right\}, gd^{1/2}(t, y^0, \{y_i\}, g)\right) \qquad (11)$$

where the scaling factor t is defined as the ratio

$$t = \left(\frac{\lambda'}{\lambda}\right)^2 \qquad (12)$$

of two normalization momenta. The normalization condition is

$$d(1, y^0, \{y_i\}, g) = 1 \qquad (13)$$

Equation (11) will be the basic equation for what follows.

3. R.G. AND SCALING: HEURISTIC CONSIDERATIONS

In this section we shall establish a connection between scaling theory and R.G. invariance by pointing out a possible mechanism reducing the complicated structure of the R.G. to the homogeneity condition required by the usual scaling hypotheses.[3]

Suppose we make the drastic assumption that all the parameters y_i, the coupling g, and the temperature variable y appearing in Eq. (11) combine together to form a new variable z and that near the critical point Eq. (11) can be written

$$d(x, z) = d(t, z_0) d\left(\frac{x}{t}, \frac{z}{t}\right) \qquad (14)$$

In other words, we have a contraction of variables which corresponds to the usual requirement of scaling theory that the dynamics near T_c be determined by only one parameter: the coherence distance ξ. If we now take the derivative of Eq. (14) with respect to t and then set $t = 1$ we have

$$x \frac{\partial d}{\partial x} + z \frac{\partial d}{\partial z} = \sigma d \qquad \sigma = \left.\frac{\partial d(t, z_0)}{\partial t}\right|_{t=1} \qquad (15)$$

The general solution of this equation has the form

$$d(x, z) = z^\sigma f\left(\frac{x}{z}\right) \qquad (16)$$

with f an arbitrary function. This form of d is the other basic requirement of scaling theory. In our context it becomes a consequence of the first, i.e., of Eq. (14). We notice that Eq. (16) is compatible with Eq. (14) only if we can choose z_0 in such a way that

$$d(t, z_0) = t^\sigma \qquad (17)$$

If, following our identification of z with the coherence distance $z = \xi^{-2}/\lambda^2$, we take $z_0 = 0$, that is, the critical point itself, Eq. (4) then requires for the function f the asymptotic behavior

$$f\left(\frac{x}{z}\right) \xrightarrow[z \to 0]{} \left(\frac{x}{z}\right)^\sigma \qquad (18)$$

However, we would like to avoid making ad hoc hypotheses such as the contraction of variables introduced in this section. A more systematic study of Eq. (11) appears desirable. This is what we shall try in the next section.

4. R.G. AND SCALING: A MORE SYSTEMATIC APPROACH

In order to exploit the full content of Eq. (11) we take the logarithmic derivatives of this equation with respect to x and y and put $t = x, y$ afterwards. We obtain the two equations

$$\frac{\partial \ln d(x, y, \{y_i\}, g)}{\partial x} = \frac{1}{x}\left[\frac{\partial}{\partial \rho} \ln d\left(\rho, \frac{y}{x}, \left\{\frac{y_i}{x}\right\}, gd^{1/2}(x, y^0, \{y_i\}, g)\right)\right]_{\rho=1} \quad (19)$$

$$\frac{\partial \ln d(x, y, \{y_i\}, g)}{\partial y} = \frac{1}{y}\left[\frac{\partial}{\partial \eta} \ln d\left(\frac{x}{y}, \eta, \left\{\frac{y_i}{y}\right\}, gd^{1/2}(y, y^0, \{y_i\}, g)\right)\right]_{\eta=1} \quad (20)$$

Equations (19) and (20) can be written in slightly different form by defining the vertex functions

$$\Gamma_1 = \frac{\partial d^{-1}}{\partial x} \quad (21)$$

$$\Gamma_2 = \frac{\partial d^{-1}}{\partial y} \quad (22)$$

We obtain

$$\frac{\partial \ln d(x, y, \{y_i\}, g)}{\partial x} = -\frac{1}{x}\tilde{\Gamma}_1\left(1, \frac{y}{x}, \left\{\frac{y_i}{x}\right\}, gd^{1/2}(x, y^0, \{y_i\}, g)\right) \quad (23)$$

$$\frac{\partial \ln d(x, y, \{y_i\}, g)}{\partial y} = -\frac{1}{y}\tilde{\Gamma}_2\left(\frac{x}{y}, 1, \left\{\frac{y_i}{y}\right\}, gd^{1/2}(y, y^0, \{y_i\}, g)\right) \quad (24)$$

where

$$\tilde{\Gamma}_1 = d\left(1, \frac{y}{x}, \left\{\frac{y_i}{x}\right\}, gd^{1/2}(x, y^0, \{y_i\}, g)\right)$$

$$\times \Gamma_1\left(1, \frac{y}{x}, \left\{\frac{y_i}{x}\right\}, gd^{1/2}(x, y^0, \{y_i\}, g)\right) \quad (25)$$

$$\tilde{\Gamma}_2 = d\left(\frac{x}{y}, 1, \left\{\frac{y_i}{y}\right\}, gd^{1/2}(y, y^0, \{y_i\}, g)\right)$$

$$\times \Gamma_2\left(\frac{x}{y}, 1, \left\{\frac{y_i}{y}\right\}, gd^{1/2}(y, y^0, \{y_i\}, g)\right) \quad (26)$$

The Eq. (23) and (24) system is compatible only if the following condition is satisfied

$$\frac{\partial}{\partial y}\left(\frac{\tilde{\Gamma}_1}{x}\right) = \frac{\partial}{\partial x}\left(\frac{\tilde{\Gamma}_2}{y}\right) \quad (27)$$

We now introduce the notion of critical point by requiring that d be singular when *both* x and y vanish. If we exclude essential singularities, Eqs. (23) and (24) yield for the two limiting cases $y = 0, x \to 0$ and $x = 0, y \to 0$

$$d(x, 0) \xrightarrow[x \to 0]{} x^{-\sigma_1} \quad (28)$$

$$d(0, y) \xrightarrow[y \to 0]{} y^{-\sigma_2} \quad (29)$$

where

$$\sigma_1 = \lim_{x \to 0} \tilde{\Gamma}_1\left(1, 0, \left\{\frac{y_i}{x}\right\}, gd^{1/2}(x, y^0, \{y_i\}, g)\right) \quad (30)$$

$$\sigma_2 = \lim_{y \to 0} \tilde{\Gamma}_2\left(0, 1, \left\{\frac{y_i}{y}\right\}, gd^{1/2}(y, y^0, \{y_i\}, g)\right) \quad (31)$$

σ_1 and σ_2 will be our basic critical indices and Eqs. (30) and (31) represent their definitions in terms of microscopic quantities. Of course one could define a critical index for any line of the (x, y) plane going into the origin. We remark that Eqs. (28) and (29) give the leading term of the logarithm of our function. In fact if we added to σ_1 and σ_2 a small correction, for example, $\sigma_1 + \sigma_1^1/\ln x$ or $\sigma_2 + \sigma_2^1/\ln y$, we would get $d \sim x^{-\sigma_1}(\ln x)^{-\sigma_1^1}$ or $d \sim y^{-\sigma_2}(\ln y)^{-\sigma_2^1}$ respectively. Our leading contribution is then defined up to logarithmic terms.

RENORMALIZATION GROUP

Having defined the critical indices we now discuss the behavior of d in the hydrodynamical and in the critical region.

4.1. The Hydrodynamical Region $y \neq 0, x \to 0$

From general arguments d is expected to be an analytic function of x in this region. A direct integration of Eq. (23) gives

$$d(x, y) = d(0, y) \exp\left[-\int_0^x dx^1 \frac{\tilde{\Gamma}_1}{x^1}\right] \tag{32}$$

From analyticity it follows that $(\tilde{\Gamma}_1/x)_{x=0}$ is finite. For small x, using Eq. (29) we then have

$$d(x, y) \sim y^{-\sigma_2}\left(1 - \frac{x}{m(y)}\right) \tag{33}$$

where we have defined

$$m^{-1}(y) = \left(\frac{\tilde{\Gamma}_1}{x}\right)_{x=0} \tag{34}$$

It is now possible to give the more precise characterization of the hydrodynamical region

$$\frac{x}{m(y)} \ll 1 \tag{35}$$

$m(y)$ as given by Eq. (34) has to be identified with the coherence distance according to

$$m^{-1}(y) = \xi^2(y)\lambda^2 \tag{36}$$

Thus we have obtained also a microscopic expression of the fundamental correlation length in terms of the vertex $\tilde{\Gamma}_1$. As the next step we calculate the behavior of m as a function of y for $y \to 0$. In order to find an equation for m we take the limit $x \to 0$ of Eq. (27). A straightforward calculation gives

$$\frac{\partial \ln m(y)}{\partial y} = -\frac{1}{y} m(y)\left(\frac{\partial \tilde{\Gamma}_2}{\partial x}\right)_{x=0} \tag{37}$$

this equation has a power law solution of the form $m \sim y^\sigma$ if

$$\lim_{y \to 0} m(y) \left.\frac{\partial \tilde{\Gamma}_2}{\partial x}\right|_{x=0} = \lim_{y \to 0} \lim_{x \to 0} \left.\frac{x}{\tilde{\Gamma}_1} \frac{\partial \tilde{\Gamma}_2}{\partial x}\right|_{x=0}$$

$$= \lim_{y \to 0} \lim_{x \to 0} \frac{\tilde{\Gamma}_2(x, y) - \tilde{\Gamma}_2(0, y)}{\tilde{\Gamma}_1(x, y)} = \text{finite} = -\sigma \qquad (38)$$

where only the relevant variables have been indicated. To evaluate the limit we have to introduce an assumption: The result of the last step does not depend on whether the $x \to 0$ limit or the $y \to 0$ limit is taken first. With this assumption, using Eqs. (31) and (32) we have

$$\sigma = \frac{\sigma_2}{\sigma_1} \qquad (39)$$

This result allows a rewriting of Eq. (33) in the form

$$d(x, y) \sim m(y)^{-\sigma_1} \left(1 - \frac{x}{m(y)}\right) \qquad (40)$$

This is exactly the prediction of scaling theory for the hydrodynamical region. The mathematical condition for its validity is given by the possibility of exchanging the limits in the last step of Eq. (38). The only explicit dynamic information which has been introduced into the R.G. equations is the analyticity of d in the region considered. Actually the only information which is being used is the fact that $(\tilde{\Gamma}_1/x)_{x=0}$ is finite and this of course is true if d is analytic. We finally remark that even if the limits in Eq. (38) cannot be interchanged we could have a power law for m with an exponent $\sigma \neq \sigma_2/\sigma_1$.

4.2. The Critical Region $x \neq 0, y \to 0$

There is an essential asymmetry between the macroscopic and the critical region. As shown by Polyakov[5] d is expected to be nonanalytic at $y = 0$ for any value of x. This means that a parallel treatment of this region is not possible. $(\tilde{\Gamma}_2/y)_{y=0}$ in general is not finite. For example, in the case of the two-dimensional Ising model $\tilde{\Gamma}_2/y$ diverges like $\ln y$. If one assumed $\tilde{\Gamma}_2/y$ finite at $y=0$ one would get

$$d(x,y) \sim x^{-\sigma_1}\left[1 - a\left(\frac{m}{x}\right)^{\sigma_1/\sigma_2}\right] \qquad (41)$$

A result of this type may exceptionally hold. It is clear however from the above discussion that the rather general assumptions introduced to this moment do not necessarily support the hypothesis that d is an homogeneous function of x and m in the whole range of values of the ratio x/m. Deviations from scaling can occur and more specific dynamical informations are needed to fill the gap between the hydrodynamical region and the critical region.

We come now to a very crucial question. To what extent the explicit expressions we obtained for the indices can be used for actual calculations? At least we would like to be able to say something about the degree of universality that can be expected for these quantities. Power laws are well known also in other fields of physics. A typical example is the behavior of the electron propagator near the mass-shell singularity once radiative corrections due to multiphoton processes are taken into account. The change from a simple pole singularity to a more complicated one is due to a phenomenon of accumulation of branch points. Another example is the Regge pole behavior in high-energy scattering and the mathematical mechanism is very much the same. An accumulation of branch points is present also at the critical point and it is natural to exploit this analogy to get some understanding. Having explicit expressions for the indices, such as Eqs. (31) and (32), the natural thing would be to apply perturbation theory. However in the limit when $x, y \to 0$ we encounter infrared divergences. This is due to the lack of any fundamental length at the critical point. One may then ask whether such a series can eventually be resummed and get a finite answer. A phenomenon of this type is known from Regge pole theory where one has examples[7] in which the expansion expressing the index diverges term by term. The series however can be resummed to give a new power series whose zero-order term depends essentially on the dimensionality of the space. On this basis we would like to conjecture that the indices σ_1 and σ_2 are obtained from two contributions, $\sigma_i = \sigma_i^{(1)} + \sigma_i^{(2)}$. The first contribution depends only on the dimensionality of the system and it is somehow universal. The second term is system dependent but smaller in value. This conjecture offers at least a definite direction for investigation.

What we have accomplished in this paper is to set up a general mathematical framework within which hypotheses leading to scaling can be given a precise form. Some exact results have been obtained but the full content of equations such as Eqs. (23) and (24) is far from being exhausted. Especially the strong nonlinearity of the equations has not been used at all. All this makes one hopeful that progress will be possible in the near future.

The ideas developed in this work will be discussed in a wider context including results on deviations from scaling in a forthcoming paper in collaboration with F. De Pasquale and C. Di Castro.

ACKNOWLEDGMENTS

This work was done to a large extent while the author was visiting the University of Marseille, Luminy, and the International Center for Theoretical Physics in Trieste. The author would like to express his gratitude to Profs. D. Kastler, A. Visconti, S. Doplicher, A. Salam, and P. Budini for the warm hospitality received in these places.

He would like to acknowledge also the continued and fruitful interaction with C. Di Castro, F. De Pasquale, M. Cassandro, D. Capocaccia, and G. Ciccotti at the University of Rome.

REFERENCES

1. Kadanoff, L. P., et al.: *Rev. Mod. Phys.,* **39**:395 (1967). Ferrell, R. A., et al.: *Ann. Phys.,* **47**:565 (1968). Halperin, B. I., and P. C. Hohenberg: *Phys. Rev.,* **177**:952 (1969). Other references can be found in these papers.
2. Halperin, B. I., and P. C. Hohenberg: *Loc. cit.*
3. This idea was first proposed by C. Di Castro and G. Jona-Lasinio, *Phys. Lett.,* **29A**:322 (1969).
4. Migdal, A. A.: *Sov. Phys. JETP,* **28**:1036 (1969).
5. Polyakov, A. M.: *ibid.,* **28**:533 (1969).
6. Since we are rather sketchy about the derivation of the R.G. functional equation, the reader may consult M. Gell-Mann and F. E. Low, *Phys. Rev.,* **95**:1300 (1954); Bogoliubov, N. N., and D. V. Shirkov, "Introduction to the Theory of Quantized Fields," chap. 8, Interscience Publishers, New York, 1959; Bonch-Bruevich, V. L., and S. V. Tyablikov, "The Green Function Method in Statistical Mechanics," chap. 11, North Holland Publishing Co., Amsterdam, 1962.
7. Cassandro, M., M. Cini, G. Jona-Lasinio, and L. Sertorio: *Nuo. Cimento,* **28**:1351 (1963).

DISCUSSION on *Paper by G. Jona-Lasinio*

G. JONA-LASINIO: I want to summarize briefly the logical structure of my paper which I think is quite simple.

The main idea is to see whether the scaling invariance properties which are expected to hold at the critical point, at least for a certain class of systems, can be embedded in a larger invariance group expressing an exact property of classical and quantum statistical mechanics. The reason for this is that we can then ask questions within a well-defined framework. For instance, we may ask which is the mechanism reducing at the critical point the larger invariance group to the simpler scaling invariance used phenomenologically. This gives also the possibility of getting some ideas about the reasons leading eventually to a breakdown of scaling.

The exact property which was taken as a basis for discussion is the multiplicative renormalization group which seems to be the most general scaling property holding for a wide class of systems. Within this framework the following points were especially discussed:

1. In the ordinary phenomenological scaling theory of the correlation functions one makes essentially the following hypotheses: There is a contraction of variables near the critical point so that the only relevant dimensional parameter is the correlation length; the correlation function is a homogeneous function of the momentum and the correlation length; the correlation function is singular only when the momentum and the inverse correlation length vanish. In this connection, it has been shown that the constraints imposed by the renormalization group invariance are strong enough to make the two last assumptions a consequence of the first.

2. If one does not want to introduce any assumption about contraction of variables, one can still obtain the scaling law $2 - \eta = \gamma/\nu$ from the renormalization group equations if two limits can be interchanged in evaluating an appropriate ratio for two vertex functions. In other words, the above scaling law can hold without the correlation function being an homogeneous function in the whole range of its arguments.

3. One should stress the fact that in this approach we obtain explicit definitions of the critical indices and of the correlation length in terms of microscopic vertex functions. This in principle gives the possibility of a microscopic calculation of the indices and we may notice certain analogies between the critical indices problem and the calculation of Regge trajectories in scattering theory. On this basis a conjecture was made pointing out a mechanism which could account for the crucial dependence of the critical indices on the dimensionality of the space.

SERIES EXPANSIONS AND THE UNIVERSALITY HYPOTHESIS

H. Eugene Stanley

Physics Department
Massachusetts Institute of Technology
Cambridge, Massachusetts

ABSTRACT

This work consists of two parts: In the first part, evidence from high-temperature series expansions is presented which supports the hypothesis that critical point exponents may not depend upon such features of the interactions Hamiltonian as (1) the spin quantum number S, (2) the degree of anisotropy, (3) the nonuniformity of the exchange interaction, and (4) the range of the interaction. In the second part, one form of the so-called "universality hypothesis" is stated, and a "universality Hamiltonian" is introduced, the study of which is expected (according to the universality hypothesis) to encompass the entire range of critical point exponents for systems with short range interactions. The predictions of this Hamiltonian (which essentially corresponds to a system of D-dimensional spins situated on a d-dimensional lattice) are discussed for all values of the parameters D and d.

1. ON WHAT FEATURES OF AN INTERACTION HAMILTONIAN DO CRITICAL POINT EXPONENTS DEPEND?

In recent years considerable experimental and theoretical attention has been directed toward the study of critical point exponents. Very recently increasing attention has been focused on the question of precisely which features of a physical system are relevant for determining the critical point exponents and which are not relevant. In this paper we discuss the evidence (provided principally by series expansion methods) which is germane to this question. We consider four particular features of model spin Hamiltonians which are thought *not* to affect critical point exponents. It is important to stress at the outset that the arguments which we shall present are by no means rigorous. On the other hand, series expansions have thus far provided essentially the only means of obtaining information about most *realistic* cooperative systems.

1.1 Dependence of Critical Point Exponents upon Spin Quantum Number S

We begin by considering the quantum-mechanical Heisenberg model with Hamiltonian given by

$$\mathcal{H}_H = -J \sum_{\langle ij \rangle}^{*} \mathbf{S}_i \cdot \mathbf{S}_j \qquad (1)$$

where the summation is restricted to nearest neighbor pairs of lattice sites $\langle i, j \rangle$ and J is the exchange parameters. In this section we discuss recent evidence which suggests that the susceptibility exponent $\gamma (\chi \sim (T - T_c)^{-\gamma})$ as $T \to T_c^+$ is independent of spin quantum number S. This result has an interesting history. When high-temperature expansion methods were first applied to this problem, it was proposed[1,2] that $\gamma(S) = 4/3$ for all values of the spin quantum number S. In 1964 Stanley and Kaplan[3] suggested that there appeared to be a weak spin dependence in which γ decreased from 1.43 to $S = 1/2$ to 1.33 for $S = \infty$ (the so-called "classical Heisenberg model") and that to a crude approximation

$$\gamma(S) \simeq 1.33 + \frac{0.05}{S} \qquad (2)$$

The estimates of Eqs. (1) and (2) were based upon the six terms in the susceptibility series which were available, at that time, for all values of the spin quantum number S. It was recognized that additional terms in the series could be calculated for the special cases $S = 1/2$[4] and $S = \infty$[5] due to certain simplifications which arise in those limits. Hence in 1966 three additional terms became available for $S = 1/2$[6] and two terms for $S = \infty$[7,8] and analysis of these extended series provided the estimates $\gamma(1/2) = 1.43 \pm 0.01$[6] and $\gamma(\infty) \simeq 1.36$[7] or $\gamma(\infty) \simeq 1.38$.[8] Thus although Eq. (2) would appear to be in error, it nevertheless seemed that there was a "spin dependence" to $\gamma(S)$. For some time workers argued whether this spin dependence was smoothly varying or abrupt: One proposal was that $\gamma(S) \simeq 1.43$ for the quantum mechanical case (finite S) and $\gamma(S) \simeq 1.38$ for the classical case $S = \infty$. However evidence slowly began accumulating to support other possibilities. Thus, e.g., Bowers and Woolf[9] argued that

$$\gamma(S) = 1.375 \pm 0.001 \simeq \frac{11}{8} \tag{3}$$

for *all* values of S, while (independently) Stanley[10] put forth evidence supporting

$$\gamma(S) = 1.38 \pm 0.02 \quad S > \frac{1}{2} \tag{4}$$

declining any conclusions whatsoever for the $S = 1/2$ case since the series for $S = 1/2$ are rather irregular. The irregular behavior of the $S = 1/2$ series has recently been interpreted[11] as arising from the presence of singularities in the complex $\mathcal{J}$ plane ($\mathcal{J} \equiv J/kT$) that are close to the physical singularity $\mathcal{J}_c \equiv J/kT_c$ which is on the positive real axis. A conformal mapping onto a transformed plane is given by

$$\mathcal{J}^* = \frac{A\mathcal{J}}{1 + B\mathcal{J}} \tag{5}$$

where A, B are real numbers whose values are chosen in such a fashion as to greatly separate the physical and the nonphysical singularities.

Using this method, Lee and Stanley[11] argue that $\gamma(1/2) = 1.36 \pm 0.04$ so that the possibility

$$\gamma(S) = \gamma_H = 1.38 \pm 0.02 \tag{6}$$

for *all* S is well within the confidence limits.

A similar situation existed for the case of the Ising model

$$\mathcal{H}_I = -J \sum_{\langle ij \rangle}^{*} S_i^z S_j^z \qquad (7)$$

for which it had been proposed[12] that

$$\gamma\left(S = \tfrac{1}{2}\right) = 1.25 \qquad (8)$$

while

$$\gamma(S = \infty) = 1.23 \qquad (9)$$

The discrepancy between Eqs. (8) and (9) has recently been explained[13] in terms of confluent singularities and it is now felt that for the Ising model

$$\gamma(S) \equiv \gamma_I = 1.25 \qquad (10)$$

for *all* S.

1.2. Dependence of Critical Point Exponents upon Isotropy of Interaction

Consider the "anisotropic" Heisenberg model

$$\mathcal{H}_A = -\sum_{\langle ij \rangle}^{*} [J_x S_i^x S_j^x + J_y S_i^y S_j^y + J_z S_i^z S_j^z] \qquad (11)$$

Clearly $\mathcal{H}_A$ reduces to the isotropic Heisenberg model when $J_x = J_y = J_z$; and to the Ising model when $J_x = J_y = 0$. Jasnow and Wortis[12] have investigated the fashion in which the Heisenberg model critical point exponents approach the Ising model critical point exponents when J_x and J_y approach zero. They argued, using extrapolations from high-temperature expansions, that the critical point exponents jump discontinuously from their Heisenberg values to their Ising values as soon as the slightest amount of anisotropy is introduced (i.e., if $J_z > J_x = J_y$).

1.3. Dependence of Critical Point Exponents on Nonuniformity of the Exchange Interactions

Another generalization of Eq. (7) is

$$\mathcal{H} = - \sum_{\langle ij \rangle}^{*} J_{ij} S_i^z S_j^z \qquad (12)$$

where, as before, two spins can interact only if they are on the neighboring sites of the lattice but *now* the interaction strength J_{ij} depends upon whether the sites i and j are horizontally or vertically disposed with respect to each other. One soluble example of Eq. (12) which has been studied is the case of a two-dimensional $d = 2$ "square lattice" of Ising spins which interact with their nearest neighbors in the horizontal and vertical directions with energies J_h and J_v respectively. Here the critical point exponents which can be obtained exactly (e.g., the magnetization exponent β, where $M \sim (T_c - T)^\beta$ as $T \to T_c^-$) are not affected as we vary the energy parameters J_h and J_v.[14]

1.4. Further Than Nearest Neighbor Interactions

A more general (and, mathematically speaking, considerably more complex) case is

$$\mathcal{H} = - \sum_{\langle ij \rangle} J_{ij} S_i^z S_j^z \qquad (13)$$

where now we allow for the very realistic possibility that spins *other* than nearest neighbors may interact with each other. Numerical calculations[15] have been carried out for the case for which $J_{ij} = J$ for spins which are either nearest neighbors, second nearest neighbors, or third nearest neighbors, with $J_{ij} = 0$ otherwise. The results of these calculations indicate that the critical point exponents may *not* change significantly from their values for the nearest neighbor case. However, when one assumes that the exchange interactions is of extremely long range, the exponents *are* found to change from their values for the case of nearest neighbor interactions. For example, the particularly tractable case

$$J_{ij} = \frac{1}{|r_i - r_j|^{d+x}} \qquad (14)$$

has been studied for both the Ising[16] and spherical[17] models. It appears that for large values of the parameter x, the critical point exponents assume their "short-range" values, while for smaller values of x, the exponents change—in the direction of the results predicted by the mean field theory, which in a sense corresponds to the case of *infinite* range interactions.[18]

2. THE UNIVERSALITY HYPOTHESIS AND A "UNIVERSALITY HAMILTONIAN"

Let us consider the following "universality hypothesis":[19] Critical point exponents for systems with short-range interactions depend upon

1. the lattice dimensionality
2. the "symmetry of the ground state"

and *not* upon properties of the system such as those discussed in Sec. 1:

1. the spin quantum number S (for example, $S = 1/2$ is the same as $S = \infty$)
2. the degree of anisotropy (for example, $J_x = J_y = 0$ and $J_z = 1$ is the same as $J_x = J_y = 0.9$ and $J_z = 1$)
3. the nonuniformity of the interactions (for example, $J_h = J_v$ is the same as $J_h = J_v$)
4. the range of the interaction (for example, the nearest-neighbors-only case is the same as allowing for second and third neighbor interactions)

According to the universality hypothesis, then, we should be able to study essentially all possible cases of short-range interactions if we study the dependence upon the lattice dimensionality d and the spin dimensionality D of the critical point exponents predicted for the Hamiltonian

$$\mathcal{H}_U = -J \sum_{\langle ij \rangle}^{*} \mathbf{S}_i^{(D)} \cdot \mathbf{S}_j^{(D)} \tag{15}$$

where here $\mathbf{S}^{(D)}$ is a D-dimensional unit vector or "classical spin"

$$\mathbf{S}^{(D)} = (\sigma_1, \sigma_2, \ldots, \sigma_D) \quad \sum_{n=1}^{D} \sigma_n^2 = 1 \tag{16}$$

TABLE 1. Special Cases of the Model Hamiltonian (Eq. (15))*

D	Hamiltonian	Name	System
1	$\mathcal{H} = -J \sum_{\langle ij \rangle} S_{ix} S_{jx}$	Ising model	One-component fluid; binary alloy, mixture
2	$\mathcal{H} = -J \sum_{\langle ij \rangle} (S_{ix} S_{jx} + S_{iy} S_{jy})$	Plane rotator model ("Vaks-Larkin" model)	λ-transition in a Bose fluid
3	$\mathcal{H} = -J \sum_{\langle ij \rangle} (S_{ix} S_{jx} + S_{iy} S_{jy} + S_{iz} S_{jz})$	Classical Heisenberg model	Ferromagnet; antiferromagnet
.			
.			
∞	$\mathcal{H} = -J \sum_{\langle ij \rangle} \left(\sum_{n=1}^{\infty} S_{in} S_{jn} \right)$	Spherical model	None

*The parameter D is the *spin* dimensionality.

Note that for $D = 1$, the Hamiltonian $\mathcal{H}_U$ reduces to the $S = 1/2$ Ising model (the spins being simply one-dimensional unit vectors); this model has come to serve as a practical model for a binary alloy, and a classical gas. From our discussion in Sec. 1.2 we might expect that the case $D = 1$ is also appropriate for anisotropic systems whose Hamiltonian is of the form of Eq. (11). For $D = 2$, $\mathcal{H}_U$ corresponds to a system of spins constrained to lie in a plane, and this "planar" model is often applied as a lattice model for the λ transition in a Bose fluid.[20] For $D = 3$ we have the classical Heisenberg model which, as discussed in Sec. 1.1, is now thought to have the same critical point exponents as the finite-spin Heisenberg model, and hence should be a realistic model for some magnetic systems with isotropic interactions (for example, EuO and EuS). For $D > 3$, the Hamiltonian $\mathcal{H}_U$ corresponds to *no* known physical system (Table 1).

When the Hamiltonian $\mathcal{H}_U$ of Eq. (15) was first proposed,[21] it was solved exactly in only two limits (Table 2):

1. $\mathcal{H}_U$ can be solved exactly for all values of *spin* dimensionality D if the lattice is one-dimensional ($d = 1$).[22]
2. $\mathcal{H}_U$ can be solved exactly for all values of lattice dimensionality d in the limit that the *spin* dimensionality approaches infinity.[23]

Concerning (1) we should remark that of course there are relatively few $d = 1$ systems in nature and hence the $d = 1$ solution has a rather restricted domain of applicability. Nevertheless we feel that the $d = 1$ results *are* worthy of interest, especially since results discovered for a $d = 1$ model frequently have their counterparts for systems of larger lattice dimensionality. For the case of $\mathcal{H}_U$, two particular results for $d = 1$ appear to be valid for *all d*:

1. As mentioned above, in the limit $D \to \infty$, the solution becomes identical to the spherical model.
2. The two-spin correlation function (and hence most thermodynamic functions) varies monotonically with spin dimensionality D.

Result (2) will be commented upon below. Result (1) was quite surprising to the author since the spherical model was originally proposed[24] as a modification of the Ising model ($D = 1$) in which the N constraints that each spin be of unit magnitude are replaced by a single constraint that the sum of the squares of the magnitudes of all N spins in the system be N (Fig. 1).

For cases other than $d = 1$ and $D = \infty$ (that is, for most realistic situations in nature) no exact solutions have been produced (Table 2)

TABLE 2. Some of the Cases in Which the Model Hamiltonian (Eq. (15)) is exactly soluble*

D	$d = 1$	$d = 2$	$d = 3$	$d > 3$
1	All H; both $n.n.$ and $1/r^{d+x}$	$H = O; n.n.$	$\cdots$	$\cdots$
2	$H = O; n.n.$	$\cdots$	$\cdots$	$\cdots$
3	$H = O; n.n.$	$\cdots$	$\cdots$	$\cdots$
.				
.				
.				
∞	All H; both $n.n.$ and $1/r^{d+x}$	All H; both $n.n.$ and $1/r^{d+x}$	All H; both $n.n.$ and $1/r^{d+x}$	Only critical point exponents are known exactly

*The notation $n.n.$ stands for "nearest-neighbor interactions only." A blank indicates that the system has not yet been solved. Here D and d denote the spin and lattice dimensionality respectively.

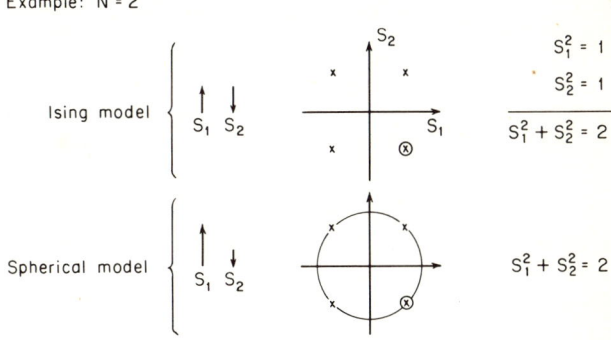

Fig. 1 *Comparison of the Ising model and the spherical model for the case of a simple system consisting of only $N = 2$ spins. We represent the $2^N = 4$ allowed states of the Ising system by the corners of an N-dimensional "hypercube." The spherical model replaces this discrete set of allowed states with a continuum of states, by replacing the N-dimensional hypercube with an N-dimensional "hypersphere." Note that the spins are still one-dimensional scalar quantities—all that we have done is to replace the "strong" constraints that each spin be of unit magnitude by the "weak" constraint that the sum of the squares of all the N spins in the system add up to N.*

and therefore we must rely upon approximation techniques. Among the more reliable techniques for determining critical properties are extrapolations based upon high-temperature expansions.[25] We have found that expansions for arbitrary D[26,27] can be carried out with little more labor than expansions for, say $D = 3$.[8,12,28] The reason is essentially that the requisite diagrams are the same regardless of the dimensionality of the spins. We have therefore calculated general

TABLE 3. Estimates of Critical Properties for D-dimensional Spins Situated on a $d = 3$ lattice (f.c.c. in this case)*

D	1	2	3	...	∞
T_c/T_M	0.816	0.804	0.793	...	0.7436
γ	1.25 $\left(\frac{5}{4}\right)$ ±0.002	1.32 $\left(\frac{4}{3}\right)$ ±0.02	1.38 $\left(\frac{7}{5}\right)$ ±0.02	...	2
α	0.125 $\left(\frac{1}{8}\right)$ ±0.010	−0.02 (0) ±0.05	−0.07 $\left(-\frac{1}{10}\right)$ ±0.05	...	−1
Δ	1.56 $\left(\frac{25}{16}\right)$ ±0.01	1.67 $\left(\frac{5}{3}\right)$ ±0.02	1.73 $\left(\frac{7}{4}\right)$ ±0.02	...	$\frac{5}{2}$
ν	0.638 $\left(\frac{5}{8}\right)$ ±0.002	0.675 $\left(\frac{2}{3}\right)$ ±0.01	0.70 $\left(\frac{7}{10}\right)$ ±0.02	...	1
"η" $= 2 - \frac{\gamma}{\nu}$	0.041 (0) ±0.006	0.04 (0) ±0.04	0.03 (0) ±0.03	...	0
$d\nu - (2 - \alpha)$	0.041 (0) ±0.10	0.04 (0) ±0.04	0.03 (0) ±0.03	...	0
"β"	0.3125 $\left(\frac{5}{16}\right)$ ±0.0010	0.33 $\left(\frac{1}{3}\right)$ ±0.02	0.35 $\left(\frac{7}{20}\right)$ ±0.05	...	$\frac{1}{2}$
"δ"	5.00 (5) ±0.20	5.00 (5) ±0.20	5.00 (5) ±0.20	...	5

*The simple fractions quoted are not intended to be the "closest rational fraction" but rather are obtained by *assuming* that each exponent varies with D according to the simple form $(a + bD)/(c + D)$, and choosing values of the parameters a, b, and c, so that the $D = \infty$ value is obtained (the spherical model exponents being exact) and the values for D finite are "best fit" (D. D. Betts and H. E. Stanley, to be published). This procedure produces one somewhat strange result: The exponent ν for $D = 1$ has the value $5/8 = 0.625$, which is smaller than the Jasnow-Wortis value 0.638 ± 0.002. The estimated error in each extrapolated estimate represents a subjective "confidence limit." Values of exponents for $3 < D < \infty$ vary smoothly, agree with the bilinear form, and are not given only due to lack of space.

D and general lattice expressions for the coefficients in the energy series, the susceptibility series, and the second moment series. With few exceptions,[29] such expansions have been available in the past only for $D = 1, 3$ and $d = 1 - 3$. Somewhat smoother series—which facilitate more reliable extrapolations—are then obtained by re-expanding all of the series in terms of the nearest neighbor spin correlation function for a $d = 1$ lattice.[30] We then study the dependence upon D and d (spin and lattice dimensionality) of the critical temperature $T_c(D, d)$ and the critical point exponents $\gamma(D, d)$, $\alpha(D, d)$, and $\nu(D, d)$ (Table 3). These critical properties are found to vary monotonically both with D and with d (Figs. 2 and 3).

The reader will note from Table 3 that the critical temperatures for a $d = 3$ lattice (in this case, a f.c.c. lattice) decrease smoothly with increasing D, suggesting the very plausible inequality

$$\langle S_0^{(D)} \cdot S_R^{(D)} \rangle \geq \langle S_0^{(D+1)} \cdot S_R^{(D+1)} \rangle \tag{17}$$

for arbitrary fixed T and R. The relation (17) has been proved rigorously by Milošević, Matsuna, and Stanley[31] only for $d = 1$, but if Eq. (17) were valid for all d, then it would clearly follow that

$$T_c^{(D)} \geq T_c^{(D+1)} \tag{18}$$

in agreement with the results of the second column of Table 3.

Not so intuitively obvious is the monotonicity displayed by the critical point exponents γ, α, and ν shown in the third, fourth, and fifth rows of Table 3. The entries in the sixth and seventh rows are predicted to be zero by various of the scaling relations. That the departure from zero appears to be a monotonically decreasing function of spin dimensionality should be viewed with caution—the errors in these small numbers are roughly as large as the numbers themselves. The quantity "β" tabulated in the eighth row is obtained by assuming the validity of the scaling relation $\alpha + 2\beta + \gamma = 2$. The fact that "$\beta$" so defined turns out to be $\gamma/4$ (regardless of whether or not we choose γ and α to be the "closest rational fraction") suggests the validity of the scaling relation $\gamma = \beta(\delta - 1)$ with $\delta = 5$. If we assume the validity of $\gamma = \beta(\delta - 1)$ (or, equivalently, of $\alpha + \beta(\delta + 1) = 2$), then we obtain the values of "δ" given in the last row. That "δ" turns out to have the value 5 independent of spin dimensionality D implies that the scaling parameter a_H in the homogeneity relation for the Gibbs potential

$$G(\lambda^{a_\epsilon} \epsilon, \lambda^{a_H} H) = \lambda G(\epsilon, H) \tag{19}$$

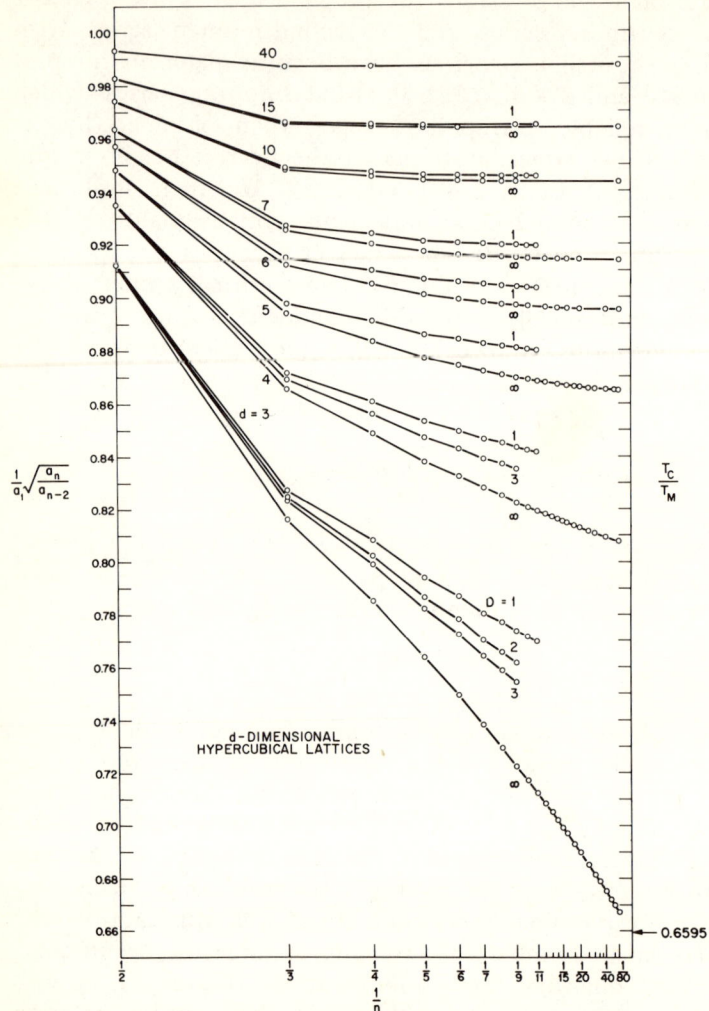

Fig. 2 *Ratio plot for the susceptibility series, indicating the variation of $T_c(D,d)$ with spin and lattice dimensionality D and d. The additional terms for the spherical model ($D = \infty$) come from a direct expansion of the analytic expressions themselves, and the additional terms for $D = 1$ come from M. E. Fisher and D. S. Gaunt: Phys. Rev., **A224**: 133 (1964). The case $D = 2$ is rather special and is treated in Fig. 3.*

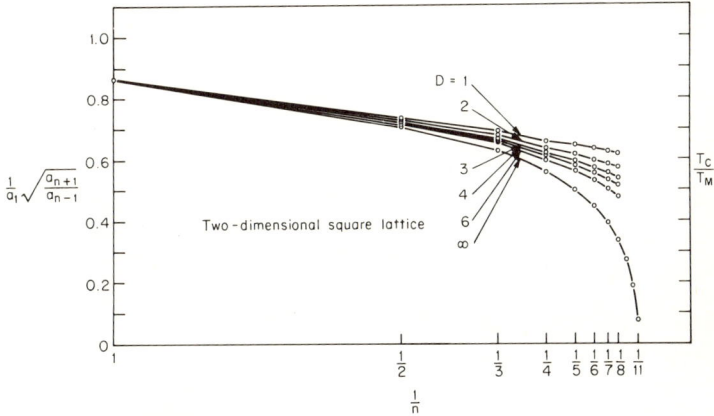

Fig. 3 *Variation of $T_c(D,2)$ with D for the plane square lattice. Shown is a conventional ratio plot, so that in the limit $n \to \infty$ the intercept should be given by T_c/T_M where T_M is the critical temperature predicted by the molecular field approximation. Curves for values of D which are not shown lie between the values shown.*

has the value (see Eq. (11.4) of Ref. 25)

$$a_H = \frac{\delta}{\delta + 1} = \frac{5}{6}$$

for *all* values of D; here $\epsilon \equiv (T - T_c)/T_c$ and H is the magnetic field.

We conclude with one cautionary remark. Although all of the critical properties appear to be monotonic functions of D and d, there is one property which may not be a *smooth* function of D. This is $T_c(D, d = 2)$, the critical temperature of a two-dimensional lattice. For the plane square lattice, e.g., it is known exactly that for $D = 1$—the Ising model—$T_c(1, 2) = 0.5673\ldots$ and also that for $D = \infty$—the spherical model—$T_c(\infty, 2) = 0$. Now the evidence from high-temperature expansions (Fig. 3) suggests that $T_c(D, 2)$ is smoothly decreasing function of D.[21,22,33] However it is known rigorously that if $D > 1$, $M(D, 2) = 0$ for all nonzero temperatures.[34] Hence it follows that *if* some sort of "phase transition" exists—as conjectured from the series expansions evidence—then it must be to a low-temperature phase with zero "infinite range order"

$$M^2 \propto \lim_{R \to \infty} \langle S_0 \cdot S_R \rangle \tag{20}$$

yet with sufficient "long-range order" that the susceptibility

$$\chi \propto \sum_{R} \langle S_0 \cdot S_R \rangle \tag{21}$$

diverges to infinity at some nonzero temperature T_c.[35] This is possible, e.g., if

$$\langle S_0 \cdot S_R \rangle \sim R^{-\lambda T} \tag{22}$$

where λ is a parameter.[35] We regard this as an open question at the present time. In particular recent experiments on a rather large class of materials which are layered magnets and therefore are "almost two dimensional" are in many cases *not* germane to answering this question (e.g., because of the presence of a small amount of anisotropy) though some evidence has been put forth to support an affirmative answer to this question.[36,37]

ACKNOWLEDGMENT

This work was supported by National Science Foundation Research Grant No. GP-15428.

REFERENCES

1. Domb, C., and M. F. Sykes: *Phys. Rev.*, **128**:168 (1962).
2. Gammel, J., W. Marshall, and L. Morgan, *Proc. Roy. Soc. (London)*, **A275**:257 (1963).
3. Stanley, H. E., and T. A. Kaplan: Comment at the 1964 Magnetism Conference; *J. Appl. Phys.*, **38**:977 (1967).
4. Domb, C.: *Adv. Phys.*, **9**:149 (1960).
5. The simplifications for $S = \infty$ were recognized independently by three groups of workers, and were published in the following order: Stanley, H. E., and T. A. Kaplan, *Phys. Rev. Lett.*, **16**:981 (1966); Wood, P. J., and G. S. Rushbrooke: *Ibid.*, **17**:307 (1966); Joyce, G. S., and R. G. Bowers: *Proc. Phys. Soc.*, **88**:1053 (1966).
6. Baker, G. A., H. E. Gilbert, J. Eve, and G. S. Rushbrooke: *Phys. Lett.*, **20**:146 (1966); *Phys. Rev.*, **164**:800 (1967).
7. Wood, P. J., and G. S. Rushbrooke: *Phys. Rev. Lett.*, **17**:307 (1966); Joyce, G. S., and R. G. Bowers: *Proc. Phys. Soc.*, **89**:776 (1966).
8. —— and ——: *Ibid.*
9. Bowers, R. G., and M. E. Woolf: *Phys. Rev.*, **177**:917 (1969).
10. Stanley, H. E.: *J. Appl. Phys.*, **40**:1546 (1969).
11. Lee, M. H., and H. E. Stanley: *Proc. 1970 Int. Conf. on Magnetism, J. Phys. (Paris)*, **32S**:352 (1971); ——: *Phys. Rev.*, **B4**:(Aug. 1971); the general idea of transforming unphysical singularities away from the physical singularity was first proposed, we believe, over a decade ago by A. Danielian and K. W. H. Stevens (*Proc. Phys. Soc.*, **B70**:326 (1957)) and, more recently, by A. J. Guttman and C. J. Thompson (*Phys. Lett.*, **28A**:679 (1969)).

12. Jasnow, D., and M. Wortis: *Phys. Rev.*, **176**:731 (1968).
13. Wortis, M.: *1969 Newport Beach Conf. on Critical Phenomena*.
14. Chang, C. H.: *Phys. Rev.*, **88**:1422 (1952).
15. Dalton, N. W., and D. W. Wood: *J. Math. Phys.*, **10**:1271 (1969); see also the recent work for arbitrary values of J_{nnn}/J_{nn}, Paul, G., and H. E. Stanley, to be published.
16. Hiley, B. J., and G. S. Joyce: *Proc. Phys. Soc.*, **85**:493 (1965).
17. Joyce, G. S.: *Phys. Rev.*, **146**:349 (1966).
18. Kac, M.: in "Brandeis Institute in Theoretical Physics," Gordon and Breach, New York, 1968; in *Proc. NATO Summer School on Mathematical Physics*, Istanbul, 1970.
19. Griffiths, R. B.: *Phys Rev. Lett.*, **24**:1479 (1970); Kadanoff, L. P.: *1970 Varenna Summer School on Critical Phenomena*; Stanley, H. E.: *Ibid.*
20. Vaks, V. G., and A. I. Larkin: *Sov. Phys. JETP*, **22**:678 (1966).
21. Stanley, H. E.: *Phys. Rev. Lett.*, **20**:589 (1968).
22. ———: *Phys. Rev.*, **179**:570 (1969); *J. Phys. Soc. Japan*, **26**:102 (1969).
23. ———: *Phys. Rev.*, **176**:718 (1968); see also Helfand, E.: *Ibid.*, **183**:562 (1969); and Thompson, C. J., and M. Kac: *Proc. Norwegian Acad. Sciences*, in press.
24. Berlin, T. H., and M. Kac: *Phys. Rev.*, **86**:821 (1952).
25. Stanley, H. E.: in W. Marshall (ed.), "Introduction to Phase Transitions and Critical Phenomena," International Series of Monographs on Physics, Oxford University Press, 1971.
26. ———: *J. Appl. Phys.*, **40**:1272 (1969); *Proc. NATO Summer Institute on Mathematical Physics*, Istanbul, 1970.
27. ——— and M. H. Lee: *Int. J. Quantum Chem.*, **4S**:407 (1970).
28. ———: *Phys. Rev.*, **158**:537, 546 (1967).
29. Bowers, R. G., and G. S. Joyce: *Phys. Rev. Lett.*, **19**:630 (1967); Stanley, H. E.: *Ibid.*, **20**:150 (1968).
30. ———: *Phys. Rev.*, **164**:709 (1967).
31. Milošević, S., K. Matsuno, and H. E. Stanley: *Phys. Stat. Solidi*, **42**:K163 (1970).
32. Moore, M. A.: To be published; Milošević, S., and H. E. Stanley: *Proc. 1970 Int. Conf. on Magnetism, J. Phys. (Paris)*, **32S**:346 (1971).
33. Moore, M. A.: *Phys. Rev. Lett.*, **23**:861 (1969).
34. Mermin, N. D., and H. Wagner: *Ibid.*, **17**:1133 (1966); Betts, D. D., C. J. Elliott, and R. V. Ditzian, *Can. J. Phys.*, **49**:1327 (1971).
35. Stanley, H. E., and T. A. Kaplan: *Phys. Rev. Lett.*, **17**:913 (1966); *J. Appl. Phys.*, **38**:975 (1967). See also V. Mubayi and R. V. Lange: *Phys. Rev.*, **178**:882 (1969), and R. M. Watson, M. Blume, and G. H. Vineyard: *Ibid.*, **B2**:684 (1970); Berezinskii, V. L., *Sov. Phys. JETP*, **32**:493 (1971); Lines, M. E., *J. Phys. (Paris)*, **32S**:477 (1971).
36. Birgeneau, R. G.: To be published.
37. Miedema, A. R.: *J. Phys. (Paris)*, **32S**:305 (1971).

DISCUSSION *on Paper by H. E. Stanley*

G. S. RUSHBROOKE: The argument of Bowers and Woolf in favor of $\gamma \sim 1.38$ for $H(1/2)$ is, of course, a strong one and I believe the series expansions for $H(1/2)$ do call for reexamination. Dr. Stanley's results are thus of great interest. But I should like to ask two things:

1. Has he similarly examined the gap-parameter Δ and does he now find the same value as for the classical Heisenberg model?

2. Are there any grounds for supposing the $H(1/2)$ indexes should be the "renormalized" $I(1/2)$ values? For this has been an odd feature of the 1967 values.

H. E. STANLEY: The answer to the first question is as follows: The gap parameter was predicted by Lee[1] to be given by

$$\Delta\left(S = \tfrac{1}{2}\right) = 1.75$$

(in contrast to the Baker-Rushbrooke prediction $\Delta = 1.81$), while Stephenson and Wood found that $\Lambda(S = \infty) = 1.73$. Hence I would say that it is likely that Δ is independent of S.

L. GUTTMAN: Were not the values of ν and η for the spin $-1/2$ Ising model on the slide those given by Fisher and Burford?

Those of Wortis et al. are rather different, and may affect the monotonicity, expecially of η.

H. E. STANLEY: The slide is an old one, and was used only because the transparency which I prepared did not work. The Wortis values are indeed slightly different, but they do not affect the monotonicity which I conjectured.

C. DOMB: This method of transformation in the complex plane was used a year or two ago effectively by Guttman, Ninham, and Thompson for the low-temperature series for the Ising model. Guttman also located all the unphysical singularities of the low-temperature Ising model.

G. A. BAKER: Could you please tell us the argument on which your belief that

$$\gamma\left(S = \tfrac{1}{2}\right) \simeq 1.38$$

is based?

H. E. STANLEY: Consider the susceptibility series

$$\chi = \sum_{n=0}^{\infty} a_n K^n \qquad (1)$$

where $K = J/kT$. Now we can reexpand Eq. (1) in terms of other functions $f(K)$

$$\chi = \sum_{n=\infty}^{\infty} A_n [f(K)]^n \qquad (2)$$

and it is trivial to determine the new coefficients A_n providing we know the expansion $f(K)$ in terms of K. Guttman et al. have pointed out (in connection with the Ising model series) that one source of irregularity of the coefficients a_n is due to the presence, in the complex K plane, of *nonphysical* singularities which are close to the *physical* singularity $K_c = J/kT_c$. We choose

$$f(K) = \frac{A}{1 + BK} \qquad (3)$$

and we find that the coefficients A_n in series (2) is smoother than the coefficients a_n in series (1). Our extrapolations yield a value of γ which is consistent with 1.40, the value that we currently believe for infinite spin. It should be emphasized that the idea of changing the expansion variable is by no means new, dating back at least to the work of Danielian and Stevens.[2] The specific choice of the "transformation function," Eq. (3), has been used very ingeniously by Betts et al.[3] to provide evidence supporting the existence of a phase transition in the XY model for the case of a $d = 2$ dimensional lattice.

REFERENCES

1. Lee, M. H.: *Phys. Rev.*, **B4**:(Aug. 1971).
2. Danielian, A., and K. W. H. Stevens: *Proc. Phys. Soc.*, **B70**:326 (1957).
3. Moore, M. A.: *Phys. Rev. Lett.*, **23**:861 (1969).

INEQUALITIES AMONG THE ISING-HEISENBERG MODEL CRITICAL INDICES

George A. Baker, Jr.
Applied Mathematics Department
Brookhaven National Laboratory
Upton, New York

Generally speaking there are two types of rigorous inequalities which are known. There are those which are of general nature, based usually on thermodynamic considerations, and those which depend on the details of the model. We will treat here some of those which belong to the second class. We will restrict our attention to the ferromagnetic Ising-Heisenberg model. That is, a system described by the Hamiltonian

$$H = -\sum_{i>j} J_{ij}[S_i^z S_j^z + \gamma_{ij}(S_i^x S_j^x + S_i^y S_j^y)] \qquad (1)$$

where the **S**'s are spin operators, and we assume

$$J_{ij} = J_{ji} \geq 0$$
$$\gamma_{ij} = \gamma_{ji} \qquad -1 \leq \gamma_{ij} \leq 1 \qquad (2)$$

The restriction on the J's insures a ferromagnetic interaction and that on the γ's makes the model interpolate between the Ising ($\gamma = 0$) and the isotropic Heisenberg models ($\gamma_{ij} = 1$).

As a generalization of previous results[1-3] we will show for this model that convergent upper and lower bounds can be given for the magnetization ($H \neq 0$) throughout the $H - T$ plane. Further, we are able to prove that the gap indices (the difference between the rates of divergence of successively higher magnetic field derivatives) satisfy the rigorous inequality

$$\Delta_{i+1} \geq \Delta_i \qquad (3)$$

and that

$$(-1)^n \Delta^n G(\tau^2) \geq 0 \qquad (4)$$

where Δ is the difference operator with respect to τ^2 and

$$\tau = \tanh\left(\frac{mH}{kT}\right)$$

$$G(\tau) = \frac{(M/mN) - \tau}{\tau(1 - \tau^2)} \qquad (5)$$

with M the magnetization, H the magnetic field, N a normalizing constant, m the magnetic moment, T the absolute temperature, and k Boltzman's constant. The spins may vary with the lattice site.

Finally, by means of the mathematical structure, and a mild extension of the proven analyticity of the free energy, we show

$$\delta \geq \frac{\Delta}{\Delta - \gamma} \qquad \delta \geq \frac{\Delta'}{\Delta' - \gamma'} \qquad \Delta \leq \Delta' \qquad \gamma \leq \gamma' \qquad (6)$$

when the relevant indices exist. (Presumably all do when $-1 < \gamma_{ij} < +1$, although there is considerable question about the low temperature side for the isotropic Heisenberg model because of spin waves.)

The starting point for our results is the theorem of Lee and

Yang.[4] They considered the partition function for the spin-1/2 ferromagnetic Ising model with interactions of unspecified range, and proved a theorem concerning the location of its zeros as a function of the fugacity. The partition function can be written as

$$Z = \sum_{n=0}^{N} \frac{Q_n}{n!} \mu^n \qquad (7)$$

where N is the number of lattice sites, and $\mu = \exp(-2mH/kT)$. As z is a finite polynomial it can be factored as

$$Z = \prod_{j=1}^{N} \left(1 - \frac{\mu}{\mu_j}\right) \qquad (8)$$

The theorem of Lee and Yang[4] states

$$|\mu_i| = 1 \qquad (9)$$

Their result was extended by Suzuki[5] to the spin 1 and 3/2 Ising models and by Griffiths[6] to arbitrary spin. Recently Asano[7] by a very nice method has also proven the Lee-Yang theorem (Eq. (9)) for the Ising-Heisenberg model described by Eqs. (1) and (2). As he points out, his result is a proof of "Conjectured Theorem B" of Suzuki.[8] On the basis of that conjecture, Suzuki then demonstrates the validity of Eq. (9) (Lee-Yang Theorem) for arbitrary spins at every lattice site, i.e., the general ferromagnetic Ising-Heisenberg model. Here N is $2\Sigma_j |S_j|$, the maximum power of μ in Z. If we now calculate the spontaneous magnetization in units which leave it between ±1, we get

$$\frac{M(\mu)}{(Nm)} = 1 - 2\mu \frac{d}{d\mu} \ln Z = \sum_{i=1}^{N} \frac{1 + \mu/\mu_i}{1 - \mu/\mu_i} \qquad (10)$$

$$\text{Re } \frac{M(\mu)}{(Nm)} \geq 0, \quad \text{if} \quad |\mu| \leq 1 \qquad (11)$$

From this form it follows by standard arguments[9] that

$$F(w) = (1 + w)^{-1/2} \frac{M[(1 + w)^{1/2} - 1]/[(1 + w)^{1/2} + 1]}{(Nm)} \qquad (12)$$

is a series of Stieltjes for all temperatures. Therefore the $[N, N](w)$ and $[N, N-1](w)$ Padé approximants form rigorous, converging upper and lower bounds to the magnetization. One can also bound the free energy and all its magnetic field derivatives by the use of generalized Padé approximants as was done for the Ising model.[1]

Thus a rigorously converging solution procedure can be given, $H \neq 0$, by means of the Padé approximants to Eq. (12) formed from the high-field series expansions. A substantial number of terms are known[10] in the spin-1/2 Ising model limit, but more generally, or even in the Heisenberg limit[11] they are quite difficult to provide. High-temperature expansions of the spin-1/2 Heisenberg model series are implicit in the results of Baker, Eve, and Rushbrooke.[12]

If we rewrite Eq. (10) as an integral over the density of roots on the unit circle and make a change of variables, we can cast it into the form

$$\frac{M}{mN} = \tau + \int_0^\infty \frac{2\tau(1-\tau^2)\omega}{1+\tau^2 w} d\phi\left(\frac{1}{1+\omega}\right) \tag{13}$$

where

$$d\phi \geq 0, \quad \int_0^1 d\phi(y) = \frac{1}{2} \tag{14}$$

Thus the function G, defined by Eq. (5), has the form

$$G(\tau) = \frac{(M/mN) - \tau}{1 - \tau^2} = \int_0^\infty \frac{d\Psi(\omega)}{1 + \tau^2 \omega} \tag{15}$$

a series of Stieltjes in τ^2. This has the consequence that if we expand

$$G(\tau) = G_0(T) - G_1(T)\tau^2 + G_2(T)\tau^4 - \cdots \tag{16}$$

then

$$D(m, n) = \det \begin{vmatrix} G_m & \cdots & G_{m+n} \\ \cdot & & \cdot \\ \cdot & & \cdot \\ \cdot & & \cdot \\ G_{m+n} & \cdots & G_{m+2n} \end{vmatrix} \geq 0 \tag{17}$$

As it is easy to show that the G_m have the same divergence as the high field derivatives, that is,

$$G_m(T) \propto (T - T_c)^{-\gamma_m} \qquad T \to T_c^+$$

it follows at once from Eq. (17) for $n = 1$, that

$$\gamma_{i+1} - 2\gamma_i + \gamma_{i-1} \geq 0$$

which is equivalent to Eq. (3). It further follows directly from Eq. (15) that Eq. (4) holds. *This result is experimentally checkable and should hold for any completely ferromagnetic Ising-Heisenberg model system.*

The remainder of my results are not quite rigorous, as they depend on a mild extension of proven results. First, one combines the results of Asano[7] and Suzuki[8] mentioned previously with that of Ginibre[13] who showed for the spin-1/2 anisotropic Heisenberg model with general interactions that there exists a neighborhood of any real positive T and $\mu = 0$ in which the free energy is analytic in μ and T. These results are sufficient to prove, by the results of Lebowitz and Penrose,[14] that the free energy is an analytic function of both T and τ for $\text{Re}(\tau) \gtrless 0$ and for any real finite temperature, just as Lebowitz and Penrose showed for the Ising model. In addition, Gallavotti, Miracle-Sole, and Robinson[15] have shown for the spin-1/2 general anistropic Heisenberg model (J_{ij} and γ_{ij} are however taken as functions of $(i - j)$ alone in their work) that we have analyticity in the neighborhood of $\tau = 0$, provided the absolute value of the complex temperature is high enough. All these results, taken together, have the consequence that the free energy is an analytic function of both temperature and magnetic field in the cut τ^2 plane

$$-\infty \leq \tau^2 \leq [R(T)]^{-1} \qquad R(T) < 0 \quad \text{for} \quad T > T_G \qquad (18)$$

for some T_G. Put another way, for T large enough there is a "window" in the distribution of singularities on the unit circle at $\mu = +1$ for T large enough, which allows the inside and outside of the unit circle to be analytically connected.

The first additional assumption we shall make is that T_G of Eq. (18) is in fact coincident with the critical temperature T_C. All available evidence points this way, at least as a function of T alone. It is in the spirit of the analysis of the Ising-Heisenberg model to date to assume the existence of critical exponents. In particular we assume

CRITICAL PHENOMENA

$$R(T) \approx \Gamma\left(\frac{T - T_c}{T_c}\right)^{-2\Delta} \qquad T \geq T_c \qquad (19)$$

Thus Eq. (15) becomes

$$G(\tau) = \int_0^{R(T)} \frac{d\Psi(\omega)}{1 + \tau^2 \omega} \qquad (20)$$

For $T > T_c$ we have the zero-field susceptibility given by

$$= 1 + \int_0^{R(T)} d\Psi(\omega)$$

$$= G(0) \approx 1 + A_0 \left(\frac{T - T_c}{T_c}\right)^{-\gamma} \qquad (21)$$

where we have assumed the existence of the critical index γ.

On the critical isotherm we see that $M \propto \tau^{1/\delta}$ implies that

$$G(\tau) \sim \tau^{(1-\delta)/\delta} \qquad (22)$$

We can obtain a lower bound on $G(\tau^2)$ directly from Eqs. (19) to (21) as

$$G(\tau) \geq \frac{A_0[(T - T_c)/T_c]^{-\gamma}}{1 + \tau^2 \Gamma[(T - T_c)/T_c]^{-2\Delta}} \qquad (23)$$

for T near, but greater than, T_c. From the property[16] that G is monotonic nonincreasing function of T and Eq. (22) for $T = T_c$, it follows that in the region $T \geq T_c$, $\tau^2 \geq 0$ there exists a constant Λ such that

$$\Lambda \geq \tau^{(\delta - 1)/\delta} G(\tau) \geq \frac{A_0 \tau^{(\delta-1)/\delta}[(T - T_c)/T_c]^{-\gamma}}{1 + \tau^2 \Gamma[(T - T_c)/T_c]^{-2\Delta}} \qquad (24)$$

The choice $\tau^2 = [(T - T_c)/T_c]^{2\Delta}/\Gamma$ immediately implies, in the limit $T \to T_c$

$$-\gamma + \frac{\Delta(\delta - 1)}{\delta} \geq 0 \tag{25}$$

or a convenient rewriting of Eq. (25) is

$$\delta \geq \frac{\Delta}{\Delta - \gamma} \tag{26}$$

We remark that it is easy to show how differentiating Eq. (20) and taking the limit as $T \to T_c$ that $\Delta \geq \Delta_{i+1} = \gamma_{i+1} - \gamma_i$, that is, it is an upper bound to the monotonic sequence of gap parameters.

Let us now turn our attention to the region $T < T_c$. Our picture for $T > T_c$ was that in the μ plane two singularities moved along the unit circle and reached $\mu = 1$ at the critical temperature. One of the fundamental results of the theory of more than one complex variable is that singular points move on trajectories and do not stop or disappear. Thus the moving singularities observed on the unit circle for $T > T_c$ must continue to be felt for $T < T_c$. Other singularities which approach $\mu = 1$ when $T \to T_c^+$, but more slowly (corresponding to smaller Δ), will be dominated by the leading one. Hence we expect to see a Δ' of at least Δ. However, we must consider also the possibility that other singularities approach the point $\mu = +1$ more rapidly than the fastest high-temperature one. They cannot do this on the physical sheet, but may on a different Riemann sheet. The dependence of G on τ instead of τ^2 below T_c for the Ising model is an example of a singularity which remains fixed at $\mu = 1$ ($\omega^{-1} = 0$ in Eqs. (15) and (20)). In the Heisenberg model, spin-wave theory predicts that the difference between the spontaneous magnetization and the magnetization vanishes like the square root of the magnetic field. This singularity would correspond to a fixed fourth root singularity at $\omega^{-1} = 0$. Consequently, we must have, taking account of more rapid singularities on other Riemann sheets

$$\Delta' \geq \Delta \tag{27}$$

More detailed arguments leading to this result are given by Gaunt and Baker.[3] These same arguments lead to the conclusion that each singularity yields

$$\beta_s = \frac{\Delta_s}{\delta_s} \qquad \gamma'_s = \left(\frac{\delta_s - 1}{\delta_s}\right)\Delta_s \tag{28}$$

We remark that a second singularity which asymptotically approaches $\omega^{-1} = 0$ at the same rate as the dominate one can make $\gamma_s < \gamma'_s$ (in accordance with Eq. (25)) so that we need not have symmetry between the high- and low-temperature indices.[3]

Let us now summarize the restriction on the possible values which the critical exponents can attain. The result can be gotten by thinking of the contribution from a set of singularities S_i. Each singularity will contribute the exponent set

$$\left\{ S_i: \Delta_i, \delta_i, \Delta'_i = \Delta_i, \gamma'_i = (\delta_i - 1)\frac{\Delta'_i}{\delta_i}, \gamma_i \leq \gamma'_i, \right.$$

$$\alpha'_i = 2 + \gamma'_i - 2\Delta'_i, \alpha_i = 2 + \gamma_i - 2\Delta_i,$$

$$\left. -\beta_i = -\frac{\Delta'_i}{\delta_i} \right\} \quad (29)$$

Singularities on the other sheet do not contribute to the high-temperature indices. Then any critical exponent E will be the maximum over the singularities S_i of E_i. (Note that we use $-\beta$ as the magnetization exponent, since the smallest β dominates the larger ones.) From this construction we can derive various relations. For example,

$$\Delta \leq \Delta' \qquad \gamma \leq \gamma' \qquad \alpha \leq \alpha'$$

$$\frac{\Delta'}{\beta} \geq \delta \geq \left(\frac{\Delta'}{(\Delta' - \gamma')} \quad \text{and} \quad \frac{\Delta}{(\Delta - \gamma)} \right)$$

$$\alpha' + 2\beta + \gamma' \geq 2 \qquad \alpha' + \beta(1 + \delta) \geq 2 \quad (30)$$

For the spin-1/2 Ising model, Gaunt and Baker[3] have estimated using the best available numerical data the bounds

$$5.13^{+0.24}_{-0.19} \geq \delta \geq 4.99^{+0.09}_{-0.08} \quad (31)$$

for δ. For the spin-1/2 Heisenberg model, we may use the results of Baker et al.[17] to give a lower bound on δ

$$\delta \geq 4.7 \pm 0.35 \quad (32)$$

which is in accord with the best estimates of $\delta \approx 5.0 \pm 0.2$ of Baker, Eve, and Rushbrooke.[12]

ACKNOWLEDGMENT

This work was performed under the auspices of the U.S. Atomic Energy Commission.

REFERENCES

1. Baker, G. A., Jr.: *Phys. Rev.,* **161**:434 (1967).
2. ———: *Phys. Rev. Lett.,* **20**:990 (1968).
3. Gaunt, D. S., and G. A. Baker, Jr.: *Phys. Rev.,* (3)**B1**:1184 (1970).
4. Lee, T. D., and C. N. Yang: *Ibid.,* **87**:410 (1952).
5. Suzuki, M.: *J. Math. Phys.,* **9**:2064 (1968).
6. Griffiths, R. B.: *Ibid.,* **10**:1559 (1969).
7. Asano, T.: *Phys. Rev. Lett.,* **24**:1409 (1970); *J. Phys. Soc. Japan,* **29**:330 (1970).
8. Suzuki, M.: *Prog. Theoret. Phys.,* **41**:1438 (1969).
9. Baker, G. A., Jr.: *Adv. Theoret. Phys.,* **1**:1 (1965).
10. Sykes, M. F., J. W. Essam, and D. S. Gaunt: *J. Math. Phys.,* **6**:283 (1965).
11. Wortis, M.: *Phys. Rev.,* **138A**:1126 (1965).
12. Baker, G. A., Jr., J. Eve, and G. S. Rushbrooke: *Phys. Rev.,* (3)**B2**:706 (1970).
13. Ginibre, J.: *Comm. Math. Phys.,* **10**:140 (1968).
14. Lebowitz, J. L., and O. Penrose: *Ibid.,* **11**:99 (1968).
15. Gallavotti, G., S. Miracle-Sole, and D. W. Robinson: *Ibid.,* **10**:311 (1968).
16. This property has been proven rigorously by R. B. Griffiths[6] for the general-spin Ising model. A more general property which would imply this one has been disproved for the Heisenberg spin-1/2 model by C. A. Hurst and S. Sherman, *Phys. Rev. Lett.,* **22**:1357 (1969). This weaker property has not yet been rigorously settled for our more general case.
17. Baker, G. A., Jr., H. E. Gilbert, J. Eve, and G. S. Rushbrooke: *Phys. Rev.,* **164**:800 (1967).

DISCUSSION *on Paper by G. A. Baker*

M. SUZUKI: Dr. Baker's argument may be extended to the fully anisotropic Heisenberg ferromagnets ($J_x \neq J_y$), because Fisher and I have proved the analyticity of the free energy for such a fully anisotropic Heisenberg ferromagnet under some ferromagnetic condition (that is, $|J_x| \leq J_z$ and $|J_y| \leq J_z$).

The case $J_x = J_y$ has been proved by Asano, as mentioned by Baker in his paper.

THEORY OF CRITICAL SLOWING DOWN

Masuo Suzuki[*]

Baker Laboratory
Cornell University
Ithaca, New York

ABSTRACT

Some general aspects of *linear* and *nonlinear* responses are discussed, particularly concerning critical slowing down. It is pointed out that the critical exponent of slowing down differs, in general, from that of the static susceptibility (or fluctuation). It is argued for ergodic systems that the singularity of the critical slowing down in the nonlinear response should be equal to that in the linear response. In connection with these properties, two specific examples are investigated: the kinetic Ising model and the linear XY model.

1. CRITICAL SLOWING DOWN IN THE LINEAR RESPONSE

The phenomenon of critical slowing down is one of general properties characteristic of dynamical critical phenomena. By critical

[*]On leave from the Institute for Solid State Physics, University of Tokyo, Tokyo, Japan.

slowing down, we mean that the relaxation time of any physical quantity becomes longer and longer as the temperature (or any other external parameter) approaches the critical point.

1.1. General Definition of Relaxation Time in Linear Response Theory

The relaxation time $\tau_A^{(l)}$ is defined[1] by the time integral of a normalized relaxation function $\hat{\Phi}(t)$ as

$$\tau_A^{(l)} = \int_0^\infty \hat{\Phi}_A(t)\,dt; \qquad \hat{\Phi}_A(t) = \frac{\Phi_A(t)}{\Phi_A(0)} \tag{1}$$

where the relaxation function is expressed by a canonical correlation according to the linear response theory[2]

$$\Phi_A(t) = \beta(\delta A, \delta A(t)); \qquad \delta A = A - \langle A \rangle \tag{2}$$

Here, the canonical correlation is defined by the equation

$$(A, B) = \frac{1}{\beta} \int_0^\beta \langle A(-i\lambda) B \rangle\, d\lambda \tag{3}$$

where $A(t)$ is the Heisenberg representation of A with the Hamiltonian $\mathcal{H}$ of the system

$$A(t) = e^{it\mathcal{H}} A e^{-it\mathcal{H}} \equiv e^{it\mathcal{L}} A \tag{4}$$

and $\mathcal{L}$ is Liouville operator defined by

$$\mathcal{L}A = [\mathcal{H}, A]; \qquad (\hbar = 1) \tag{5}$$

1.2. Reduction of Relaxation Time

It can be shown that the relaxation time $\tau_A^{(l)}$ of any physical quantity A is composed of two factors,[1] that is,

$$\tau_A^{(l)} = \frac{\chi_A}{\beta R_A(\xi)} \qquad (6)$$

where ξ represents T, H, etc., and

$$R_A^{-1}(\xi) = \int_0^\infty \exp\left(\sum_{n=1}^\infty \frac{i^n}{n!} \lambda_n x^n\right) dx \qquad (7)$$

where λ_n are cumulants defined by

$$\lambda_1 = \mu_1, \quad \lambda_2 = \mu_0 \mu_2 - \mu_1^2,$$
$$\lambda_3 = \mu_0^2 \mu_3 - 3\mu_0 \mu_1 \mu_2 + 2\mu_1^3 \quad \text{etc.} \qquad (8)$$

in terms of the moments μ_n

$$\mu_n = (\delta A, \mathcal{L}^n \delta A) \qquad (9)$$

The critical slowing down arising from the first factor χ_A in Eq. (6) is called "thermodynamical critical slowing down." It is caused simply by the equilibrium thermal fluctuations of the relevant quantity A. The remaining factor $R^{-1}(\xi)$ represents the "kinetic critical slowing down (or speeding up)," which is characteristic to the Liouville operator or kinetics of the system. Consequently, only when $R^{-1}(\xi)$ is nonvanishing and finite at the critical point, one obtains the usual "Van Hove" results $\tau_A^{(l)} \propto \chi_A$. However, when the cumulants $\lambda_n (n \geq 2)$ are divergent, it is expected that $R^{-1}(\xi)$ may be singular near the critical point. Then, the critical exponent of slowing down is different from that of the susceptibility or fluctuation. Such examples have been shown theoretically by Suzuki, Yahata, and Kubo[3] and experimentally by Hatta.[4]

1.3. Critical Slowing Down in Kinetic Ising Model

We have found that the critical exponent $\Delta_M^{(l)}$ of slowing down for the magnetization is different from that of the static susceptibility γ in the kinetic Ising model. This model is described by a master equation (as a stochastic process) with a transition probability which depends on the configuration of surrounding spins and satisfies the condition of detailed balance. We have obtained[3]

$$\Delta_M{}^{(l)} = 2.00 \pm 0.05 \quad \text{whereas} \quad \gamma = \frac{7}{4} \quad (10)$$

in two dimensions, where $\Delta_M{}^{(l)}$ is defined by

$$\tau_M{}^{(l)} = \int_0^\infty \frac{\langle M(t) M \rangle}{\langle M^2 \rangle} \, dt \sim (T - T_c)^{-\Delta_M{}^{(l)}} \quad (11)$$

It should be remarked that the critical exponent $\phi_M \, (= \Delta_M - \gamma)$ of the second factor $R_M{}^{-1}$ in Eq. (6) is equal to $\eta\nu$ ($= 1/4$; $\nu = 1$ and $\eta = 1/4$) in this case where ν is the critical exponent of the correlation length and η is an index parameter indicating a deviation of the correlation function from the Ornstein-Zernike type. Furthermore, we have confirmed that the dynamical susceptibility can be scaled in the form[3]

$$\chi(\omega) = \epsilon^{-\gamma} f\left(\frac{i\omega}{\epsilon^\Delta}\right); \quad \epsilon = \frac{T - T_c}{T_c} \quad (12)$$

1.4. Critical Slowing Down in Linear XY Model

The Hamiltonian to consider here is given by

$$\mathcal{H} = -J \sum_j [(1 + \delta) S_j^x S_{j+1}^x + (1 - \delta) S_j^y S_{j+1}^y] - \mu_B H \sum_j S_j^z \quad (13)$$

where S_j^x, S_j^y, and S_j^z denote the spin operators for $S = 1/2$ at the jth site, and δ is a constant satisfying $1 \geq \delta > 0$. This system has a critical field H_c, where the susceptibility exhibits a logarithmic singularity (that is, $\gamma = 0$) with respect to the variable $(H - H_c)$ near the critical field at zero temperature. In the present model the relaxation time of the magnetization is defined by[1]

$$\tau_M{}^2 = -\int_0^\infty \frac{(\delta M, \delta M(t))}{(M, M)} \, t \, dt \quad (14)$$

which has the following asymptotic form[1]

$$\tau_M \sim (H - H_c)^{-\Delta_M^{(l)}}; \quad \Delta_M^{(l)} = 1 \tag{15}$$

near the critical field at $T = 0$, where the logarithmic singularity has been neglected at Eq. (15). The derivation of this result will be illustrated in Sec. 2 in connection with the nonlinear response. (It should be noted that the definition (Eq. (1)) of the relaxation time is inappropriate in the present model because it vanishes identically.) This is the first rigorous example in which the exponent of critical slowing down Δ_M is different from that of the susceptibility.

2. CRITICAL SLOWING DOWN IN THE NONLINEAR RESPONSE

We discuss here a nonlinear situation in which the system is in equilibrium described by an initial (time-independent) Hamiltonian $\mathcal{H}_i$ (or $\mathcal{H}_0$) and at a certain time (say $t = 0$), the system is abruptly forced into a different condition described by a new (time-independent) Hamiltonian $\mathcal{H}$.

The relaxation time for this nonlinear response is defined by

$$\tau_A^{(n.l.)} = \int_0^\infty \frac{A(t) - \langle A \rangle}{\langle A \rangle_i - \langle A \rangle} dt = \langle \delta A \rangle_i^{-1} \int_0^\infty \langle e^{it\mathcal{L}} \delta A \rangle_i dt \tag{16}$$

where $\langle \cdots \rangle_i$ indicates an average taken over the initial Hamiltonian $\mathcal{H}_i$.

2.1. Relation between Singularity of Relaxation Times in Nonlinear and Linear Responses

2.1.1. Ergodic Systems.
An ergodic system in the nonlinear response approaches more and more closely its equilibrium situation as time increases. Then, it seems to be reasonable in general, to assert that both critical exponents should be equal in *ergodic systems*

$$\Delta_A^{(n.l.)} = \Delta_A^{(l)} \tag{17}$$

because the differences in the initial (or intermediate) stages of the relaxation are expected not to affect the singularity of the relaxation time, and because anomalous (or critical) fluctuation will appear dominant in (or very close to) equilibrium, which will be attained by the final stage of the relaxation.

The kinetic Ising model has been investigated by using perturbation expansions in the resolvent form

$$\frac{1}{\mathcal{L}} = \frac{1}{\mathcal{L}_0 - \mathcal{L}_1} = \frac{1}{\mathcal{L}_0} + \frac{1}{\mathcal{L}_0}\mathcal{L}_1\frac{1}{\mathcal{L}_0} + \cdots \quad (18)$$

For simplicity, we discuss a relaxation process in which the relevant system is completely ferromagnetic (or with an infinite field) at the initial time and approaches its equilibrium situation in the absence of the magnetic field. The relaxation time thus obtained in the plane square lattice is

$$\frac{\tau_M^{(n.l.)}}{\tau_0} = 1 + \frac{16}{3}v + \frac{64}{3}v^2 + \frac{208}{3}v^3 + \frac{608}{3}v^4 + \frac{15{,}664}{27}v^5$$

$$+ \frac{681{,}664}{405}v^6 + \cdots \quad (19)$$

where $v = \tanh(J/kT)$ and J is the exchange interaction between nearest neighbor spins. Applying the ratio method to this result (Eq. (19)), we find that

$$\Delta_M^{(n.l.)} \simeq 2.209, 2.314, 2.039, 1.843, 1.929, 2.210, \cdots \quad (20)$$

as the successive approximations for the critical exponent of slowing down of the total magnetization. We believe that the estimates for the critical exponent are oscillating around *the value 2* and will approach this value in the limit of the infinite series. Hence, we conclude that the value of critical exponent of slowing down of the nonlinear response is equal to that in the linear response for this ergodic kinetic Ising model. This illustrates nicely the general property (Eq. (17)).

2.1.2. Nonergodic Systems. It is difficult to predict the general behaviors of the relaxation of a nonergodic system. In fact, a nonergodic system does not approach its equilibrium situation, and the initial condition may sometimes be crucial in discussing the singularity of the relaxation time, so that both critical indices may, in general, be different.

We have calculated rigorously the relaxation time in the onedimensional XY model (Eq. (13)) for the nonlinear response. That is, the system is in equilibrium under a magnetic field H_0 (or $h_0 = \mu_B H_0/J$) for $t < 0$. At time $t = 0$, the field is suddenly changed to the value H (or $h = \mu_B H/J$) close to the critical field H_c. Then, how does the magnetization relax to its limiting value as $t = \infty$? This situation can be characterized by the relaxation time[1]

$$\tau_M^{(2)(\text{n.l.})} = \left[\lim_{\epsilon \to +0} \int_0^\infty e^{-\epsilon t} \frac{M(t) - M(\infty)}{M(\infty) - M(0)} t\, dt \right]^{1/2} \quad (21)$$

Our problem is solved, in principle, by the following observations

$$M(t) = \langle M \rangle_t = Tr M \rho(t)$$

where

$$\rho(t) = e^{-i\mathcal{H}t} \rho(0) e^{i\mathcal{H}t} \quad \text{for} \quad t > 0$$

and

$$\rho(t) = \frac{e^{-\beta \mathcal{H}_0}}{Tr\, e^{-\beta \mathcal{H}_0}} \quad \text{for} \quad t \leq 0 \quad (22)$$

First, we transform the Hamiltonian (Eq. (13)) into the form[5]

$$\mathcal{H} = \sum_k \left\{ \alpha_k a_k^\dagger a_k + \frac{1}{4} i \beta_k (a_k^\dagger a_{-k}^\dagger + a_k a_{-k}) \right\} \quad (23)$$

with $\alpha_k = J(\cos k - h)$ and $\beta_k = -J_\delta \sin k$, by using the well-known transformation

$$S_n^- = S_n^x - i S_n^y = a_n^\dagger \prod_{m < n} (1 - 2 a_m^\dagger a_m) \quad \text{etc.} \quad (24)$$

where $a_n^\dagger$ and a_n are, respectively, the Fermion creation and annihilation operators and also we have made use of the Fourier transforms a_k and $a_k^\dagger$ of the Fermi operators a_n and $a_n^\dagger$, respectively.

Here, instead of computing Eq. (22) directly, it is more convenient to use the equation of motion for the operators $a_k^\dagger$ and a_k. Thus, we find the differential equation

$$\frac{d^3}{dt^3} M_k(t) + \omega_k^2 \frac{d}{dt} M_k(t) = 0; \quad \omega_k^2 = 4(\alpha_k^2 + \beta_k^2) \quad (25)$$

for the Fourier component $M_k(t) \equiv \langle a_k^\dagger a_k + a_{-k}^\dagger a_{-k} - 1 \rangle$ of the magnetization. It is easy to compute the initial values of $d^n M_k(t)/dt^n$ for $n = 0, 1,$ and 2, by diagonalizing the Hamiltonian (Eq. (23)). The results are given by

$$\dot{M}_k(0) = 0 \quad \text{and} \quad \ddot{M}_k(0) = 4J^3\delta^2 \sin^2 k (1 - 2n_k^0)(\epsilon_k^0)^{-1}(h_0 - h) \quad (26)$$

where $\epsilon_k = (1/2)\omega_k$, $n_k = [\exp(\beta\epsilon_k) + 1]^{-1}$, and the superscript 0 denotes the values of the relevant quantities corresponding to the initial Hamiltonian $\mathcal{H}_0$. Thus, we arrive at the final expression of the magnetization, namely

$$M(t) = \frac{1}{2\pi} \int_0^\pi M_k(t)\, dk = M(0) + \frac{1}{\pi} \int_0^\pi [\cos(\omega_k t) - 1] g(k)\, dk \quad (27)$$

where

$$g(k) = \left(\frac{2J^3 \gamma^2 \sin^2 k}{\epsilon_k^0 \omega_k^2} \right) T_0(k)(h_0 - h) \quad (28)$$

and

$$T_0(k) = \tanh\left(\frac{\epsilon_k^0}{2k_B T} \right) \quad (29)$$

It should now be remarked that this system is *nonergodic* even at $T = 0$ in the sense that

$$\lim_{t \to \infty} M(t) \neq M_{\text{eq.}}(H) \quad (30)$$

Now, as the relaxation time defined by Eq. (21) is easily rewritten as

$$\tau_M^{(2)(\text{n.l.})} = \left\{ \frac{\int_0^\pi dk\, [T_0(k) \sin^2 k / 4\epsilon_k^0 \epsilon_k^4]}{\int_0^\pi dk\, [T_0(k) \sin^2 k / \epsilon_k^0 \epsilon_k^2]} \right\}^{1/2} \quad (31)$$

This is a rigorous result valid for any temperature and any magnetic field. It is easily shown to exhibit singularity of the form

$$\tau_M^{(2)(\text{n.l.})} \sim \delta |H - H_c|^{-1/2} \qquad \Delta_M^{(\text{n.l.})} = \frac{1}{2} \qquad (32)$$

near the critical field at $T = 0$. From Eq. (31), one can obtain Eq. (15) in the limit of the linear response; i.e., for $H_0 \to H$ in Eq. (31). As anticipated, *the critical exponent of the relaxation time of the nonlinear response is different from that in the linear response.* The origin of the difference is explained mathematically as follows. For the linear response, the quantity ϵ_k^0 of Eq. (31) becomes equal to ϵ_k and consequently one obtains a stronger singularity as is shown in Eq. (15). For the nonlinear response, ϵ_k^0 remains nonzero even for $k \neq 0$ and at the critical field. Consequently, a weaker singularity has been obtained. Physically speaking, the initial condition affects the final relaxation of the nonlinear response in this nonergodic system.

ACKNOWLEDGMENTS

The author would like to express his sincere thanks to Professor M. E. Fisher for helpful comments. The support of the National Science Foundation is gratefully acknowledged.

REFERENCES

1. Suzuki, M.: *Prog. Theoret. Phys.*, **43**(4):882 (1970).
2. Kubo, R.: *J. Phys. Soc. Japan*, **12**:570 (1957).
3. Suzuki, M., H. Yahata, and R. Kubo: *Ibid.*, **26** Suppl.:153 (1969); Yahata, H., and M. Suzuki: *Ibid.*, **27**:1421 (1969).
4. Hatta, I.: *J. Phys. Soc. Japan*, **28**:1266 (1970).
5. Lieb, E., T. Schultz, and D. Mattis: *Ann. Phys.*, **41**:407 (1961); Katsura, S.: *Phys. Rev.*, **127**:1508 (1962); Pikin, S. A., and V. M. Tsukernik: *Soviet Phys. JETP*, **23**:914 (1966).

DISCUSSION on *Paper by M. Suzuki*

M. KAC: Could some of the difficulties which are attributable to nonergodicity of this standard *XY* model be removed by considering a model in which the interactions between nearest neighbors are *not* the same? For example, the interaction can assume two distinct values with complementary probabilities.

B. M. McCOY: E. Barouch has the problem of the ergodicity of randomized XY model under consideration.

J. VILLAIN: Have you found any simple explanation for the fact that τ^{-1} goes to zero more strongly than χ^{-1}?

M. SUZUKI: Not yet. However, I think that my discussion by using cumulant expansions for the relaxation time gives some explanation for our numerical results.

Part Three
DEVELOPMENT OF SPATIAL ORDERING

SPATIAL ORDER

Lester Guttman
Solid State Sciences Division
Argonne National Laboratory
Argonne, Illinois

ABSTRACT

The developments in the field of atomic order in alloys will be reviewed. In a few cases, the transition from long-range order to disorder seems to occur without a latent heat, although the majority of such transitions are of first-order. The higher-order transitions involve simple cubic or body-centered cubic crystals, whereas first-order transitions involve close-packed structures. The systems showing higher-order transitions are suitable for testing theories of phase transitions, especially since the rigid Ising model is a physically reasonable starting point. When the compressibility of real crystals is taken into account, the character of the transition is modified, but it does not become first-order, at least for one model that has been solved exactly.

The pair-correlation function has been measured near T_c for two alloys, CuZn and Fe_3Al. In nearly all respects, the Ising model predictions are borne out quantitatively. This model seems also to work for the heat capacity of CuZn. However, the dependence of T_c

on pressure and composition for CuZn and CoFe is not consistent with the assumption of nearest neighbor forces only.

The principal gaps in our information are: (1) The critical properties of a wider variety of systems, including covalent or ionic crystals, (2) measurements of the compressibility and thermal expansion coefficients near the ordering temperature, and (3) studies of solid solutions showing critical solution temperatures.

The conspicuous features missing from current theory are: (1) Treatments of more general interactions, such as those with more distant neighbors, or the dependence of nearest neighbor potentials on distance, (2) treatments of compositions other than 1:1, especially in close-packed crystals, and (3) theory of transitions accompanied by small finite changes in volume, or by small distortions from cubic symmetry.

1. SPATIAL ORDERING

The purpose of an introductory talk like this I take to be to give a survey of the field, particularly for nonspecialists, so that later speakers do not have to spend so much valuable time on generalities. It is natural for me to do this with reference to a review[1] that I wrote some fifteen years ago, and I find it instructive to see how our present view of this field differs from the one current at that time. For instance, although the subject of "second-order phase changes" was discussed at length, I never referred to "critical phenomena" *per se*, simply because no relevant experiments had been done, nor was there any theory beyond the classical theory. Nevertheless, in what follows, I may raise some questions that are still unanswered, in spite of the renascence of which this Conference is a part.

Before getting down to specific problems, I want to remark that calling this session of the conference "Spatial Ordering" implies a constraint that I am reluctant to accept; we are surely aware that the common features of various types of transitions are still of more interest than the differences between them. So when I talk of atomic ordering, I trust that I may be allowed to touch on the parallels with other kinds of cooperative phenomena. In any such comparison of diverse transitions, the Ising model dominates the discussion. The qualitative features of Onsager's solution have become the norm for continuous transitions, replacing those of classical second-order transitions. In three dimensions, the Ising model has been the least intractable, and more of its properties have been computed with higher precision than for any other model. For the experimenter in the field of ordering, the Ising model is physically reasonable as well, so he naturally turns to it first in the course of interpretation of data. In fact, the combination of attractions just mentioned has been so nearly irresistible that properties of magnetic systems and even of fluids have been compared with the Ising model, which could be only a crude picture in these cases. In this general context, it becomes clear why so much attention has been paid to order-disorder transitions despite their rarity: These phenomena serve as a testing ground for theories that are at once simple enough to be manageable, and at the same time fairly realistic. One must hasten to add that alloys are interesting for other reasons, and that the study of ordering or clustering in solid solutions is presumably the best way to assess the configurational contributions to equilibrium properties of alloys, to the extent that they are separable from dynamical effects.

2. THE THERMODYNAMIC CHARACTER OF ORDER-DISORDER TRANSITIONS

At this Conference, one assumes that only higher-order transitions are being considered. Actually, the majority of order-disorder transitions in binary alloys are first-order, i.e., have a nonzero latent heat. On the basis of very scanty evidence, I suggested[1] that the exceptions occurred in solid solutions that are body-centered cubic (b.c.c.) when disordered. Since then, a little more information has been gathered, and I would state the following: Ordering takes place by a first-order transition unless (1) the disordered form is b.c.c., and the ordered form is $L2_0$ with the same unit cell size (examples: β-CuZn, CoFe, FeAl) or (2), the disordered form is $L2_0$ and the ordered form is DO_3 with the lattice parameter twice as large (Fe_3Al is the only case). For those unfamiliar with the Strukturbericht names, I might add that in case (1) the ordered form has one type of atom at the corners and the other kind at the centers of the unit cells, while the distribution is random among all sites of the b.c.c. lattice in the disordered state. In case (2), the atoms involved in the transition occupy a simple cubic lattice at random in the disordered state, but in the ordered state the two types alternate along rows parallel to the cube axes; an equal number of atoms (Fe atoms in this case) occupies the remainder of the sites of the crystal, and takes no essential part in the transition. This modified rule excludes the case of Fe_3Si, in which the disordering goes in a single step[2] from DO_3 to b.c.c., by a first-order transformation, with coexisting ordered and disordered phases of (generally) different compositions. With so few examples, one hesitates to use the term "rule," but it does seem that higher-order transitions occur only between structures that are loose-packed in a topological sense. This picture is supported by some examples, studied by Muldawer,[3] of ternary alloys such as $AgAuZn_2$, which resemble Fe_3Al, and seem also to disorder in two steps, each of which is a higher-order transition.

As far as I know (and again, there is not much data), the separation of ordering transitions into two classes is clean-cut. That is, there are no borderline examples of transitions that look like higher-order transitions until $\Delta T/T_c$ is of the order of 10^{-2} or so, and then show a small latent heat (contrast ferroelectrics). It seems to me now, as it did some years ago, that the crystallographic classification is suggestive, and that it may be rewarding to look into this aspect of the theory of the Ising model on a rigid lattice. I will return to this point at the end.

Another problem on which some progress has been made recently is that of the influence of the finite compressibility of real systems on their behavior near transitions where the density does not play a major role, as in ordering of binary alloys. We have already heard something about this from Rice, and will hear more from Griffiths. In earlier publications from these authors, we learned first[4] that if the constant volume heat capacity C_V were infinite along a line in the $p-T$ plane, the system would become mechanically unstable (negative compressibility); the conclusion drawn by Rice[4] was that such transitions would become first order. Wheeler and Griffiths proved[5] the converse, namely, that a system that was mechanically stable could not exhibit a line of infinite C_V. Quite recently, the situation has been further clarified by a model calculation by Baker and Essam.[6] They were able to express the thermodynamic functions of a certain compressible Ising system exactly in terms of those of the rigid system. Since statistical mechanics and thermodynamics are equivalent when done properly, their results were consistent with the Wheeler-Griffiths restriction, namely, C_V turned out to be always finite for this system. (The upper bound might be of the order of $15 k_B$ per spin in a real system.[7]) On the other hand, the upper limit for C_p was found to be inversely proportional to the applied pressure, and, except very close to T_c, C_p was quite indistinguishable[7] from that of the Ising model for values of the parameters typical of real systems. The transition was not found to become first order, however; C_p reaches its maximum with infinite slope both from above and from below, but the compressibility is never negative, and there is no latent heat.

The effects of allowing the interatomic distances to vary are certainly of some importance in real systems, and useful results have come from attempts to deal with them, as just mentioned. However, it seems to me that it is legitimate to ask, "Why should we care if C_V is constrained by thermodynamics to be finite?" Two sorts of negative answers can be given. First, this may be only a pseudo-problem: Until Baker and Essam's paper appeared only the *rigid* Ising model was at all tractable, and because this model happens to correspond also to a system at constant volume, we said, fallaciously, "Only C_V of real systems should be compared with theory." Of course, even at constant volume, the interatomic distances fluctuate as the atoms vibrate, and it is by no means clear that by fixing the average value we are closer to the rigid lattice case than by allowing it to vary. It turns out, as Baker and Essam have shown, that just the reverse is sometimes true. (See also Fisher.[8]) Another answer is thermodynamic. Consider first-order transitions of pure substances, which we think we understand. The latent heat appears only at constant pressure; at constant volume, the transition is spread out over a finite temperature range. By extension, when the latent-heat is zero, it is the behavior of C_p that should characterize the transition, not that of C_V.

3. THE PAIR-CORRELATION FUNCTION NEAR T_c

Although we are accustomed to thinking about critical properties primarily in thermodynamic terms, such as the behavior of the heat capacity, there are distinct advantages, both experimentally and theoretically, to considering the pair-correlation function as equally fundamental. We would like to study distribution functions relating to more than two atoms, but still do not know how to measure them (cf., however, Egelstaff *et al.*[9]). Furthermore, not all aspects of the pair-correlation function have a quasi-thermodynamic aspect, but only the values for distances that are large on an atomic scale. In other words, we are interested in the correlation, if any, at infinity—the long-range order—and in the asymptotic behavior, which might be called "intermediate-range order," as distinct from the conventional short-range order.

When the long-range order is not zero, i.e., below the critical temperature, the crystal symmetry is lower, and certain diffraction maxima are present with an intensity that is proportional to the square of the long-range order parameter, suitably defined. Moreover, for finite interatomic distances, the pair-correlation function differs from its value at infinite distance, i.e., the crystal is not perfect; hence there is coherent scattering at other than the Bragg angles, the angular distribution of which can be transformed into the desired pair-distribution function.

I want to emphasize somewhat the difference between the state of order prevalent in alloys showing critical behavior, and in those that have first-order transitions. There are a number of studies[1,10] of the diffuse scattering above T_c from systems having first-order transitions. The details differ, but in all cases the order is accurately represented by a small number of short-range order parameters (one for each kind of neighbor) that are not strongly temperature dependent, even close to T_c, and the correlation practically vanishes for interatomic distances greater than about three times the nearest-neighbor distance. When, as I recall, Walker and Keating[11] set out to measure the short-range order in β-CuZn by neutron diffraction they expected much the same situation. However, nature had prepared a little surprise for them: When they attempted a Fourier analysis of their diffuse scattering data they found it impossible to extract a finite number of short-range order parameters. They were forced by the long range of the correlation, even well above T_c, to a *continuum* description, namely, the Ornstein-Zernike formula as it applies to an ordering alloy. One can, of course, dismiss this difference between the two types of systems as merely "quantitative," but it indicates to me that one would be well advised to adopt a rather different starting point in the two cases for any theoretical treatment that falls short of complete rigor.

There have now been diffraction investigations of two ordered systems showing critical behavior—a later and much more complete study of CuZn by Als-Nielsen and coworkers,[12-15] and an X-ray study of Fe_3Al by my associates and myself.[16,17] We have unfortunately been deprived of the chance to hear at first hand from Als-Nielsen, but you will shortly hear a discussion by Walker of his own recent X-ray work on CuZn and its relation to some of the neutron scattering results. I will presume to describe the other parts of the neutron scattering measurements, as well as our own work.

The experimental results can be concisely summarized by saying that they are in general very close to the Ising model predictions (Table 1). The only discrepancy that seems to exceed the combined errors is that between the theoretical and experimental values of η in the simple cubic case, for which a number of reasons might be suggested.

TABLE 1. Critical Parameters for Correlations*

	β	ν	γ	F^+	η
CuZn, expt.	0.305 ± 0.005	0.65 ± 0.02	1.25 ± 0.02	3.0 ± 0.2	0.077 ± 0.067
Ising, b.c.c.	0.307 to 0.314†	0.643 ± 0.003‡ $0.638 {+0.002 \atop -0.001}$§	1.250¶	2.24‡	0.056 ± 0.008‡ $0.041 {+0.006 \atop -0.003}$§
Fe$_3$Al, expt.	0.307	0.649 ± 0.005	1.246 ± 0.01	2.34 ± 0.05	0.080 ± 0.005
Ising, s.c.	0.307 to 0.314†	0.643 ± 0.003‡ $0.638 {+0.002 \atop -0.001}$§	1.250¶	2.09‡	0.056 ± 0.008‡ $0.041 {+0.006 \atop -0.003}$§

*For notation, see M. E. Fisher: *Rep. Prog. Phys.*, 30:615 (1967).
†Baker, Jr., G. A., and D. S. Gaunt: *Phys. Rev.*, 155:545 (1967).
‡Fisher, M. E., and R. J. Burford: *Ibid*, 156:583 (1967).
§Moore, M. A., D. Jasnow, and M. Wortis: *Phys. Rev. Lett.*, 22:940 (1969).
¶Guttmann, A. J., B. W. Ninham, and C. J. Thomson: *Phys. Rev.*, 172:554 (1968).

Als-Nielsen[14] has also measured the scattering from CuZn in the difficult region just below T_c where the diffuse peak and the Bragg "superlattice" peak are superimposed. He has concluded that $\gamma' = \gamma$, with an uncertainty of about 1 percent, and that the ratio of the susceptibility amplitudes, $C+/C-$, is 5.46 ± 0.05, in good agreement with the static scaling laws.

4. OTHER CRITICAL PROPERTIES

Some other critical properties of ordering systems have been measured and can be compared with theory. Equilibrium appears to be established so rapidly in β-CuZn that the heat capacity can be determined by an A.C. method.[18] The results for C_p above the transition are in fairly good agreement with those calculated by Baker and Essam[19] for a compressible Ising model; below T_c, the agreement is not so good.

The dependence of T_c on pressure and on composition near the 1:1 atomic ratio for Cu-Zn alloys and for Co-Fe alloys has been analyzed by Bienenstock and coworkers.[20-22] The results are inconsistent with the predictions of the simple Ising model—i.e., with nearest neighbor interactions and a potential that is dependent on volume but independent of composition. The assumptions underlying the analysis have been clearly set forth by these authors. I would add only that we have already seen, in treating the effect of compressibility, that constant volume is by no means equivalent to constant interatomic distance. Therefore, it may be premature to conclude that there is evidence for the inadequacy of the Ising model; when the dependence of the ordering energy on interatomic distance is properly taken into account these discrepancies may disappear.

In addition to refinements in techniques for the measurement of the more obvious critical properties of ordering alloys, it seems to me that it might be worthwhile to try to measure expansion coefficients and compressibilities near T_c, even though the ordering transitions are not primarily connected with density changes (in contrast to the gas-liquid critical point). The motivation comes in part from a desire to fill the gap that is pointed to by the Baker-Essam theory. However, we might also consider* the Buckingham-Fairbank relations[23]

---

*I am indebted to Henry A. Kierstead for numerous discussions on this subject.

$$\frac{C_p}{T} = \left(\frac{\partial S}{\partial T}\right)_t + P'_c\left(\frac{\partial V}{\partial T}\right)_p$$

$$\left(\frac{\partial V}{\partial T}\right)_p = \left(\frac{\partial V}{\partial T}\right)_t - P'_c\left(\frac{\partial V}{\partial P}\right)_T$$

where $t = T - T_c(P)$, and $P'_c = (dP/dT)_c$. These equations show that if the transition temperature is marked by an infinite heat capacity C_p behaving as $(t)^{-\alpha}$, then for small enough t, the expansion coefficient α_p and compressibility K_T also become infinite with the same exponent α.

There may be experimental advantages to determining α_p rather than the heat capacity. Especially when equilibrium times are long or at high temperatures, these advantages may outweigh the difficulties inherent in the fact that one measures lengths, which must be differentiated numerically (of course, C_p is a quantity of the same kind), as well as the expectation that the anomaly in α_p may be relatively harder to find than that in C_p.

We should perhaps be warned by the experience of Cadieu and Douglass,[24] who found that α_p of gadolinium near its Curie temperature became temperature-independent for $|t/T_c| \lesssim 10^{-3}$. They suggested that this rounding of the transition was due to electron scattering by impurities, and that the maximum coherence length observable would be of the order of the electron-impurity mean free path. Such a limitation is probably not general, however, and a narrower region of t/T_c is almost certainly accessible in other types of transitions.[25]

5. SOLID SOLUTIONS

On general grounds, we would expect binary solid solutions exhibiting a critical solution temperature to behave like their liquid analogs—i.e., to show critical fluctuations of composition and the attendant singularities in thermodynamic properties and in coherent scattering. However, an argument to the contrary has been made by Cahn,[26] and the question remains unsettled theoretically. There are not many examples of systems that are suitable for experimental investigation. The alloys of Al and Zn have been studied by Münster and Sagel,[27,28] who claimed that the scattering was in agreement with the classical formula, both as to its dependence on temperature and on angle. These data are not, however, of the quality we have now come to expect. The work by Alefeld[29] on the critical elastic properties of metallic solutions of hydrogen would seem to be in conflict with Cahn's conjecture; we shall hear more about this later from Alefeld himself. It seems to me that there is room for more work on all aspects of this area.

I have cited all the examples known to me of studies of critical properties of solid solutions, whether of the ordering or the segregating type. They have all, you will note, been of metallic systems, and the short list of possibilities is now nearly exhausted. It would seem that we ought to look around for examples of other kinds of crystals that might behave similarly. In the midst of really profound ignorance, we can hardly afford to overlook clues as to which aspects of critical phenomena are sensitive to the form of the interaction energy and which are not. Studies of ionic, covalent, or molecular solid solutions that order or cluster might supply this kind of information.

6. STATUS OF THEORY

Up to this point I have confined myself to a description of the phenomenology of critical points in solid solutions, digressing only to comparisons with the Ising model. Within the limits set by the nominal subjects of this Conference and of this session in particular, nothing further would be appropriate. However, I cannot resist taking this opportunity to give some gratuitous advice to my theoretical friends.

To the extent that a rigorous statistical theory can be said to exist for a class of phase transitions, the same model and essentially the same averaging procedure should apply to both of the phases involved. Otherwise, the transition point is merely that at which the free energies of the two phases happen to be equal, and we cannot learn much more by studying the system in the vicinity of the transition than in other regions of temperature, pressure, field, etc. By this criterion, we may claim presently to have a theory of order-disorder transitions in some alloys, and with less justice, perhaps, a theory of some Curie and Néel points, and even of the gas-liquid critical point. On the other hand, no theory could be said to exist of the gas-liquid transition well away from the critical point, or of melting, or of nearly all other first-order transitions.

This is all well known, I am sure. What I want to contribute is the suggestion that a start might be made towards bridging the enormous gap that undoubtedly separates theories of continuous transitions and theories of first-order transitions, and that the "thin place in the wall"* is at order-disorder transitions in close-packed alloys. I have already touched on the experimental evidence for two crystallographically distinct classes of alloys in this respect. There has been very little theory in this area—for good reason, no doubt. For what they are worth, approximate treatments[1,30] of the rigid Ising model support the notion that in close-packed alloys the transition from order to disorder takes place with a latent heat, and that the equilibrium phases are at least metastable near T_c, one being still nearly completely ordered, the other nearly random. A rigorous theory would have to have one feature that is absent from the calculations that have been made up to now. If one wants to stick to the composition 1:1 then, to be realistic, it would be important to include the variation of the ordering energy with distance, since the ordered form of CuAu and many similar alloys is *tetragonal*, with $c/a \simeq 0.95$, consisting of planes perpendicular to the c axis alternately of one or the other type of atom. In this structure, the twelve sites holding the equidistant nearest neighbors of each atom in the cubic disordered state split into a group of eight at one distance, occupied by the opposite kind of atom, and a group of four at slightly greater distance, occupied by the same kind of atom. On the Ising model, the net ordering energy is that due to gaining two $A-B$ bonds per atom, but if there is no change in atomic volume (as is the case, to a good approximation) the $A-B$ bonds in the ordered state are shorter, and presumably stronger, than those in the disordered state, and it would be a mistake to ignore this difference.

The other common ordered structure in f.c.c. solutions is that typified by Cu_3Au. To fix the composition at the 3:1 ratio, one must deal either with the appropriate grand ensemble,[31] or with the equivalent antiferromagnet in a finite magnetic field. I need not stress the difficulties of these problems. Here, too, it would be desirable to include the distance dependence of the ordering energy, since the bond lengths in the ordered and disordered forms at T_c are different, although both structures have cubic symmetry.

None of the preceding is new, to say the least, but it is worth recalling if, as I hope, it challenges the theorists to take up the problem. In the last few years the only contributions to theory that have any relevance have been made by Clapp and Moss[32,33] and by Bienenstock and Lewis.[20]

---

*Einstein is said to have expressed his disdain for the kind of research that consists of looking for the thin places in the wall and drilling a few holes.

ACKNOWLEDGMENT

The preparation of this paper was performed under the auspices of the U.S. Atomic Energy Commission.

REFERENCES

1. Guttman, L.: *Sol. State Phys.*, 3:146 (1956).
2. Glaser, F. W., and W. Ivanick: *Trans. AIME*, 206:1290 (1956).
3. Muldawer, L.: *J. Appl. Phys.*, 37:2062 (1966).
4. Rice, O. K.: *J. Chem. Phys.*, 22:1935 (1954).
5. Wheeler, J. C., and R. B. Griffiths: *Phys. Rev.*, 170:249 (1968).
6. Baker, G. A., Jr., and J. W. Essam: *Phys. Rev. Lett.*, 24:447 (1970). Similar results seem to have been obtained by H. Wagner: *Phys. Rev. Lett.*, 25:31 (1970), and H. Wagner and J. Swift, unpublished.
7. Guttman, L.: *Phys. Rev.*, B2:1432 (1970).
8. Fisher, M. E.: *Ibid.*, 176:257 (1968).
9. Egelstaff, P. A., D. I. Page, and C. R. T. Heard: *Phys. Lett.*, 30A:376 (1969).
10. Cohen, J. B., and J. E. Hilliard (eds.): "Local Atomic Arrangements Studied by X-ray Diffraction," Gordon and Breach Science Publishers, Inc., New York, 1966.
11. Walker, C. B., and D. T. Keating: *Phys. Rev.*, 130:1726 (1963).
12. Dietrich, O. W., and J. Als-Nielsen: *Ibid.*, 153:711 (1967).
13. Als-Nielsen, J., and O. W. Dietrich: *Ibid.*, 153:706, 717 (1967).
14. ———: *Ibid.*, 185:664 (1969).
15. Norvell, J. C., and J. Als-Nielsen: *Ibid.*, B2:277 (1970).
16. Guttman, L., H. C. Schnyders, and G. J. Arai: *Phys. Rev. Lett.*, 22:517 (1969).
17. ——— and ———: *Ibid.*, 22:520 (1969).
18. Ashman, J., and P. Handler: *Phys. Rev. Lett.*, 23:642 (1969).
19. Baker, G. A., Jr., and J. W. Essam: *J. Chem. Phys.*, 55:(1971) (in press).
20. Bienenstock, A., and J. Lewis: *Phys. Rev.*, 160:393 (1967).
21. Yoon, D. N., and A. Bienenstock: *Phys. Rev.*, 170:631 (1968).
22. ——— and R. N. Jeffery: *J. Phys. Chem. Solids*, 31:2635 (1970).
23. Buckingham, M. J., and W. M. Fairbank: *Prog. Low Temp. Phys.*, 3:80 (1961).
24. Cadieu, F. J., and D. H. Douglass, Jr.: *Phys. Rev. Lett.*, 21:680 (1968).
25. See also Philp, J. W., and E. D. Adams: *J. Low Temp. Phys.*, 2:309 (1970).
26. Cahn, J. W.: *Acta Met.*, 9:795 (1961); 10:179 (1962).
27. Münster, A., and K. F. Sagel: *Mol. Phys.*, 1:23 (1958).
28. ——— and C. Schneeweiss: *Z. Physik. Chem.*, 37:369 (1963).
29. Alefeld, G., G. Schaumann, J. Tretkowski, and J. Völkl: *Phys. Rev. Lett.*, 22:697 (1969).
30. Muto, T., and Y. Takagi: *Solid State Phys.*, 1:194 (1955).
31. Newell, G. F., and E. W. Montroll: *Rev. Mod. Phys.*, 25:353 (1953).
32. Clapp, P. C., and S. C. Moss: *Phys. Rev.*, 142:418 (1966); *ibid.*, 171:754 (1968).
33. Moss, S. C., and P. C. Clapp: *Ibid.*, 171:764 (1968).

DISCUSSION *on Paper by L. Guttman*

P. C. CLAPP: Do you feel that the loose-packed binary alloys show higher order phase transitions because of their special configurational properties or because of special elastic behavior?

L. GUTTMAN: The configurational properties.

L. ONSAGER: I think that the question of whether a transition is continuous or discontinuous was first treated by Landau. He stated that for any continuous transition the symmetry elements of order two must be lost. If the higher order symmetry elements are lost, you are almost certainly dealing with a discontinuous transition.

G. ALEFELD: In regard to the question about first-order transitions in close-packed lattices and second order in non-close-packed lattices, I will give in the last talk this morning examples of second order transitions in close-packed as well as in non-close-packed lattices, namely, transitions of hydrogen in Pd and Nb.

J. D. LITSTER: I would like to add to Prof. Onsager's remarks on the role of symmetry in phase transitions. Tomorrow I shall present experimental results on a liquid crystal which would like to undergo a second order phase transition but is required by the symmetry of the ordered phase to have a small discontinuity in the change to the ordered state.

G. A. BAKER: The preprint of Wagner and Swift given to me by Dr. Guttman which he referred to in his address, indicates that, in their approximations, C_p along a path of constant volume should be rather close to the rigid Ising model functions for suitable systems. This result is in accord with those of Dr. Essam and myself. However, in fact neither C_p or C_v is precisely the rigid Ising function. Furthermore, to compare with experiments one must compute the thermodynamic functions along a path of constant pressure and not of constant volume. Our calculations for β brass based on the adiabatic or Domb model where the exchange energy depends only on the volume, as is fairly appropriate for a highly shear resistant material, show good agreement with the data of Als-Nielsen on the order parameter.[1] The data on the specific heat of Ashman and Handler[2] do not agree with these Ising results, particularly for $T < T_c$.

C. DOMB: Could you please explain in more detail how AB ordering takes place in a close-packed lattice, e.g., f.c.c.?

L. GUTTMAN: The typical structure consists of (100) planes alternately all A or all B. This selects a particular (001) direction, which becomes a tetragonal axis.

P. C. MARTIN: Are there not some experiments on solutions at M.I.T. in which the order of the transition changes as a function of pressure although there seems to be no change in symmetry?

O. K. RICE: Garland's work was on ammonium halide crystals. The order-disorder phenomenon had to do with orientation of ammonium ions. These crystals showed a first-order transition (believed to arise from the tendency of C_v to increase toward infinity) at low pressures. At least in some cases, the first-order transition seems to disappear (go over to higher order transition) at a higher pressure.

REFERENCES

1. Norvell, J. C., and J. Als-Nielsen: *Phys. Rev.*, **B2**:277 (1970).
2. Ashman, J., and P. Handler: *Phys. Rev. Lett.*, 23:642 (1969).

EARLY-STAGE CLUSTERING AND ORDERING KINETICS IN BINARY SOLID SOLUTIONS*

D. de Fontaine
School of Engineering and Applied Science
University of California at Los Angeles
Los Angeles, California

H. E. Cook
Product Development Group
Ford Motor Company
Dearborn, Michigan

1. INTRODUCTION

The purpose of this study is to present, in a unified manner, a theoretical model governing the early-stage kinetics of clustering and ordering in binary substitutional solutions. The present treatment includes and generalizes the continuum theory of Cahn[1,2] applicable to spinodal decomposition, the discrete or lattice theory of Hillert[3] and of Cook, de Fontaine, and Hilliard,[4] applicable also to

*This and the next paper are discussed together on page 287.

order-disorder reactions, the microscopic theory of elasticity of Cook and de Fontaine[5] and the influence of fluctuations as given by Cook.[6,7] In some of its aspects the present treatment parallels earlier ones by Krivoglas[8] and Khachaturyan.[9]

Let us consider a binary solid solution defined by its Bravais lattice, each site of which is associated with an A or with a B atom. Given an initial distribution of A and B atoms, allow some rearrangement to take place by diffusion during a time interval Δt. The concentration variable $c(p)$ is then defined as the probability, with respect to the initial distribution, of finding a B atom at site (p) after the time interval Δt. It is convenient to define also a local concentration variation function, or deviation from the mean

$$v(p) = c(p) - c_0 \qquad (1)$$

where c_0 is the average concentration in the crystal. The above viewpoint leads to the concept of "average atoms" which, because of local strain fields, may be displaced from their parent sites by the vectors $\mathbf{u}$ of Cartesian components u_i. With each site (p) we therefore associate four variables: $v(p)$ and $u_i(p), (i = 1, 2, 3)$. The thermodynamics of the system are thus characterized by $4N$ state variables, N being the total number of atoms in the crystal.

2. CHEMICAL AND ELASTIC ENERGY

On the atomic scale the distinction between "chemical" and elastic energies is a rather artificial one but can be justified as follows: Consider (Fig. 1) the schematic potential energy curves for an $A-A$ and an $A-B$ pair in a solid. In general the energy minima for

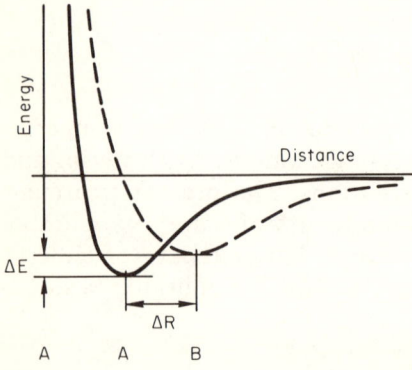

Fig. 1 *Hypothetical potential energy curves for an $A-A$ (full curve) and an $A-B$ (dotted curve) pair illustrating "chemical" (ΔE) and "elastic" (ΔR) effects.*

these two curves will differ both in height ΔE in distance ΔR from the origin atom A. Thus the substitution of an A by a B atom will give rise to chemical effects related to ΔE and elastic effects related to ΔR. The total (Helmholtz) free energy F_t of the system is then given by the sum of chemical G and elastic F free energy contributions

$$F_t = G + F$$

(The symbols G and F used here do not have their usual classical thermodynamic meaning).

The simplest expression for the chemical energy is a quadratic expansion in the concentration variations about the state of uniform concentration $v(p) = 0$ (all p), from which state the solid solution is assumed to deviate only slightly. Allowing interatomic interactions to take place to arbitrarily distinct neighbors, we obtain the phenomenological expression

$$G = \frac{1}{2} \sum_{p,p'} g(p' - p) v(p) v(p') \qquad (2)$$

in which the double sum extends over all pairs (p, p') of lattice sites, the chemical energy coefficients g depending on the lattice distance $(p' - p)$ between neighbors of the pair. The model is thus a nonlocal harmonic one or, equivalently, a pairwise interaction model which incorporates the entropy of the solution in a highly approximate manner. It will be shown later that the second derivative of the configurational entropy appears in the "self-pair" term corresponding to $(p') = (p)$.

In a like manner, the elastic free energy can be written as a quadratic expansion in the state variables $u_i(p), v(p)$ since, obviously, the elastic energy must depend on the displacements of atoms from their parent lattice sites and also on the local distortions resulting from the nature of the atoms at lattice sites, the atoms A and B being either "too large" or "too small" to fit in the average lattice of the solution. The required expansion has been given previously[5] and has the expression

$$F = \frac{1}{2} \sum_{p,p'} [\phi_{ij}(p' - p) u_i(p) u_j(p')$$

$$+ 2\phi_i(p' - p) u_i(p) v(p') + \phi(p' - p) v(p) v(p')] \qquad (3)$$

in which summation over repeating Cartesian subscripts $i, j = 1, 2, 3$ is implied. Just as in the theory of lattice vibrations, $\phi_{ij}(p' - p)$ represents the force at (p) in direction i due to a displacement at (p') in direction j. The coupling parameter $\phi_i(p' - p)$ represents the force at (p) in direction i due to the homogeneous expansion caused by a concentration change at (p'), finally the parameter $\phi(p' - p)$ represents the work required to bring the pair (p', p) to its ideal lattice location against the homogeneous strain fields due to the concentration changes at (p) and (p'). The displacement variables u_i can be eliminated from the energy expression by assuming that the relaxation time to reach mechanical equilibrium is much smaller than the time required for effecting an appreciable local concentration change, i.e., by imposing the condition

$$\frac{\partial F}{\partial u_i} = 0 \qquad (4)$$

The calculations, however, are best performed in Fourier space in which the representations of Eqs. (2) and (3) become diagonalized.

3. FOURIER TRANSFORMATION

Consider a monatomic crystal containing $N = N_1 \times N_2 \times N_3$ unit cells, assume periodic boundary conditions and define the discrete Fourier transform operator by

$$\mathcal{F} \equiv \frac{1}{N} \sum_p \exp[-i k(h) \cdot x(p)]$$

with x and k being real space and k-space vectors respectively. The symbol h denotes an allowed site in k-space just as the symbol p denotes a lattice point in real space. Using upper case symbols for the transforms of the corresponding lower case symbols, we can write

$$\begin{aligned} V &= \mathcal{F} v \\ U_i &= \mathcal{F} u_i \end{aligned} \qquad (5)$$

and

$$\begin{aligned} G^* &= N \mathcal{F} g \\ \phi^* &= N \mathcal{F} \phi \end{aligned} \quad \text{(all combinations of subscripts)} \qquad (6)$$

in which the star denotes the complex conjugate.

The energy quadratics transform to the following diagonal forms

$$G = \frac{N}{2} \sum_h G(h) Q(h) \qquad (7)$$

$$F = \frac{N}{2} \sum_h F(h) Q(h) \qquad (8)$$

in which the summations extend over all points h in the first Brillouin zone. We have defined, further

$$Q = VV^*$$

and

$$F = \phi - \phi_i \tilde{\phi}_{ij} \phi_j \qquad (9)$$

where the quantities $\tilde{\phi}_{ij}$ are elements of the matrix inverse to that of the ϕ_{ij}'s. Equation (9) is arrived at by imposing condition (4), solving for the displacement amplitudes U_i and substituting back into the elastic energy expression.[5] The function $Q(h)$, the "intensity" of the harmonic concentration waves, is the Fourier transform of the concentration covariance

$$q(p) = \frac{1}{N} \sum_{p'} v(p') v(p' + p)$$

The normalized covariance, i.e., $q(p)$ divided by the variance

$$q(0) = c_0(1 - c_0)$$

is the Warren short-range order parameter α for shell (p).[10]

The total free energy of the solution can now be expressed in the compact Fourier representation

$$F_t = \frac{N}{2} \sum_h \psi(h) Q(h) \qquad (10)$$

with

$$\psi(h) = F(h) + G(h)$$

The advantage of the Fourier representation is that (1) the double sums in Eqs. (2) and (3) are replaced by a combined single sum in Eq. (10) and that (2) the displacement variables are eliminated. Thus, the free energy of the solution can, in principle, be evaluated when the probable distribution of A and B atoms is known yielding, by Fourier transformation, the intensities $Q(h)$. The coefficients $\psi(h)$ are known since they arise from well-defined operations on pairwise interaction coefficients $g(p' - p)$, $\phi_{ij}(p' - p)$, ... which are assumed given.

Equation (10) can be used to test the stability of a solid solution with respect to particular harmonic concentration waves: The solution will be unstable with respect to a wave of vector $k(h)$ if the corresponding coefficient $\psi(h)$ is negative and conversely. Instabilities in the long wavelength region of the concentration spectrum give rise to clustering effects, whereas instabilities in the short wavelength region give rise to ordering effects.[4] Nucleation problems and all cases involving large local concentration changes must be handled by a more complete theory including higher order terms in the expansions of Eqs. (2) and (3). A few attempts have been made along these lines in the case of spinodal decomposition[11,12] but the general case, for which a full many body approach is required, remains presently untractable.

4. FLUCTUATIONS

The Fourier representation of the free energy is precisely that required for the Landau theory of fluctuations above the transformation temperature.[13] By direct substitution, we immediately obtain the following expression for the expectation value of the intensity $Q(h)$

$$<Q(h)> = \frac{k_B T}{N\psi(h)} \tag{11}$$

where $k_B T$ has its usual meaning. A detailed account of fluctuation theory based on this approach is given by Krivoglaz[14] with a somewhat different formulation for the order-disorder case.

5. DIFFUSION EQUATION

Through the averaging procedure described in the Introduction, it is possible to consider a probability density flux of B atoms $J_r(p)$ from site $p + r$ to site p.[4] If the driving force for diffusion were of purely chemical nature, we would naturally assume the flux J to be proportional to the difference in potentials w at the two sites

$$J_r(p) = Mm(r)[w(p + r) - w(p)] \qquad (12)$$

with

$$w(p) = \frac{\partial G}{\partial v(p)} = \sum_r g(r) v(p + r)$$

In Eq. (12), M is a positive quantity related to the atomic mobility and $m(r)$ takes into account differences in jump frequencies of atoms to neighboring sites in noncubic crystals. The quantities M and $m(r)$ will be determined in the next section by comparing the above discrete treatment with the continuum formulation given by Fick's law. The rate of change of concentration variation $\dot{v}$ at p must be proportional to the sum of fluxes J_r from neighboring sites

$$\dot{v}(p) = \sum_r J_r(p) \qquad (13)$$

We shall henceforth adopt the convention that a summation over r indicates a sum over those nearest neighbors which are assumed to contribute to the flux at p.*

The system of linear differential equations (Eq. (13)) can be uncoupled by a Fourier transformation to obtain

$$\dot{V}(h) = \alpha(h) V(h) \qquad (14)$$

with the "amplification factor" α given by

$$\alpha(h) = -M\beta(h) G(h) \qquad (15)$$

---

*Minus signs are not required in the flux and continuity equations (Eqs. (12) and (13)) because of our convention regarding the definition of J_r and because of our use of finite differences rather than derivatives.

in which the coefficient β is given by

$$\beta(h) = \sum_r m(p)\{1 - \cos[k(h) \cdot x(r)]\}. \qquad (16)$$

If elastic effects are taken into account, the amplification factor takes the more general form

$$\alpha(h) = -M\beta(h)\psi(h). \qquad (17)$$

The solution of Eq. (17) is given by

$$V(h) = V_0(h) \exp[\alpha(h)t] \qquad (18)$$

and can be interpreted as follows: In the harmonic approximation used here, the Fourier components of the concentration variation are independent so that the amplitude of the N composition waves $V(h)$ of initial values V_0 change with time t according to the exponential law given above. The amplitudes increase for positive values of the amplification factor, decrease for negative α. The concentration waves which grow and those which decay are separated by a critical surface in h space defined by

$$\alpha(h) = 0$$

According to Eq. (17), this relation defines two loci: The first is given by the equation $\beta(h) = 0$ which, because the cosines form a complete set, is equivalent to the vanishing of the expression enclosed in braces in Eq. (16), yielding

$$h_1 r_1 + h_2 r_2 + h_3 r_3 = \text{integer}$$

The amplification factor thus vanishes at the nodes of the reciprocal lattice. The other locus is given by

$$\psi(h) = 0$$

and represents the true critical surface h_c, say. The critical wavelength λ_c, introduced by Hillert[3] for the one-dimensional discrete case, and by Cahn[1] for the case of spinodal decomposition is then given by

$$\lambda_c = \frac{2\pi}{k(h_c)}$$

The critical surfaces in h space behaves very much like Fermi surfaces in that, in the absence of strain effects, they are nearly spherical in cubic crystals for large critical wavelengths (small h_c). For systems in which the critical wavelengths are progressively shorter, the critical surfaces in h space are progressively distorted as they approach the Brillouin zone boundaries. The existence of strain energy contributions will, however, distort a spherical critical surface close to the origin even in cubic systems.

Within the range of validity of the harmonic approximation, the Fourier component receiving maximum amplification is determined by seeking the extremum of the amplification factor α and is found by solving the system

$$\frac{\partial \alpha}{\partial h_\gamma} = 0 \, (\gamma = 1, 2, 3)$$

These extreme values will not necessarily give rise to maximum amplification: One must first check whether these points in h space represent maxima rather than minima or saddle points, and furthermore if the corresponding amplification factor is positive; for $\alpha < 0$, we have maximum rate of decay rather than maximum amplification. By geometrical considerations we see that extreme values of the amplification factor must exist when a direction of high symmetry in h space intersects the Brillouin zone boundary; if these points correspond to maxima, then ordering will take place since maximum amplification occurs at the superlattice points. These concepts were illustrated for the case of cubic crystals, in the absence of strain energy contributions in a previous paper.[4]

The amplitude equation 14 can readily be transformed into an "intensity" equation

$$\dot{Q}(h) = 2\alpha(h) Q(h) \tag{19}$$

There is a distinct advantage in using the intensity rather than the amplitude whenever a source term need be added to the continuity equation (Eq. (13)). In particular, the fluctuation intensity source term can be taken from Eq. (11) whereas no corresponding expression is available for the expectation value of the amplitudes $V(h)$. It was shown elsewhere[6] that the complete intensity equation must read

$$\dot{Q}(h) = 2\alpha(h)[Q(h) - <Q(h)>]$$

with solution at time t, for each (h), given by

$$Q = [Q_0 - <Q>] \exp(\alpha t) + <Q>$$

in which Q_0 is the initial intensity $V_0 V_0^*$. This equation has been used successfully to describe the kinetics of short-range ordering above the transformation temperature[6] and to the early stages of spinodal decomposition.[7]

6. COMPARISON AT LONG WAVELENGTHS

In a previous paper[5] we established a relation between the coupling parameters and the elastic constants of the material by imbedding the lattice in an elastic continuum and by comparing the microscopic and macroscopic theories of elasticity at long wavelengths of the displacement and composition variation spectra. We shall now follow essentially the same procedure in order to derive relations linking the phenomenological coefficients $g(p)$ to the Helmholz free energy of the solution and to gradient energy coefficients of the type introduced by Cahn and Hilliard.[15]

Let us define a continuously varying concentration function by the equation[16]

$$v(x) = \sum_h V(h) \exp[i\mathbf{k}(h) \cdot \mathbf{x}]$$

in which the Fourier coefficients V are those defined by Eq. (5). We can then transform the summation in Eq. (2) by an integration over the volume of the crystal to obtain the continuum expression of the chemical free energy

$$G = \frac{1}{2\Omega} \int v \sum_p g(p) \left[v + \frac{1}{2!} x_i(p) x_j(p) v_{,ij} \right.$$

$$\left. + \frac{1}{4!} x_i(p) x_j(p) x_k(p) x_l(p) v_{,ijkl} + \cdots \right] dx \quad (20)$$

in which Ω is the atomic volume. The subscripts in Eq. (20) refer to a Cartesian coordinate system imbedded in the elastic continuum and whose axes have directions determined by the eigenvectors of the matrix of elements $(\mathbf{a}_\alpha \cdot \mathbf{a}_\beta)$, the scalar products of the lattice translation vectors. A comma followed by subscripts indicates a derivative with respect to space coordinates as follows:

$$v_{,i} = \frac{\partial v}{\partial x_i} \text{ etc.}$$

Let us now compare Eq. (20) to a phenomenological quadratic expansion of the Helmholtz free energy in terms of concentration and concentration gradients

$$G = \int \left[\frac{1}{2} f_0'' v^2 + \kappa_{ij}^{(1)} v v_{,ij} + \kappa_{ij}^{(2)} v_{,i} v_{,j} \right.$$

$$\left. + \kappa_{ijkl}^{(1)} v v_{,ijkl} + \kappa_{ijkl}^{(2)} v_{,i} v_{,jkl} + \kappa_{ijkl}^{(3)} v_{,ij} v_{,kl} + \cdots \right] dx \quad (21)$$

Terms which are odd in the number of indices are not included since they must vanish for a centrosymmetric crystal. Note that our coefficient $\kappa_{ij}^{(2)}$ is identical to the gradient energy coefficient introduced by Cahn and Hilliard[15] but that our coefficient $\kappa_{ij}^{(1)}$ differs from theirs by the factor v. By expanding the Helmholtz free energy $f(c)$ in a Taylor's series about $c = c_0$ one can show that the coefficient v^2 must be the second derivative $f(c)$ evaluated at $c = c_0$

$$f_0'' = \left[\frac{\partial^2 f}{\partial c^2} \right]_{c=c_0} \quad (22)$$

It is essential that the comparison between two equivalent expressions for G be performed in reciprocal space: In direct space, the quantities v, $v_{,ij}$, ... are not independent and the coefficients of like terms in the expression under the integral sign cannot be equated since this would lead to long wavelength relations which have incorrect symmetry properties. Comparison of the two h space representations of G then yield the required long wavelength relations

$$\sum_p g(p) = \Omega f_0'' \quad (23a)$$

CRITICAL PHENOMENA

$$\frac{1}{2} \sum_p g(p) x_i(p) x_j(p) = 2\Omega(\kappa^{(1)}_{(ij)} - \kappa^{(2)}_{(ij)}) \tag{23b}$$

$$\frac{1}{4!} \sum_p g(p) x_i(p) x_j(p) x_k(p) x_l(p) = 2\Omega(\kappa^{(1)}_{(ijkl)} - \kappa^{(2)}_{(ijkl)} + \kappa^{(3)}_{(ijkl)}) \tag{23c}$$

The right hand sides of Eqs. (23b) and (23c) involve the completely symmetric parts of the indicated tensors, as denoted by the parenthesis notation, for example

$$\kappa_{(ij)} = \frac{1}{2!}(\kappa_{ij} + \kappa_{ji})$$

The long wavelength relations corresponding to the elastic energy contribution were given in a previous paper.[5]

Let us now compare the discrete and continuum expressions of the diffusion equation. By expanding the cosine terms in Eq. (16) we obtain, by Eqs. (14), and (15), the expression

$$\dot{V}(h) = -\frac{MN}{2} \sum_r m(r) [\mathbf{k}(h) \cdot \mathbf{x}(r)]^2$$

$$\sum_p g(p) [1 + i\mathbf{k}(h) \cdot \mathbf{x}(p) - \cdots] V(h) \tag{24}$$

Fick's second law

$$\dot{v}(\mathbf{x}) = D_{ij} v_{,ij}(\mathbf{x}) \tag{25}$$

has the h space formulation

$$\dot{V}(h) = -D_{ij} k_i(h) k_j(h) V(h)$$

where the D_{ij} are the (concentration independent) Darken interdiffusion coefficients for anisotropic crystals. By comparing Eqs. (24) and (25) we obtain, noting that D_{ij} is a symmetric tensor, the long wavelength relation

$$\frac{1}{2} \Omega M f_0'' \sum_r m(r) x_i(r) x_j(r) = D_{ij} \tag{26}$$

In this way, the parameter $m(r)$ can be determined from a knowledge of the interdiffusion coefficients, given a suitable expression for the mobility term M.[4]

7. CONTRIBUTION FROM CONFIGURATIONAL ENTROPY

Let us express the free energy $f(c, T)$ per unit volume of solution as the sum of an excess free energy $f_x(c, T)$ plus the ideal configurational entropy

$$f(c, T) = h(c) - Ts_x(c) + \left(\frac{k_B T}{\Omega}\right)[c \ln c + (1 - c) \ln(1 - c)]$$

h and s_x being respectively the enthalpy and excess entropy of the system. Hence, the second derivative of f (Eq. (22)) can be written

$$f_0'' = h'' - Ts_x'' + \frac{k_B T}{\Omega c_0(1-c_0)}$$

the accents denoting differentiation with respect to concentration.

We now assume, in accordance with a derivation given by Cahn and Hilliard,[15] that the ideal configurational entropy has a purely local character and further that the excess free energy can be represented by a sum of pairwise interaction energies. The first long wave equation (Eq. (23a)) can then be split into two parts

$$g(0) = \frac{k_B T}{c_0(1 - c_0)} \qquad (27)$$

$$\sideset{}{'}\sum_p g(p) = \Omega(h'' - Ts_x'') \qquad (28)$$

The accent on the summation sign indicates that the origin $(p) = (000)$ is to be excluded. By Eqs. (5), (27), and (28) we may then write the Fourier transform of $g(p)$ as

$$G(h) = \frac{k_B T}{c_0(1 - c_0)} + G^\circ(h) \qquad (29)$$

where $G°$ is the transform of the strictly pairwise $g(p)$ function, origin excluded, related to the excess free energy as shown. The solute intensity A expressed in Laue units is given by

$$A(h) = \frac{NQ(h)}{c_0(1 - c_0)} \qquad (30)$$

and is identical to the Fourier transform of the Warren short-range order parameters. Neglecting the elastic energy contribution in Eq. (11), we obtain by Eqs. (29) and (30)

$$\langle A(h) \rangle = \frac{1}{1 + c_0(1 - c_0) G°(h)/k_B T}$$

which is, in a different notation, the expression derived by Clapp and Moss[17] for the transform of the correlation function above the transition temperature except for a normalization constant in the numerator and a factor of two due to a different definition of pair energies. An alternate method of satisfying the integrated intensity condition consists in introducing an additive constant in the denominator of the expression for $\langle Q \rangle$ given above. This method[18] is computationally awkward but has the advantage of yielding a bounded value of Q at the critical point.

8. CUBIC CRYSTALS

In cubic crystals, second-rank tensors must be scalars so that we may define the gradient-energy and diffusion coefficients K and D as

$$\kappa_{(ij)}^{(2)} - \kappa_{(ij)}^{(1)} = \delta_{ij} K$$

$$D_{ij} = \delta_{ij} D$$

Equation (26), with the sum over (r) extended to the equidistant neighbors in the first coordination shell, yields

$$\frac{1}{2} \Omega M f_0'' m \sum_r [x_1(r)]^2 = D$$

With the mobility M defined as in a previous paper[4]

EARLY-STAGE CLUSTERING AND ORDERING   271

$$M = \frac{D}{f_0'' \Omega}$$

we obtain

$$m = \frac{1}{a^2}$$

since the sum over x_1^2 is equal to $2a^2$ for both b.c.c. and f.c.c. crystals, a being the lattice parameter. With this determination of m, the expression for $\beta(h)$, given by Eq. (16), agrees with the one given previously.[4]

Let us denote by g_i ($i = 0,1,2,3 \ldots$) the pair interaction coefficient for the ith coordination shell. The coefficient g_0 is given by Eq. (27) and the explicit forms of the long wave relations (b) and (c) to the third shell are: For b.c.c.

$$8g_1 + 6g_2 + 12g_3 = \frac{a^3}{2}(h'' - Ts_x'')$$

$$g_1 + g_2 + 4g_3 = -aK$$

for f.c.c.

$$12g_1 + 6g_2 + 24g_3 = \frac{a^3}{4}(h'' - Ts_x'')$$

$$2g_1 + 2g_2 + 12g_3 = -aK$$

Explicit forms of the elastic long wave relations for b.c.c. and f.c.c. crystals are given elsewhere.[5]

9. THE GOLD-NICKEL SYSTEM MODULATED STRUCTURE

Several studies[19-21] have shown that coherent precipitation in the Au-Ni system appears to take place by a highly anomalous spinodal mechanism. Woodilla and Averbach[19] observed that a well-defined portion of the concentration Fourier spectrum $V(h)$ received maximum amplification, although the resulting maximum in intensity $Q(h)$ did not gradually shift to lower values of (h) with prolonged aging as would be expected from the normal coarsening reaction following spinodal decomposition. Indeed, the intensity close to the origin of k space was observed to decrease during aging. Furthermore, the onset of decomposition, detected by the appearance of satellite intensity corresponding to an 8Å concentration modulation along $< 100 >$ directions, occurred about 220°C above the spinodal temperature calculated by Golding and Moss.[22] The continuum theory, valid for modulation wavelengths larger than 15–20Å, is not expected to offer satisfactory explanations for such short wavelength modulations as those reported for the Au-Ni system. We have therefore sought an explanation based on the microscopic theory presented above.

Since the thermodynamic properties of the Au-Ni system are largely determined by the substantial strain energy in the solid solution, our attention has centered on the possible deviations of the

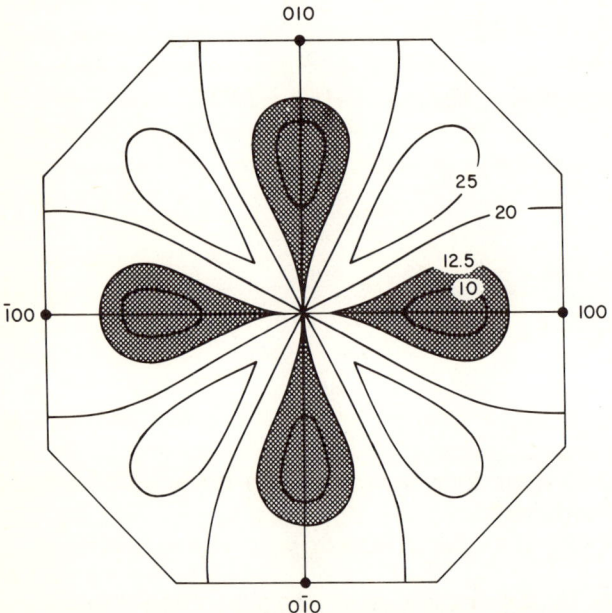

Fig. 2 *Contour plot of effective modulus $M(h)$ (in units of 10^{12} dynes/cm^2) in a (100) section of the first Brillouin zone for Au-Ni. The following constants were used: $\alpha_1 = \alpha_2 = 1.0149 \times 10^4$ dynes/cm, $\beta_1 = \beta_2 = 1.0547 \times 10^3$ dynes/cm, $\gamma_1 = 2.1979 \times 10^4$ dynes/cm, $\alpha_1 = 6.5896 \times 10^{-5}$ dynes, $\alpha_2 = 0$, $\alpha_1 = 2.0444 \times 10^{-13}$ dynes·cm, $\alpha_2 = -6.7923 \times 10^{-14}$ dynes·cm. For the meaning of symbols $\alpha_1 \ldots$, see Ref. 5.*

effective elastic modulus[5]

$$M(h) = \frac{F(h)}{2\Omega\eta^2}$$

from its value given by the continuum theory.[2] In the above equation, valid only for cubic crystals, $F(h)$ is the quantity defined by Eq. (9) and η is an atomic misfit parameter defined by[1]

$$\eta = \frac{d(lna)}{dc}$$

Figure 2 is a contour plot of the effective modulus $M(h)$ in a (100) section of an f.c.c. Brillouin zone. The contours were calculated for the equiatomic composition by using a first and second neighbor force model. The force constants were fitted to the elastic constant data[23] through the use of the long wave relations given previously.[5] The values of the remaining undetermined constants were adjusted so as to create minima or "soft" regions of $M(h)$ centered around the 8Å positions along < 100 > directions. With the position so adjusted, it was then observed that the depth of the minimum, which varied with its position, caused the 8Å Fourier components of the $V(h)$ spectrum to become unstable at temperatures some 200–300°C above the spinodal calculated by Golding and Moss[22] in agreement with the observations of Woodilla and Averbach.

The force constant model used for Fig. 2 displays very strong nearest neighbor solute-lattice coupling[5] and equal lattice coupling between first and second neighbors. Using a spring (forces) and ball (ion) model, nearest neighbor gold ions are thus connected by highly compressed springs whereas second neighbor ions are connected by unstretched springs of similar spring constant. Interestingly enough, this simplified model simulates the very strong nearest-neighbor repulsive force resulting from core interactions between gold ions in the solid solution.[23]

If the postulated elastic contribution dominates the "chemical" one, it can be shown that the amplification factor $\alpha(h)$ will exhibit a maximum in the vicinity of the 8Å position, midway between the fundamental and superlattice positions. The resulting < 100 > modulations are not expected to coarsen, i.e., the 8Å satellites will not tend to more towards the fundamental since, according to Fig. 2, the energy minima themselves are located close to the satellite positions. By contrast, the (chemical) energy minima are always located at the origin of k space in normal clustering systems. Intensity maxima occur some distance away from the origin in the usual spinodal systems for kinetic reasons only, maxima in the amplification factor α occurring typically in the range of 50–100Å. The coherent phase transformation in Au-Ni, on the other hand, can be described as the decomposition of a quasi-random solid solution to a short wavelength modulated structure intermediate between the clustered and ordered states. The precise nature of this structure has yet to be determined.

10. SUMMARY

The main features of the theoretical model outlined above are: (1) The free energy expansion is limited to terms which are quadratic in the state variables v (concentration) and u_i (displacement components); (2) the model is a nonlocal one, i.e., the thermodynamic properties of a small neighborhood of lattice site p depend, in principle, on the probability of finding a given atom with given displacements at all other sites p' of the crystal; (3) a lattice, rather than continuum model is used, leading to results valid on an atomic scale; (4) the treatment, though "microscopic" and "discrete," is classical throughout. Thus, quantum mechanical considerations intervene only indirectly through the pair-wise interaction coefficients if one were to evaluate these from first principles.

Hypothesis (1) was introduced for convenience. Higher order terms could, of course, be retained, but this would lead to a many body problem, the solution of which is untractable in the general case. Higher order terms must be considered, however, if such events as nucleation or particle coarsening are to be treated successfully. As it stands, the diffusion equation given here can only be used to describe the kinetics of local concentration changes in solutions deviating slightly from the quasi-random state. Amplification of concentration waves in the long wavelength portion of the spectrum gives rise to clustering effect (spinodal decomposition), amplification in the short wavelength portion gives rise to ordering.[4]

The nonlocal hypothesis (2) is equivalent to the introduction of gradient terms in an otherwise local theory. Cahn and Hilliard[1,5] showed that a positive chemical gradient energy term in the continuum approximation corresponded to an interfacial free energy and Hillert[3] and Cook, de Fontaine, and Hilliard[4] showed that a negative gradient term in the discrete case characterized a solution whose atoms tended to surround themselves with unlike neighbors.

The lattice hypothesis (3) takes into account the discrete nature of the lattice and, in conjunction with the nonlocal approach (2), allows one to treat correctly the short wavelength portion of the concentration variation spectrum. The early stages of spinodal decomposition and of certain order-disorder transformations can thus be treated within the same mathematical framework, a result which is conceptually pleasing but which becomes essential for the case of multicomponent solid solutions whose early-stage decomposition can manifest ordering and clustering behavior simultaneously.[24,25] Likewise, the strain-ordering of b.c.c. solutions[4] and strain-induced clustering can be rationalized on the basis of the concentration of solute intensity $Q(k)$ in regions of k space for which the elastic contribution to the function $\psi(h)$ takes on minimum values. A tentative explanation of the short wavelength clustering effects in Au-Ni solutions is offered on this basis. A study of the chemical and elastic effects and their influence on the shapes of GP zones is currently in progress.

REFERENCES

1. Cahn, J. W.: *Acta Met.*, **9**:795 (1969).
2. ———: **10**:179 (1962).
3. Hillert, M.: *Ibid.*, **9**:525 (1961).
4. Cook, H. E., D. de Fontaine, and J. E. Hilliard: *Ibid.*, **17**:765 (1967).
5. ——— and ———: *Ibid.*, **17**:915 (1969).
6. ———: *J. Phys. Chem. Solids*, **30**:2427 (1969).
7. ———: *Acta Met.*, **18**:297 (1970).
8. Krivoglaz, M. A., and E. A. Tikhonova: *Ukr. Fiz. Zh.*, **3**:297 (1958).
9. Khachaturyan, A. G.: *Sov. Phys. Solid State*, **9**:2040 (1968).
10. de Fontaine, D. *J. Appl. Cryst.*, in press.
11. Cahn, J. W.: *Acta Met.*, **14**:1687 (1966).
12. de Fontaine, D.: Doctoral dissertation, Northwestern University, Evanston, Ill. (1967).
13. Landau, L. D., and E. M. Lifshitz: "Statistical Physics," Addison-Wesley, Reading, Mass., 1958.
14. Krivoglaz, M. A.: "Theory of X-Ray and Thermal Neutron Scattering by Real Crystals," Plenum Publishing Corp., 1969.
15. Cahn, J. W., and J. E. Hilliard: *J. Chem. Phys.*, **28**:258 (1958).
16. Krumhansl, J. A.: in R. F. Wallis (ed.), "Lattice Dynamics," p. 627, Pergamon Press, 1965.
17. Clapp, P. C., and S. C. Moss: *Phys. Rev.*, **171**:754, 764 (1968).
18. de Fontaine, D.: This symposium, p. 277.
19. Woodilla, J. E., and B. C. Averbach: *Acta Met.*, **16**:255 (1968).
20. Moss, S. C., and B. C. Averbach: *Proc. Conf. Small Angle X-ray Scattering, Syracuse, N.Y.* edited by H. Brumberger, Gordon and Breach, Science Publishers, Inc., 1965.
21. Kimball, O. F., and J. B. Cohen: *Trans. AIME*, **245**:661 (1969).
22. Golding, B., and S. C. Moss: *Acta Met.*, **15**:1239 (1967).
23. Golding, B., S. C. Moss, and B. C. Averbach: *Phys. Rev.*, **158**:637 (1967).
24. de Fontaine, D.: *Acta Cryst.*, **A25**:S229 (1969).
25. Morral, J. E.: Doctoral dissertation, Massachusetts Institute of Technology, 1969.

BOSE-EINSTEIN CONDENSATION OF CONCENTRATION FLUCTUATIONS IN BINARY SOLID SOLUTIONS

D. de Fontaine

School of Engineering and Applied Science
University of California at Los Angeles
Los Angeles, California

1. INTRODUCTION

For many practical purposes the Landau theory of critical fluctuations[1] has proved to be a very useful one. Its application to the scattering of radiation by real crystals has been covered extensively by Krivoglaz.[2] Capp and Moss[3] have successfully used a somewhat different approach, easily shown[4] to be equivalent to the Landau formulation, in their analysis of short-range order of binary solid solutions. Among other shortcomings, discussed for example by Kadanoff et al.,[5] the Landau theory, applied to clustering and ordering in solid solutions, fails to yield a constant for the concentration fluctuation intensity spectrum integrated over the Brillouin zone. Furthermore, the predicted intensity increases without limit at the transition temperature and has physically unacceptable values below. These deficiencies can be eliminated by introducing an additive parameter in the denominator of the Landau formula as will be demonstrated presently. By contrast, multiplying the Landau expression by a normalization constant[3] does not take care of the improper behavior of the expected intensity at and below the transition temperature.

2. BOSE-EINSTEIN ANALOGY

It was shown elsewhere[4] that the free energy of a binary substitutional solid solution deviating slightly from the disordered state can be written as

$$F = \frac{N}{2} \sum_k \psi(\mathbf{k}) Q(\mathbf{k}) \quad (1)$$

where N is the number of atoms in the crystal (to which we impose periodic boundary conditions), ψ is an energy coefficient obtained by Fourier transforming the pairwise interaction coefficients, Q is the "intensity" VV^*, V being the Fourier transform of

$$v(p) = \begin{cases} 1 - c_0 & \text{if} \quad B \text{ atom at}(p) \\ -c_0 & \text{if} \quad A \text{ atom at}(p) \end{cases}$$

where c_0 is the atomic fraction of B in the solution. The sum in Eq. (1) is extended over all points $\mathbf{k}$ in the first Brillouin zone. If we define a normalized intensity

$$\bar{Q}(\mathbf{k}) = NQ(\mathbf{k}) \quad (2)$$

we obtain, for the expected intensity $<\bar{Q}>$ above the transition temperature according to the Landau theory, the equation[4]

$$<\bar{Q}> = \frac{k_B T}{\psi(\mathbf{k})} \quad (3)$$

where $k_B T$ has its usual meaning. Equation (3) is reminiscent of the high-temperature approximation of the Bose-Einstein expression for the expected number of particles $<n_i>$ occupying energy level ϵ_i

$$\left[\exp\left(\frac{\epsilon_i}{k_B T}\right) - 1\right]^{-1} \simeq \frac{k_B T}{\epsilon_i}$$

If the number of particles is to be conserved, one must introduce an appropriate Lagrangian multiplier

$$\lambda = e^\alpha = e^{-\epsilon^*/k_B T} \qquad (4)$$

Following the same procedure in the present case, one would obtain, by analogy, the modified expression

$$<\bar{Q}(k)> = \frac{k_B T}{\psi(k) - \psi^*} \qquad (5)$$

the parameter ψ^*, independent of k, being determined by the integrated intensity condition

$$\sum_k <\bar{Q}(k)> = Nc_0(1 - c_0) \qquad (6)$$

The above analogy can be justified by the arguments of the next section.

3. IDEAL GAS OF PSEUDOPARTICLES IN K SPACE

Let $n = Nc_0$ be the number of B atoms in the solid solution. For given c_0, the number of ways of distributing A and B atoms among the available lattice sites is given by

$$w = \frac{N!}{(N - n)! n!}$$

a large, but finite number. To each distribution of atoms will correspond a set of $\bar{Q}(k)$ values so that, for given k, the intensities $\bar{Q}$ can only take a finite, discrete set of values. It is therefore permissible to regard a given $Q(k)$ as made up of a sum of arbitrarily small units, each unit corresponding to a fictitious pseudoparticle in k space. The "quantum of concentration intensity" need not be defined further since the final result is quite independent of the precise nature of the postulated pseudoparticles; we must merely recognize that, as in the very similar case of phonons, these particles are indistinguishable. Furthermore, these intensity units can be regarded as particles of an ideal gas since the Fourier intensities do not interact in the approximation of Eq. (1).

There is, of course, no restriction as to the number of particles occupying a given level, i.e., there is no restriction on the value of $\bar{Q}(k)$ corresponding to energy level $\psi(k)$. It therefore follows that Bose-Einstein statistics must be used to describe the distribution of intensity $\bar{Q}$ over the available energy levels at equilibrium. Taking into account the constraint of Eq. (6), we then obtain for the expectation value of the intensity $\bar{Q}_k$

$$<\bar{Q}_k> = \frac{g_k}{\exp[(\psi_k - \psi^*)/k_B T] - 1}$$

where g_k is the statistical weight of level k,[6] ψ^* being related to the Lagrangian multiplier as in Eq. (4). The evaluation of ψ^* is not a simple matter. With certain simplifying assumptions however, the problem can be rendered formally identical to the one treated by London in his theory of superfluidity.[6]

4. ANALOGY WITH LONDON'S THEORY OF SUPERFLUIDITY

As an example, let us evaluate ψ^* by choosing the following simple form for the energy coefficient ψ_k, valid for cubic crystals[7]

$$\psi(k) = f'' + \frac{1}{4} a^2 \omega \beta(k) \qquad (7)$$

in which the first term f'' is the second derivative with respect to concentration of the Helmholtz free energy of the solution evaluated at the average concentration c_0. The second term represents a "gradient energy" which takes into account the nonlocal character of the free energy of a nonuniform solution.[4] In Eq. (7), a is the lattice parameter and ω is a coefficient depending on the nature of the system. Let us consider specifically the case of b.c.c. crystals, for which a second-order phase transition is possible at the equiatomic composition[8] and for which we have[7]

$$\beta(k) = \left(\frac{8}{a^2}\right)\left[1 - \cos\left(\frac{k_1 a}{2}\right)\cos\left(\frac{k_2 a}{2}\right)\cos\left(\frac{k_3 a}{2}\right)\right] \quad (8)$$

We can expand the geometrical factor $\beta(k)$ about the harmonic wave describing the ground k_0 state of a solid solution:

k_0 = any reciprocal lattice vector for the phase-separated (or clustered) state
k_0 = a vector from the origin to any superlattice position for the ordered state.

By thus expanding $\beta(k)$ we obtain, by Eqs. (7) and (8), the following expression for the k dependent part of the gradient energy:

$$\epsilon \simeq |\omega|\pi^2 N^{-2/3}(h_1^2 + h_2^2 + h_3^2) \quad (9)$$

where the h_i are the Cartesian components of the vector $(aN^{1/3}/2\pi)(k - k_0)$. The h_i are integers if the cyclic lattice is assumed to be a cube containing N lattice sites. The expression for the gradient energy expanded about the ground state turns out to be always positive even though ω is positive for clustering and negative for ordering.[4]

Equation (8) is essentially identical to that giving the kinetic energy levels of particles of an ideal gas. This allows us to use the results of London's theory of superfluidity based on the concept of an ideal Bose-Einstein gas.[6] All of his results will now be taken over with only a slight change in notation; the reader is referred to London's monograph for additional details and proofs. If we assume that the states with k close to the ground state k_0 dominate in the intensity spectrum, we can replace the summation in Eq. (6) by an integral after performing a change of variables from h_i to ϵ through the use of Eq. (9). The result is

$$\sum_k <\bar{Q}(k)> \simeq \frac{2N}{(\pi\omega)^{3/2}} \int_0^E \frac{\epsilon^{1/2} d\epsilon}{\exp[(\alpha + \epsilon)/k_B T] - 1} \quad (10)$$

in which

$$\alpha = \frac{(f'' + \delta - \psi^*)}{k_B T}$$

with $\delta = 0$ for clustering and $\delta = 4\omega$ for ordering. Assuming that high-energy states contribute negligibly to the integrated intensity, the upper limit E of the integral can be allowed to go to infinity. The integral can then be evaluated,[6] and we obtain the approximate integrated intensity condition

$$2\left(\frac{k_B T}{\pi \omega}\right)^{3/2} F_{3/2}(\alpha) = c_0(1 - c_0) \qquad (11)$$

in which $F_{3/2}(\alpha)$ is a monotonically decreasing function of α defined by London.[6] In Eqs. (10) and (11), only the magnitude of ω should be used, in accordance with Eq. (9).

It can be shown that as T decreases the nonnegative parameter α must also decrease until a critical point is reached for which $\alpha = 0$. We then have, by Eq. (11), the following expression for the critical temperature

$$k_B T_0 = \pi |\omega| \left[\frac{c_0(1 - c_0)}{2F_{3/2}(0)}\right]^{2/3} \qquad (12)$$

For $c_0 = 1/2$, and with $F_{3/2}(0) = 2.612$, we obtain

$$\omega = \pm 2.4\, k_B T_0 \qquad \text{(+: cluster, −: order)}$$

Recall that, for the regular solution model, the interaction coefficient ω_R is given by

$$\omega_R = \pm 2\, k_B T_0$$

showing that our gradient energy coefficient $\omega a^2/8$ is related to ω_R in accordance with the derivation of Cahn and Hilliard.[9]

Above the critical temperature T_0, the value of α is given by

$$\frac{F_{3/2}(\alpha)}{F_{3/2}(0)} = \left(\frac{T}{T_0}\right)^{3/2} \qquad \text{(for } T > T_0\text{)}$$

In converting the sum of Eq. (6) to an integral (10), the ground state, $h_1 = h_2 = h_3 = 0$, has been treated incorrectly. The ground state contribution must then be added to Eq. (10), yielding

$$(e^\alpha - 1)^{-1} + Nc_0(1 - c_0)\left(\frac{T}{T_0}\right)^{3/2}\frac{F_{3/2}(\alpha)}{F_{3/2}(0)} = Nc_0(1 - c_0) \qquad (13)$$

(for $T < T_0$)

This equation has a solution for α of order $1/N$. The ground state population, or long-range order intensity is then approximately given by

$$\bar{Q}_0 = Nc_0(1 - c_0)\left[1 - \left(\frac{T}{T_0}\right)^{3/2}\right] \qquad \text{(for } T < T_0\text{)}$$

whereas the contribution to the integrated intensity from the next higher states is of order $N^{2/3}$ only, which amounts to an infinitesimal fraction of the total intensity in the limit of large N. At $T = 0°$K, all of the intensity is in the ground state.

5. LIMIT OF STABILITY

It is interesting to note that Eq. (12) defines the transition temperature T_0 as a function of the average concentration. Thus Eq. (12) represents the limit of stability of the solution with respect to the following two diffusional processes: The spinodal for clustering systems, the ordering-start temperature for ordering systems. We have accordingly reserved the symbol T_0 for the temperature defined by Eq. (12); the symbol T_c is reserved for the true critical temperature which must be located at the maximum of the T_0 vs. c_0 curve of Eq. (12). If we assume, for the moment, that the gradient energy parameter ω is concentration independent, we see that the limit of stability has the following functional dependence

$$T^p \sim c(1 - c)$$

with $p = 3/2$. For a regular solution model, however, the exponent of T would be $p = 1$ (in this section, temperature and concentration are "running" variables so that the subscripts (0) on T and c can be dropped).

For a symmetric system, with critical concentration at $c = 1/2$ and with constant ω, the limit of stability has the equation

$$\tau^p = 1 - u^2 \qquad (14)$$

with

$$\tau = \frac{T}{T_c}$$
$$u = 2c - 1 \qquad (15)$$

$$p = \begin{cases} 1 & : \text{regular solution model} \\ 3/2 & : \text{present model} \end{cases}$$

Equation (14) was chosen because of its correct behavior at the critical point ($\tau = 1, u = 0$ i.e., $c = 1/2$) and at absolute zero ($\tau = 0$, $u = \pm 1$ i.e., $c = 0$ and 1) regardless of the (positive) value of p. The model with the 3/2 law gives a limit of stability which lies completely outside the one given by $p = 1$ and has infinite slope at $c = 0$ and 1.

In attempting to check these models against experimental data, we must first modify Eq. (14) since, in the real world, critical points seldom do us the favor of being located at $c = 1/2$ The simplest way of introducing the required flexibility into Eq. (14) is to modify Eq. (15)

$$u = \frac{c - c_c}{c_c(1 - c_c)} [2(c_c - 1)c + (1 - c_c)] \qquad (16)$$

where c_c is the critical concentration corresponding to T_c One easily verifies that this modified expression yields correct behavior at the critical and absolute zero temperatures.

The models were tested against the experimentally determined spinodals of two systems: AlZn and AuNi with $c_c = 0.39$ and 0.70 respectively. In Figs. 1 and 2, taken from the work of Cook and Hilliard,[10] we have plotted points calculated from Eqs. (14) and (16) with $p = 1$ (filled circles) and $p = 3/2$ (filled triangles). It is apparent

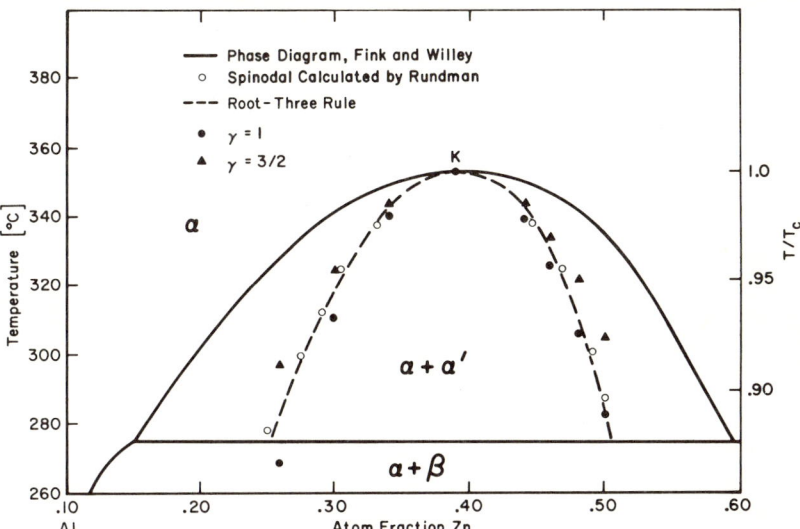

Fig. 1 Al-rich side of Al-Zn phase diagram from Ref. 10 with filled symbols added, as calculated from Eqs. (14) and (16).

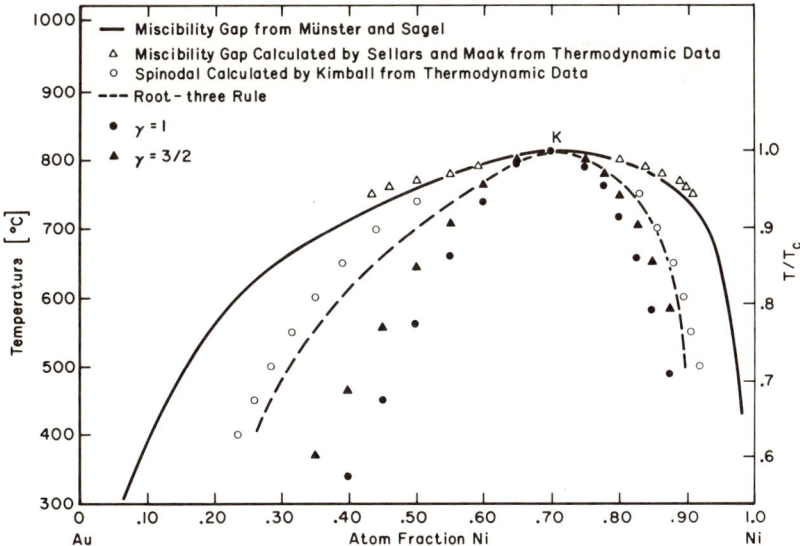

Fig. 2 Au-Ni phase diagram from Ref. 10 with filled symbols added, as calculated from Eqs. (14) and (16).

that the proposed analytical models give poorer fit to the data than does the "root three rule" of Cook and Hilliard. It must be recalled, however, that the root three rule uses the full information of the given miscibility gap, the present model merely uses, as input, the location of a single point, the critical point K. The models with $p = 1$ and $p = 3/2$ appear to bracket the thermodynamic data in the AlZn case; in the AuNi case the fit, though generally poor, is somewhat better with $p = 3/2$.

6. DISCUSSION

Normalization of the integrated intensity $\bar{Q}$ can be accomplished by the introduction of an additive parameter in the denominator of the Landau expression for the expectation value of $\bar{Q}$, as indicated in Eq. (5). This equation can be regarded as the high-temperature approximation of a Bose-Einstein equation describing the statistics of an ideal gas of harmonic concentration waves in Fourier space. The parameter responsible for normalization thus appears as a Lagrangian multiplier required by the constraint of constant integrated intensity, a consequence of the impossibility of creating or of destroying atoms of the solution. By analogy with the London theory of superfluidity, it was shown that the energy levels populated by concentration waves can be obtained by expanding the gradient energy about the ground or long-range ordered state of the solution. The entropy and free energy of the postulated ideal gas of pseudoparticles in k space can then be evaluated from the partition function according to standard procedures.

Bose-Einstein condensation of these noninteracting particles occurs at a critical temperature which is proportional to the 2/3 power of $c(1 - c)$, in contrast to the familiar parabolic law of the regular solution model. The 2/3 power law of Eq. (12) suggests that surface-to-volume-ratio effects have been indirectly taken into consideration. Such effects become important near the critical point where spatial correlations of the order of the crystal size tend to develop. Following London's treatment, we find that the number of pseudoparticles in the ground state just below the critical point is proportional to N, the number of atoms in the crystal, while the next higher states have population of order $N^{2/3}$ only, proportional to the number of surface atoms. At the critical point itself, the intensity remains finite, as required.

The assumption of ideality of the k space gas of pseudoparticles precludes obtaining correct scaling laws for the system. Furthermore, the temperature dependence of the ground state population, or long-range order intensity (see Eq. (13) and also Fig. 21 of London's monograph[6]), though qualitatively acceptable cannot, for the reason just given, yield the proper quantitative behavior.

ACKNOWLEDGMENTS

The author wishes to thank Professor J. E. Hilliard for permission to use his previously published figures of the Au-Ni and Al-Zn diagrams.

REFERENCES

1. Landau, L. D., and E. M. Lifshitz: "Statistical Physics," p. 366, Addison-Wesley Publishing Co., Inc., Reading, Mass., 1958.
2. Krivoglaz, M. A.: "Theory of X-Ray and Thermal Neutron Scattering by Real Crystals," Plenum Publishing Corp., 1969.
3. Clapp, P. C., and S. C. Moss: *Phys. Rev.,* **171**:754, 764 (1968).
4. de Fontaine, D., and H. E. Cook: This symposium.
5. Kadanoff, L. P., W. Gotze, D. Hamblen, R. Hecht, E. A. S. Lewis, V. V. Palciauskas, M. Rayl, and J. Swift: *Rev. Modern Phys.,* **39**:395 (1967).
6. London, F.: "Superfluids," vol. 2, p. 40, J. Wiley & Sons, New York, 1954.
7. Cook, H. E., D. de Fontaine, and J. E. Hilliard: *Acta Met.,* **17**:765 (1967).
8. Lifshitz, E. M.: *J. Phys.,* **4**:251 (1942).
9. Cahn, J. W., and J. E. Hilliard: *J. Chem. Phys.,* **28**:258 (1958).
10. Cook, H. E., and J. E. Hilliard: *Trans. AIME,* **233**:142 (1965).

DISCUSSION *on Papers Presented by D. de Fontaine and H. E. Cook*

G. ALEFELD: Have you compared your assumptions about elastic constants of Au-Ni with measured dispersion curves?

H. E. COOK: To determine the several force constants it is necessary to measure the dispersion curves of the alloy and also the size-effect scattering at small wavelengths in the short-range order intensity. The first set of measurements would yield $\alpha_1, \alpha_2, \beta_1, \beta_2,$ and γ_1. The second set would yield $\hat{\alpha}_1$ and $\hat{\alpha}_2$ as shown by Ref. 5 of our paper. Neither of these experiments has been carried out for the Au Ni system.

S. C. MOSS: It should be emphasized that at the maximum unmixing temperature in the Au-Ni alloys ($\sim$890°C) there does not appear (and should not appear) any critical scattering or build-up in fluctuations. This is due simply to the fact that the atom size disparity is so large ($r_0(Au) \gg r_0(Ni)$) that any composition fluctuation is accompanied by an elastic wave which costs so much energy that the composition wave is essentially damped out. In fact, the phase-separation at ~890°C is completely incoherent, characterized by nucleation and growth of the new phases out of the parent single phase matrix and starting at, say, grain boundaries. It is only when the supercooling is sufficient to overcome the elastic energy contribution that coherent phase separation occurs and then it obviously occurs first for those waves with the smallest elastic energy. Thus the Cook-de Fontaine calculation of a possible minimum in $M(h)$ near 8Å^{-1} is important as an explanation of the appearance of an 8Å wavelength in Au-Ni as the first decomposition effect below the coherent spinodal.

L. GUTTMAN: Are there in fact measurements of the diffuse scattering at high temperature (i.e., near the critical temperature) which prove the chance of critical opalescence?

S. C. MOSS: There are indeed the observations reported some years ago by Muenster and Sagel of critical X-ray opalescence above 890°C in polycrystalline Au-Ni samples. It was subsequently shown, however, that double Bragg scattering could more than account for the scattered intensity of Muenster and Sagel, but that particular measurement was never redone properly on a single crystal.

Regarding Alefeld's earlier question on phonon dispersion curves, I would only add that in concentrated alloys of large mass disparity (i.e., Au-Ni) the phonon lifetimes become vanishingly small and the measured neutron groups consequently too broad to measure. The Cook-de Fontaine method then becomes nearly the only way of estimating the "average" force constants.

LONG-RANGE ORDER IN BETA BRASS

C. B. Walker and D. R. Chipman
Army Materials and Mechanics Research Center
Watertown, Mass.

ABSTRACT

The temperature dependence of long-range order in β-CuZn is being determined from X-ray single crystal measurements of the integrated intensities of the (100) and (300) superlattice reflections made over the entire range of temperatures below T_c. Improvements over previous studies include allowance for the appreciable difference between the amplitudes of thermal vibration of the Cu and Zn atoms. The results obtained for the long-range order parameter at temperatures outside the critical region are consistently greater than those predicted by both "rigid" and "compressible" Ising model theories.

1. INTRODUCTION

Beta brass, the CuZn alloy of approximately 1:1 composition, offers one of the classical examples of order-disorder transformations in binary alloys.[1] At room temperature its structure is nearly perfectly ordered, with one sublattice, e.g., the corner sites of the cubic unit cells, occupied almost exclusively by Zn atoms and the other, the body center sites, occupied by Cu atoms. As the temperature is raised, the atomic arrangement begins to disorder, each sublattice being occupied by an increasing fraction of atoms of the "wrong" type, and this varies continuously at an ever-increasing rate up to the critical temperature $T_c = 468°C$ where both sublattices contain just the average composition of the alloy, and only short-range atomic correlations remain. This order-disorder transformation is apparently a true second-order (or higher order) phase transformation,[2] and recent experiments[3-7] have shown that at temperatures near T_c a number of its characteristics are described quite well by the Ising model.

The object of the present experiment is to determine accurately the temperature dependence of the long-range order in β-CuZn over the entire range of temperatures below T_c using X-ray diffraction techniques. The long-range order, a measure of the mean sublattice composition, is specified by the Bragg-Williams long-range order parameter S defined as

$$S = r_\alpha - w_\beta$$

where r_α is the fraction of sites on one sublattice "rightly" occupied, and w_β is the fraction of sites on the other sublattice that are "wrongly" occupied. This parameter can range in value from unity, for perfect order, to zero, for $T \geq T_c$. It is determined from measurements of intensities of Bragg superlattice reflections, for which we can write approximately

$$I \propto S^2 \cdot |f_{Zn} e^{-M_{Zn}} - f_{Cu} e^{-M_{Cu}}|^2 \qquad (1)$$

where f_x is the scattering factor of an atom of type X at rest, and M_X is its Debye-Waller exponent, proportional to its mean square amplitude of thermal vibration, which we initially treat as depending only on the identity of the atom.

Chipman and Warren[8] investigated this same problem in 1950, but the advances in diffraction equipment, theory, and techniques since that time offered several opportunities for improvement on that work. Norvell and Als-Nielsen[9] have now just done this same experiment using neutron diffraction techniques, but their results contain significant systematic errors, as will be discussed below.

Our measurements in the critical region (that is, $T \geq 0.97 T_c$) are still incomplete, so the discussion here will be restricted to results for temperatures below this limit. Our major conclusion is that the experimental values for the long-range order parameter are appreciably greater than those calculated from Ising models. To support this we shall discuss several experimental features in some detail.

2. EXPERIMENT

Our experiment used more-or-less conventional X-ray diffraction equipment and techniques, with only a few unusual features. The primary beam was monochromatic Cu Kα radiation obtained with a doubly bent LiF monochromator from an X-ray tube operated at 20 ma and 14 kvp. The specimen was a single crystal containing approximately 51.9 at. % Cu, with a mosaic spread of only a few minutes, in the form of a cylinder 3/4 in. in diameter, 3/16 in. thick, with flat faces parallel to (100) planes. Duplicate scintillation detector counting systems measured the diffracted radiation and monitored the primary beam power. The diffractometer and the sample furnace and cryostat were standard commercial instruments. A helium atmosphere was used in the furnace.

We have measured the integrated intensities of the (100) and (300) superlattice reflections at temperatures ranging from liquid helium up to T_c using an omega scan technique. Our procedure has been to measure the integrated intensity of each reflection at room temperature both before and after a short set of measurements at high (or low) temperatures, and then to normalize the integrated intensity of a reflection at temperature T in terms of its room temperature value. In the upper range of temperatures the experimental runs were shortened in several steps to minimize possible effects due to zinc loss by evaporation.

The question of zinc loss effects was checked in two ways: By comparing the room temperature "before" and "after" measurements bracketing each high-temperature run, and by comparing integrated intensities for similar temperatures from samples with different thermal histories. The comparison of "after" with "before" values, each reproducible to 0.2 percent, for our runs above 300°C generally showed a small decrease, with a r.m.s. difference of 0.6 percent and a maximum change of 1.4 percent. The checks for samples with different histories, though less accurate, showed similar results: in one case the crystal was held at 425°C for 30 minutes for

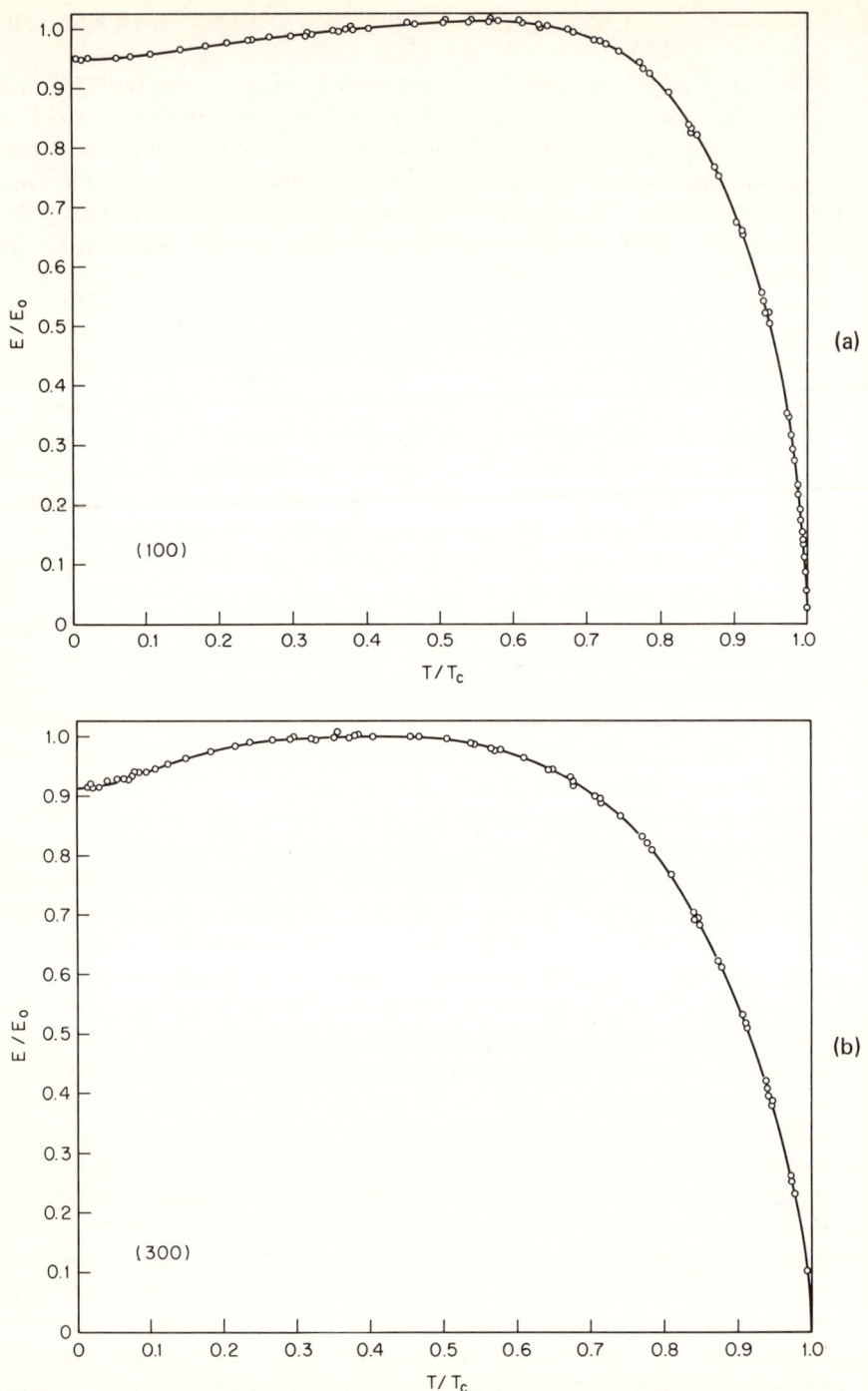

Fig. 1 *(a)The measured integrated intensity E of the (100) reflection, normalized in terms of its room temperature value E_0 as a function of T/T_c. (b) As in (a), but for the (300) reflection.*

one set of (100) and (300) measurements, then heated up to T_c and brought back to 425°C during 10 minutes, and then held at 425°C for another 30 minutes for a second set of (100) and (300) measurements; the second value for the (100) intensity at 425°C was 0.4 percent smaller than the first, and the second (300) intensity was 2.0 percent smaller than the first, with an experimental precision for each intensity of approximately 1.0 percent. We have concluded from these checks that the errors in our intensity data due to zinc loss effects are less than 1 percent even at the highest temperatures.

3. RESULTS

The raw data—measured integrated intensities E normalized in terms of room temperature values E_0—for the (100) and (300) reflections are plotted in Fig. 1a and 1b respectively as a function of the reduced temperature T/T_c. Room temperature corresponds to $T/T_c = 0.40$. Most of the points below $T/T_c = 0.85$ correspond to an average of six measurements and show a reproducibility varying from approximately 1/4 percent below room temperature to 1/2 percent at $T/T_c = 0.85$. The remaining points are for fewer and/or shorter measurements, with correspondingly larger errors.

These raw data plots show two general characteristics: At high temperatures the intensity of each reflection decreases with increasing temperature, going apparently continuously to zero at T_c; and at temperatures below room temperature the intensity of each reflection *increases* with increasing temperature. The high-temperature behavior is the result to be expected as the long-range order decreases to zero. The behavior at low temperatures, where the long-range order is effectively constant, shows unambiguously from Eq. (1) that since $f_{Zn} > f_{Cu}$ at these temperatures the Cu atoms in this alloy must have an appreciably larger mean square amplitude of thermal vibration than that of the Zn atoms. This difference was not recognized in any previous study of this alloy.

The effect of this difference in the amplitudes of vibration is displayed more clearly after the intensities have been corrected for the usual factors of extinction, included thermal diffuse scattering and critical scattering, and a Debye-Waller factor representing an average mean square amplitude of vibration. The largest correction, that for the mean Debye-Waller factor, was determined in a separate experiment[10] from measurements of the (310) fundamental reflection from a polycrystalline specimen over this same range of temperatures. The included thermal diffuse scattering was calculated using an improved method[11] that correctly includes anisotropic effects. The included critical scattering, negligible except near T_c, was calculated on the basis of the Ising model. The extinction correction, at its worst only 3½ percent for the (100) and 2 percent for the (300), was calculated using the theory of Zacharaisen[12] together with absolute intensity measurements of the (200) and (310) fundamental reflections from our crystal. To reduce computations, these corrections and all subsequent steps in the analysis were carried out for points on smooth curves drawn through the data rather than on the individual data points.

The square root of each corrected normalized intensity value then yielded a normalized amplitude value for that reflection and temperature. Our results for these normalized amplitudes for the (100) and (300) reflections as a function of temperature are shown in Fig. 2 by the A_1 and A_3 curves respectively. If all atoms had had the same amplitude of vibration, these two curves would be identical, giving directly the normalized long-range order S/S_0 as a function of temperature. The observed differences between them are quite marked, and they allow one to estimate that at room temperature

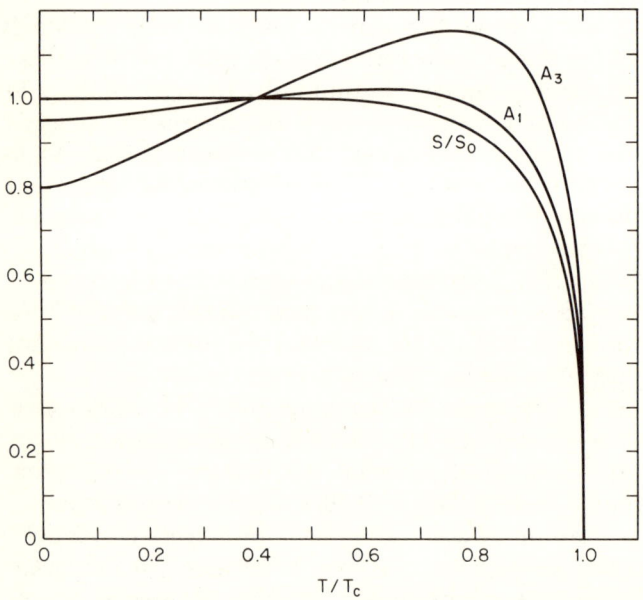

Fig. 2 *Normalized reflection amplitudes—A_1 for the (100), A_3 for the (300)—and the long-range order S/S_0 as a function of temperature.*

the mean square amplitude of vibration of a Cu atom is approximately 12 percent greater than that of a Zn atom. This difference, which may seem rather large, has been confirmed by Nicklow and Dolling[13] from computer calculations using the interatomic force constant values measured by Gilat and Dolling.[14]

A method of analysis has been developed that allows us to determine the long-range order as a function of temperature from these two normalized amplitude curves without requiring accurate values for the scattering factors or the difference between the amplitudes of vibration of the two atoms. Details of this method will be given elsewhere. We note here only two points: (1) Exactly the same results are obtained from the alternative approach[15] of associating the different amplitudes of vibration with the different sublattices rather than with the two kinds of atoms, and (2) two normalized amplitude curves are required to obtain one long-range order curve, leaving us with no consistency check. Our results for the long-range order parameter, normalized in terms of its room temperature value, are plotted as the curve labelled S/S_0 in Fig. 2 for comparison with the two amplitude curves.

In Fig. 3 we compare this experimental long-range order curve with recent Ising model calculations. Our experimental result is shown by the solid curve. The short-dashed curve is that given by the [10,10] Padé approximant of Essam and Fisher,[16] and the long-dashed curve gives the results of the recent "compressible lattice" Ising calculation of Baker and Essam.[17] Our results are in marked disagreement with these calculations. The random error in our measured long-range order parameter increases with temperature, but it is only approximately 1 percent at $T/T_c = 0.9$, which is roughly an order of magnitude smaller than the discrepancy between our results and the Baker-Essam calculation.

Norvell and Als-Nielsen have recently made this same measurement using neutron diffraction techniques and obtained results that were in excellent agreement with the Baker-Essam calculation. However, their experiment must be criticized on several points—no allowance was made for different amplitudes of vibration for the two kinds of atom; their self-consistent procedure for evaluating the large correction for the mean Debye-Waller factor gave results quite different from our measured values for this correction, and they had to deal with quite large extinction corrections. We suggest that these factors have led to serious systematic errors in their results.

Recent experiments[3-7] on various properties of β-CuZn at temperatures in the critical region have generally shown reasonably good agreement with Ising model calculations. Our results show that

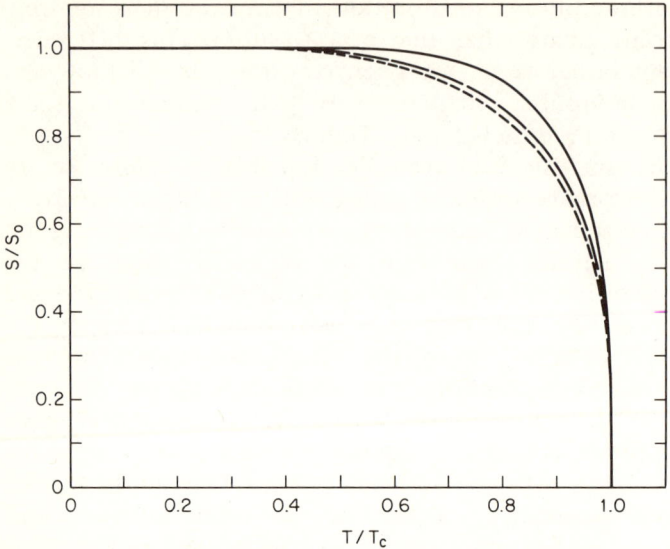

Fig. 3 *Comparison of the measured long-range order with Ising model calculations. The solid curve gives the experimental results; the short-dashed and long-dashed curves give respectively the "rigid" lattice calculation of Essam and Fisher and the "compressible" lattice calculation of Baker and Essam.*

outside of this critical region the measured long-range order differs considerably from the Ising calculations. Our alloy is not stoichiometric, however, and it is a metal in which interactions often extend beyond first neighbors. It thus seems quite possible that the differences between our results and the calculated values are due to the effects of one or both of these limitations in the present Ising theories.

REFERENCES

1. Nix, F. C., and W Shockley: *Rev. Mod. Phys.*, **10**:1 (1938).
2. Guttman, L.: in F. Seitz and D. Turnbull (eds.), "Solid State Physics," vol. 3, Academic Press, Inc., New York, 1956.
3. Walker, C. B., and D. T. Keating: *Phys. Rev.*, **130**:1726 (1963).
4. Dietrich, O. W., and J. Als-Nielsen: *Ibid.*, **153**:711 (1967).
5. Als-Nielsen, J., and O. W. Dietrich: *Ibid.*, **153**:717 (1967).
6. Als-Nielsen, J.: *Ibid.*, **185**:664 (1969).
7. Ashman, J., and P. Handler: *Phys. Rev. Lett.*, **23**:642 (1969).
8. Chipman, D. R., and B. E. Warren: *J. Appl. Phys.*, **21**:696 (1950).
9. Norvell, J. C., and J. Als-Nielsen: *Phys. Rev.*, **B2**:277 (1970).

10. Chipman, D. R., and C. B. Walker: *Bull. Am. Phys. Soc.*, **14**:379 (1969); to be published.
11. Walker, C. B., and D. R. Chipman: *Acta Cryst.*, **A26**:447 (1970).
12. Zacharaisen, W. H.: *Acta Cryst.*, **23**:558 (1967).
13. Nicklow, R. M., and G. Dolling: Private communication.
14. Gilat, G., and G. Dolling: *Phys. Rev.*, **138**:A1053 (1965).
15. Schwartz, L. H., and J. B. Cohen: *J. App. Phys.*, **36**:598 (1965).
16. Essam, J. W., and M. E. Fisher: *J. Chem. Phys.*, **38**:802 (1967).
17. Baker, G. A., and J. W. Essam: *Phys. Rev. Lett.*, **24**:447 (1970).

DISCUSSION on *Paper Presented by C. B. Walker*

S. C. MOSS: Is it possible that the nonstoichiometry ($Cu_{0.52}Zn_{0.48}$) influences the behavior (when compared to the prediction for Cu Zn) much as an applied field might shift an exponent in an antiferromagnet?

C. DOMB: I think it unlikely that a difference of this order could be due to this small deviation in composition. One must remember that the Ising interaction is very simple, and one should not be surprised if it fails to provide exact agreement with metallic atom interactions. In the critical region the changes are extremely rapid, and insensitive to the details of the interaction; hence one may get agreement. But this is not true over the whole range, and these experimental data provide a challenge to modify the interaction to a more sophisticated character.

H. E. COOK: A long-range force which could be responsible for the deviation of the experiment from Ising theory results from the elastic interaction between ions of different size. This interaction favors ordering in the CuZn system and is roughly 20 percent of the ordering energy.

G. A. BAKER: The apparent disagreement of your experimental data with Ising model predictions is in the same temperature range as the marked disagreement of Ashman and Handler's specific heat data. Since you recently have had an exchange of views with Als-Nielsen, could you summarize his views on your experimental disagreements?

C. B. WALKER: I believe that our evidence for the difference between the amplitudes of vibration of the two atoms, together with the rather large discrepancies between the values for the correction for the mean Debye-Waller factor in the two experiments, have led Als-Nielsen to agree that their results may be in error.

W. MARSHALL: Your criticism of Als-Nielsen's experiments depends heavily upon the assumption that their Debye-Waller factors should be the same as yours. However, the nuclear Debye-Waller factors need not necessarily be equal to those for X-rays. It would therefore be useful to measure these by neutron techniques. It remains true that, since your experiments show that the X-ray factors unambiguously are different, then surely so also should be the neutron factors—and hence there *is* a systematic error in Nielsen's results. Nevertheless to get a quantitative estimate of their error it is necessary to check if the two types of Debye-Waller factor differ.

C. B. WALKER: The room temperature values that we have obtained for the mean Debye-Waller exponent and for the difference between the exponents for the two atoms agree rather well with the corresponding values calculated by Nicklow and Dolling using the force constants determined in the neutron inelastic scattering experiment of Gilat and Dolling. This indirect evidence indicates that there should not be much difference between neutron and X-ray Debye-Waller factors in this alloy.

MULTISITE CLUSTER PROBABILITIES IN ISING LATTICES

Philip C. Clapp

Ledgemont Laboratory
Kennecott Copper Corporation
Lexington, Massachusetts

ABSTRACT

If it can be assumed that the configurational energy of the lattice depends only upon the pair correlations (i.e., the pair-wise interaction model) then multisite correlations can be inferred from the pair correlations by finding that value of the multisite probability distribution that maximizes the entropy within the microcanonical ensemble defined by the pair correlations.

A variational technique for carrying out this procedure will be described and applied to several examples.

1. INTRODUCTION

The first descriptions of order-disorder transformations around the turn of the century were in terms of a single long-range order parameter, and for temperatures above the critical temperature T_c where this parameter was zero the configurational state of the system was considered to be essentially random. It soon became apparent from specific heat measurements in the vicinity of T_c that the configurational entropy did not reach the value appropriate to a completely random lattice until the temperature was very far above T_c. This residual order of a "disordered" lattice has been called short-range order (SRO) and the first attempts to quantify and describe it were in terms of a parameter that measured the nearest neighbor pair correlation.

The Ising square lattice results of Onsager[1] and Onsager and Kaufmann[2] were vitally important not only because they finally resolved the question of whether a system having only short-range interactions could achieve a cooperative transition to a state of long-range order, but because it also gave the first exact and detailed description of the propagation of spatial ordering in the SRO state in terms of the range and magnitude of pair correlations.

Characterizing the SRO state of a lattice by means of the pair correlations is now a widespread practice, partly because these are the quantities which can be measured by diffuse SRO neutron or X-ray scattering data, and partly because they are such useful theoretical functions with which to evaluate other thermodynamic properties of the system.

However, there are a number of important questions about the nature of a partially ordered lattice that cannot be answered by the pair correlation description. A frequent and surprisingly frustrating question of this kind is "What does a typical region of the SRO lattice look like?" If one tries to construct the "average" configuration of a region containing more than two sites by adjusting the occupation of sites to satisfy the average nearest neighbor correlation it will then be found that the second and higher neighbor pair correlations are, in general, not correct.

A second question is "What is the distribution in size and frequency of regions in the SRO phase that have a configuration identical to that of the perfectly ordered state?" This question is relevant to our understanding of nucleation processes as well as providing evidence for (or against) microdomain models of the SRO state.

A third question that can be important for real systems where the configuration of the ordered state is not known (usually because atomic mobility has become too sluggish at T_c) is "What is the ordered state configuration that the system is trying to achieve?"

These are all questions that, in general, require a knowledge of the many body correlations present in the SRO state and yet all the information that is normally available is in the form of pair correlations. As we shall show, this information implies a set of sum rules[3] which place definite limits on the range of values that the many body correlations can assume. By using a postulate of Information Theory[4] we can take the further step of determining the "most probable" value of the many body correlations within the range allowed by the known pair correlations and calculate multisite probability distributions for various cases of interest.

The calculation is exact if two conditions are satisfied:

(1) The configurational energy of the lattice is determined by central pair-wise interactions whose range does not exceed the dimensions of the multisite cluster.

(2) The Information Theory choice corresponds to the maximal entropy point in the configurational phase space bounded by the sum rules.

2. THEORY

Consider a cluster of n sites in some fixed geometrical relationship (e.g., a tetrahedral nearest neighbor group, a site and its nearest neighbor sites, etc.). If there are N sites in the lattice and sites are regarded as distinguishable, then there will be N different n site clusters, ignoring boundary effects. The occupation of each site is specified by $\sigma_i (= \pm 1)$. This provides the information of spin alignment (up or down) in magnetic lattices or of atomic configuration in binary alloy lattices. In any case, there will be 2^n possible configurations of the cluster, although many may be symmetrical equivalents if the geometry of the cluster possesses a symmetry.

Denote each configurational type by an index k, and the fraction of n site clusters in the lattice of one type by P_k. P_k can be interpreted equally well as the probability that a cluster chosen at random in the lattice will be of configuration type k. It is obvious that the P_k's must be positive semidefinite and satisfy the simple sum rule

$$\sum_{\{k\}} P_k = 1 \qquad (1)$$

The lattice average $<\sigma>$ of the occupation numbers σ_i (i.e., the composition or magnetization) provides n sum rules of the form

$$\sum_{\{k\}} \sigma_j^{(k)} P_k = <\sigma> \quad (j = 1, n) \qquad (2)$$

where $\sigma_j^{(k)}$ corresponds to the occupation number of the jth cluster site when it is in configuration k.

In addition, the contribution that each configurational type will make to the lattice averaged pair correlations can be calculated by inspection and leads to other sum rules, $n(n-1)/2$ in number, as follows

$$\sum_{\{k\}} \sigma_j^{(k)} \sigma_{j+p}^{(k)} P_k = <\sigma_j \sigma_{j+p}> \quad (j = 1, n; p > j) \qquad (3)$$

In essence, these sum rules confine the possible P_k values to a reduced volume in P_k space. The range of P_k's within this subspace depends upon the composition and pair correlations, being largest when the pair correlations are zero and shrinking to a single point (in most cases) when the pair correlations are those of a perfectly ordered configuration. The P_k's then become completely determined.

Again regarding P_k as the probability that an arbitrarily chosen n site cluster will have the configuration k, a theorem of Information Theory[5] can be used to find the most probable values of the P_k's which are consistent with the information (the sum rules) that we do possess. The theorem states that the least biased choice for the P_k's is that set which maximizes within the allowed subspace the "ignorance function" $I(P_k)$ given by

$$I(P_k) = -\sum_k P_k \ln P_k \qquad (4)$$

This theorem is based on a demonstration that any other set of P_k's would imply more information than has been given. More information of a very complicated kind may, in fact, be present due to overlapping cluster conditions that we shall now speak about.

The "ignorance function" can also be interpreted as an entropy function for a particular microcanonical ensemble. Consider the configurational entropy $W(P_k)$ obtained by placing NP_1 clusters of type 1, NP_2 clusters of type 2, etc., on a lattice of N sites in all possible ways, disregarding overlaps. This is given by the combinatorial factor

$$W(P_k) = \frac{N!}{\prod_k (NP_k)!} = {}^N\left[-\sum_k P_k \ln P_k\right] \quad (5)$$

and is obviously identical to the "ignorance function" above, apart from the factor of N.

$W(P_k)$ is identical to the combinatorial factor used in the Cluster Variation Method developed by Kikuchi and others[6] using this particular cluster as a basis. The limitations of this expression for calculating the entropy of the lattice are well known since it counts not only all the possible lattice configurations but many more that are impossible because of incompatible overlaps. However, it is clear for our purposes that it is only the location of the maximum and not its magnitude that is of importance.

We shall propose then a postulate which, if true, would make our determination of the most probable set of P_k's an exact procedure. The postulate is: If the value of the P_k's are changed such that $W(P_k)$ is increased (decreased) then the subset of possible lattice configurations will also increase (decrease).

We cannot offer a mathematical proof of this conjecture at present, but we can say that for all the examples where the exact answer is known by other means, the postulate is apparently correct. These include four site clusters on an Ising linear chain, triplet clusters for an equi-composition three-dimensional binary lattice and nearest-neighbor triplet clusters in Cu_3Au. The latter case was determined by Gehlen and Cohen[7] using a computer simulation technique for the SRO state, and our results are compared to theirs and to the predictions of the Kirkwood superposition approximation below.

In summary, if the configurational energy depends only upon the pair correlations spanned by the n site cluster then the sum rules define a microcanonical ensemble of lattice configurations that contain many different sets of P_k's. As N is allowed to increase indefinitely the statistical law of large numbers assures us that the fraction of lattice configurations having any set of P_k's other than the most probable one becomes vanishingly small. Thus, if our postulate about the correspondence of entropy maxima is correct then we have determined exactly the distribution of multisite probabilities characteristic of the system.

3. RESULTS

A computer simulation of a typical lattice configuration in a region of 4000 sites has been carried out by Gehlen and Cohen[7] for Cu_3Au at 450°C (10 percent above T_c) based upon Moss's experimental data[8] for the first, second, and third neighbor pair correlations. Starting from either a perfectly random or a perfectly ordered configuration they used a computer to shuffle atomic configurations, accepting the new configuration only if it was more consistent with the experimental data, until a configuration was reached that had essentially the same pair correlations as the measured values. A number of properties of the final computer configurations were then measured including the various fractions of nearest neighbor triplets. Their results are given in the second and third columns of Table 1. We have calculated from the sum rules the range for the triplet probabilities allowed by the pair correlations and this is given in the first column. The most probable set of triplet probabilities according to our method is shown, as are the predictions of the Kirkwood superposition approximation[9] which is frequently used to calculate triplet probabilities. The latter method approximates a triplet probability as a product of the three pair probabilities as follows:

$$P(\sigma_1 \sigma_2 \sigma_3) = CP(\sigma_1 \sigma_2) P(\sigma_2 \sigma_3) P(\sigma_3 \sigma_1) \tag{6}$$

where C is a normalization constant adjusted to make the sum of the triplet probabilities unity.

It is apparent from the table that the agreement between our results and the values of Gehlen and Cohen is very good. It suggests that our answer may well be the exact result for the infinite lattice ensemble.

TABLE 1. Nearest Neighbor Triplet Probabilities in Cu_3Au at 450°C

Triplet type	Range allowed by sum rules at $T = 450°$	Computer Simulation		This calculation	Superposition approx.
		Ordered start	Random start		
Cu-Cu-Cu	0.302–0.328	0.3273	0.3275	0.3271	0.3474*
Cu-Cu-Au	0.672–0.594	0.5962	0.5956	0.5964	0.5676*
Cu-Au-Au	0.000–0.078	0.0758	0.0761	0.0759	0.0840*
Au-Au-Au	0.026–0.000	0.0007	0.0007	0.00067	0.0011

*Note that these values are outside the allowed range.

TABLE 2. Cu-Centered Cluster Distribution

Cu₃Au T = 450°C (405°C)

$a_1 = -.195(-.218)$ $a_2 = +.215(+.286)$ $a_3 = +.003(-.012)$ $a_4 = +.077(+.122)$

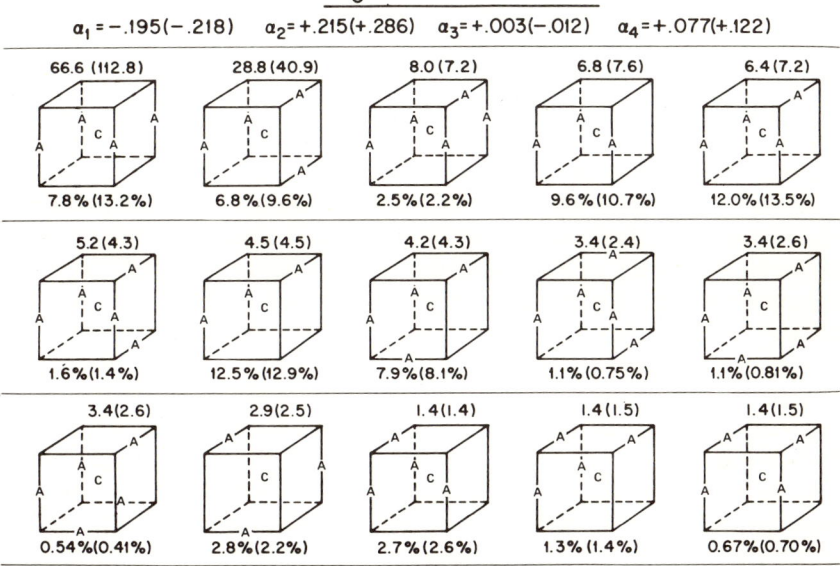

TOTAL 71.6% (80.5%)

Although the superposition approximation would appear to give a reasonable estimate of the triplet probabilities, the first three values are impossible because they lie outside the range allowed by the sum rules. One may conclude that an arbitrary choice for the triplet probabilities within the allowed range would actually be better than using the superposition approximation, for this at least would produce a set of values consistent with the known alloy data.

TABLE 3. Au-Centered Cluster Distribution

Cu₃Au T = 450°C (405°C)

$a_1 = -.195(-.218)$ $a_2 = +.215(+.286)$ $a_3 = +.003(-.012)$ $a_4 = +.077(+.122)$

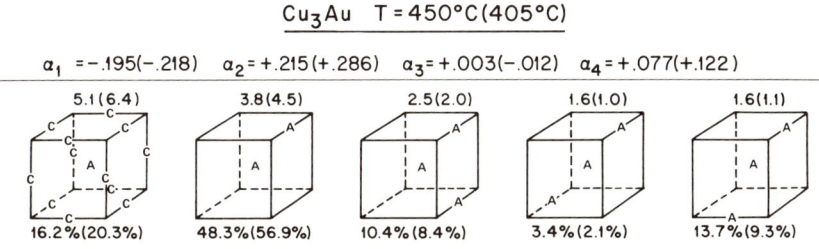

TOTAL 92% (97%)

We have also employed the variational method to calculate the probability distribution for the configurations of a cluster consisting of a site and its twelve nearest neighbors in Cu_3Au. It is impractical to display the entire distribution since this would involve over 300 distinct configurations. We have instead shown in Tables 2 and 3 only the most enhanced clusters in order of their enhancement factors (EF). EF is the ratio of a cluster's population in the SRO state to its population in the completely random state.

Moss[8] has measured the SRO parameters at two temperatures above the ordering temperature in this system. Therefore it is possible to study the evolution of the cluster distribution as it approaches the long range ordered (LRO) state. The spectrum for Cu-centered clusters is shown in Table 2 and that for Au-centered in Table 3. The numbers at the lower temperature (3 percent above T_c) are shown in parentheses and those at the higher temperature (10 percent above T_c) without. The numbers above the clusters are the EF's and those below are the actual lattice fraction for each cluster (including its symmetrical equivalents). The total lattice fraction covered by the displayed clusters is given at the bottom of the figures so that the fraction of the lattice that is distributed over the rest of the 300 or so clusters is apparent.

Several interesting aspects of the distribution are immediately apparent. There is a surprising degree of "order" in the nearest neighbor (n. n.) environment even at 10 percent above T_c. In fact, 34 percent of the Cu atoms and 64 percent of the Au atoms have one mistake or less in their n. n. shell (by comparison with the n. n. environment in the perfectly ordered Ll_2 structure). These fractions increase to 42 and 77 percent respectively at 3 percent above T_c.

The second most enhanced Cu-centered cluster is a particularly interesting type of mistake because it corresponds to the perfectly ordered environment of a different LRO arrangement variously referred to as the DO_{22} structure, the Ni_3V structure or the Cu_3Au long period (M = 1) superlattice. Clapp and Moss[10] have previously demonstrated that the pair interactions in Cu_3Au are such that the normal Ll_2 Cu_3Au structure is only slightly more favored energetically than the DO_{22} structure and this is undoubtedly the source of the "confusion" in the SRO state.

It may be added that as one goes to the lower temperature, although the other Cu-centered clusters either remain at the same lattice fraction or decrease, both Ll_2 and DO_{22} clusters increase appreciably, so that this competition apparently persists into the LRO state and may be regarded as the source of the very common (100) antiphase boundaries.

4. CONCLUSION

We have provided a set of exact sum rules that are applicable to the probability distribution of any n site cluster. These sum rules provide upper and lower bounds for the n site configuration probabilities that will ensure compatibility with the known composition and pair correlation parameters.

In addition, we have described a variational method for calculating the most probable n site distribution. This method rests upon two postulates; one physical, in the sense that it is an assumption about the nature of the atomic interactions, and the other mathematical, in that it is an assumption about the density of states in two closely related configuration spaces.

The variational method has been used to find the distribution of nearest neighbor 3-site clusters and nearest neighbor 13-site clusters in Cu_3Au. The three-site distribution could be compared to the same distribution previously determined by Gehlen and Cohen by their computer simulation technique and the results were found to be essentially identical. This provides significant support for our mathematical postulate, but not for the physical postulate because it is implicitly assumed in their procedure also.

In fact, if it becomes possible to measure n body correlations directly, any departure from the values calculated within the present scheme will provide important evidence about the relative strength of n body and pair interactions in real systems.

REFERENCES

1. Onsager, L.: *Phys. Rev.*, **65**:117 (1944).
2. Kaufmann, B., and L. Onsager: *Phys. Rev.*, **76**:1244 (1949).
3. Clapp, P. C.: *Phys. Rev.*, **164**:1018 (1967).
4. ———: *J. Phys. Chem. Solids*, **30**:2589 (1969).
5. Katz, A.: "Principles of Statistical Mechanics," p. 18, W. H. Freeman and Co., San Francisco, 1967; also L. Brillouin: "Science and Information Theory," Academic Press Inc., N. Y., 1962.
6. Kikuchi, R.: *Phys. Rev.*, **81**:988 (1951); *J. Chem. Phys.*, **19**:1230 (1951).
7. Gehlen, P. C., and J. B. Cohen: *Phys. Rev.*, **139**:A844 (1965).
8. Moss, S. C.: *J. Appl. Phys.*, **12**:3547 (1964).
9. Hill, T. L.: "Statistical Mechanics," pp. 195, 278, McGraw-Hill Book Co., Inc., N. Y., 1956.
10. Clapp, P. C., and S. C. Moss: *Phys. Rev.*, **171**:754 (1968), and *Phys. Rev.*, **171**:764 (1968).

DISCUSSION *on Paper by P. C. Clapp*

B. McCOY: While you discussed three-particle correlations in detail your method also applies to four-particle correlation. Then, as a nontrivial check on your method, you can use the exact computation on the four-particle correlation[1] of the two-dimensional Ising model in zero magnetic field.

P. C. CLAPP: I was not aware of this work and I thank you for bringing it to my notice.

C. DOMB: If you are not concerned with the immediate neighborhood of the critical point, you can certainly use series expansions for any of the multiple correlations. It is only when it is necessary to assess critical behavior (say within 1 percent of the critical point) that difficulties arise in assessing the precise asymptotic form of the coefficients. Elsewhere the Padé approximant, for example, should give an approximation accurate enough for testing the hypothesis.

L. ONSAGER: In the Ising model that I solved, one could not only obtain the four-particle correlations but also the three-particle correlations below the critical temperature (where they are nonzero). These could be used to test your own method.

REFERENCES

1. Computed by Y. Y. Stephenson, *J. Math. Phys.*, (1964).

THE "GESTALT" OF LOCAL ORDER

J. E. Gragg, Jr.
Department of Metallurgy and Materials Science
Carnegie-Mellon University
Pittsburgh, Pennsylvania

P. Bardhan and J. B. Cohen
Department of Materials Science
The Technological Institute
Northwestern University
Evanston, Illinois

ABSTRACT

It is shown that the first few measured pair probabilities restrict the values for higher order terms and higher order configurations. Computer programs generating atomic arrangements from these values are described, based on this principle. Tests of these programs are illustrated, and the configurations in the real systems examined to date are summarized.

1. INTRODUCTION

Local order (clustering or short-range order) expresses itself in diffuse scattering of short-wave length radiation. From this, for a binary system at least, we can directly obtain information on the conditional pair probability of finding an A atom at the end of some interatomic vector if there is a B atom at the origin of the vector $P_{lmn}^{A|B}$. $X_B P_{lmn}^{A|B}$ is the total probability (lmn represents the interatomic vector $l\vec{a}_1 + m\vec{a}_2 + n\vec{a}_3$). Theoretical equations for many porperties of materials can be derived in terms of the pair probabilities, but the derivations require the assumption that only pair interactions are involved. The *measured* pair probabilities, however, are the result of *all* the many body interactions in the system. It would, therefore, be intrinsically best to produce the actual local configurations from the measured pair probabilities, and to derive the behavior directly from these arrangements. One fact makes it possible to do this: The pair probabilities for various vectors are related and strongly restrict higher order configurations. We will demonstrate this in Sec. 2. Thus one can hope to proceed with, say, the values of $P_{lmn}^{A|B}$ for a few of the nearest-neighbor vectors. The procedure is very limited (due to computational time) as to the size of the configuration that can be computed.* P. Gehlen and J. B. Cohen first devised a more general procedure[1,2] and there are two other versions of the same approach.[3,4] Large numbers of atoms are placed in a three-dimensional array in a computer and pairs of A and B atoms are selected at random and interchanged until the configuration has the same pair probabilities for a few interatomic vectors. No energetics are involved. It is then possible to search for certain patterns of arrangements that one believes may be present from other evidence, such as the fully ordered structure, or the equilibrium precipitate that occurs at lower temperatures and longer times, or the configurations suggested by Dr. Clapp's calculations. This pattern recognition is often highly subjective in nature, involving all of the trials and tribulations of the psychology of good patterns and bad—what someone tends to "see" as a "good" pattern. Fortunately many of the patterns are simple. The programs are described in Sec. 3 and tests of the practical uniqueness of the structures are considered in Sec. 4. Finally, results for those real systems which have been examined in detail with this method will be reviewed.

---

*Dr. P. Clapp has described *his* procedures[11] for this in this volume; these involve some assumptions. Ours described in Sec. 2, do not, but the sizes he can deal with are somewhat larger.

2. THE INTERRELATIONSHIP OF THE PAIR PROBABILITIES AND HIGHER ORDER CONFIGURATIONS

The measured pair probability for a given interatomic vector is averaged over all such vectors underneath the incident beam of radiation. Nonetheless, because of the connectivity of the lattices of most solid solutions of metals and ceramics, the averages for different vectors are related. Consider, for example, the interatomic vectors shown in Fig. 1a. The bar graph in Fig. 1b serves as the basis for deriving the necessary relationships. Each of the three bars represents the total atomic fraction, i.e., unity. The thick region is the fraction of A atoms; this "thick" region has the same length in all three bars—but is distributed in different ways, depending on the probability of finding an A atom next to an A for a specific interatomic vector. (These are total probabilities *not* conditional probabilities—and we ignore end effects for a finite solid.) With this figure, the seven listed equations can be written immediately. There are seven equations with eight unknowns—the eight triplet probabilities. The measured values are the α's (the Fourier coefficients of the diffuse scattering intensity) and these are included through the relationships

$$\alpha_{ij} \equiv 1 - \frac{P_{ij}^{B|A}}{x_B} \tag{1a}$$

where x_B is the atomic fraction, and i, j the atomic positions at the ends of vector lmn

$$P_{ij}^{B|A} + P_{ij}^{A|A} = 1 \tag{2}$$

and

$$x_A P_{ij}^{A|A} = P_{ij}^{AA} = x_A(x_A + x_B \alpha_{ij}) \tag{3}$$

Now α_{23} for a large group of atoms would be the same as the *measured* value of α_{14}.[5-7] Thus knowing α_{12}, α_{13} we can say that we wish to know the range of α_{14} given the constraints of the first six equations in Fig. 1. As the probabilities are equal to or greater than zero, the mathematics of linear programming can be employed to accomplish this.[8-10] Some of the results were presented in Ref. 9 for a six-point cluster. Confinement was clearly indicated. The possible range was reduced to 70 percent of its maximum value when only short-range order was present, and to 50 percent when long range was appreciable.

In Fig. 1c a four-point cluster is shown with the matrix of coefficients for the probabilities. The last three rows show how cubic symmetry can be added; nearest-neighbor triplet probabilities must be equivalent. To consider long-range order, bar graphs are drawn for each sublattice. For example, Eqs. (2) to (4) of Fig. 1b are, for each sublattice, equal to the fraction of A atoms on the particular sublattice—which is obtained from the long-range order parameter. Equation (2) would thus be equal to the fraction of A atoms on the sublattice associated with site 1, because the equation is the sum of all the A atoms in row 1. Summing the new Eq. (2) for each sublattice gives x_A. However, no effect of long-range order was found, nor was there any effect of symmetry as far as the ranges of α's were concerned.

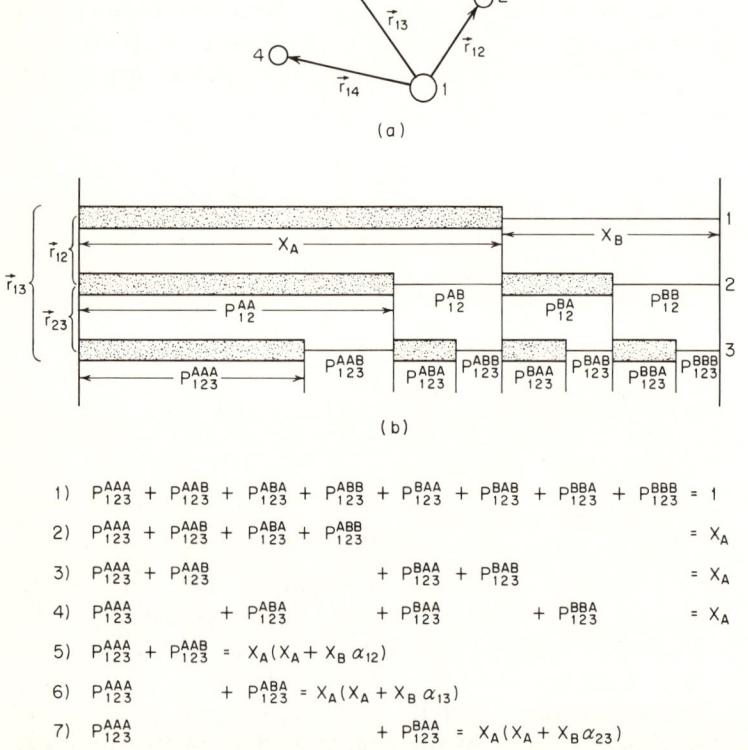

Fig. 1 (a) A four-point cluster of interatomic vectors. (b) Each of the three horizontal bars represents the total atomic fraction, i.e., unity. The thick portion represents A atoms. P_{12}^{AA} is the probability of an A atom at the end of $\vec{r}_{12}$ if an

"GESTALT" OF LOCAL ORDER 313

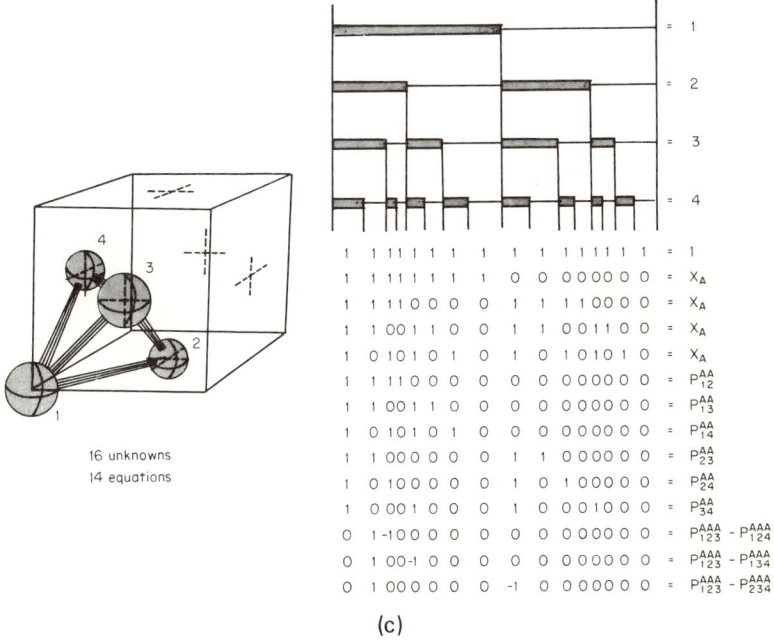

(c)

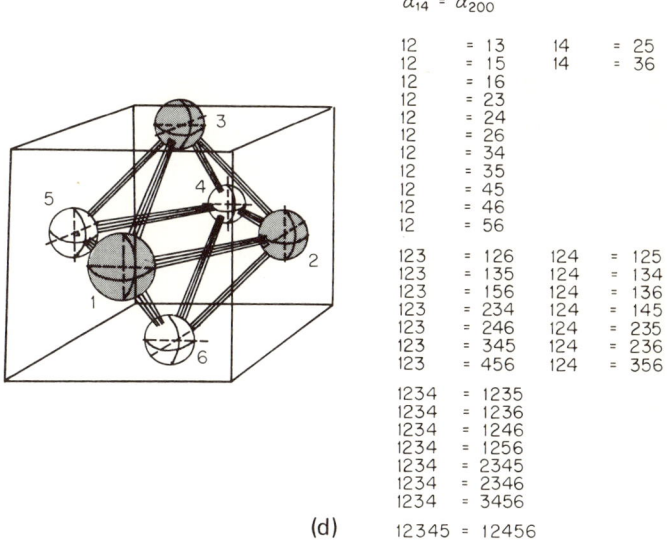

$\alpha_{12} = \alpha_{110}$
$\alpha_{14} = \alpha_{200}$

12	= 13	14	= 25
12	= 15	14	= 36
12	= 16		
12	= 23		
12	= 24		
12	= 26		
12	= 34		
12	= 35		
12	= 45		
12	= 46		
12	= 56		
123	= 126	124	= 125
123	= 135	124	= 134
123	= 156	124	= 136
123	= 234	124	= 145
123	= 246	124	= 235
123	= 345	124	= 236
123	= 456	124	= 356
1234	= 1235		
1234	= 1236		
1234	= 1246		
1234	= 1256		
1234	= 2345		
1234	= 2346		
1234	= 3456		

(d)   12345 = 12456

A is at the origin. Thus it appears as the region of overlap of the thick bars in the first two rows. The definition of the other terms and the equations follow from these "overlaps." [9] *(c) A four-point cluster. (d)–(f) Three six-point cluster, showing equivalent configurations. The minima and maxima of the shaded configuration was sought. (In (e) and (f) the range for first-neighbor triplets was also sought.)*

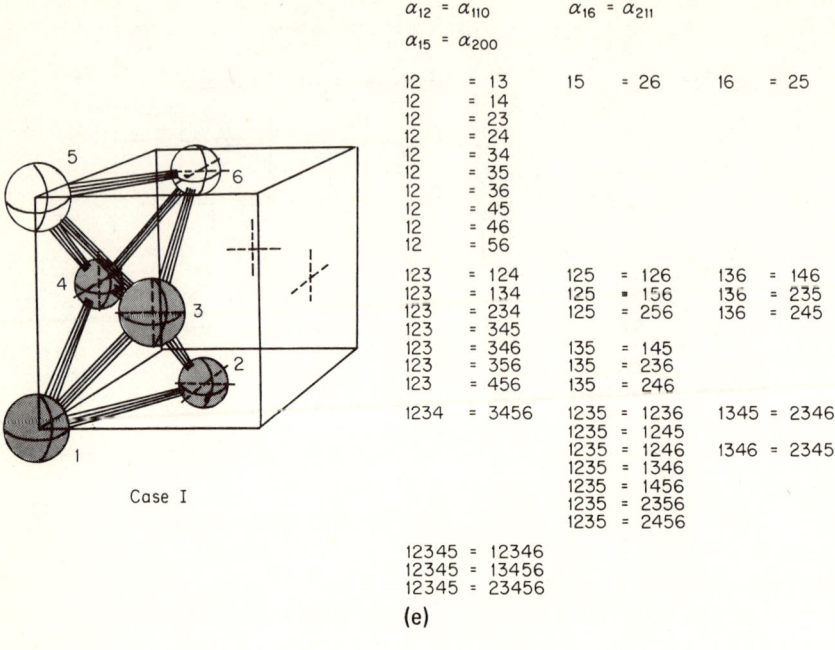

$\alpha_{12} = \alpha_{110}$ $\alpha_{16} = \alpha_{211}$
$\alpha_{15} = \alpha_{200}$

12	= 13	15	= 26	16	= 25
12	= 14				
12	= 23				
12	= 24				
12	= 34				
12	= 35				
12	= 36				
12	= 45				
12	= 46				
12	= 56				

123	= 124	125	= 126	136	= 146
123	= 134	125	= 156	136	= 235
123	= 234	125	= 256	136	= 245
123	= 345				
123	= 346	135	= 145		
123	= 356	135	= 236		
123	= 456	135	= 246		

1234	= 3456	1235	= 1236	1345	= 2346
		1235	= 1245		
		1235	= 1246	1346	= 2345
		1235	= 1346		
		1235	= 1456		
		1235	= 2356		
		1235	= 2456		

12345 = 12346
12345 = 13456
12345 = 23456

(e)

Case I

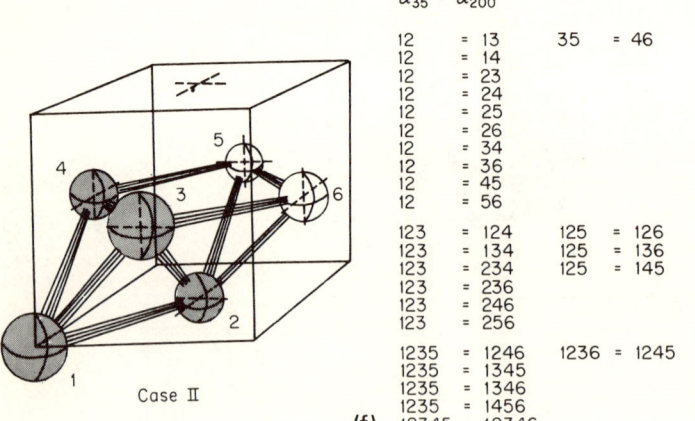

$\alpha_{12} = \alpha_{110}$ $\alpha_{15} = \alpha_{211}$
$\alpha_{35} = \alpha_{200}$

12	= 13	35	= 46	15	= 16
12	= 14				
12	= 23				
12	= 24				
12	= 25				
12	= 26				
12	= 34				
12	= 36				
12	= 45				
12	= 56				

123	= 124	125	= 126	135	= 146
123	= 134	125	= 136		
123	= 234	125	= 145		
123	= 236				
123	= 246				
123	= 256				

1235	= 1246	1236	= 1245	2345	= 2346
1235	= 1345			2345	= 2356
1235	= 1346			2345	= 2456
1235	= 1456				
12345	= 12346				

(f)

Case II

Fig. 1 *(Continued)*

From the bar graph in Fig. 1c it is clear that $P^{AAA}_{123} = P^{AAAA}_{1234} + P^{AAAB}_{1234}$. This equation can be solved in the same manner for minimum and maximum values. The fourth and fifth columns of Tables 1a and b give the results for Cu_3Au above and below T_c (columns 2 and 3 are the ranges due to composition alone). Figure 1d shows a six-point cluster which contains first- and second-nearest neighbor vectors; the equivalent pairs of triplets, quadruplets, and quintuplets are given. Solutions were carried out for the shaded triplet and are given in columns six and seven using α_1 and α_2. Figures 1e and f show six-point clusters including third nearest neighbors. Case I includes a third-neighbor vector between *two* sublattices while case II has third-neighbor vectors between *three* sublattices. The remaining columns in Table 1 are for these six-point groups. The last column gives the value obtained from the computer simulations to be described in the next section.

(Dashed entries in Table 1 mean narrowing the range was not narrower than for the results given in previous columns.)

Note how tightly the triplet and quadruplet probabilities are confined. Above T_c, α_3 adds nothing for the triplets, whereas α_2 adds nothing below; α_1 and α_3 are controlling. For the quadruplets, case I of adding third neighbors is most important above T_c, but case II is more important below. Note also that the results from the computer simulation are near the extremes of the ranges and not in the center; the position is the end unaffected by additional pair probabilities. Small clusters, like 136 and 12345 in Fig. 1d, involving first and second neighbors, were also well confined.

It should be pointed out that, in contrast to the solutions for higher order α's, it is the additional symmetry considerations that cause α_2 and α_3 to reduce the ranges of the multipoint configurations when only α_1 is specified. α_2 and α_3 have some effect, but nowhere near that due to symmetry even when only α_1 alone is specified. Finally, all the symmetry conditions are *not* necessary. Only some play a role; the rest are redundant.

This kind of mathematics can also be particularly useful in determining if a set of α's is internally consistent. If they are not, no solutions are possible.

3. PRINCIPLES OF THE COMPUTER PROGRAMS

3.1. Gehlen and Cohen's Original Program[1,2]

We shall speak of 1 and 0 instead of A and B atoms, to become accustomed to the notation used in the printouts to be shown later.

TABLE 1a. First-neighbor Triplet and Quadruplet Distributions in Cu_3Au above T_C ($S = 0.00$)

	Min/Max		$\alpha_{110} = -0.195$*	$\alpha_{200} = +0.215$		$\alpha_{211} = +0.003$		Computer† Simulation
						Case I	Case II	
pAuAuAu	0	1/4	0.0259	0.0	0.0143	—	—	0.0008
pAuAuCu	0	1/4‡	0.0777	0.0348	0.0777	—	—	0.0758
pAuCuCu	0	3/4	0.6723	0.5943	0.6372	—	—	0.5963
pCuCuCu	1/4	3/4	0.3278	0.3135	0.3278	—	—	0.3272
pAuAuAuAu	0	1/4	0.0259	—	—	—	0.0 0.0141	0.0
pAuAuAuCu	0	1/3	0.0520	—	—	0.0 0.0476	—	0.0030
pAuAuCuCu	0	1/2§	0.1554	—	—	0.0126 0.1554	—	0.1470
pAuCuCuCu	0	1	0.8964	—	—	0.6888 0.8316	—	0.6970
pCuCuCuCu	0	3/4	0.0778 0.1556	—	—	0.1069 0.1556	—	0.1530

*α's from S. C. Moss, *J. Appl. Phys.*, 35:3547 (1964).
†Starting from the fully ordered structure using the first three α's; 4,000 atoms.[1]
‡This value can range up to 3/8 if symmetry is not considered.
§This value can range up to 6/10 if symmetry is not considered.

Table 1b. First-neighbor Triplet and Quadruplet Distributions in Cu$_3$Au Below T_c ($S = 0.80$)

	Min/Max		$\alpha_{110} = -0.265$*		$\alpha_{200} = +0.715$	$\alpha_{211} = -0.209$	
						Case I	Case II
pAuAuAu	0	1/4	0.0	0.0128	—	—	0.0	0.0090
pAuAuCu	0	1/4†	0.0	0.0384	—	—	0.0114	0.0384
pAuCuCu	0	3/4	0.6732	0.7116	—	—	0.6732	0.7002
pCuCuCu	1/4	3/4	0.2756	0.2884	—	—	0.2794	0.2884
pAuAuAuAu	0	1/4	0.0	0.0128	—	—	0.0	0.0089
pAuAuAuCu	0	1/3	0.0	0.0256	—	—	—	—
pAuAuCuCu	0	1/2‡	0.0	0.0768	—	—	—	—
pAuCuCuCu	0	1	0.8460	0.9488	—	—	0.8460	0.9316
pCuCuCuCu	0	3/4	0.0384	0.0769	—	—	0.0471	0.0769

*α's from L. H. Schwartz and J. B. Cohen: *J. Appl. Phys.*, **36**:598 (1965).
†This value can range up to 3/8 if symmetry is not considered.
‡This value can range up to 6/10 if symmetry is not considered.

The original program involved 4,000 or 16,000 atoms in a three-dimensional array, with periodic boundary conditions. It will be best in what follows to consider a specific example, cubic Cu_3Au (or 0_31). For the ordered alloy, there are three simple cubic sublattices of Cu, β, γ, δ, with origins at centers of a face-centered cube, and one of Au-α, with its origins at the corner of the face-centered cube. We rewrite the definition of α_k here in terms of 1 and 0 as the species, and for shells of atoms rather than specific vectors

$$\alpha_k = 1 - \frac{P_k^{0|1}}{x_0} = 1 - \frac{P_k^{1|0}}{x_1} \tag{1b}$$

k represents the kth shell; as the materials we will be dealing with are cubic, all permutations of lmn are equivalent, and so we can speak of these lmn as all in the kth shell. Hence, with C_k the coordination number of the kth shell around an atom and N the total number of atoms in the model, the number of 1-0 pairs that should be in the "final" configuration (designated with superscript f) is

$$(1,0)_k^f = (1 - \alpha_k) x_1 x_0 C_k N \tag{4}$$

We define the long-range order parameter S as

$$S = \frac{3}{4} \frac{(r_\beta - x_0)}{(1 - x_0)} + \frac{1}{4} \frac{(r_\alpha - x_1)}{(1 - x_1)} \tag{5}$$

where r_α and r_β are the fractions of α and β sublattices correctly occupied ($r_\beta = r_\gamma = r_\delta$).

Equation (5) and a mass balance for 1 atoms enables one to calculate r_α and r_β and hence the number of 1 atoms that *should be* on each sublattice. We then define (LRO_j) ($j = \alpha, \beta, \gamma, \delta$) as the difference between the initial and final numbers of 1 atoms on each sublattice. A typical computer run was initiated as follows.

For a fully ordered alloy, the occupation of each site is known, as is $(LRO)_j$, and $(m_{11})_k$, $(m_{01})_k$ respectively the number of 1 atoms in the kth shell around each 1 atom and the number of 1 atoms in the kth shell around each 0 atom. LRO_j, $(1,0)_k^i$, $(1,0)_k^f$, $(m_{11})_k^i$, $(m_{01})_k^i$ (where i is the initial ordered state) are all input to the computer (for k up to 3) as well as the identity of each atom at each site. Coordinates of a possible 1 atom are selected at random. If $(LRO)_j > 0$, 1 atoms are to be removed from the sublattice j specified by the coordinates, to satisfy long-range order, and the sublattice is accepted. (If not, another set of coordinates is chosen.) A scan is then made of the sublattice to *locate* a 1 atom, for at the coordinates there *may not* be a 1. This is done by scanning along a row in the x direction starting from the chosen coordinates and incrementing x and then, if necessary, incrementing y and again scanning along the x-axis; and if still no 1 has been found, incrementing z and scanning x, y again. An 0 atom is similarly chosen. Now, if superscript 2 represents the situation after an exchange, 1 before it

$$(1,0)_k^2 = (1,0)_k^1 + 2[(m_{11})_k^1 - (m_{01})_k^1 + \delta] \qquad (6)$$

where δ is unity if the sites are kth neighbors, zero if they are not. These numbers are computed for up to $k = 3$ (i.e., for up to the third shell) and an interchange carried out (and data updated), only if *each* of the three $(1,0)_k$'s comes closer to the final values. If the exchange is not possible, a new 0 atom is found; if necessary, the whole crystal is scanned. If a match is still not found, a new 1 atom is chosen.

Certain self-teaching schemes aid the process. Suppose $(1,0)_k^f > (1,0)_k^2$. Then $(m_{11})_k > (m_{01})_k$ is required for an exchange from Eq. (6). But if $(\overline{m}_{01})_k^1 = 4, (\overline{m}_{11})_k^1 = 0.5$, it will be difficult to find a suitable pair for exchange. Speed can be increased by specifying, for example, that the 1 atom must have $(m_{11})_k^1 > 2$. Suitable values for such conditions can be judged from the final values of m_{11}, m_{01}. In the program these values are automatically "tightened" if it becomes more and more difficult to find a suitable pair. Finally, if after a period of time no atom can be found, a few random jumps are made, taking into account only the long-range order S.

If a different initial state is desired, say a near-random alloy, the ordered state could be used to start a run to force $S = \alpha_1 = \alpha_2 = \alpha_3 = 0$.

The run is terminated for $\sum_{k=1}^{3} |(1,0)_k^f - (1,0)_k^2| \leq 10$, equivalent to about 10^{-2} percent of the total number of pairs in the first three shells being different than required by the measured α's. In addition to counts of various quantities and the printouts of the final structures, the program allows one to hold two of the α's constant and vary the other to its maximum or minimum value; this can be done to test the effects on the results of experimental errors in the alphas.

Some *2,000* jumps per minute is currently typical, for a 4,000 atom model. The cost for a typical run (30 seconds) is about *$5*—quite inexpensive even if you consider this procedure nothing more than larger model making than one can ever hope for by hand calculations!

3.2. Gragg's Modifications[3]

The size of the model was increased to 108,000 atoms, in any three-dimensional array, (b.c.c. or f.c.c.). With such a size it is possible to study large clusters, such as Guinier-Preston zones, and to deal with situations closer to a critical temperature. At the largest size *any four* α's can be controlling, but at smaller sizes even more can be employed; for any size, any number of α's may be calculated from the final model. The operator can specify when the configuration is satisfying the α's to within some estimated experimental error. An initial state, satisfying only composition, can be generated, as a starting point, internally, without a separate run. The use of the program was greatly simplified for the operator, minimizing his need to remember symbols, formats, etc.

The first atom chosen is accepted if it is *either* a 1 *or* a 0, depending on which species must be removed to satisfy $(LRO)_j$ for that sublattice. The second atom must be on the sublattice with the maximum error in $(LRO)_j$ for the opposite atom type. The x, y, z coordinates from a random number generator are mixed in a random fashion—that is, they become x, z, y for example. The first direction scanned (x or y or z) is also chosen at random. The "self-teaching" was altered somewhat. A "test point" is chosen as the average of the instantaneous values $(\bar{m}_{01})_k$ and $(\bar{m}_{11})_k$. Suppose there are currently too few 1 atoms in shell k, around a 1 atom and that the atom chosen initially is a 1 atom. This atom is accepted for a try at an interchange with an 0 only if $(m_{11})_k$ for this atom is less than the test point—for all shells for which agreement with α's is desired. After a certain number of tries for an exchange (specified by the user) two "remedial corrections" are applied:

1. If the difficulties have been evenly divided among all the shells being considered, a certain number of random jumps is carried out (i.e., without considering S or the α's). The operator can choose the number before the run;

2. If the problem is primarily associated with one shell, its test point is altered. For the example considered above, the test point for the shell would be increased by one.

The test procedures for a successful interchange are the same as for A, and running times and number of jumps per minute are about the same.

3.3. Williams' Program[4]

A simpler program for 8,000 atoms has recently been developed by R. O. Williams. Instead of interchanging atoms he merely changes the identity one at a time. The condition for a switch is that the overall composition should not exceed specified limits and that the change in the controlling alphas should be a *vector* toward the final values. Self-teaching schemes are not included in Williams' version, and no special effort is made to randomize the selection process. Thus, several runs in succession are required for good sampling. Equivalent sublattices are not "forced" to be equivalent because in small models some statistical fluctuations are expected.*

Indeed, in our own work (programs A and B) we initially found that this condition has little effect on the results anyway. Williams' program is faster and more efficient than the previous two as a result of the simpler coding, and the more open decision on the accepted change. As the results are generally obtained with extremely few jumps, the initial configuration can exert a large influence, another reason for multiple runs—or a run should be started with a state far from the final one. Close to a critical temperature or for large clusters, the composition limits may have to be large, which could alter the overall composition; this point has not been tested.

4. TESTS OF THE PRACTICAL UNIQUENESS OF THE PATTERNS

4.1. Starting from Two Initial States

One way of testing the uniqueness of the results is simply to approach "equilibrium" as an experimenter might do—by coming at it from two directions. As an example, consider some results for Cu-14 at. percent Al (Table 2).

As we shall see later, this f.c.c. alloy consists of rods of second-neighbor Al atoms ("1" atoms) in <100> directions. It is easy then to compare the number of atoms in rods of a given length, starting from a near-random alloy and from an ordered alloy. There is very excellent agreement, despite the fact that the number of jumps to the final configuration was vastly different for the two

---

*Williams gives the formula for the expected standard deviation of α_k for the kth shell (with 1 as the minor species) as

$$\sigma_{\alpha_k} = \left[\frac{x_1}{NC_k(x_0)}\right]^{1/2}$$

For the first shell of an f.c.c. structure ($C_k = 12$) with $N = 4,000$ atoms and with $x_1 = 0.25$, $\sigma \simeq 0.025$.

TABLE 2. Distribution of Al Atoms in Domains in Cu-14.5 at %Al.*[1]

	Numbers of domains containing between					Total number of domains	Number of jumps
	1 and 5 atoms	5 and 10 atoms	10 and 15 atoms	15 and 20 atoms	20 and 25 atoms	25 and 31 atoms		
From order	163	10	7	3	2	2	187†	10,288
From random	162	11	7	4	0	2	186	1,527

*The numbers are related to a 4,000-atom model.
†This corresponds to 3.88×10^{21} domains/cc.
Data from C. R. Houska and B. L. Averbach: *J. Appl. Phys.*, **30**:1525 (1959).

cases. We can look at much more detail, but we must be careful to pick a system where the numbers we wish to compare can be trusted. The Cu-Al system is not a good one because the data was taken many years ago, and is probably not very accurate. A better system would be Cu_3Au because good data exists for several temperatures—both above and below T_c. In addition to higher-order α's, we will compare

1. near-neighbor triplets and quadruplets
2. amount of ordered (or clustered) material, surface/volume ratio of such regions
3. distribution of neighbors around atoms in various shells

Results above T_c for Cu_3Au ($\approx 395°C$ [12]) for (1) and (2) are given in Table 3 and for (3) in Fig. 2.

Considering the aforementioned statistical errors associated with a 4,000 atom model, all values of higher order α's are quite well fixed by the first two α's, more so than indicated by the calculations in Sec. 2. The values from the two starting states are in excellent agreement. The distribution plots in Fig. 2 are also in excellent agreement for the two states.

Two minor words of caution are necessary.

1. For the smaller model sizes we find agreement within statistical error for many higher order α's. Lin, Spruiell and Williams[13] have shown that if, say, five α's are used as input rather than three, a better match to the details of the diffuse scattering pattern occurs. These additional constraints, however, do not seem to materially affect the atomic pictures. To show this, we have held two α's constant and varied the third (in several cases) to its maximum and minimum. For variations that change the sign of a small α but not its magnitude much, the changes are very minor. In some cases we have also worked out the changes in shapes necessary to bring all α's much closer to experiment than the statistical errors expected with the computer models. Again the changes are small.[1,2]

2. A real problem in pattern recognition occurs in trying to define the "ordered atoms" such as those listed in Table 3. Subjectively, you quite naturally look for regions like any fully ordered phase at low temperatures, if such a phase is known. Even then, defining the region is difficult unless the shape is simple and there is not much overlap—which is fortunately the case in most systems we have examined! In Cu_3Au for example, as we shall see, it is not. For the fully ordered phase each Au atom is surrounded by 12 Cu's. But Au atoms in boundaries of any ordered regions in a partially ordered state will not have twelve Cu neighbors, so a relaxed definition is needed for quantitative searches—somewhere between 9 (the random

TABLE 3. Results for Cu_3Au at $450°C$.*[1]

	Experimental values	4,000-atom model							16,000-atom model[†]
		Results obtained with the computer							
		From initially ordered configuration using				From initially random configuration			
		α_1	α_1, α_2	$\alpha_1, \alpha_2, \alpha_3$	α_1	α_1, α_2	$\alpha_1, \alpha_2, \alpha_3$	
α_1	-0.195	-0.1942	-0.1949	-0.1947	-0.1958	-0.1951	-0.1942	-0.1949
α_2	$+0.215$	$+0.2942$	$+0.2168$	$+0.2160$	$+0.2689$	$+0.2168$	$+0.2147$	$+0.2152$
α_3	$+0.003$	$+0.0045$	$+0.0365$	$+0.0031$	$+0.0161$	$+0.0379$	$+0.0030$	$+0.0029$
α_4	$+0.077$	$+0.1395$	$+0.0627$	$+0.1329$	$+0.0858$	$+0.0511$	$+0.1313$	$+0.1407$
α_5	-0.052	-0.0704	-0.0800	-0.0662	-0.0855	-0.0729	-0.0671	...
α_6	$+0.028$	$+0.0610$	-0.0173	$+0.0487$	-0.0093	-0.0296	$+0.0507$	$+0.0577$
Cu-Cu-Cu		10494	10466	10472	10442	10467	10481	10468
Cu-Cu-Au		19022	19094	19080	19150	19087	19061	19088
Cu-Au-Au		2474	2414	2424	2374	2425	2435	2421
Au-Au-Au		10	26	24	34	21	23	23
Cu-Cu-Cu-Cu		1242	1220	1224	1204	1223	1229	1225
Cu-Cu-Cu-Au		5526	5586	5576	5626	5575	5565	5578
Cu-Cu-Au-Au		1222	1168	1176	1136	1181	1183	1177
Cu-Au-Au-Au		10	26	24	34	21	23	22
Au-Au-Au-Au		0	0	0	0	0	0	1
No. of ordered Au atoms		558	473	473	544	461	462	465
No. of ordered Cu atoms		857	915	1102	1045	948	1121	1113
Number of jumps		750	1536	2273	3693	3473	4127	14705

*Data from S. C. Moss: *J. Appl. Phys.*, **12**:3547 (1964).
†The results have been reduced to a 4,000-atom model.

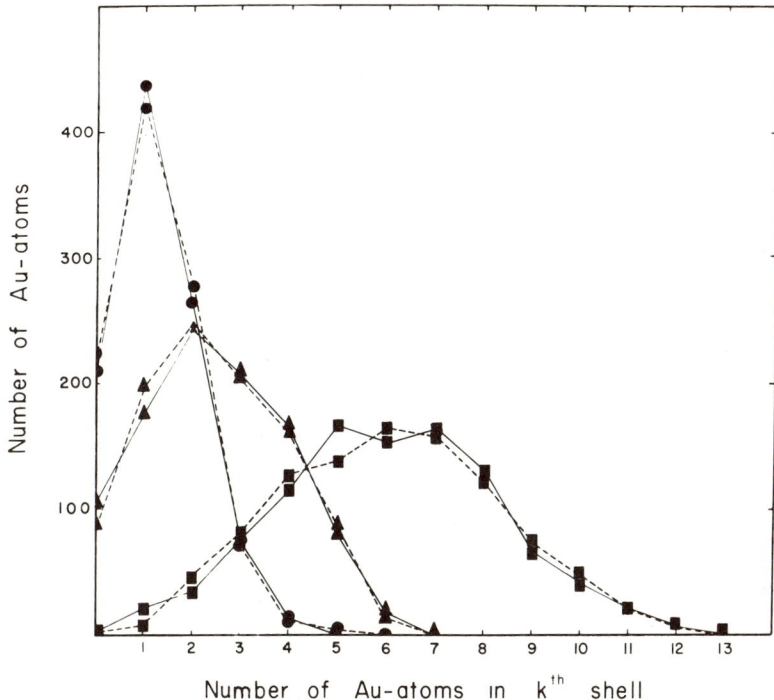

Fig. 2 *Atomic distributions in the first three shells for Cu_3Au at $450°C$.*[1]
● = *1st shell;* ▲ = *2nd shell;* ■ = *3rd shell;* ——— *initially ordered;* ----- *initially near-random.*

value) and 12. This leads to some arbitrariness in the shape and the best one can do it seems, is just to try different definitions in complex situations.

4.2. Synthesized Structures

As a final test, we constructed a series of different alloys with clusters or ordered regions of specific shape and composition, calculated the pair probabilities and used these to generate configurations. One configuration is illustrated in Fig. 3. The printouts were examined with masks of a certain shape—that of the original region *or*, in some cases, another shape. Similar procedures were followed for a random alloy. The numbers and size of circles in the figure give the fraction of sites correctly occupied by 1 atoms in a number of such regions. A lower limit of the number of 1 atoms to be included

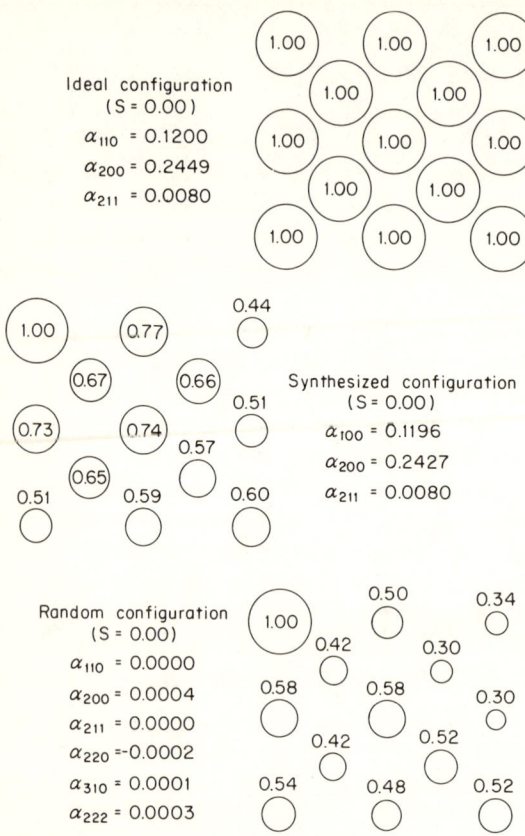

Fig. 3 *Average domain configurations for simulated f.c.c. 1_30 alloy. In (a) is shown one of the 50 regions inserted on $\{100\}$ planes of a model of 4,000 atoms. The numbers and the size of the circles indicate the probability of finding a 1 atom. In (b) the probabilities are given for the sites of the configurations found in a simulation with the first three α's and S. In (c) the probabilities for these configurations are given for a near-random alloy ($\alpha_1 = \alpha_2 = \alpha_3 = S = 0.0$).*

as a region was used to preclude counting background—this number was 4 or 5 for regions containing 9 or more 1s, 3 for random alloys. There is a very clear trend toward the original configuration. Models involving up to 108,000 atoms were tried, with regions up to 256 atoms, with similar success.[3]

All of these tests give us the warm feeling that we can indeed trust this approach to provide us with some good information on the "gestalt" of local order.

5. RESULTS FOR REAL SYSTEMS

5.1. Ordering Systems

We shall be concerned with both short-range order above T_c, and partial long-range order below T_c. What is short-range order? Is it local regions of high-long-range order suspended in a near-random matrix, a mass of small contiguous ordered regions, or is it something else? Below T_c when order is not perfect, how are atoms on wrong sites arranged? The presence of diffuse scattering[14-16] indicates the arrangement is not random.

5.1.1. The Cu-Au System.[1, 2, 14-16]

We shall first discuss results above T_c. Much of the quantitative data for 450°C has been presented in Table 3, where it can be seen that about 46 percent of the atoms are in ordered regions. This increases by 20 percent as the temperature decreases to 405°C—about 10°C above T_c. These regions have a size of 1 to 8 cells (3.7–30.0Å) along $<100>$ directions. The average size, about 12Å, is in agreement with that from the half-breadth of the diffuse scattering. The order is like Cu_3Au below T_c—Au atoms at corners, Cu at faces of a cubic cell. A typical region is shown in Fig. 4. These volumes are quite irregular and often (as shown by the shaded regions) have antiphased portions, associated with a shift of $\frac{1}{2}<110>$ in the plane of the boundary, so that Au and Cu interchange sublattices across a boundary. The regions are strikingly like those found by Beeler and Delaney[17] in Monte-Carlo experiments to simulate the initial stages of ordering below T_c. $CuAu_3$ was found to be quite similar to Cu_3Au. For CuAu the regions are small cuboidal shapes with many boundaries; some are of the "shift" variety described above, and some are associated with a 90° rotation. The ordered structure consists of layers (stacked in a $<100>$ direction) of Au then Cu, then Au, etc., so that such boundaries become possible.

Some typical results below T_c for Cu_3Au and $Cu_{72}Au_{28}$ are shown in Fig. 5. We will first discuss the stoichiometric alloy Cu_3Au, heat-treated to give $S = 0.8$, 10–15°C below T_c. Some 300 atoms out of 4,000 are on wrong sites. Most are in two-dimensional plates on $\{100\}$ planes with $<100>$ edges. There are 11×10^{20} per cc of these, containing an average of five atoms. (About one percent of the Au atoms on Cu sites are isolated.) These two-dimensional regions have mostly Cu atoms in the boundaries.

Two nonstoichiometric alloys have also been studied, $Cu_{72}Au_{28}$ (see Fig. 5b,c) and $Cu_{77}Au_{23}$. When both alloys are heat-treated for maximum long-range order* the sublattice of the species in excess is

---

*For $Cu_{72}Au_{28}$ $S = 0.88$ at maximum order, for $Cu_{77}Au_{23}$ $S = 0.97$.

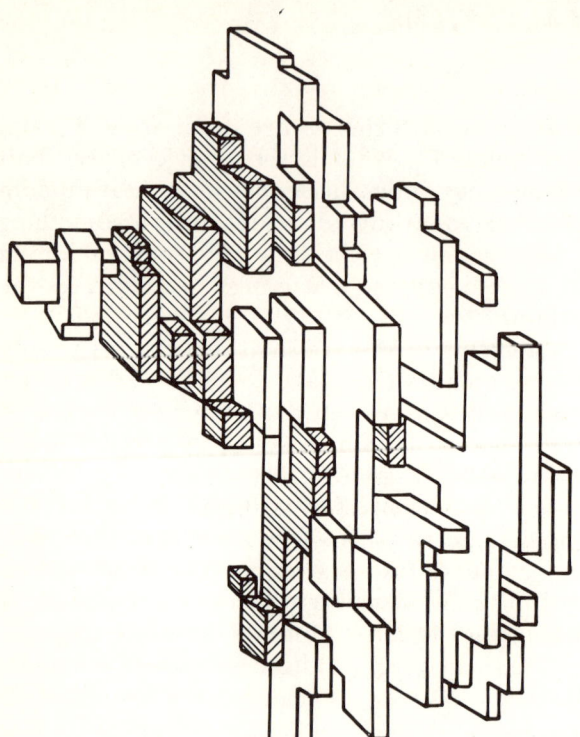

Fig. 4 *An ordered region in Cu_3Au above T_c (at $450°C$). $\alpha_1 - \alpha_3$ used in simulation from S. C. Moss, J. Appl. Phys., 12:3547 (1964). The shaded volume has its Au atoms on a different sublattice than the unshaded volume.*[1]

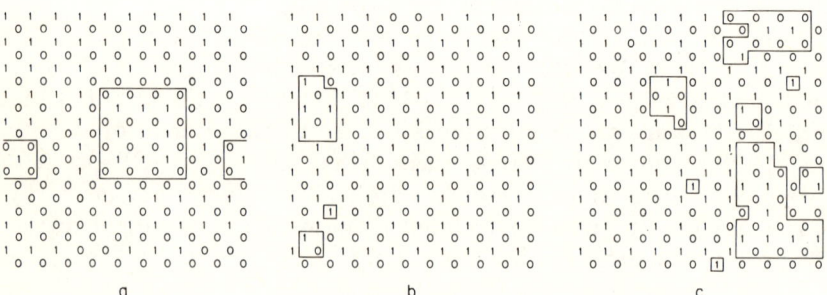

Fig. 5 *Cu_3Au below $T_c - \{100\}$ sections—model of 4,000 atoms. Data for $\alpha_1 - \alpha_3$ from L. H. Schwartz and J. B. Cohen, J. Appl. Phys., 36:598 (1965).*[13] *(a) Cu_3Au, $S = 0.80$; (b) $Cu_{72}Au_{28}$, $S = 0.83$; (c) $Cu_{72}Au_{28}$, $S = 0.71$, $1 = Au$.*

filled and the remaining excess atoms are distributed at random on the other sublattices. As $Cu_{72}Au_{28}$ is equilibrated at higher temperatures, two-dimensional ordered regions develop on {100} planes with <100> edges. A large preponderance of the excess Au atoms appear in the boundaries of these regions. (For $S = 0.83$ there are 6×10^{20} regions per cc containing an average of five atoms.) As the temperature is raised to within 10 to 15°C of T_c, where $S = 0.71$, there are 18×10^{20} planar regions containing an average of four atoms—but 70 percent of the atoms are in clustered platelets, i.e., most of the misplaced atoms are on three-dimensional regions like those above T_c. Some {111} boundaries appear.

For Cu_3Au $S = 0.8$, $Cu_{72}Au_{28}$ $S = 0.71$, and $Cu_{77}Au_{23}$ $S = 0.87$, there are the same number of wrongly occupied Au sites. This can be ascertained by solving Eq. (5) and a mass balance for r_α, r_β. Thus, a comparison of all three configurations is interesting. Of the 225 wrongly located Cu atoms in a 4,000 atom model of the Cu-rich alloy, about a quarter were in small Cu clusters, another quarter were isolated (but there were *no* isolated Au atoms) and the remaining half were in two-dimensional and three-dimensional ordered regions (4×10^{20}/cc) containing an average of six atoms, about the size for the other alloys.

5.1.2. **The Cu-Pt System.** At low temperatures there are two well-known ordered phases in this system, one like Cu_3Au (Ll$_2$) and the Ll$_1$ phase, CuPt, a rhombohedral structure with alternating (111) layers of Cu and Pt. Kaplow[18] obtained data on a series of alloys at different temperatures, all above T_c. At the Cu_3Pt composition we find the ordered regions more platelike than for Cu_3Au, and much smaller (five or six atoms), even though the measurements were made in a range above T_c similar to that for Cu_3Au. For CuPt there were three-dimensional regions like the ordered phase. The ordered volume was always much less than for Cu_3Au. This volume varied with temperature but less for Cu_3Pt. For compositions between these two, both regions were observed, generally contiguous.

5.1.3. **Rodlike Systems.** The Cu-Al and Ni-Pd systems are continuous solid solutions to room temperature, but both exhibit local order.[19,20] They are strikingly similar in atomic configurations consisting of rods of second neighbors in <100> directions as shown in Fig. 6 for Cu-14 at. percent Al. The 1 atoms are Al. (In the Ni-Pd system, Ni-base alloys with 25, 50, 75 at. percent Pd were examined. For the first alloy, order was quite small and the configuration was not as certain.)

For the Cu-Al alloy, about 85 to 90 percent of the atoms are in such rods. However, these rods overlap and about 54 percent of the

Fig. 6 Cu-14.5 at. percent Al. (100) section—model of 4,000 atoms. 1 = Al. Data for $\alpha_1 - \alpha_3$ from C. R. Houska and B. L. Averbach, J. Appl. Phys., 30:1525 (1959).[1]

Al are in tetrahedra made up of Al atoms connected along <100> and <211> directions.[13] This is like a model first suggested by Borie and Sparks based on the Al atom arrangement in metastable (β_1) Cu_3Al.[21] In view of the larger percentage of atoms in rods rather than in connected tetrahedra and the absence of *any* isolated tetrahedra in a 4,000 atom model, this description is clearly not as good as the one involving rods. Work on more dilute alloys however, might remove any lingering doubts.

Still a third alloy system exhibits this rodlike character in solid solutions. This is Ni-10 at. percent W, despite the fact that there is an intermetallic Ni_4W in the system. The rods are shorter: 20 percent of the W atoms are in rods only two atoms long, with 31 percent in longer sizes. Only a few containing ten atoms were found. *No* regions like Ni_4W could be found. (In a random alloy of the same composition, 37 percent of the W were in two-atom rods, with 15 percent in longer rods.)

5.1.4. The Au-Pd System. Lin, Spruiell and Williams[13] have recently examined the diffuse scattering from an alloy of 40 at. percent Pd in this system of continuous solid solutions. The variation of the α's did not seem to fit well with ordered structures like Cu_3Au, Ni_4Mo, $CuAu$, or an A_3B alloy of the DO_{22} structure. They decided to search for the regions that Clapp's calculations indicated to be most likely; shown in Fig. 7, it is a long-period structure of two unit cells, with an antiphase boundary caused by a ½<110> shift on the (100) plane between the two cells. In 8,000 atoms there were 54 such regions, five times the number found in a near-random alloy, in agreement with the ratio for these two states calculated by Clapp (see Sec. 4). Thus, like Cu_3Au above T_c, there is a tendency for some periodic antiphased regions in this alloy system. This particular study

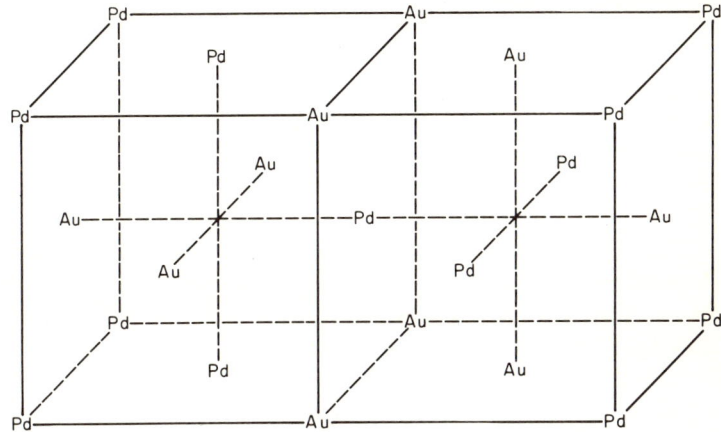

Fig. 7 *Typical region found in a simulation of 8,000 atoms with $\alpha_1 - \alpha_5$ of an Au—40 at. percent Pd alloy, after W. Lin, J. E. Spruiell, and R. O. Williams.*[13]

also shows how helpful Clapp's procedure can be in the problem of pattern recognition.

In *all* the systems examined, short-range order consists of ordered regions in a near-random matrix.

5.2. Clustering Systems

5.2.1. The Al-Ag System. Gragg[3] has recently examined in detail the diffuse scattering from an alloy of Al-5 at. percent Ag alloy aged at 110°C (below the solvus) for 14 hours. The scattering thought to be due to internal ordering in the zones was shown to be due to displacements of atoms from lattice sites. The computer simulation was the first to employ 108,000 atoms. Some 35 zones were found in the model, corresponding to a zone density of $\approx 2 \times 10^{19}$/cc. The zones were all octahedra with an average composition of 68 at. percent Ag and an average size along <100> directions of 19 Å. The matrix contained no Ag. This octahedral shape explained many unusual features of the diffuse scattering that have been reported in the literature, but not interpreted.

5.2.2. The Fe-Mo System. T. Ericsson and co-workers have examined two aged alloys, Fe with 3.9 and 6.1 at. percent Mo.[22,23] The clusters in both these alloys are irregular and quite dilute in Mo. For 3.9 pct. Mo, the median number of Mo atoms in a zone is 8, the maximum is 29, and the average concentration of Mo in zones is ≈10 at. percent. For the alloy of 6.1 at. percent Mo these numbers are 12, 66, and 10; the zones are just larger and more numerous in the more concentrated alloy and are not higher in Mo content. The smallest zones bear some resemblance to Fe_2Mo, the equilibrium phase. The very dilute Mo concentration in the zones seems to explain why aging produces no hardening, a fact reported in Ref. 24.

The Mössbauer spectrum for aged Fe-6 at. percent Mo was determined by Marcus, Fine, and Schwartz.[25] In Table 4a the relative areas of the peaks found (last column) are compared to that expected for a random alloy (second column) and that from a count of computer simulations with 16,000 atoms (third column). The decision to consider the peaks as due to Fe atoms with none, one, or two atoms in the first two shells is that of Marcus et al.[25] Consider just the fraction of Fe atoms with no Mo neighbors in the first shell. The measured fraction is much *less* than a random alloy or the computer simulation. But, the location of diffuse scattering under Bragg peaks clearly indicates this alloy is clustering, so that there should be *more* Fe atoms with no Mo neighbors than for a random alloy. A much better interpretation is given in Table 4b, where

TABLE 4. Comparison of Atomic Distributions with Interpretation of Mössbauer Spectra in Ref. 25 for Fe-6.1 at. percent Mo. (Table from Ref. 23.)

	a		
Fraction of Fe atoms with indicated number of Mo neighbors in first two shells	Random (Calculated)	Clustered, from computer simulation	From Mössbauer-pattern[25]
0	0.438	0.498	0.384
1	0.402	0.296	0.561
2	0.160	0.133	0.055
>2	...	0.072	...
	b		
Fraction of Fe atoms with indicated number of Mo neighbors in first three shells	Random (Calculated)	Clustered, from computer simulation	From Mössbauer-pattern[25]
0	0.25	0.36	0.384
1 or 2 (with 1 in each of two shells)	0.64	0.55	0.561
2 in either first, second, or third shell.	0.11	0.09	0.055

three-neighbor shells are employed. It appears that these simulations can be particularly useful in understanding the complex Mössbauer spectra from alloys.

5.2.3. **The Cu-Ni System.** This alloy system is a continuous series of solid solutions with some indication of a possible miscibility gap near room temperature.[26] The magnetic and optical properties have attracted considerable attention of late. (See Refs. 27–31 for recent work and summaries of the literature.) In particular, Cu-rich alloys exhibit superparamagnetism and, near the equi-atomic composition, experiments with neutron diffraction indicate large magnetic spin clusters, qualitatively similar to what is found for dilute solutions.[27] There is considerable debate as to whether these clouds are solely due to magnetic interactions or whether clustering in the solid solution is the cause (or perhaps both are involved). In view of the fact that the magnetic behavior is affected by heat treatment, plastic deformation, and radiation, there seems to be no question but that some effect of chemical concentration is present. Also, attempts to calculate the properties assuming a random alloy are not entirely satisfactory.[30]

The justification used by some authors for neglecting the effect of local order in this system is that measurements of diffuse scattering indicate that only α_1 is different from zero for $Cu_{0.525}Ni_{0.475}$;[26] thus many concluded that as the polarization clouds contained 30 to 50 atoms[27,28] the absence of a chemical effect beyond the first shell of 12 neighbors precludes such an explanation. However, underlying this thinking is the assumption that the regions are spherical! Computer simulation looked like it might prove interesting.

The results for the distribution of Ni neighbors is shown in Fig. 8. The alloy consists of Ni-rich rods and two and three-dimensional platelike clusters of Ni on {111} planes with two such regions often crossing. The average size of the crossed planes was 17 (including only Ni atoms with more than 8 Ni first neighbors) compared to nine or ten atoms in the random alloy. The spacing was 12Å (compared to $\approx$ 13.5Å from neutron scattering[27]). It is because of the platelike nature of many of the regions that higher order α's are zero! These regions are clearly ample in size to explain the observed effects. And, it is not surprising that only Cu-rich (not Ni-rich) alloys show unusual magnetic effects. It would be interesting confirmation of these configurations if anisotropy in magnetic behavior, or streaking in small-angle scattering patterns, could be detected. Interestingly, for Ni-20 at. percent Cu, using the data of Ref. 3, there are also crossed {111} plates and some rods—but of Cu in this case. Their density is about the same as for the more Ni-rich alloy, but the

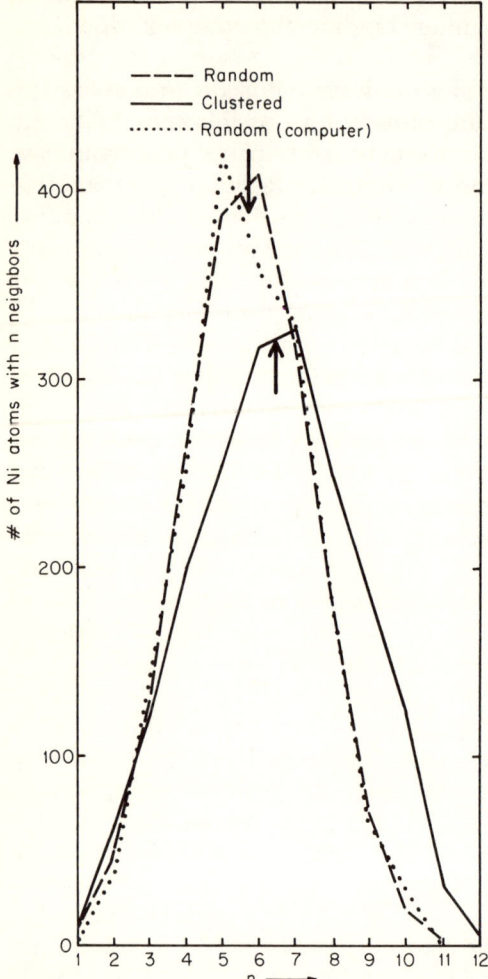

Fig. 8 *Atomic distribution in the first-neighbor shell for $Cu_{0.525}Ni_{0.475}$. Computer simulation of 4,000 atoms using $\alpha_1 - \alpha_3$ from B. Mozer, D. T. Keating, and S. C. Moss: Phys. Rev., 175: 868 (1968). Note the agreement between the distribution for a random alloy produced in the computer, and calculated from the binomial distribution. Arrows indicate the average values calculated from α_1.*

plates generally contain only six or seven Ni atoms (and the rods four or five atoms). No such regions are apparent in the corresponding random alloy.

Incidentally, even in dilute alloys with large "magnetic clouds" the possibility of chemical clustering should not be ignored. Also, the results we have presented on partially ordered alloys indicate that in these cases too, sizeable chemical regions may play a key role in understanding the behavior.

5.3. An Oxide

$FeLiO_2$ has two ordered phases,[32] a tetragonal phase Q_I consisting of (001) planes of an ordered arrangement of Fe^{2+} and Li^+, and a metastable phase Q_{II} appearing in the transition from the disordered cubic NaCl phase to Q_I. Q_{II} is like Q_I, with antiphase boundaries on {100} planes due to ½<110> shifts. From the measured short-range order parameters[32] we find the disordered state to be about 80 percent of small regions of Q_{II} (in agreement with the authors' suggestion), 2 to 10 planes thick, containing 7 to 34 atoms.

It would be interesting to examine the computer models generated from theoretical models, especially to compare the size and/or spacing of any regions with calculated correlation lengths. This is now underway.

ACKNOWLEDGMENTS

Support initially by ARPA, through Northwestern University's Materials Research Center, and now by the NSF is gratefully acknowledged. ONR kindly supplied funds to make the trip to this conference possible. This effort was initiated by Dr. P. Gehlen as part of his doctoral dissertation. It was a pleasure to have him as a colleague. Messrs. H. Berg and A. Krawitz helped develop some of the computer programs and made some of the runs. Finally, we thank Prof. J. Spruiell for letting us see some of his work prior to publication.

REFERENCES

1. Gehlen, P. C., and J. B. Cohen: *Phys. Rev.*, **139**:A844 (1965).
2. Gehlen, P. C.: Doctoral dissertation, Northwestern University, Evanston, Ill., 1966.

3. Gragg, J. E., Jr.: Doctoral dissertation, Northwestern University, Evanston, Ill., 1970.
4. Williams, R. O.: ORNL-TM-2866, March 1970.
5. Buerger, M. J.: "Vector Space and its Applications in Crystal Structure Investigation," John Wiley & Sons, Inc., New York, 1959.
6. ——: in C. N. Ramachandran (ed.), "Advanced Methods of Crystallography," Academic Press, Inc., New York, 1964.
7. ——: *Proc. Nat. Acad. Sci., U.S.,* **136**:376 (1950).
8. Gass, S. I.: "Linear Programming," McGraw-Hill Book Co., New York, 1964.
9. Cohen, J. B.: in H. I. Aaronson (ed.), "Phase Transformations'" p. 561, ASM, Novelty Park, Ohio, 1970 (App. II by R. W. Bednarz, J. Gragg, Jr., and J. B. Cohen).
10. Bednarz, R. W.: Series in Applied Mathematics, Rep. 70-9, Northwestern University, Evanston, Ill., April 1970.
11. Clapp, P.: *J. Phys. Chem. Solids,* **30**:2589 (1969).
12. Keating, D. T., and B. E. Warren: *J. Appl. Phys.,* **22**:286 (1951).
13. Lin, W., J. E. Spruiell, and R. O. Williams: *J. Appl. Cryst.,* **3**:297 (1970).
14. Chipman, D. R.: *J. Appl. Phys.,* **27**:739 (1956).
15. Schwartz, L. H., and J. B. Cohen: *Ibid.,* **36**:598 (1965).
16. Gehlen, P. C., and J. B. Cohen: *Ibid.,* **40**:5193 (1969).
17. Beeler, J. R., and J. A. Delaney: *Phys. Rev.,* **130**:962 (1963).
18. Kaplow, R.: Doctoral dissertation, Massachusetts Institute of Technology, Cambridge, Mass., 1958.
19. Houska, C. R., and B. L. Averbach: *J. Appl. Phys.,* **30**:1525 (1959).
20. Lin, W., and J. E. Spruiell: *J. Appl. Phys.,* to be published.
21. Borie, B. S., and C. J. Sparks: *Acta Cryst.,* **17**:827 (1964).
22. Ericsson, T., and J. B. Cohen: *Ibid.,* **A27**:97 (1971).
23. ——, S. Mourikis, and J. B. Cohen: *J. Mat. Sci.,* **5**:901 (1970).
24. Urakami, A., H. L. Marcus, M. Meshii, and M. E. Fine: *Trans. ASM,* **60**:344 (1967).
25. Marcus, H. L., M. E. Fine, and L. H. Schwartz: *J. Appl. Phys.,* **38**:4750 (1967).
26. Mozer, B., D. T. Keating, and S. C. Moss: *Phys. Rev.,* **175**:868 (1968).
27. Hicks, T. J., B. Rainford, J. S. Kouvel, G. G. Low, and J. B. Comly: *Phys. Rev. Lett.,* **22**:531 (1969); Kouvel, J. S., and J. B. Comly: *Ibid.,* **24**:598 (1970).
28. Robbins, C. G., H. Claus, and P. A. Beck: *Ibid.,* **22**:1307 (1969).
29. Cohen, J. B.: *J. Mat. Sci.,* **4**:1012 (1969).
30. Perrier, J. P., B. Tissier, and R. Tournier: *Phys. Rev. Lett.,* **24**:313 (1970).
31. Cable, J. W., E. O. Wollan, and H. R. Child: *Ibid.,* **22**:1256 (1969).
32. Brunel, M., and F. de Bergevin: *J. Phys. Chem. Solids,* **30**:2011 (1969).

DISCUSSION on *Paper Presented by J. B. Cohen*

H. E. COOK: Have you Fourier-transformed your resulting atomic distribution to see how well its intensity distribution corresponds to the experimental intensity data from which you obtained the first three SRO parameters? The SRO peak should be wider than the experimental peak since only three SRO parameters, i.e., three Fourier coefficients, were obtained from the original data.

J. B. COHEN: We have not done this very seriously as yet, but I am not sure that one can say the distribution will *always* be broader. The reason I say this is that the "linear-programming" procedures we have described in our paper indicate there is a *range* of solutions for higher neighbor pair probabilities, given the first three. Thus, the simulation can give a sharper or broader pattern.

C. N. YANG: If the system has only nearest-neighbor interactions, your procedure stops when the correct energy is reached. Have you investigated

1. Whether the ensemble of all final states reached by your procedure is the microcanonical ensemble?

2. If the answer to (1) is yes, is your procedure the most economical way of reaching such an ensemble?

3. Did you discuss the variance of the distributions you obtained?

J. B. COHEN: We are not sure what kind of ensemble we are dealing with, as this is a very subtle question. It would be nice to know the answer, or how to adjust the jump procedure to arrive at a better defined ensemble, so that the multisite correlations would be more useful for comparison with theoretical calculations. We can only say that for certain alloys we have made several runs from different initial states with different numbers of atoms and have obtained the same answer so that there is some indication we have a physically highly-probable configuration. It is the topology—the physical picture of what local order is—that has been our primary motivation in the past. As in the explanation of the complex Mössbauer pattern in Fe-Mo alloys, this can be very helpful. Our jump procedure is definitely not the most efficient for two reasons. The first is a practical one. It is necessary to make many jumps to destroy the influence of the initial state. Secondly, by making many jumps we can "maximize the entropy," i.e., attempt to obtain the most probable configuration. It certainly would be nice to have an algorithm that assures a particular ensemble and would provide more definitive numbers for comparison with theoretical approaches.

Professor Cowley's group has available some information on variances of multisite configurations.

D. de FONTAINE: In your computer simulation of the CuNi system, you show domains consisting of intersecting platelets. How frequent is this configuration compared to configurations consisting of noninteresting platelets?

J. B. COHEN: I should say that the simulation indicates rods as well as plates. Concerning the plates, I can only say here that these were *generally* crossed. If you wish a number, or fraction, we can supply this.

FERROELASTICITY DUE TO HYDROGEN IN METALS

G. Alefeld [*]

Institut für Festkörperforschung
Kernforschungsanlage Jülich, Germany

INTRODUCTION

The phase transitions of most alloys are barely understood today. In general, phase diagrams are established purely empirically. Rarely a detailed understanding for the physical reason for the transitions exists. In this paper it will be shown that the new concept of ferroelasticity[1-3] provides access to the understanding of a class of relatively complicated-looking phase diagrams, namely, for metal-hydrogen systems. First, analogies between electric and elastic phenomena will be demonstrated. Then the temperature dependence of the elastic susceptibilities is calculated. The results are compared with experiments. Furthermore, it is shown that the relaxed elastic constants violate elastic stability criteria at characteristic temperatures. In Sec. 1 the average elastic interaction energy is calculated in mean field approximation. As an example, the phase diagram for dipoles which are pure dilatation centers is calculated. Finally the

---

[*] New address: Physics Department, Technische Universität München, D-8046 Garching, Germany.

phase diagrams for general dipoles are given schematically and compared with experimental phase diagrams.

1. ELECTRIC AND ELASTIC ANALOGS

For a long time in the last century, there was a strong effort to get more insight into electromagnetic phenomena by using analogies to an elastic continuum. In the meantime the understanding of electromagnetism has proceeded further than that for elasticity, especially if we consider crystals with defects. Therefore it is useful today to apply the opposite procedure, namely, to start from e.g., electric phenomena and to translate the electric concepts into elastic language.[4,5] By establishing electric and elastic analogs, not only a conceptual unification in the description of material properties can be reached but furthermore deeper insight into physical mechanisms can be gained. The concepts of diaelasticity and paraelasticity have been formulated by Kröner.[4,5] An extension to ferroelasticity is possible quite naturally.

Table 1 shows the correspondence of elastic and electric concepts and equations. The undistorted harmonic elastic continuum or lattice corresponds to the vacuum. Applying forces k_i to the lattice, e.g., by introducing defects corresponds to creating electric charges q. The elastic displacements u_i correspond to the electric potential U. The energy of a monopole in the corresponding field is therefore qU or $k_i u_i$ respectively (row (9)). The two columns for elasticity contain in principle identical equations, but written such that either strain or stress corresponds to the electric field. Below we will give arguments in favor of one or the other analogy. In row (8) the dipole moments are defined as the first moment of the charges and forces resp. The sum n extends over all forces which the defect exerts on the neighboring atoms. Since the torque $M_{ij} = \sum_n (k_i^{(n)} x_j^{(n)} - x_i^{(n)} k_j^{(n)})$ of forces exerted by a defect in general is zero, the elastic dipole moment is a symmetric tensor. In Fig. 1a and 1b simple examples of elastic dipoles are given. Analogous definitions as in row (8) hold for electric and elastic quadrupoles and so on. Kröner[4,5] usually worked with the quantity P_{ij}, whereas Nowick and Heller[13] preferred P_{ij}^*. The energies and forces are given in rows (10) and (11). The polarization is the dipole moment per unit volume. The polarizations are written for the case in which the dipoles have discrete orientations ν. The susceptibilities are defined in row (13). In row (14) σ_{ij} means the real measurable stresses. Only for homogeneous distribution of polarization is σ_{ij} identical with stresses due to external forces.[7] In

FERROELASTICITY DUE TO HYDROGEN IN METALS   341

this case these equations have a simple interpretation which is symbolically indicated in Fig. 1c and 1d and which demonstrates the analogy between surface forces (stress) and surface charges. Let us first consider defects of the sort in Fig. 1b. For homogeneous distribution over the sample and the three orientations all forces in the sample cancel on the average. If now external forces are applied, the dipoles will orient. The net forces of all dipoles now are forces on the surfaces as indicated in Fig. 1d, which act like stresses p_{kl} and add on to the external stress σ_{kl} to give the total strain $\epsilon_{ij} = S_{ijkl}(\sigma_{kl} + p_{kl})$. For dipoles of the type in Fig. 1a the forces never cancel completely. Even for completely statistical distribution and without external stresses a surface layer of forces exists (Fig. 1c), causing a strain $\epsilon_{ij} = S_{ijkl}p_{kl}$, which is a homogeneous expansion of the lattice. It should be pointed out that due to the vector character of forces in contrast to charges, an elastically "polarized" state may have surface forces on all surfaces of the sample. Spontaneous polarization manifests itself macroscopically as strain or stress, depending on the boundary conditions. D-E-hysteresis of ferroelectrics corresponds to $\sigma - \epsilon$ hysteresis of ferroelastic material.

In Table 2 the correspondence of the connection between mean field and local field is demonstrated. In determining the mean field for a given external field it is important to consider boundary conditions. Per definition, a paraelectric sample in vacuum corresponds to an elastic specimen, loaded with dipoles but imbedded into an infinite elastic continuum with identical elastic constants. Under these conditions in both cases a polarized specimen is surrounded by complicated stray fields. On the other hand if the elastic specimen is cut out of the continuum, arbitrary forces and displacements can be applied at the surface without causing any reaction of the now surrounding vacuum. No stray field exists. Analogous electric boundary conditions are given for a dielectric between metallic condenser plates. Arbitrary charges and potentials can be applied at the condenser plates without creating a stray field. If we move a dielectric between two open condenser plates the average field decreases from E_i^{ext} to $E_i^{ext} - 1/\epsilon_0 P_i$ due to surface charges. Analogously, the average strain of a homogeneously loaded specimen is given by the strain due to external forces ϵ_{ij}^{ext} and the strain caused by polarization p_{ij}, which now has the same direction as ϵ_{ij}. The average stress is the stress due to external forces only.

The local field is that field which has to be used if one calculates the energy of a dipole in the field of other dipoles and external forces. Therefore in computing the local field the contribution of the

TABLE 1. Analogies between Electric and Elastic Concepts and Equations.[5–8,13]

Electrostatics	Elasticity	
Vacuum	Harmonic elastic continuum or lattice	
Charges q Charge densities ρ	Forces k_i Force densities f_i	(1) (2)
Potential $U(r)$	Displacement $u_i(r)$	(3)
Electric field E_i $E_i = -\nabla_i U$	Strain ϵ_{ij} $\epsilon_{ij} = \text{sym } \nabla_i u_j = \frac{1}{2}(\nabla_i u_j + \nabla_j u_i)$	(4)
Dielectric displacement D_i $D_i = \epsilon_{ij} E_j$	Stress σ_{ij} $\sigma_{ij} = C_{ijkl} \epsilon_{kl}$	(5)
Energy density e $e = \frac{1}{2} D_i E_i = \frac{1}{2} \epsilon_{ij} E_i E_j$	Strain ϵ_{ij} $\epsilon_{ij} = S_{ijkl} \sigma_{kl}$	(6)
	Energy density e $e = \frac{1}{2} \sigma_{ij} \epsilon_{ij} = \frac{1}{2} C_{ijkl} \epsilon_{ij} \epsilon_{kl}$ $e = \frac{1}{2} \epsilon_{ij} \sigma_{ij} = \frac{1}{2} S_{ijkl} \sigma_{ij} \sigma_{kl}$	
Potential equation $\nabla_i D_i = \rho$ (7) + (5) + (4) yields $\nabla_i \epsilon_{ij} \nabla_j U = -\rho$	Equilibrium condition $\nabla_i \sigma_{ij} = -f_j$ (7) + (5) + (4) yields $\nabla_i C_{ijkl} \nabla_k u_l = -f_j$ for a lattice $\phi_{jl}^{mn} u_l^n = k_j^m$	(7)
Dipole moment p_i $p_i = \sum_n x_i^{(n)} q^{(n)}$	Dipole moment P_{ij}^* $P_{ij}^* = S_{ijkl}^o \sum_n x_k^{(n)} k_l^{(n)}$	(8)

Monopole energy
$$U^m = k_i u_i$$ (9)

Dipole energy
$$U^d = -P^*_{ij}\epsilon_{ij}$$ (10)

Force on dipole
$$K_j = P^*_{il}\nabla_j \sigma_{il}$$ (11)

Polarization p^*_{ij}
$$p^*_{ij} = \sum_\nu \rho^{(\nu)} p^{*(\nu)}_{ij}$$ (12)

Susceptibility χ
$$\chi_{ijkl} = \frac{\partial p^*_{ij}}{\partial \sigma_{kl}} \text{ or } \frac{\partial p_{ij}}{\partial \sigma_{kl}}$$ (13)

$$\epsilon_{ij} = S^\circ_{ijkl}\sigma_{kl} + p^*_{ij}$$ (14)

Monopole energy
$$U^m = qU$$

Dipole energy
$$U^d = -p_i E_i$$

Force on dipole
$$K_j = p_i \nabla_j E_i$$

Polarization P_i
$$P_i = \sum_\nu \rho^{(\nu)} p_i^{(\nu)}$$

$\rho^{(\nu)}$ = dipole density (cm^{-3}) with orientation ν

Susceptibility χ
$$\chi_{ij} = \frac{\partial P_i}{\partial E_j}$$

$$D_i = \epsilon^\circ_{ij} E_j + P_i$$

$$\sigma_{ij} = C^\circ_{ijkl}\epsilon_{kl} - p_{ij}$$

TABLE 2. Mean Field and Local Field.[5,7]

Electrostatics		Elasticity	
Average field E_i	$E_i = E_i^{ext} - \alpha_{ij} P_j$	Average strain ϵ_{ij}	$\epsilon_{ij} = \epsilon_{ij}^{ext} + S^o_{ijkl} p_{kl}$
E^{ext} = field due to external charges		ϵ_{ij}^{ext} = strain due to external forces	
α_{ij} = depolarization factor		$\epsilon_{ij}^{ext} = S^o_{ijkl} \sigma_{kl}^{ext}$	(15)
		Average stress	$\sigma_{ij} = \sigma_{ij}^{ext}$
		σ_{ij}^{ext} = stress due to external forces	
Local field	$E_i^{loc} = E_i + \lambda_{ij} P_j$	Local strain	$\epsilon_{ij}^{loc} = \epsilon_{ij} + \lambda_{ijkl} p_{kl}$
λ_{ij} = Lorentz field factor		$\lambda_{ijkl} = F\delta_{ik}\delta_{jl} + G\delta_{ij}\delta_{kl}$	
		for cubic systems	
$\lambda_{ij} = \dfrac{1}{3\epsilon_0} \delta_{ij}$		$F = -\dfrac{4-5\nu}{15\mu(1-\nu)} \quad G = \dfrac{1}{30\mu(1-\nu)}$	
(isotropic and cubic)		for isotropic systems	
		Local stress	$\sigma_{ij}^{loc} = \sigma_{ij} + \lambda^*_{ijkl} p^*_{kl}$
		$\lambda^*_{ijkl} = C^o_{ijkl} + \epsilon_{ijmn} \lambda_{mn\mu\nu} C^o_{\mu\nu kl}$	(16)

FERROELASTICITY DUE TO HYDROGEN IN METALS

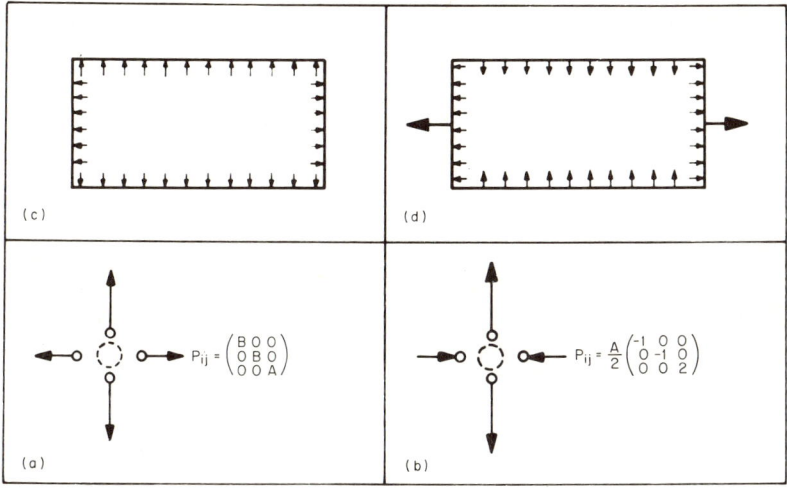

Fig. 1 *Examples for elastic dipoles and elastic polarization.*

dipole considered to the average field must be subtracted. We will sketch very briefly the computational procedure: The average strain is given by the external strain and the contribution ϵ^d of all dipoles in the volume V

$$\epsilon = \epsilon^{\text{ext}} + \sum_V \epsilon^{(d)} \tag{1}$$

The local strain is given by

$$\epsilon^{\text{loc}} = \epsilon^{\text{ext}} + {\sum_V}' \epsilon^{(d)} \tag{2}$$

(Σ' implies a sum over all but the dipole considered.)

In Eq. (2) the total volume V of the body is divided into two regions $V - v$ and v, where v is a small volume with the dipole considered in the center. In the region $V - v$ the dipoles are smeared out to a continuous dipole distribution so that the summation can be replaced by integration. Inside v the summation is done explicitly. Therefore (in symbols!)

$$\epsilon^{\text{loc}} = \epsilon^{\text{ext}} + \int_{V-v} dV + {\sum_V}' \epsilon^{(d)} \tag{3}$$

In Eq. (1) the sum is converted into an integral as well and divided up into integrals over the two regions $V - v$ and v. Thus

$$\epsilon = \epsilon^{\text{ext}} + \int_{V-v} dV + \int_{v} dV \qquad (4)$$

The volume integral over the small volume can be converted into a surface integral over the small volume. Therefore, combining Eqs. (3) and (4), ϵ^{loc} can be written as

$$\epsilon^{\text{loc}} = \epsilon - \int_{f} df + \sum_{v}{}' \epsilon^{(d)} \qquad (5)$$

This formula describes exactly the same method as used by Lorentz in calculating the local electric field: The procedure outlined above is explicitly or implicitly used in papers by Kröner[4,5] and by Siems.[9] The functions $F(c_{11}, c_{12}, c_{44})$ and $G(c_{11}, c_{12}, c_{44})$ in Table 2 are defined in Ref. 9. The equation in row (16) holds only if $\sum_{v}{}' \epsilon^{(d)} = 0$. It should be pointed out that the local electric field is larger than the average electric field by $P/3\epsilon_0$ ($= 4\pi P/3$) since for the local field the negative (depolarizing) contribution of the dipole considered is missing. In the elastic case the local field is smaller than the average field ($\lambda < 0$) since here the positive strain contribution of the dipole considered is subtracted from the mean field.

The analogy with $E_i \to \epsilon_{ij}$, which is favored by Kröner, gives elastic equations which look less complicated. Furthermore the calculation of the local field is more transparent. Yet Eqs. (14) and (15) and Eqs. (21) and (22) below will show that the susceptibilities defined as $\partial p/\partial \sigma$ have the same temperature dependence and a singularity at a phase transition as the electric susceptibility $\partial P/\partial E$, whereas $\partial p/\partial \epsilon$ has no singularity at the phase transitions. The derivative $\partial p/\partial \epsilon$ actually shows the same temperature dependence as the derivative $\partial P/\partial D$ for the electric case.

2. THE PARAELASTIC REGION

The energy of an elastic dipole, characterized by the dipole moment tensor P^{ν}_{ij} ($\nu = 1 \ldots n$, n = number of possible sites) in an

FERROELASTICITY DUE TO HYDROGEN IN METALS

external strain field plus the average strain field of all the other dipoles is given by

$$U^\nu = -P^\nu_{ij}\epsilon_{ij}^{loc} \qquad (6)$$

We assume that $\omega\tau_i \ll 1$ (τ_i = characteristic relaxation times). For small defect concentration as compared to the maximum possible concentration and with the conditions that $U^\nu < kT$ and $\Sigma_\nu \int U^\nu dV = 0$ the defect concentrations ρ^ν are connected with the energies U^ν by the relations[3]

$$\rho^\nu - \frac{\rho_0}{n} = -\frac{\rho_0 U^\nu}{nkT} \qquad (7)$$

The total average strain is given by

$$\epsilon_{ij} = S_{ijkl}\left[\sigma_{kl} + \left(\rho^\nu - \frac{\rho_0}{n}\right)P^\nu_{kl}\right] \qquad (8)$$

Here the strain $(\rho_0/n)S_{ijkl}\sum_\nu P^\nu_{kl}$, which one gets without stress by introducing the defects into the lattice, has been subtracted. If we ignore for the moment the Lorentz correction (Table 2) and replace ϵ^{loc} by the average field ϵ we can immediately combine Eqs. (6) to (8) and solve for the stress σ with the following result

$$\sigma_{ij} = \left(C_{ijkl} - \frac{\rho_0}{nkT}P^\nu_{ij}P^\nu_{kl}\right) = C'_{ijkl}\epsilon_{kl} \qquad (9)$$

For a set of dipole moment tensors with

$$P^\nu_{ij} = \begin{pmatrix} A & 0 & 0 \\ 0 & B & 0 \\ 0 & 0 & B \end{pmatrix} \text{ and cyclic } (\nu = 1, 2, 3) \qquad (10)$$

which apply for the octahedral and tetrahedral sites in b.c.c. lattices as well as for the octahedral sites in f.c.c. lattices (with $A = B$), we get from Eq. (9) the following relaxed elastic moduli C'_{ij}

$$C'_{11} = C_{11} - \frac{\rho_0(A+2B)^2}{9kT} - \frac{2\rho_0(A-B)^2}{9kT} \qquad (11)$$

$$C'_{12} = C_{12} - \frac{\rho_0(A+2B)^2}{9kT} + \frac{\rho_0(A-B)^2}{9kT} \tag{12}$$

$$C'_{44} = C_{44} \tag{13}$$

The minus sign indicates that for decreasing temperature elastic stability criteria are violated. Before we discuss this point we will state the corresponding equations which one gets taking into account the Lorentz correction for the local field. After a straightforward but somewhat lengthy calculation one finds

$$C'_{11} = C_{11} - \frac{\rho_0(A+2B)^2}{9k\left(T - T_G \frac{\gamma_G - 1}{\gamma_G}\right)} - \frac{2\rho_0(A-B)^2}{9k\left(T - T_S \frac{\gamma_S - 1}{\gamma_S}\right)} \tag{14}$$

$$C'_{12} = C_{12} - \frac{\rho_0(A+2B)^2}{9k\left(T - T_G \frac{\gamma_G - 1}{\gamma_G}\right)} + \frac{\rho_0(A-B)^2}{9k\left(T - T_S \frac{\gamma_S - 1}{\gamma_S}\right)} \tag{15}$$

$$C'_{44} = C_{44} \tag{16}$$

The numerical factors γ_G and γ_S are given by

$$\gamma_G = 1 + \frac{F + 3G}{S_{11} + 2S_{12}} \tag{17}$$

$$\gamma_S = 1 + \frac{F}{(S_{11} - S_{12})} \tag{18}$$

The temperatures T_G and T_S stand for

$$kT_G = \tfrac{1}{3} \gamma_G \rho_0 (A + 2B)^2 (S_{11} + 2S_{12}) \tag{19}$$

$$kT_S = \tfrac{1}{3} \gamma_S \rho_0 (A - B)^2 (S_{11} - S_{12}) \tag{20}$$

Instead of elastic moduli C_{ij} the elastic coefficients S_{ij} are often measured. Using the relations $c_{11} - c_{12} = (s_{11} - s_{12})^{-1}$, $c_{11} + 2c_{12} = (s_{11} + 2s_{12})^{-1}$, $c_{44} = s_{44}^{-1}$ for cubic systems, we can solve Eqs.

(14)–(16) for s_{ij} yielding

$$S'_{11} = S_{11} + \frac{\rho_0 (A + 2B)^2 (S_{11} + 2S_{12})^2}{9k(T - T_G)} + \frac{2\rho_0 (A - B)^2 (S_{11} - S_{12})^2}{9k(T - T_S)} \quad (21)$$

$$S'_{12} = S_{12} + \frac{\rho_0 (A + 2B)^2 (S_{11} + 2S_{12})^2}{9k(T - T_G)} - \frac{\rho_0 (A - B)^2 (S_{11} - S_{12})^2}{9k(T - T_S)} \quad (22)$$

$$S'_{44} = S_{44} \quad (23)$$

Eqs. (21) to (23) can also be derived directly by solving Eqs. (7) and (8) for $\epsilon_{ij} = S'_{ijkl}\sigma_{kl}$, as has recently been done by Siems.[9] The temperature dependence of the elastic coefficients is Curie Weiss-type. T_G and T_S define temperatures, at which the elastic coefficients s_{11} and s_{12} approach infinity. The physical meaning of the three terms in Eqs. (14) to (16) and (21) to (23) respectively is as follows: The first term characterizes the response of the dipole-free lattice. The second term is due to the recently discovered Gorsky effect [1,10,11] (long-range diffusion of centers of dilatation with the strength $A + 2B$ under the action of a gradient of dilatation), whereas the third term is due to the Snoek effect (local response of elastic dipoles to a strain, i.e., parallel alignment). It is significant to note that the characteristic temperatures T_S and T_G for Snoek and Gorsky relaxation are determined by the same components of the dipole moment tensor as are the individual relaxation strengths (susceptibility). Except for the numerical values γ the temperature T_S has previously been derived by Zener[12] and Nowick and Heller.[13] A general expression for T_G, which also applies for large concentrations is given in Ref. 3.

3. COMPARISON WITH EXPERIMENTS

The Gorsky relaxation of hydrogen and deuterium in V, Nb, and Ta has been investigated recently.[1,10,11] The predicted Curie-Weiss temperature dependence of the elastic susceptibility has been verified very well, even close to the critical point of H in Nb (see Sec. 6 and 7). Since the relaxation strength as well as T_G have been measured a check of Eq. (19) which is based on the assumption of purely elastic

TABLE 3. Experimental Results for the Trace of the Dipole Moment Tensor and for T_G.

	V–H	V–D	Nb–H	Nb–D	Ta–H	Ta–D
$(A + 2B)/3$ eV	2.4 ± 0.2	2.5 ± 0.2	3.8 ± 0.3	3.8 ± 0.3	3.0 ± 0.1	2.9 ± 0.2
T_G[°K/1 at %]	20.5	20.5	25	25	20 ± 2	20 ± 3
γ (exp)	0.41 ± 0.11	0.38 ± 0.11	0.29 ± 0.08	0.29 ± 0.08	0.42 ± 0.07	0.45 ± 0.12

FERROELASTICITY DUE TO HYDROGEN IN METALS

interaction is possible. In Table 3 the experimental values for $(A + 2B)/3$ and for T_G are listed. From these informations one can deduce experimental values for γ which are listed in the third row. According to Siems[22] the theoretical values for γ are found in the region of 0.25 (Nb) and 0.35 (Ta), which is on the lower limit but still within the error bars of the experimental values. The agreement therefore can be considered quite satisfactory especially since mean field approximation has been applied.

4. VIOLATION OF STABILITY CRITERIA

The elastic stability criteria are derived on the basis of the postulate that the free energy

$$F(\epsilon_{ij}) = \frac{1}{2} C_{11} (\epsilon_{11}^2 + \epsilon_{22}^2 + \epsilon_{33}^2) + C_{12}(\epsilon_{11}\epsilon_{22} + \epsilon_{22}\epsilon_{33} + \epsilon_{33}\epsilon_{11})$$
$$+ 2C_{44}(\epsilon_{12}^2 + \epsilon_{23}^2 + \epsilon_{31}^2) \qquad (24)$$

must be positive for every combination of strain components. From the eigenvalues of the matrix of coefficients one derives

$$c_{11} + 2c_{12} > 0 \qquad c_{11} - c_{12} > 0 \qquad c_{44} > 0 \qquad (25)$$

The eigenvectors, which diagonalize the coefficient matrix correspond to those modes of deformation for which the lattice becomes unstable, when the corresponding stability criteria is violated. Applying Eq. (25) to Eqs. (14) to (16) we find

$$C_{11} + 2C_{12} - \frac{\rho_0 (A + 2B)^2}{3k\left(T - T_G \dfrac{\gamma_G - 1}{\gamma_G}\right)} > 0 \qquad (26)$$

$$C_{11} + C_{12} - \frac{\rho_0 (A - B)^2}{3k\left(T - T_S \dfrac{\gamma_S - 1}{\gamma_S}\right)} > 0 \qquad (27)$$

$$C_{44} > 0 \qquad (28)$$

The temperatures, at which the stability criteria are violated, are T_G and T_S of Eqs. (19) and (20), T_G being connected with the compressibility (Eq. (26)) and T_S with the shear modulus (Eq. (27)). The violation of stability criteria signalizes phase transitions, which will show up for decreasing temperature. T_G and T_S are not necessarily the transition temperatures, but manifest the so-called stability curve (spinodal) as a function of ρ_0. At a critical point the stability curve touches the phase boundary.

The following mode of deformation ϵ is connected with Eq. (26)

$$\epsilon_{ij}^{(1)} = \epsilon \begin{pmatrix} 1 & 0 & 0 \\ 0 & 1 & 0 \\ 0 & 0 & 1 \end{pmatrix} \qquad (29)$$

For $\epsilon_{ij}^{(1)}$ the cubic symmetry of the lattice is preserved. The lattice is unstable against compression or dilatation, the corresponding phase transition therefore is precipitation. Locally the density increases ($\epsilon > 0$) on account of other areas ($\epsilon < 0$). The Gorsky effect therefore is intimately connected with that phase transition, in which elastic dipoles precipitate (= gas-liquid transition in the space of the host lattice[14]). This transition is observed, e.g., in H in Pd,[15] PdAg–alloys[16] as well as for H in Nb.[17] The violation of the second stability criterion (Eq. (27)) indicates instability against the following modes of deformation (or combinations of these modes)

$$\epsilon_{ij}^{(2)} = \epsilon \begin{pmatrix} -1 & 0 & 0 \\ 0 & -1 & 0 \\ 0 & 0 & 2 \end{pmatrix} \quad \epsilon_{ij}^{(3)} = \epsilon \begin{pmatrix} 1 & 0 & 0 \\ 0 & -1 & 0 \\ 0 & 0 & 0 \end{pmatrix} \qquad (30)$$

$\epsilon_{ij}^{(2)}$ yields a tetragonal lattice, $\epsilon_{ij}^{(3)}$ yields an orthorhombic lattice. The corresponding phase transitions are ordering processes in which for $\epsilon_{ij}^{(2)}$ one site is populated on account of both the others (parallel alignment of elastic dipoles). For $\epsilon_{ij}^{(3)}$ one site is populated on account of one other site, whereas the population of the third site stays constant. These ordering processes are intimately connected with the Snoek-type relaxation. Using statistical thermodynamics it can be shown that the ordering process $\epsilon_{ij}^{(2)}$ has the lower free energy. Therefore, a tetragonal lattice is expected, as has been observed for H in Ta.[18]

In the Appendix it is shown that for interstitials occupying the four 1/4 (111) positions in b.c.c. lattices, instead of $c_{11} - c_{12}$ the condition $c_{44} > 0$ is violated. From Eq. (24) it is immediately evident, that with $c_{44} \to 0$ the lattice is unstable against the deformations ϵ_{12}, ϵ_{23}, or ϵ_{31} or any combination. The resulting lattice symmetries are orthorhombic ($\epsilon_{12} \neq 0$, $\epsilon_{23} = \epsilon_{31} = 0$), monoclinic ($\epsilon_{12} = \epsilon_{23} \neq 0$ $\epsilon_{31} = 0$) or trigonal ($\epsilon_{12} = \epsilon_{23} = \epsilon_{31}$). These three phase transitions correspond to different ordering processes, e.g., orthorhombic ($\rho_1 = \rho_2 > \rho_3 = \rho_4$) or trigonal ($\rho_1 > \rho_2 = \rho_3 = \rho_4$).[19] The orthorhombic transition (with $\epsilon_{12} \neq 0$, not with $\epsilon_{11} = -\epsilon_{22}$!) has been observed for H in Nb.[20] The stability criteria can make tentative predictions about the symmetry of the low-temperature phases. The phase boundaries must be calculated by statistical thermodynamics. Since apparently mean field approximation works fairly well, we will in Sec. 6 first calculate the elastic interaction energy and then demonstrate on an example how to determine phase diagrams.

5. THE ELASTIC INTERACTION ENERGY

In mean field approximation the total elastic interaction energy can be written as

$$U = \frac{1}{2} \int \rho^\nu U^\nu \, dV \tag{31}$$

with

$$U^\nu = -P^\nu_{ij} \epsilon_{ij}^{\text{loc}} = -P^\nu_{ij}(S_{ijkl} + \lambda_{ijkl}) \rho^\mu P^\mu_{kl} \tag{32}$$

Combining Eqs. (31) and (32) yields

$$U = -\frac{1}{2} \int \rho^\mu \rho^\nu M^{\mu\nu} \, dV \tag{33}$$

with the coupling matrix

$$M^{\mu\nu} = P^\mu_{ij}(S_{ijkl} + \lambda_{ijkl}) P^\nu_{kl} \tag{34}$$

With the dipole moment tensors (Eq. (10)) the matrix M splits up into two parts which are given by

$$M^{\mu\nu} = \frac{1}{3}\gamma_G(A + 2B)^2(S_{11} + 2S_{12})\begin{pmatrix} 1 & 1 & 1 \\ 1 & 1 & 1 \\ 1 & 1 & 1 \end{pmatrix}$$

$$+ \frac{1}{3}\gamma_S(A - B)^2(S_{11} - S_{12})\begin{pmatrix} 2 & -1 & -1 \\ -1 & 2 & -1 \\ -1 & -1 & 2 \end{pmatrix} \quad (35)$$

The interaction energy therefore can be written as

$$U = -\frac{1}{2}\int \left\{ \frac{1}{3}\gamma_G(A + 2B)^2(S_{11} + 2S_{12})(\rho_1 + \rho_2 + \rho_3)^2 \right.$$

$$+ \frac{1}{3}\gamma_S(A - B)^2(S_{11} - S_{12})[(\rho_1 - \rho_2)^2$$

$$\left. + (\rho_2 - \rho_3)^2 + (\rho_3 - \rho_1)^2] \right\} dV \quad (36)$$

From this interaction the following phase transitions, which have partly been sketched in the previous chapter can be expected:

1. The system can lower the energy by increasing locally the average density $(\rho_1 + \rho_2 + \rho_3)$. With decreasing temperature one expects a phase transition, in which the total system decays into two subsystems with different dipole density (precipitation). The interaction energy responsible for this phase transition is determined by the same quantities as the Gorsky effect and T_G (Eq. (19)).

2. According to the second term in Eq. (34) the energy can be lowered by increasing the population of one site on account of the other sites. This is an ordering process. The interaction energy is determined by the parameters entering into Snoek relaxation. Yet the following point is important: In contrast to the Snoek process this ordering process is always accompanied by a precipitation process. Not only the relative occupation of the sites enters into Eq. (36) but the absolute differences. Therefore, the system can lower the energy by separating into two subsystems of which one is higher ordered and denser than the other.

We therefore find two precipitation processes with different characteristic interaction energies. The condensation process, accompanying the parallel alignment, constitutes an essential difference to spin ordering processes and results from the fact that elastic dipoles which we deal with (H in metals) in addition to the orientation can also change the local density.

Ferromagnetism in liquids may show similar properties. If the precipitation of the elastic dipoles is compared with the "gas-liquid" transitions, then the first transition corresponds to the ordinary transition "gas-liquid", whereas the second transition is similar to the transition "liquid isotropic" to "liquid anisotropic", as observed for many organic liquids.

In the following chapter we will, as an example, determine the phase diagram for systems for which $A - B = 0$ so that the second precipitation process is missing. In the subsequent chapter we sketch the diagram for the case $(A + 2B) = 0$, so that the first precipitation process is missing. The general diagram is the combination of both.

6. THE GAS-LIQUID PHASE TRANSITION

Hydrogen or deuterium in Pd and Pd alloys occupy the octahedral sites and therefore $P^{\nu}_{ij} = P\delta_{ij}$. Thus only the first precipitation process is expected. We now apply lattice-gas theory: The system is divided in N^o cells with the volume v. Each cell can be occupied by one particle. N is the total number of particles. At low temperatures the system consists of two phases with N_1^o and N_2^o cells, populated with N_1 and N_2 particles. The entropy of a subsystem is given by

$$S = k \ln \frac{N^o}{(N^o - N)! N!} \quad (37)$$

Ignoring interaction between the two subsystems the total free energy is given by

$$F = -\frac{\gamma_G P^2}{2\kappa v} \left[\frac{N_1^2}{N_1^o} + \frac{N_2^2}{N_2^o} \right] - T(S_1(N_1^o, N_1) + S_2(N_2^o, N_2)) \quad (38)$$

where κ is the bulk modulus. F must be minimized with respect to the four variables N_1, N_1^o, N_2, and N_2^o, taking into account the relations

$$N_1 + N_2 = N$$

$$N_1^o + N_2^o = N^o \quad (39)$$

Differentiation of F yields

$$2(n_2 - n_1) = \frac{kT2\kappa v}{\gamma_G P^2} \ln \frac{n_2(1 - n_1)}{n_1(1 - n_2)} \tag{40}$$

$$n_2^2 - n_1^2 = \frac{kT2\kappa v}{\gamma_G P^2} \ln \frac{1 - n_1}{1 - n_2} \tag{41}$$

n_1 and n_2 stand for

$$n_2 = \frac{N_2}{N_2^o} \quad n_1 = \frac{N_1}{N_1^o} \tag{42}$$

Equations (40) and (41) must be fulfilled simultaneously and yield the densities n_1 and n_2 as a function of T. If we now introduce an order-parameter M as follows

$$n_1 = \frac{1 - M}{2} \quad n_2 = \frac{1 + M}{2} \tag{43}$$

we find that Eqs. (40 and (41) both reduce to

$$M = \frac{kT2\kappa v}{\gamma_G P^2} \ln \frac{1 + M}{1 - M} \tag{44}$$

or

$$M = \operatorname{tgh} \frac{\gamma_G P^2 M}{2\kappa v k T} \tag{44a}$$

Equations (44) and (44a) are well known from ferromagnetism. The order-parameter $M = n_2 - n_1$ corresponds to the magnetization. The solution of Eq. (44) is plotted in Fig. 2 which represents indeed (at least schematically) the phase diagrams as found for H or D in Pd or Pd alloys.[15,16]

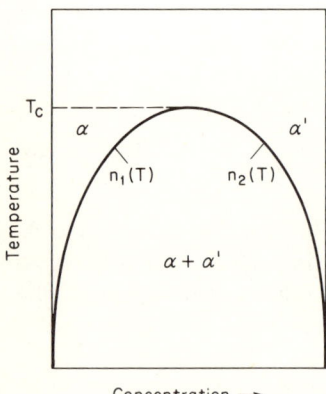

Fig. 2 *Solution of Eq. (44) (schematic phase diagram for H in Pd).*

The critical temperature T_c is given by

$$kT_c = \frac{\gamma_G P^2}{4\kappa v} \qquad (45)$$

For temperature $T < T_c$, for which $n_1 < 1$ the temperature dependence of the concentration n_1 is exponential as

$$n_1 = \exp\left[-\frac{\gamma_G P^2}{kT 2\kappa v}\right] = \exp\left[-\frac{2T_c}{T}\right] \qquad (46)$$

as is well known for the low-temperature magnetization of ferromagnets without spin waves. r may define the number of sites per host atom. Experimentally the maximum concentration is found as $c_m = 58$ at%.[15] Therefore, r must be chosen as $r = 0.58$. v is for a f.c.c. lattice given by $(a^3/4r)$. The concentration c_1 and c_2 (hydrogen atoms per host atom) follow from n_1 and n_2 by multiplying with r.

The critical temperature has experimentally been found as 290°C.[15] Using the data of Wicke and Nernst,[21] a plot of the phase boundary at low concentrations yields indeed an exponential temperature dependence $C_1 \sim \exp(-a/kT)$ with $a = 0.095$ eV. With $2kT_c = a$ we find $T_c = 570°$K in good agreement with the measured critical temperature. Unfortunately, P is not well known for Pd. From lattice parameter changes one estimates $P \approx 3$ eV, which may be wrong by 50 percent. Using $\gamma_G = 0.3$[22] one derives from the critical temperature (Eq. (45)) $P \approx 4$ eV. The order of magnitude appears to be correct.* According to Eq. (45) the experimentally observed difference in the critical temperature for H and D may be attributed to different dipole moments as a result of different zero-point oscillations. A 20°C difference in T_c corresponds to only about 2 percent difference in P.

7. THE GENERAL PHASE DIAGRAM

Figure 3 (dashed) shows schematically the result for the phase diagram when $A + 2B = 0$ and $A - B \neq 0$.[23] The general phase diagram is in principle a superposition of Figs. 2 and 3 (dashed). In superimposing both diagrams the temperature is scaled according to the two characteristic energies in Eq. (36). Therefore, a diagram like Fig. 4 is possible which indeed represents schematically the topological features of the experimental phase diagram for H in

---

*Recent Gorsky-effect measurements by J. Völkl et al. (unpublished) yield $P = (3.5 \pm 0.4)$ eV. With $\gamma_G = 0.3$ one predicts a value for T_c between 550 and 650°K.

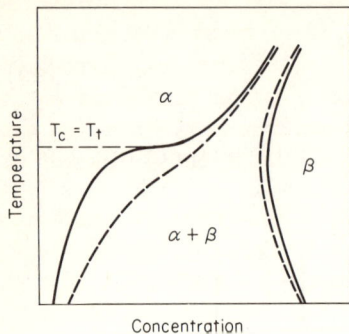

Fig. 3 Phase diagram for $A + 2B = 0$, $A - B \neq 0$ (dashed lines) (schematic phase diagram for H in Ta (full lines)).

Nb.[17] By lowering the ratio $U_1/U_2 = T_G/T_S$ the triple point of Fig. 4 disappears and the diagram Fig. 3 is found. Figure 3 represents schematically the diagram for H in Ta.[18]

Experimentally the critical point for H in Nb has been found at $\approx 450°K$. Using Eq. (39) in Ref. 3 and the experimental values of Table 3 one finds $T_s = 460°K$. Furthermore, Table 3 shows that the critical point for H in Ta should be lower in temperature by about 20 percent as is roughly observed experimentally as well.

8. CONCLUSIONS

From the previous section the following can be concluded:

1. There is strong evidence that the attractive interaction responsible for the condensation and ordering phenomena of interstitial hydrogen in metals can predominantly be attributed to the elastic dipole-dipole interaction.

2. This interaction causes two characteristic phase transitions which can be called "ferroelastic transitions." Both transitions are

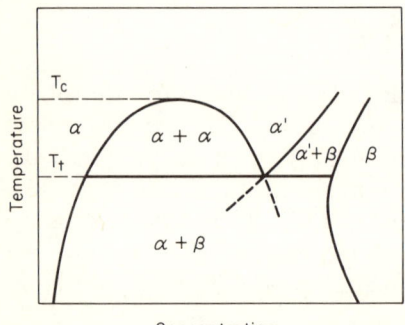

Fig. 4 Schematic phase diagram for H in Nb.

preceded by paraelastic regions with Curie-Weiss temperature dependence of the elastic susceptibilities. The general phase transitions for hydrogen in metals are combinations of the two ferroelastic transitions. The richness of phenomena must be attributed to the fact that in our case the elastic dipoles can change position and thus increase locally their density and that the elastic dipole moment is a tensor.

3. Although we have considered only diagrams for interstitial hydrogen, the results should in principle be applicable to other solid solutions like O, N, C, and B in metals (or eventually also to, e.g., Li in Ge, etc.). Yet the traces of the dipole moment tensors, e.g., of O, N, C in b.c.c. metals are about two to three times larger; thus T_G is four to ten times larger. Therefore analogous phenomena are shifted to much higher temperatures and may be even beyond the melting point. Nevertheless, for many solid-state phase transitions the long-range elastic interaction will have to be taken into account in addition to commonly known sources of interaction. There are rarely systems which in the unordered state are free of lattice distortions.

4. Besides ferroelastic order antiferroelastic order may be possible. For this case one would expect a periodic arrangement of interstitials with alternating orientations. Phases of this type are claimed to have been observed (e.g., $Ta_{64}C$ with C periodically arranged[24]) although no careful investigation exists whether these phases are equilibrium phases.

5. One can not expect that the interaction energy Eq. (36) is still applicable for high concentrations of interstitials. Alloying with hydrogen changes considerably the electronic structure. The elastic coefficients are a function of the hydrogen content (order of magnitude: 0.1 percent change per 1 at% H). Furthermore, the dipole strength may depend on concentration as well. Therefore, for large concentration the qualitative picture of the phase transition remains correct. For quantitative results these diaelastic and nonlinear properties of the host lattice must be taken into account.

REFERENCES

1. Alefeld, G., G. Schaumann, J. Tretkowski, and J. Völkl: *Phys. Rev. Lett.*, 22:697 (1969).
2. Aizu, K.: *J. Phys. Soc. Japan*, 27:387 (1969).
3. Alefeld, G., J. Völkl, and G. Schaumann: *Phys. Stat. Sol.*, 37:337 (1970).
4. Kröner, E.: *Phys. Kondens. Materie*, 2:262 (1964).
5. For example, E. Kröner in "Theory of Crystal Defects," *Proc. Summer School in Hrazany*, edited by B. Gruber, Academia Prague, 1966.

6. Eshelby, J. D.: *Solid State Phys.*, 3:79 (1956).
7. Siems, R.: Wechselwirkung zwischen Defekten in Kristallen, JÜL-Bericht 545-FN.
8. Muggli, J.: Thesis ETH, Zürich, 1970.
9. Siems, R.: *Phys. Stat. Sol.*, 42:105 (1970).
10. Schaumann, G., J. Völkl, and G. Alefeld: *Phys. Rev. Lett.*, 21:891 (1968).
11. ———: *Phys. Stat. Sol.*, 42:401 (1970).
12. Zener, C.: "Elasticity and Anelasticity of Metals," Chicago University Press, Chicago, 1948.
13. Nowick, A. S., and W. R. Heller: *Adv. Phys.*, 12:251 (1963).
14. Alefeld, G.: Phys. Stat. Sol., 32:67 (1969).
15. For example, D. P. Smith, "Hydrogen in Metals," University of Chicago Press, Chicago, 1948.
16. For example, H. Brodowsky and E. Poeschel: *Z. Phys. Chem., Neue Folge*, 44:143 (1965).
17. Walter, R. J., and W. T. Chandler: *Trans. AIME*, 233:762 (1965).
18. Zierath, H.: Thesis, Universität Münster, 1969.
19. Pick, M., and G. Alefeld: unpublished.
20. Brauer, G., and R. Hermann: *Z. anorg. und allgem. Chemie*, 274:11 (1953).
21. Wicke, E., and G. H. Nernst: *Berichte der Bunsengesellschaft*, 68:224 (1964).
22. Siems, R.: Private communication.
23. Alefeld, G., and R. Amadori: To be published.
24. Villagrana, R. E., and G. Thomas: *Phys. Stat. Sol.*, 9:499 (1965).

APPENDIX

In the hexahedral site, for the ¼ (111) position in b.c.c. lattices the strain field has trigonal symmetry. The four dipole moment tensors can be written as:

$$P_{ij}^1 = \begin{pmatrix} P & Q & Q \\ Q & P & Q \\ Q & Q & P \end{pmatrix} \qquad P_{ij}^{(2)} = \begin{pmatrix} Q & -Q & -Q \\ -Q & P & Q \\ -Q & Q & P \end{pmatrix}$$

$$P_{ij}^{(3)} P_{ij}^4 \quad \text{minus sign cyclic} \quad \text{(A1)}$$

Evaluating the relaxed elastic coefficient, we find

$$S'_{11} = S_{11} + \frac{\rho_0 P^2 (S_{11} + 2S_{12})^2}{k(T - T_G)} \tag{A2}$$

$$S'_{12} = S_{12} + \frac{\rho_0 P^2 (S_{11} + 2S_{12})^2}{k(T - T_G)} \tag{A3}$$

$$S'_{44} = S_{44} + \frac{\rho_0 Q^2 S_{44}^2}{k(T - T_S)} \tag{A4}$$

with
$$kT_G = \gamma_G 3\rho_0 P^2 (S_{11} + 2S_{12}) \tag{A5}$$

$$kT_S = \gamma_S Q^2 S_{44} \tag{A6}$$

γ_G has the same meaning as in Eq. (17), whereas γ_S stands for

$$\gamma_S = 1 + \frac{2F}{S_{44}} \tag{A7}$$

For the stability criteria we now find

$$C_{11} + 2C_{12} - \frac{3\rho_0 P^2}{k\left(T - T_G \dfrac{\gamma_G - 1}{\gamma_G}\right)} > 0 \tag{A8}$$

$$C_{11} - C_{12} > 0 \tag{A9}$$

$$C_{44} - \frac{\rho_0 Q^2}{k\left(T - T_S \dfrac{\gamma_S - 1}{\gamma_S}\right)} > 0 \tag{A10}$$

For this type of position the shear constant c'_{44} violates the stability criterion. The interaction energy U can be written as

$$U = -\frac{1}{2} \int \Big\{ \frac{1}{3} \gamma_G P^2 (S_{11} + 2S_{12})(\rho_1 + \rho_2 + \rho_3 + \rho_4)^2$$

$$+ \gamma_S Q^2 S_{44} [(\rho_1 - \rho_2)^2 + (\rho_2 - \rho_3)^2 + (\rho_3 - \rho_4)^2 + (\rho_4 - \rho_1)^2$$

$$+ (\rho_1 - \rho_3)^2 + (\rho_2 - \rho_4)^2] \Big\} dV \tag{A11}$$

DISCUSSION on Paper by G. Alefeld

P. C. CLAPP: How do you distinguish between the hydrogenic diffusive hops responsible for elastic dipole reorientations and those that produce diffusive motion of the dipoles?

G. ALEFELD: You decide on the basis of the corresponding relaxation time. The orientation relaxation time for a hydrogen dipole in Nb, V, or Ta is typical of the order of magnitude of 10^{-11} sec. For diffusion over the distance of a sample size (1mm) about $10^{12}-10^{13}$ jumps are required, yielding relaxation times of the order 10–100 sec.

H. E. COOK: Have you used continuum theory throughout?

G. ALEFELD: I have indeed used continuum theory and I am aware that improvement may be reached by using lattice theory, especially since the hydrogen atoms come so close. Nevertheless there are recent measurements by Beyerle[1] on the interaction of OH⁻ dipoles in KCl, by means of EPR technique. He finds that dipoles with a distance $a\sqrt{3/2}$ interact as you calculate from continuum theory[2] which is certainly surprising.

REFERENCES

1. Beyerle, W. W.: Thesis ETH Zürich, 1970.
2. Muggli, J.: Thesis ETH, Zürich, 1970.

Agenda Discussion:
DEVELOPMENT OF SPATIAL ORDERING

S. C. Moss[*]
Energy Conversion Devices
Troy, Michigan

P. C. Gehlen[†]
Metal Science Group
Battelle Memorial Institute
Columbus, Ohio

The agenda session was opened with some introductory remarks by the chairman because it became apparent during the morning session that some review of the alloy case might be useful for those participants unfamiliar with the general outlook of most of the work on spatial ordering. The following points were stressed:

1. The fundamental measured quantity is the diffracted intensity (either diffuse above T_c or Bragg-like below T_c). It is only

[*]Chairman
[†]Secretary

occasionally that interest is centered on the critical point itself (i.e., in CuZn-Walker and Chipman and Fe_3Al-Guttman). In the absence of atom size disparities this intensity is written in the usual form

$$I(k) = \sum_{ij} \alpha_{ij} \exp(i\mathbf{k} \cdot \mathbf{r}_{ij}) \qquad (1)$$

where k is a general diffraction vector and $I(k)$ shows the periodicity of the reciprocal space. $\mathbf{r}_{ij}$ is a vector between sites i and j in the real lattice and α_{ij} is the conventional short-range order parameter (pair correlation function) first expressed in this particular form by B. E. Warren, but essentially the same, in the AB alloy, as $\langle \sigma_i \sigma_j \rangle$, the spin-spin correlation function in the Ising model.

2. There are almost no alloys in which even the grossest features of the diffuse scattering can be attributed solely to nearest neighbor interactions. This statement refers to a representation of the intensity $I(k)$ in a simple thermodynamic form which follows from any of several mean field type derivations (Krivoglaz, Clapp-Moss, Elliott and Marshall, Zernike, etc.):

$$I(k) \propto \frac{1}{1 + V(k)/kT} \qquad (2)$$

where

$$V(k) = \sum_{ij} V_{ij} \exp(i\mathbf{k} \cdot \mathbf{r}_{ij}) \qquad (3)$$

$$V_{ij} = V_{ij}^{AA} + V_{ij}^{BB} - 2V_{ij}^{AB} \qquad (4)$$

It is therefore most usual in alloy work to confine attention to temperatures well above T_c and attempt to extract information on the pairwise interaction energies V_{ij} (for several examples see Moss and Clapp[1]) rather than to look directly in the vicinity of T_c where that information might be lost.

3. An important approach is the treatment of the α_{ij}'s directly via the methods outlined by J. B. Cohen and P. C. Clapp to obtain a more pictorial representation of the atomic arrangements and a better appreciation of higher than two-particle configurations. This is especially important in relating local order to various properties as Cohen has pointed out on several occasions.

4. Finally, for alloys in which the elastic interactions are large, we again restrict ourselves to the Fourier or reciprocal space representation and discuss the scattering patterns not only in terms of the pairwise chemical energies but also in terms of the elastic energy coefficients of Krivoglaz and de Fontaine and Cook, which are similar, for obvious reasons, to the atomic force constants of lattice dynamics.

After these general remarks P. C. Clapp presented a simple derivation of the Cowley theory which in 1950 was the first of the approximate expressions for short- and long-range order which incorporated the essential aspects of criticality and was valid for compositions other than AB. The AB derivation can be done in a few steps and demonstrates clearly the approximation for the factorization of multiparticle correlations that yields the Cowley relation. Because there seem to be systems which obey his equation for long-range order, it seemed useful to introduce the Cowley expression to this audience. Clapp noted, much as he had done in his earlier paper,[2] that a relationship for the averages of spin operators previously reported in work on the Ising model of magnetism by Callen[3] and Doman and ter Haar[4] can be used to derive the expression of Cowley[5] for the short-range order parameters of an AB binary alloy.

Clapp proceeded with the standard Ising description of an AB alloy with the operators $\{\sigma_i\}$ where $\sigma_i = +1$ if there is an A atom and $\sigma_i = -1$ if there is a B atom at site i; then, as usual, the configurational energy of any state of the alloy can be written as

$$H = -\frac{1}{4}\sum_{ij} V_{ij}\sigma_i\sigma_j \qquad (5)$$

where

$$V_{ij} \equiv \frac{1}{2}(V_{ij}^{AA} + V_{ij}^{BB} - 2V_{ij}^{AB}) \qquad (6)$$

For a Hamiltonian of this form, Callen has pointed out that the two-particle correlation function is given by

$$\langle \sigma_f \sigma_g \rangle = \langle \sigma_f \tanh(\beta E_g) \rangle \qquad (7)$$

where

$$E_g = -\frac{1}{2}\sum_h V_{hg}\sigma_h \qquad (V_{gg} = 0) \qquad \beta = \frac{1}{kT}$$

and the brackets denote a thermodynamic average over all the allowable states of the system.

If f is taken to be the origin 0 then $\langle \sigma_0 \sigma_g \rangle$ is Cowley's short-range order parameter α_g for an AB alloy. Returning to Eq. (7) we see that it may be written as

$$\langle \sigma_f \sigma_g \rangle = \langle \tanh(\beta E_g \sigma_f) \rangle \tag{8}$$

since $(\sigma_f)^{2n+1} = \sigma_f$ and a power series expansion of $\tanh(\beta E_g)$ contains only odd terms.

If we assume (and this will turn out to be equivalent to Cowley's approximation) that

$$\langle (\beta E_g \sigma_f)^{2n+1} \rangle = \langle \beta E_g \sigma_f \rangle^{2n+1}$$

then we have

$$\langle \sigma_f \sigma_g \rangle = \tanh \langle \beta E_g \sigma_f \rangle \tag{9}$$

or

$$\langle \sigma_f \sigma_g \rangle = -\tanh \left\{ \frac{\beta}{2} \sum_h V_{hg} \langle \sigma_h \sigma_f \rangle \right\} \tag{10}$$

Using the relation $\tanh^{-1} x = 1/2 \ln[(1 + x)/(1 - x)]$ Eq. (10) becomes

$$\ln \frac{(1 + \langle \sigma_f \sigma_g \rangle)^2}{(1 - \langle \sigma_f \sigma_g \rangle)^2} = -2\beta \sum_h V_{gh} \langle \sigma_h \sigma_f \rangle \tag{11}$$

and again taking f to be the origin 0 and replacing $\langle \sigma_0 \sigma_g \rangle$ by Cowley's parameter α_g we have Cowley's result

$$\ln \frac{(1 + \alpha_g)^2}{(1 - \alpha_g)^2} = -2\beta \sum_h V_{gh} \alpha_n \tag{12}$$

To understand the nature of the approximation more clearly, let us return to the exact expression for the order parameter given by Eq. (8) and expand the right-hand side

$$\langle \sigma_f \sigma_g \rangle = \beta \langle E_g \sigma_f \rangle - \frac{\beta^3}{3} \langle (E_g \sigma_f)^3 \rangle + \frac{2\beta^5}{15} \langle (E_g \sigma_f)^5 \rangle \cdots \tag{13}$$

The β^3 term may be written as

$$\langle (E_g \sigma_f)^3 \rangle = -\frac{1}{2^3} \sum_{l,m,n} V_{lg} V_{mg} V_{ng} \langle (\sigma_l \sigma_f)(\sigma_m \sigma_f)(\sigma_n \sigma_f) \rangle \quad (14)$$

and for the moment let us assume that V_{ij} decreases very rapidly with increasing separation between i and j. Then the important terms in Eq. (14) will be those for which l, m, and n are very close to g. As pointed out above, Cowley's approximation reduces to replacing $\langle (\sigma_l \sigma_f)(\sigma_m \sigma_f)(\sigma_n \sigma_f) \rangle$ by $\langle \sigma_l \sigma_f \rangle \langle \sigma_m \sigma_f \rangle \langle \sigma_n \sigma_f \rangle$ which would be a fair approximation if one could say that σ_l, σ_m and σ_n are more strongly correlated to σ_f than to each other. However, this will probably not be true for most choices of f since for f distant from g, the sites l, m, and n must be much closer to each other (since they are clustered around g) than they are to f, and Cowley's approximation for this term should be considerably in error. On the other hand, the theory does work quite well for some systems. The reason for this is probably the fact that at temperatures sufficiently above the ordering temperature, only the term linear in β gives a significant contribution.

If we consider the long-range ordered state, then $\alpha_g \to \langle \sigma_f \sigma_g \rangle \to \langle \sigma_f \rangle \langle \sigma_g \rangle$ which is equivalent to the square of the Bragg-Williams parameter S^2. Hence we obtain the square of the familiar mean field result $S^2 = \tanh(T_c/T) S^2$. The exponent β for the temperature dependence of the vanishing of long-range order or magnetization [$S \propto (T_c - T)^\beta$] thereby goes from one-half in mean field to one-quarter in this case, even though in many respects the Cowley theory falls within the general class of mean field treatments. The system DAG (dysprosium aluminum garnet) shows a sublattice magnetization which seems to be very closely given by the Cowley expression,[6] and in DAG it is thus very tempting to search for a dipolar interaction scheme which permits the factoring of the correlations shown above.

During the presentation by Clapp, H. E. Stanley and others questioned the exactness of Eq. (7)—the identity pointed out by Callen. It was discussed briefly and agreed that the identity was all right but that it certainly was not a solution for the Ising model which, of course, had never been claimed for it. It was also noted by Clapp and reemphasized by Stanley that we could in general get a hierarchy of equations: $S^n = \tanh(T_c/T) S^n$, where as n gets large we approach a perfect first-order transition. The interest in the treatment however, was mainly to place it within the more conventional framework and not to suggest any hierarchy of approximations.

Several people including W. Marshall, B. McCoy, and P. C. Martin said that the class of mean field theories (including Cowley's) represent crude solutions that do not usually include even the proper dependence on dimensionality. Furthermore, the expression $I(k) \propto 1/1 + \beta V(k)$ is only good at high temperatures and it was quite reasonably asked by Martin why the V_{ij}'s are of interest and what relevance the whole pursuit had.

The chairman indicated again that the theory is used only as a framework to interpret gross features of the scattering which seem not to be temperature dependent—such as the ratio of the interaction energies for several coordination shells. For example, the early Cowley data showed a pronounced disk-like scattering in the 100 plane of the reciprocal lattice which could *only* be accounted for by a ratio $V_2/V_1 \simeq -.20$. This scattering is very nonisotropic (even neglecting elastic effects) and is qualitatively different from the more familiar isotropic scattering near T_c. It is to gain some insight into alloying energetics that we have studied these effects.

L. Guttman also pointed out that nearly all of these ordering alloys are close-packed (again, aside from β-CuZn and Fe₃Al) and are known to have first-order transitions. For this reason, it is less interesting to explore the scattering at temperatures near the transition point, because one does not expect any catastrophic build-up in fluctuations.

As an example of the derivation of interaction energies from X-ray measurements on short-range order, J. M. Cowley gave a short presentation of the work of S. W. Wilkins.[7] The relation he used was

$$I(\mathbf{k}) = \frac{G_2(T)}{1 + G_1(T) V(\mathbf{k})} \tag{15}$$

which is an equation of more general validity between the transforms of the short-range ordering parameters α_{ij}, and interatomic energies V_{ij}, where the functions of temperatures $G_1(T)$ and $G_2(T)$ depend on the approximation used. To eliminate the slight uncertainty of the G functions, Eq. (15) is transformed by using the critical temperature relationship, that $1 + G_1(T_c) V(\mathbf{k}_m) = 0$ where $\mathbf{k}_m$ is the value of k for which $V(\mathbf{k})$ is a minimum.

For convenience of solution, the equation is then transformed into real space to give

$$\left[\frac{V(k_m)}{V_{0i}}\right] X(T) \frac{T}{T_c} f(\alpha_{0i}) - \sum_j \alpha_{0j} \frac{V_{ij}}{V_{0i}} = 0 \qquad (16)$$

where $(T/T_c) X(T) = G_1(T_c)/G_1(T)$. For the mean field approximation $f(\alpha_{0i}) = \alpha_{0i}/m_A m_B$.

Wilkins then used a least-squares technique for deriving the quantities

$$X \quad \frac{V_2}{V_1} \quad \frac{V_3}{V_1} \quad \frac{V_4}{V_1} \cdots \frac{V_S}{V_1}$$

from the experimental α_{ij} values for Cu_3Au at $T/T_c = 1.09$ obtained by Moss.[8] A weighting factor was applied to reduce the importance of the α_1 and α_2 values (first- and second-neighbor parameters) for which the experimental errors are expected to be greatest. The resulting V_i/V_1 are plotted in Fig. 1, together with values calculated

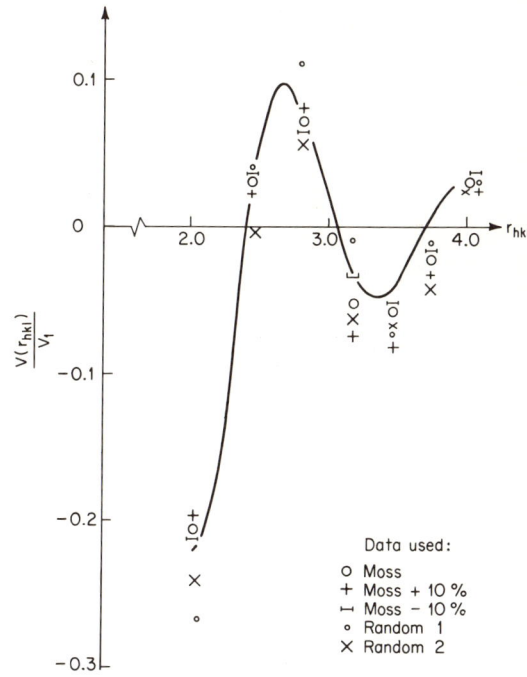

Fig. 1 *Values of the first seven pair interaction ratios for various choices of the weighting factors, obtained from the data of Moss at $T/T_c = 1.091$.*

on the assumption of various systematic or random errors in the α_{ij}. The curve drawn is the best fit of the form

$$V(r) = \frac{A \cos(2k_F r + \phi)}{r^3} \qquad (17)$$

which is the well-known long-range oscillatory potential due to free-electron screening effects. The fit implied an electron/atom ratio of 0.98 and a phase factor $\phi = -0.7$ and is perhaps the first serious attempt to fit a long-range oscillatory pair potential to scattering data, although there had earlier been reported evidence for a long-range oscillatory interaction which revealed itself directly in the reciprocal lattice via Kohn-type anomalies at $I(2k_F)$.[9]

Following Cowley's exposé, Stanley asked if this were a Rudermann-Kittel interaction within the mean field approximation and the chairman replied that it was indeed the Friedel oscillatory potential and the mean field aspect was only in the scattering formula used at temperatures well above T_c to obtain V_{ij}.

H. Cook then raised the question of the contribution of elastic energy terms to V_{ij} or, actually, to V_k. In other words, as Krivoglaz first noted, the intensity expression should have explicit contributions from V_k (electronic), V_k (elastic) and so forth. Clapp pointed out that, as the $I(k)$ is usually written, these terms are grouped together in V_k (total) and that an evaluation of V_k via scattering methods thus includes the elastic part. Cook replied that, while this may be true in a formal way, the elastic interaction is often so long ranged that it requires many terms in V_{ij} to account for it. It would thus be more efficient to approach the elastic effect from the lattice statics point of view, obtain its contribution to V_k, and leave the rest of V_k to the domain of experimental measurement. This he showed to be especially relevant in clustering systems where, with a few interatomic force constants, the anisotropic elastic response of the alloy can be characterized straightforwardly and gives rise to diffuse scattering patterns for the composition fluctuation which may be, say, extended along a <100> direction outward from the origin of k space. Since in mean field theory the scattering about $k = 0$ is essentially isotropic, even with several neighbor interactions, to account for this highly anisotropic profile would require an unusually large number of Fourier coefficients or V_k's. While in the ordering case this problem is not really so important, in principle it should be considered.

D. de Fontaine stressed that the microscopic elastic calculation could be used together with a more practical number of V_{ij}'s of chemical or electronic origin to account for the entire pattern and his comments are given below:

It is instructive to inquire whether the elastic energy contribution in the microscopic formalism given here can be written as the sum of pairwise interactions as in the purely "chemical" case. The answer is a guarded yes: In the discrete case with periodic boundary conditions considered, it is formally possible, by Fourier inversion, to rewrite the elastic energy quadratics given by Eq. (8) (in paper III-2) in its real space representation

$$F = \frac{1}{2} \sum_{p, r} f(r) v(p) v(p + r) \qquad (18)$$

in which we have defined the pair interaction coefficient by

$$f(r) = \frac{1}{N} \sum_{h}{}' F(h) \exp[i\mathbf{k}(h) \cdot \mathbf{x}(r)] \qquad (19)$$

the prime on the summation sign indicating that the origin of h space (reciprocal space) is to be excluded. It is essential to exclude this point since the value of $F(h)$ at $(h) = (000)$ is not defined, the value depending on the way the origin is approached. This is because the expression for $F(h)$, Eq. (9), of paper III-2 contains the inverse of a matrix which is singular at the origin.

Although it is thus possible to define elastic pairwise interaction coefficients according to Eq. (19), it is not fruitful to do so since the coefficients $f(r)$ have appreciable values for large pair separations r. It follows that the expansion in r of Eq. (18) limited to a few near neighbors will fail to converge because of the singular nature of $F(h)$ at the origin.

As a consequence, we expect that experimental data will not be adequately fitted by a pairwise interaction model with any reasonable number of coefficients in cases where the elastic contribution to the total free energy is an important one. Note, however, that the predicted long-range elastic interactions can occur even if only near neighbor coupling parameters are retained in the elastic energy expansion of Eq. (3) of paper III-2. It thus appears preferable, when elastic effects are large, to split the total energy into "chemical" and "elastic" parts and to fit experimental data to a few $g(R)$ and to a few $\phi_{ij}(r)$, $\phi_i(r)$ parameters (see paper III-2), relying on the microscopic elasticity model presented here to take care of a long-range interactions observed.

Marshall inquired into the origin of the anisotropy and wondered why, if all that was being considered could be treated within the pairwise formalism, the anisotropy could not still be perfectly well described by the pair energies. The distinction was made between anisotropic scattering and anisotropic elasticity and it was pointed out that the coupling between composition and displacement waves gave in real space certain soft directions along which the composition could fluctuate more easily. This in turn yielded scattering curves which streak along these preferred directions.

Marshall then returned briefly to an earlier point about the applicability of a mean field expression. His point simply was that, if these force fields are long-ranged and elastic interactions are so important, it makes little sense to use anything but a mean field treatment for the correlations—which was exactly what had been somewhat piously hoped for at the outset of the discussion.

Cowley concluded this topic with some additional remarks on the ordering case. Above T_c in an ordering alloy the elastic effects are not nearly so important and he emphasized that the extraction of important information on the electronic part of V_{ij} was a quite legitimate and useful procedure.

The discussion was then directed to the results of G. Alefeld on ferroelasticity where, because the interactions are so long-ranged, it was in a sense not surprising that the exponents seemed to obey the mean field predictions.

B. Widom put this succinctly when he asked if, given a $1/r^3$ elastic dipole interaction driving the transition in, say, H-Pd, the coexistence curve had a parabolic shape near T_c. Alefeld replied that the measurements were not that accurate but that the exponent, β_s was perhaps closer to 1/2 than to 1/3. Alefeld had more to say about exponents via measurements of the Van der Waals isotherm in his hydrogen-containing metals. From plots of chemical potential of the lattice gas vs. composition he showed that one can measure the isotherms above T_c and get γ where the system followed the Landau theory very closely. This is of course a mean field behavior and in fact one of the difficulties which Alefeld mentioned in obtaining good values for the coexistence curve was a hysteresis effect (undercooling and superheating behavior on cycling through the transition). While an approximate value of $\beta = 1/2$ was thus mentioned, this value can be attributed to the fact that the measured phase boundary was an incoherent one and that, as J. W. Cahn has suggested, all but $K = 0$ fluctuations are suppressed. In a sense, then, crossing the coexistence curve involves large macroscopic fluctuations. The question of coherent vs. incoherent phases and phase-boundaries was then picked up by Cook and de Fontaine especially as it relates to Alefeld's result. It was argued that the coexistence curve in these hydrogen alloys is either a coherent boundary (as it would be, say, in order-disorder reactions) or an incoherent boundary as has been suggested, for example, for Au-Ni where the elastic effects are large because of a large difference in atom size. Some time was devoted to understanding how the boundary was determined and how the phase separation proceeded within the gap. Alefeld said that the coexistence curve was determined via X-ray measurements of the lattice parameters. That would establish the boundary as the true (final) equilibrium boundary. The discussion then turned to the question of whether the elastic energy which would accompany the possible fluctuations above T_c (single phase) was sufficient to suppress these fluctuations.

Actually Alefeld was not measuring the path from the disordered to the phase-separated state and the problem of coherence was not encountered in his study. However, in general it can be stated that if the lattices of the separating phases are continuous through those phases then the separation is coherent. If the lattices are discontinuous at the phase boundary—which is thereby defined—then the phases are incoherent. It is likely that the measurement by Alefeld of mean field exponents for these alloys is due to the elastic dipolar strains. The more interesting question would thus be, as Cook emphasized, whether there exists within the measured coexistence curves a metastable gap which allows the hydrogen-rich and hydrogen-poor phases to decompose coherently on a fine scale. The reported hysteresis may be an indication that there is indeed a coherent and incoherent gap.

It was noted with regret that time did not permit more discussion of C. B. Walker's results and of the general question of how well alloys can be treated by Ising calculations. The influence of nonstoichiometry and atom-size effect on equiatomic Ising predictions was mentioned and it was suggested that these issues receive more attention, at least in private discussions. The general consensus was certainly that the result of Walker and Chipman on long-range order in βCuZn represented perhaps the most complete and accurate study of an Ising system in the ordered range and that attention should be directed to the deviations from calculated Ising behavior that they observed.

REFERENCES

1. Moss, S. C., and P. C. Clapp: *Phys Rev.*, **171**:764 (1968).
2. Clapp, P. C.: *Phys. Lett.*, **13**:305 (1964).
3. Callen, H. B.: *Ibid.*, **4**:161 (1963).
4. Doman, B. G. S., and D. ter Haar: *Ibid.*, **2**:15 (1962).
5. Cowley, J. M.: *Phys. Rev.*, **77**:669 (1950).
6. Frankel, N. E., and A. E. C. Spargo: *Phys. Lett.*, (1970).
7. Wilkins, S. W.: *Ibid.*, **B2**:3935 (1970).
8. Moss, S. C.: *J. Appl. Phys.*, **35**:3547 (1964).
9. ———: *Phys. Rev. Lett.*, **22**:1108 (1969).

Part Four

MAGNETS AND SUPERCONDUCTORS

CRITICAL POINTS DEPENDENT ON PARAMETERS

Robert B. Griffiths

Department of Physics
Carnegie-Mellon University
Pittsburgh, Pennsylvania

ABSTRACT

Some examples of magnetic systems are discussed in which the critical point depends on a parameter, thus leading to a line of critical points. It is reasonable to assume that the singular part of the free energy (giving rise to the usual critical point divergences) is a smooth function of this parameter, provided one makes a proper choice of thermodynamic variables. The consequences of such an assumption are worked out for (1) a ferromagnet under pressure, for which one obtains the Pippard relations; (2) an antiferromagnet in an external magnetic field, for which one obtains Fisher's relation between C_H and $d\chi/dT$ at the Néel point; and (3) Ising (or Heisenberg) models with a suitable perturbation added to the Hamiltonian, for which one can infer the temperature dependence of pair correlation functions near the critical point. Brief reference is made to "tricritical" points, e.g., when the first-order and lambda transitions meet for a metamagnet, where the singular part of the free energy is not expected to depend smoothly on a parameter.

1. INTRODUCTION

The critical "points" observed in the laboratory are often particular points on critical lines (or surfaces) which can be obtained by varying one or more thermodynamic parameters. Thus the Curie "point" of a ferromagnet becomes a line of Curie or critical points if the ferromagnet is subjected to hydrostatic pressure as well as temperature variations, and the Néel transition in an antiferromagnet becomes a line of critical points in an external magnetic field. A pure fluid exhibits a single liquid-vapor critical point, but fluid mixtures in general have critical lines or surfaces.

It is intuitively reasonable that one should be able to extend our present ideas on what happens to thermodynamic functions near a critical point to the more general critical lines (and surfaces) and this paper contains a discussion of one procedure for attacking this problem. The approach is quite phenomenological and consists of making a "smoothness postulate" to the effect that the singular part of the free energy, that which gives rise to divergent susceptibilities, heat capacities, etc., is a smoothly varying function along a line of critical points. The notion will be made more precise in Sec. 2 with reference to the simple case of a ferromagnet subject to hydrostatic pressure. The more interesting case of an antiferromagnet is taken up in Sec. 3 where it is shown that certain published relationships near the zero-field Néel point can be obtained rather quickly from the smoothness postulate by "turning the thermodynamic crank." Also included are some brief remarks on "tricritical points" where one expects strong departures from smooth behavior, and which thus possess considerable experimental and theoretical interest. Finally, in Sec. 4 it is shown how the smoothness postulate may be used to make plausible guesses as to the temperature dependence of correlation functions near the critical point of an Ising or Heisenberg ferromagnet.

At this point let me insert a disclaimer: Few if any of the ideas presented in this paper are my own original contribution; many of them have been "floating around" among workers in the field for some time. I am attempting to set them down in fairly explicit fashion and point out their broad applicability. In advance I apologize to anyone who feels reference to his own contributions should have been included; I have made little effort to track down original sources of the ideas here presented, though the footnotes will indicate where I myself learned some of them.

2. FERROMAGNETS UNDER PRESSURE

A schematic phase diagram for a ferromagnet under hydrostatic pressure is shown in Fig. 1. The magnetic field H is the "internal" magnetic field—the external field corrected for sample demagnetization—parallel to the easy axis of magnetization. Note that as H passes through zero for temperatures $T < T_c(p)$, the Curie or critical temperature as a function of pressure, the spontaneous magnetization changes sign. We shall say, therefore, that the cross-hatched surface at $H = 0$ in the pT plane is a coexistence surface, a surface along which two phases, characterized by spontaneous magnetization in opposite directions, coexist.

The basic smoothness postulate is that the singular part of the thermodynamic potential, i.e., the part which gives rise to the characteristic critical-point singularities, is essentially the same everywhere along the critical line except for a smooth variation of parameters describing the amplitudes of the various singularities, and that the critical line is a smooth curve, except for points where the first-order phase transition changes its basic character. In particular the critical indices[1] α, β, γ, δ, etc. should remain unchanged as one moves along the critical line.

A way of expressing the smoothness postulate in mathematical form is to suppose that the thermodynamic potential near the critical line has approximately the form

$$\Phi(p, T, H) = \Phi_0(p, T, H) + \phi(p) f(\tau, h) \tag{1}$$

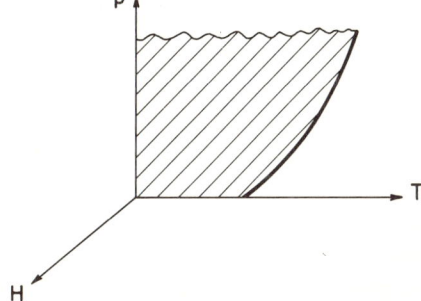

Fig. 1 *Phase diagram (schematic) of a ferromagnet under hydrostatic pressure. The coexistence (first-order phase transition) surface, shown cross hatched, terminates in a line of critical points as T increases at constant p, or p decreases at constant (and sufficiently large) T.*

where

$$\tau = \theta(p)[T - T_c(p)]$$
$$h = \eta(p)H \qquad (2)$$

and where Φ_0, ϕ, θ, η, and T_c are assumed to be analytic, or at least very smooth, functions of their respective arguments, and all the singular behavior (divergent susceptibilities, spontaneous magnetization, and the like) is contained in the function f. We assume that ϕ, θ, and η are positive and further, without loss of generality, that they are equal to 1 at $p = 0$. Then at this pressure the susceptibility is

$$\chi = -\frac{\partial^2 \Phi}{\partial H^2} = \chi_0 - \frac{\partial^2 f}{\partial h^2} \sim (T - T_c)^{-\gamma} \qquad (3)$$

for $T > T_c$ and $H = 0$; here χ_0 is the contribution coming from Φ_0 and hence a smooth function of the temperature, while the divergence comes from the second derivative of f. It is obvious that the exponent γ is independent of p, and the same holds for α, β, etc.

Let us now examine the choice of variables in Eqs. (1) and (2) a little more carefully. There is (1) a variable h which measures distance perpendicular to the coexistence surface. It is natural to choose h proportional to H, since Φ is unchanged if H is replaced by $-H$, a consequence of time reversal symmetry. In cases where one does not have such a symmetry—as, for example, in binary fluid mixtures—the appropriate requirement is that h measure distance from the coexistence surface in a direction not parallel to this surface[2] (at least not parallel near that portion of the critical line which is of interest). Then there is (2) a variable[3] τ which measures distance from the critical line in a direction parallel to the coexistence surface but not parallel to the line itself, and (3) a variable which measures distance along the critical line, in this case p.

Were we to consider a situation involving four independent thermodynamic variables—for example, a ferromagnet subject to both isotropic and anisotropic stress—there would be a coexistence volume terminating in a critical surface. Variables in categories (1) and (2) would be chosen as before, and category (3) would contain two variables indicating position along the critical surface.

Actually, Eq. (2) represents but one possible choice of variables. Provided dT_c/dp is not zero, we could just as well define

$$\tau = \hat{\theta}(T)[p_c(T) - p]$$
$$h = \hat{\eta}(T)H \qquad (4)$$

and replace ϕ in Eq. (1) by $\hat{\phi}(T)$. Here the appropriate measure of distance from the critical line is $p_c(T) - p$ (the sign is chosen consistent with $dT_c/dp > 0$, as in Fig. 1), and T serves as the parameter measuring distance along the critical line. The choice of Eq. (4) in place of Eq. (2) serves to emphasize the fact that while one customarily uses temperature to measure distance from a critical point, there is no reason (apart from experimental convenience) why pressure would not serve equally well, and the critical indices could be defined in terms of $p - p_c$ rather than $T - T_c$. (The same remark applies to liquid-vapor critical points). If our smoothness postulate is correct, pressure indices are identical with the corresponding temperature indices; e.g., a decrease in pressure at fixed temperature (assuming $dT_c/dp > 0$) would lead to a spontaneous magnetization vanishing as $(p - p_c)^\beta$. This result, while it is obvious with the choice of Eq. (4), can be obtained immediately from Eqs. (1) and (2).

Both Eqs. (2) and (4) are special cases of a more general assumption:[4] that ϕ, τ, and h are analytic (or very smooth) functions of p, T, and H with the requirements that ϕ and τ be even and h odd functions of H, that ϕ is positive, that τ is zero but has a nonvanishing gradient along the critical line, with $\tau < 0$ in the coexistence region and > 0 in the disordered state, and that dh/dH be positive at $H = 0$. [The symmetry or antisymmetry in H is imposed because Φ is unchanged when H is replaced by $-H$, and we very naturally impose the same requirement on Φ_0 as a function of H and f as a function of h.] This additional generality may be necessary in some circumstances, but for most purposes it would appear that Eqs. (2) or (4) should be quite adequate, and they have the great advantage of being (comparatively!) explicit.

It must be emphasized that there is one sense in which our choice of variables is not arbitrary: We have deliberately chosen as independent thermodynamic variables the intensive quantities or "fields" (variables which are the same in two coexisting phases) in preference to extensive quantities of "densities" such as volume, entropy, magnetization, and the like. Reasons for this choice have been stated elsewhere,[2,4] and let me add that in my own thinking about phase transitions I have found the use of "field" variables in preference to the other possibilities results in a very considerable conceptual simplification for systems with three or more independent thermodynamic variables.

We have already mentioned one application of the smoothness postulate: Critical exponents for the same quantity should be the same whether the variable is $|T - T_c|$ or $|p - p_c|$. A second type of result, obtained many years ago by Pippard[5] on the basis of a smoothness postulate slightly less general than that which we outlined above, is that with $H = 0$ (so that Φ is simply the ordinary Gibbs potential) the compressibility, isobaric thermal expansion coefficient, and isobaric heat capacity should all have the same type of singular behavior (as a function of temperature or, alternatively, as a function of pressure) at a critical point, with the coefficients of the most singular parts of these quantities related through factors of dT_c/dp (assumed neither zero nor infinite).

3. ANTIFERROMAGNETS

3.1. Uniaxial Antiferromagnets

Additional applications of the smoothness postulate arise if we consider a two-sublattice uniaxial antiferromagnet in an external field parallel to the easy axis for temperatures near the Néel temperature. The phase diagram is indicated schematically in Fig. 2, and once again H is the "internal field" obtained by correcting the applied field for demagnetization. The quantity η giving rise to the third axis in Fig. 2 is a fictitious staggered magnetic field which points "up" at the sites on one sublattice and "down" on the other. This is really the analog for the antiferromagnet of an ordinary (uniform) magnetic field for the ferromagnet, and the two "phases" which coexist at $\eta = 0$ in the cross-hatched region are obtained from each other by interchanging the two sublattices.

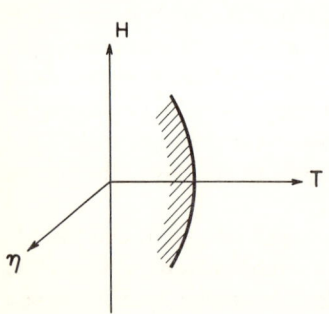

Fig. 2 *Phase diagram for an antiferromagnet near the Néel point. The coexistence surface (cross hatched) is bounded by a line of critical points, the boundary between antiferromagnetic and paramagnetic "phases", as T increases.*

It is my opinion that the conceptual advantages of introducing the "unphysical" field η far outweigh any complaint that its use by theorists is unfair because the physical counterpart is unavailable in the laboratory. In particular, we see upon comparing Figs. 1 and 2 that the "phase boundary" which at $\eta = 0$ separates the paramagnetic and antiferromagnetically ordered states in the HT plane in Fig. 2 is in reality a line of critical points. This conclusion is supported by the fact that the heat capacity singularity as a function of temperature for $H > 0$ has the same singular character as for $H = 0$.[6]

Time reversal symmetry implies the free energy is unchanged if H changes sign, and thus the critical temperature $T_c(H)$ is a symmetric function of H. Apart from this symmetry there is no reason to expect $H = 0$ to be a special point on the critical curve, so our smoothness postulate suggests a parabola to lowest order

$$T_c(H) = T_N + \frac{1}{2} T_c''(0) H^2 + \cdots \tag{5}$$

where $T_c(0) = T_N$ is the Néel temperature.

In analogy with Eq. (1), we suppose that near this curve in the $\eta = 0$ plane the free energy has the form

$$\Phi(H, T) = \Phi_0(H, T) + \phi(H) f\{\theta(H) [T - T_c(H)]\} \tag{6}$$

where H replaces p in Eq. (1) as the parameter measuring position along the critical line, and Φ_0, ϕ, θ, and T_c are all assumed to be even functions of H and analytic (or very smooth) functions of their arguments. As η, the analog of H in Eq. (1), will not be used in our subsequent discussion, we simply omit it. We assume further that ϕ and θ are positive and, without loss of generality, equal to one at $H = 0$.

The character of the function f can be immediately related to experiment by noting that the zero field heat capacity

$$T^{-1} C_{H=0} = -\frac{\partial^2 \Phi}{\partial T^2}$$

$$= -\frac{\partial^2 \Phi_0}{\partial T^2} - f''\{T - T_N\} \tag{7}$$

consists of a smooth term plus a singular term arising from the second derivative of f.

The susceptibility $\chi = -\partial^2 \Phi / \partial H^2$ has the form at $H = 0$

$$\chi = -\frac{\partial^2 \Phi_0}{\partial H^2} - \phi''(0) f\{T - T_N\}$$

$$-(T - T_N) \theta''(0) f'\{T - T_N\} + T_c''(0) f'\{T - T_N\} \tag{8}$$

Upon comparing Eqs. (7) and (8) it is evident that even if at $H = 0$ C_H has a weak divergence or a sharp maximum or cusp, the same will not be true of χ. However

$$\frac{d\chi}{dT} = T_c''(0) f''\{T - T_N\} + \cdots \qquad (9)$$

will have the same singular behavior as C_H (note that $T_c''(0) < 0$ and we have included only the most singular term in Eq. (9)), a result first predicted by Fisher[7] on the basis of rather detailed arguments involving the behavior of correlation functions near T_N. Further, the ratio of the singular parts of $d\chi/dT$ and C_H is simply related to the curvature of the critical line at the Néel point. Both of these conclusions have been confirmed experimentally for at least one material,[8] which provides some support for the smoothness postulate, at least at $H = 0$.

In addition to the foregoing results at $H = 0$, Eq. (6) permits us to draw various conclusions about thermodynamic properties for $H > 0$. In particular, the singularity in C_H should retain the same basic character for $H > 0$ as for $H = 0$, while χ exhibits a singularity similar to C_H as soon as $dT_c/dH \neq 0$. The relation between the singular parts of χ and C_H is, of course, given by the appropriate Pippard relation[5] with p replaced by H. The absence of the C_H singularity from χ at $H = 0$ in no way contradicts the smoothness postulate but, as pointed out elsewhere,[2] it simply reflects the fact that the line of critical points is parallel to the H axis; similar "anomalies" may be expected whenever a critical line is parallel, whether by "accident" or (as in the present case) by symmetry, to one of the thermodynamic coordinate axes.

3.2. Metamagnets and Tricritical Points

Antiferromagnets illustrate another aspect of the smoothness postulate which states that critical properties vary smoothly along a critical line *except* where the nature of the underlying first-order phase transition undergoes some change. In many antiferromagnets, changes do occur in the vicinity of the critical line as the temperature is lowered and the magnetic field increases. A not uncommon occurrence is a spin-flop transition, but we shall discuss the conceptually simpler (though experimentally less common) case of a simple metamagnet, such as $FeCl_2$,[9] for which it is found that the line of critical points (or "lambda transition") in the HT plane becomes a line of first-order phase transitions below a certain temperature which we call the tricritical temperature T^*.

Molecular-field calculations on simple models and the Landau phenomenological theory[10,11] suggest that the phase diagram viewed in ηHT space has the form shown schematically in Fig. 3 for $H > 0$ (the situation for $H < 0$ is, of course, symmetric). As H increases for $T < T^*$, the coexistence surface, which lies initially at $\eta = 0$, bifurcates into two surfaces extending out into the experimentally inaccessible regions $\eta < 0$ and $\eta > 0$. These surfaces themselves terminate in critical lines which connect with the experimentally observable critical line at $T = T^*$ (which suggests the name "tricritical"). The line where bifurcation takes place at $T < T^*$ is a line of triple points, and thus corresponds to a first-order phase transition in the experimentally accessible $\eta = 0$ plane.

It is well to emphasize that the preceding description is at the present time somewhat speculative. An obvious criticism is that "classical" theories of the Landau and molecular-field type are known to yield incorrect predictions on thermodynamic phenomena near simple critical points.[1] However, they nonetheless correctly predict, in many circumstances, the *qualitative* nature of the phase diagram (presence or absence of first-order transitions, existence of critical points), and this is the use to which they are being put in the present instance.

If our preceding description of the phase diagram is accepted, it is obvious that the tricritical point is an excellent candidate as a point where smooth dependence of critical properties on a parameter should break down. Already there is experimental evidence that

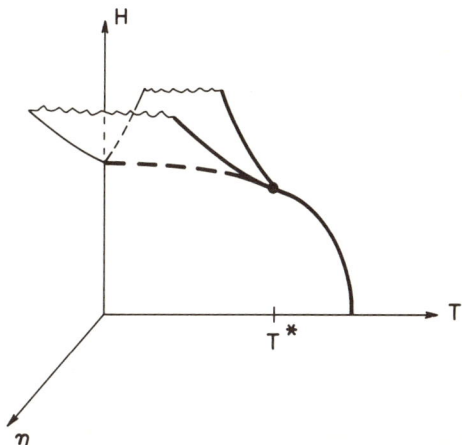

Fig. 3 *Phase diagram for a simple metamagnet. The three coexistence surfaces meet along the dotted line, observed experimentally as a first-order transition, for $T < T^*$, the tricritical temperature, and the three critical lines bounding these surfaces on the high temperature side meet at the tricritical point.*

thermodynamic quantities at tricritical points in metamagnets[12] and in the very analogous situation in the He^3-He^4 mixtures,[13] show a very different behavior from that found at "ordinary" critical points. Further experiments and better calculations should help to clarify the situation.

4. SPIN CORRELATION FUNCTION IN ISING FERROMAGNETS

As a final application of the smoothness postulate, let us consider the temperature dependence of pair correlation functions in Ising ferromagnets near the critical point. It is at first surprising that our postulate, which is essentially "thermodynamic," could have any implications for "microscopic" quantities such as correlation functions. A relationship can, however, be established with the proper choice of a parameter.

Consider a Bravais lattice of N sites with periodic boundary conditions. The pair correlation is

$$\Gamma_\delta = \langle \sigma_r \sigma_{r+\delta} \rangle = N^{-1} \sum_r \langle \sigma_r \sigma_{r+\delta} \rangle$$

$$= -\left(\frac{\partial \Phi}{\partial \epsilon}\right)_{\epsilon=0} \tag{10}$$

where

$$\Phi = -(N\beta)^{-1} \ln Tr[e^{-\beta \mathcal{H}}] \tag{11}$$

$$\mathcal{H} = \mathcal{H}_0 - \epsilon \sum_r \sigma_r \sigma_{r+\delta} \tag{12}$$

and we have added to the Hamiltonian of interest, $\mathcal{H}_0$, a small perturbation. Here σ_r is the Ising spin variable at the site r, and $\beta = (kT)^{-1}$ the inverse temperature.

With $\mathcal{H}_0$ an appropriate ferromagnetic Hamiltonian, one would expect the critical temperature to increase approximately linearly near $\epsilon = 0$

$$T_c(\epsilon) = T_c(0) + b\epsilon + \cdots \tag{13}$$

The smoothness postulate for Φ in the form of Eq. (1), with p

replaced by ϵ and $H = 0$ (and omitted as an argument of Φ) yields

$$\Phi = \Phi_0 + \phi(\epsilon) f\{\theta(\epsilon)[T - T_c(0) - b\epsilon]\} \qquad (14)$$

and, upon differentiation, we find the most singular term in $d\Gamma_\delta/dT$ to be

$$\frac{d\Gamma_\delta}{dT} = + b\phi\theta^2 f''\{\theta[T - T_c(0)]\} + \cdots$$

$$= -b\left(\frac{\partial S}{\partial T}\right)_{\epsilon=0} + \cdots \qquad (15)$$

That is to say, near the critical point the singular part of $d\Gamma_\delta/dT$ should be proportional to the singular part of the heat capacity, but with opposite sign, a result known to hold for the Ising model on a square with nearest-neighbor interactions.[14] For the model last mentioned one finds, as well, that b is an increasing function of δ, which agrees with the intuitive idea that ferromagnetic pair interactions should be more effective in increasing the critical temperature the longer their range. (Thus the mean-field Ising ferromagnet, in which each spin interacts equally with every other spin, has a Curie temperature in excess of the nearest-neighbor Ising ferromagnet with the same ground state energy.[15]

The foregoing considerations could equally well be applied to a Heisenberg ferromagnet, replacing σ_r by the vector spin operator S_r. We must assume, of course, and this is less certain than in the Ising case, that adding a small perturbation will not change the character of the underlying first-order phase transition.

When considering scattering processes, one is less concerned with Γ_δ than with its Fourier transform

$$\beta^{-1}\chi(k) = \sum_\delta \Gamma_\delta e^{i\delta \cdot k} = \sum_\delta \Gamma_\delta \cos(\delta \cdot k) \qquad (16)$$

where we assume the origin is a center of inversion for the lattice, and k is a wavevector allowed by the periodic boundary conditions. In place of Eq. (12) we introduce a perturbation, with $k \neq 0$

$$\mathcal{H} = \mathcal{H}_0 - \epsilon \sum_r \cos(k \cdot r) \sigma_r \qquad (17)$$

which is, effectively, a spatially varying field. Now Φ is an even function of ϵ [strictly speaking, this requires that there be some r in the lattice such that $k \cdot r$ is an odd multiple of π; this is a rather trivial restriction when N is large] and at $\epsilon = 0$

$$\frac{-\partial^2 \Phi}{\partial \epsilon^2} = \beta N^{-1} \sum_{r} \sum_{r'} \cos(k \cdot r) \cos(k \cdot r')$$

$$\times \{\langle \sigma_r \sigma_{r'} \rangle - \langle \sigma_r \rangle \langle \sigma_{r'} \rangle \} = \frac{1}{2} \chi(k) \qquad (18)$$

where the second equality is obtained assuming that $\mathcal{H}_0$, and consequently the correlation functions, have the translational symmetry of the lattice [the coefficient ½ in front of χ becomes 1 if 2k is a reciprocal lattice vector].

For $k \neq 0$, a finite ϵ corresponds to applying a "staggered" field to the ferromagnet which might be expected (for small ϵ) to have an effect analogous to applying a uniform field to an antiferromagnet, i.e., a reduction in critical temperature of order ϵ^2. We thus have an analog of the situation discussed in Sec. 3.1 above, as illustrated in Fig. 2, but with ϵ playing the role of H. Hence the left side of Eq. (18) should resemble the antiferromagnetic susceptibility at $H = 0$. This means that for $k \neq 0$, $\chi(k)$ should be finite at the ferromagnetic critical temperature, and $d\chi(k)/dT$ should have the same type of singularity (and, unlike dT_s/dT, the same sign) as the zero field heat capacity. Calculations by Fisher and Burford[14] based on series expansions seem to be consistent (or at least not inconsistent) with such behavior. Note that an analogous calculation for the quantum Heisenberg ferromagnet runs into difficulties in that the first equality in Eq. (18) fails to hold, due to noncommutativity of operators, with σ_r replaced by S_r.

5. CONCLUSION

The preceding examples by no means exhaust the ways in which an examination of the geometry of phase diagrams may serve as an intuitive aid to a phenomenological understanding of critical phenomena. There are, for example, critical phenomena in binary, ternary, etc. fluid mixtures in which the coexistence surface is in general curved (in contrast to magnetic systems, where flatness is often insured by symmetry), giving rise to some interesting effects.[2] There also seems to be some profit in considering the phase diagrams of models in which the critical indices are thought to change suddenly at particular values of a parameter.[16] Needless to say, phenomenological generalizations, such as the smoothness postulate, are no substitute for the detailed and often rather arduous direct calculations which have provided us with so much insight into critical phenomena, or for the experimental measurements against which our ideas must ultimately be tested. However, they may serve to organize the knowledge we possess and at the same time suggest other approaches to outstanding problems. My own belief is that the smoothness postulate will continue to serve both functions, at least until it is replaced by something better.

ACKNOWLEDGMENTS

It is a pleasure to acknowledge many helpful discussions with Professors M. E. Fisher, S. A. Friedberg, and J. C. Wheeler.

REFERENCES

1. Fisher, M. E.: *Rep. Prog. Phys.*, **30**:615 (1967).
2. Griffiths, R. B., and J. C. Wheeler: *Phys. Rev.*, **A2**:1047 (1970).
3. Buckingham, M. J., and W. M. Fairbank: in C. J. Gorter (ed.), "Progress in Low Temperature Physics," vol. 3, p. 80, North Holland Publishing Co., Amsterdam, 1961.
4. Fisher, M. E.: *Phys. Rev.*, **176**:257 (1968).
5. Pippard, A. B.: "Elements of Classical Thermodynamics," chap. 9, Cambridge University Press, New York, 1957.
6. Schelleng, J. H., and S. A. Friedberg: *Phys. Rev.*, **185**:728 (1969).
7. Fisher, M. E.: *Phil. Mag.*, **7**:1731 (1962).
8. Skalyo, J., Jr., A. F. Cohen, S. A. Friedberg, and R. B. Griffiths: *Phys. Rev.*, **164**:705 (1967).
9. Jacobs, I. S., and P. E. Lawrence: *Ibid.*, **164**:866 (1967).
10. Landau, L.: *Phys. Z. Sowjet.*, **11**:26 (1937).
11. Landau, L. D., and E. M. Lifschitz: "Statistical Physics," p. 452, Pergamon Press, London, 1958.
12. Keen, B. E., D. Landau, B. Schneider, and W. P. Wolf: *J. Appl. Phys.*, **37**:1120 (1966).
13. Griffiths, R. B.: *Phys. Rev. Lett.*, **24**:715 (1970).
14. Fisher, M. E., and R. J. Burford: *Phys. Rev.*, **156**:583 (1967).
15. Griffiths, R. B.: *Commun. Math. Phys.*, **6**:121 (1967).
16. ———: *Phys. Rev. Lett.*, **24**:1479 (1970).

DISCUSSION on *Paper by R. B. Griffiths*

P. HELLER: In connection with your discussion of the wavelength-dependent susceptibility $\chi(k)$, let us say for an antiferromagnet, may we assume that only for k *not* equal to the staggered wave vector, that the curve of the type you draw, having a temperature derivative which behaves like the specific heat, would be obtained? As k tends to the staggered wave vector, however, the maximum value of $\chi(k)$ gets larger and larger, the resulting curve tending to the expected singular curve which is described by the indices γ and γ'.

R. GRIFFITHS: Yes, that is what one expects, although I do not want to swear to that by smoothness.

G. A. BAKER: Gallavotti, Miracle-Sole, and Robinson have done considerable work on the smoothness postulate and have proved analyticity as a function of several complex variables in a restricted region for various parameters. Coopersmith and others have suggested that we assume analyticity in all available variables except in the face of explicit singularities. May this work be thought of as a precursor to your more extensive treatment?

R. GRIFFITHS: As I said before, I don't want to claim originality. I don't know where I have learnt it.

G. A. BAKER: What do you mean by intermediate states of smoothness?

R. GRIFFITHS: By that I mean if analyticity fails and one can get away with twenty continuous derivatives, I think this is enough smoothness.

R. A. FERRELL: In your first slide you introduced some functions. Would you have any difficulties if you said that all these functions were analytic except the one that contains the variable τ?

R. GRIFFITHS: No, that is my basic assumption, that they are all analytic functions in the vicinity of the real axis.

O. K. RICE: Is not the proportionality of $\partial \chi/\partial T$ to c_H the same thing thermodynamically as the proportionality of $(\partial V/\partial T)_p$ to c_p in the liquid helium case?

R. GRIFFITHS: Yes, it is derived by the same method. The only difference is that here we are talking about the pair correlation function rather than the coefficients of expansion.

M. J. COOPER: Your postulate seems very reminiscent of what is known as the parametrical representation of Schofield and Josephson. Is this a valid analogy?

R. GRIFFITHS: No, I think this analogy is not correct. If you wish, Schofield and Josephson have proposed a very nice form for the singular function. I don't really make much of an assumption as to just what form that has.

L. ONSAGER: I might point out that the correlation function is not smooth through the critical point but behaves like the energy. This is reasonable only for finite distances but not for the limit of infinite distances. In this limit the correlation function is zero above the critical point and rises steeply below. In the Ising model the type of singularity of the correlation function seems to be the same for all finite distances, but the coefficient gets bigger and bigger with increasing distance.

R. GRIFFITHS: I think what you are saying is, if I differentiate, I find a logarithmic behavior and the amplitude of this increases, as the spins are taken farther apart.

R. E. MILLS: In the case of metamagnets, mean field and RPA calculations both suggest that the staggered susceptibility is singular from T^* to T_N along a path in the $H - T$ plane. Since this might be difficult to sort out experimentally, would you comment on what your smoothness postulate predicts?

R. GRIFFITHS: The smoothness postulate says that the staggered susceptibility would be infinite until you get to T_N itself where the curve is parallel to the H axis.

L. KADANOFF: R. Griffiths has been far too modest in ascribing most of the points in his talk to other authors. As far as I know at least the following points he made are original and significant contributions to our field:

  1. The use of "fields" to characterize critical behavior (along with Wheeler).
  2. The geometric characterization of Fig. 3.
  3. The concept of a tricritical point.
  4. Very pretty new arguments for the appearance of the specific heat in temperature derivatives of the ferromagnetic $\chi^{-1}(k)$, $\sigma_r \sigma_{r+s}$, and the antiferromagnetic χ.
  5. The particular algebraic statements of Eq. (1).[1] (This was also stated independently and roughly simultaneously by myself.[2])

REFERENCES

1. Paper presented for Newport Beach Conference on phase transitions (1970). Also, R. Griffiths: *Phys. Rev. Lett.*, **24**:1479.
2. Discussion at Newport Beach Conference on phase transitions. Invited paper at Washington meeting of APS (1970).

CRITICAL PROPERTIES OF FERROMAGNETS AND LIQUID CRYSTALS

J. D. Litster

Department of Physics
Massachusetts Institute of Technology
Cambridge, Massachusetts

ABSTRACT

Light scattering studies of the nematic-isotropic transition in p-methoxy benzylidene p-n-butylaniline show divergences similar to those observed near critical points. These divergences are interpreted using a mean field model and the properties of nematic liquid crystals are compared to those of ferromagnets. General properties of nematic liquid crystals and fluctuations in the ordered phase are also discussed.

1. INTRODUCTION

Although phase transitions are common occurrences in matter and have been studied for the past century, the detailed explanation of what happens to most substances undergoing a change of phase remains one of the most challenging problems in statistical mechanics and many body physics. Our greatest advances have come in understanding second-order (continuous) phase transitions, where experiments have shown that quite different materials exhibit divergences closely analogous to those observed near the critical point of a pure fluid.[1] The classical phenomenological theories of van der Waals, Weiss, and Bragg and Williams provided an attractive general point of view which was clearly stated by Landau in his theory of second-order phase transitions.[2] As more detailed experimental information became available, it was apparent that the classical theories did not provide a correct quantitative description of the behavior of many materials near the critical point. An indication of the theoretical way to proceed was provided by Onsager's exact solution of the two-dimensional Ising model which showed nonclassical critical behavior, and much subsequent theoretical effort has been devoted to extending these ideas to real three-dimensional systems.

Compared to critical phenomena, first-order phase transitions are rather poorly understood. This is undoubtedly a result of the fact that second-order transitions occur with ample warning and take place continuously. As a result one may carry out experimental studies of all stages of the phase transition; consequently we have more detailed information on critical properties than we do about first-order phase transitions which occur with little, if any, pretransitional phenomena and take place discontinuously. In this paper, I shall discuss the order-disorder transition that takes place in nematic liquid crystals. Although this is a first-order transition in all known liquid crystals,[3] I shall present evidence that over a wide temperature range nematic liquid crystals behave as if they were going to undergo second-order phase transition at a critical point. In particular there is a critical slowing down and a divergence in amplitude of fluctuations as well as a diverging "generalized susceptibility" in the vicinity of the transition.[4]

The order-disorder transition in liquid crystals is a change from an isotropic liquid state to a state (also liquid) with long-range orientational order of the anisotropic molecules that constitute the compound. The transition from the isotropic to the ordered state is discontinuous, as shown by a latent heat and density change. The simplest type of ordering occurs in nematic liquid crystals, and I shall confine my discussion to materials of this structural classification. In nematics, the centers of mass of the molecules are randomly distributed in both the ordered and disordered phases. The ordered state is characterized by a long-range parallel alignment of the molecules, without a preferred direction, analogous to the parallel alignment of spins in an isotropic ferromagnet. The molecules usually have uniaxial symmetry with both the electric polarizability and the diamagnetic susceptibility greater parallel to the long axis. An order parameter analogous to the magnetization may be used to specify the degree of alignment of the molecules, and because of the anisotropy in polarizability the order parameter may be directly determined by measuring the dielectric constant tensor of the liquid crystal. The name liquid crystal comes from this effect; a liquid crystal is a liquid with the optical anisotropy of a crystal. The order (and hence the optical properties) of liquid crystals is readily altered by external fields and this has led to much interest in using these materials for optical display devices.

2. A MODEL FOR NEMATIC LIQUID CRYSTALS

Since ordering in liquid crystal mesophases is analogous to ordering in magnetic systems it is natural to compare their similarities and differences. This can most easily be done within the framework of a specific model, and I will use the Landau model of phase transitions[2] (which is identical to the mean field model). This model is mathematically simple; it provides an adequate qualitative description of ferromagnets, and, as we shall see, an even better description of nematic liquid crystals.

To begin, let us write the cartesian dielectric constant tensor in a nematic liquid crystal as

$$\epsilon_{\alpha\beta} = \bar{\epsilon}\delta_{\alpha\beta} + Q\left(\frac{\Delta\epsilon}{3}\right)(3n_\alpha n_\beta - \delta_{\alpha\beta}) \tag{1}$$

where n_α, n_β are cartesian components of a unit vector parallel to the optic axis. An order parameter specifying the degree of alignment of each molecule is Q; this is analogous to the magnetization in a ferromagnet. Since there are no observable physical consequences to reversing the direction of a single molecule, the order parameter must have quadrupolar symmetry. (The order parameter in a magnet has dipolar symmetry.) Therefore an appropriate form for Q is[5] $Q = (1/2)\langle 3\cos^2\theta - 1\rangle$, where θ is the angle between the long axis of a molecule and the local optic axis. If ϵ_p and ϵ_t are the dielectric constants parallel and transverse to the optic axis of a perfectly ordered ($Q = 1$) liquid crystal, then $\bar{\epsilon} = (\epsilon_p + 2\epsilon_t)/3$ and $\Delta\epsilon = \epsilon_p - \epsilon_t$.

In the mean field approximation, the free energy density in the vicinity of the phase transition is written

$$\Phi = \Phi_0 + \frac{A}{2} Q^2 - \frac{B}{3} Q^3 + \frac{C}{4} Q^4 - \frac{\Delta\chi}{3} H^2 Q + D(\nabla Q)^2 + \cdots \quad (2)$$

with $A = a(T - T_c^*)$, and $\Delta\chi = \chi_p - \chi_t$ the anisotropy in the diamagnetic susceptibility of the liquid crystal. Except for the presence of the cubic term in Q and the magnetic energy varying as H^2, this is identical to the mean field expression for the free energy of a ferromagnet near its Curie point. The term Q^3 arises because of the quadrupolar symmetry of the ordered state and reflects the fact that positive and negative values of Q correspond to quite different physical arrangements of the molecules. (If $Q > 0$, the molecules order with their long axes parallel to the optic axis. When $Q < 0$, the molecules are randomly aligned with their long axes orthogonal to the optic axis.)

The equilibrium value of Q is the one that provides the lowest possible free energy. Minimizing Eq. (2) with respect to Q gives the equation of state (for a uniform system)

$$AQ - BQ^2 + CQ^3 - \frac{\Delta\chi H^2}{3} = 0 \quad (3)$$

When Eq. (3) has three real roots, two of them correspond to minima in the free energy. At temperatures well above T_c^* and in zero field the free energy is lowest when $Q = 0$, but at $T = T_K = T_c^* + 2B^2/9aC$ the free energy becomes lower for the minimum at finite Q. A first-order transition then occurs to the ordered state. The discontinuity in order parameter is $Q_K = 2B/3C$ and the latent heat is $aT_K Q_K^2/2$. This behavior is a consequence of the cubic term in Eq. (2); if B were zero, the liquid crystal would behave exactly as a ferromagnet and have a second-order phase transition at T_c^*.

In the isotropic phase well above T_K, the Q^3 term in Eq. (2) will be dominated by the quadratic term and the liquid crystal will behave in a similar manner to a ferromagnet. For example, we may use Eqs. (1) and (2) to calculate the birefringence induced by a magnetic field. The optic axis is along the magnetic field and the birefringence is

$$\Delta n = (\epsilon_{zz})^{1/2} - (\epsilon_{xx})^{1/2} \simeq \frac{\Delta\epsilon \Delta\chi H^2}{6(\bar{\epsilon})^{1/2} a(T - T_c^*)} \quad (4)$$

The magnetic birefringence therefore diverges in an analogous manner to the susceptibility of a ferromagnet.

We may also calculate the fluctuations in Q; for long wavelengths the mean squared fluctuations of wave vector q are[6]

$$\langle \delta Q^2(q) \rangle = \frac{KT}{V} \frac{1}{A(1 + \xi^2 q^2)} \quad (5)$$

where $\xi^2 = \xi_0^2 (T - T_c^*)^{-1} = 2D/A$ is the square of the correlation length for the fluctuations. We see there is also a divergence of the fluctuations in order parameter analogous to that observed near critical points.

The Landau model has been extended by DeGennes to include the time dependence of the fluctuations. In DeGennes' model[7] a restoring force $-\partial\Phi/\partial Q$ is equated to a damping force $\nu \partial Q/\partial t$, where ν is a transport coefficient. The resulting equation of motion yields exponentially damped fluctuations wth a relaxation time

$$\tau = \frac{\nu}{A + 2Dq^2} = \frac{\nu}{a(T - T_c^*)(1 + \xi^2 q^2)} \quad (6)$$

and so there is critical slowing of the fluctuations as the phase transition is approached.

3. EXPERIMENTAL STUDIES OF MBBA

I have used the mean field model to discuss the behavior of a nematic liquid crystal near its phase transition and to compare this behavior with the critical properties of a ferromagnet. Now I shall present the results of experimental studies of the isotropic phase of the nematic crystal *p*-methoxy benzylidene *p-n*-butylaniline (MBBA). These experiments were done in collaboration with Thomas W. Stinson at the Massachusetts Institute of Technology.[4]

As we have seen in the previous discussion optical methods are ideally suited to study ordering in liquid crystals. We have measured the magnetic birefringence of MBBA as a function of temperature in the isotropic phase and our results are shown for two different samples[8] in Fig. 1. Over a wide temperature range $H^2/\Delta n$ varies

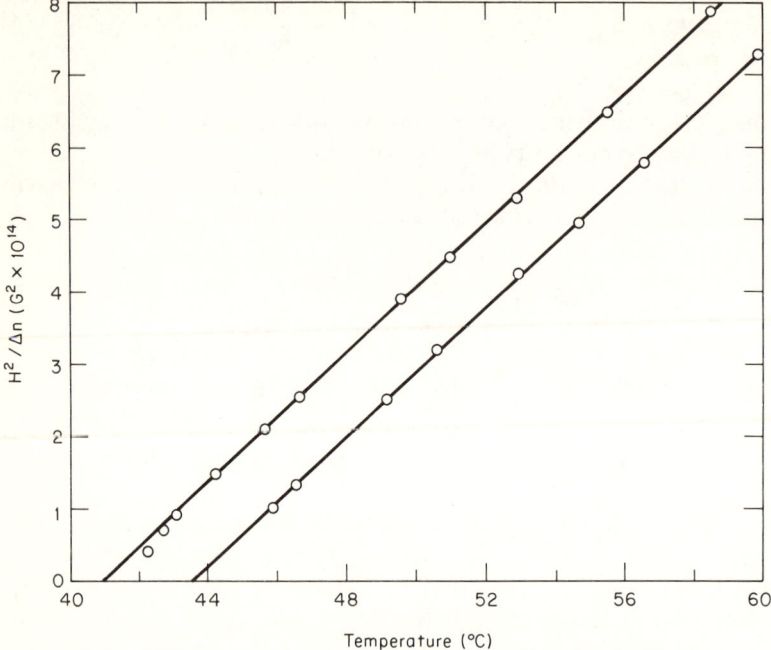

Fig. 1 *Temperature dependence of the magnetic birefringence in the isotropic phase of two different samples of MBBA. The solid lines are a fit to Eq. (4).*

linearly, and the mean field result Eq. (4) is in excellent agreement with our measurements. The solid lines extrapolate to zero at T_c^*, and for both samples we observed the first-order phase transition to occur at $T_K = T_c^* + 1.0°K$. There is a slight departure from the mean field result at temperatures very close to T_K, and this will be discussed later. The values of $\bar{\epsilon}$, $\Delta\epsilon$, and $\Delta\chi$ are all known for MBBA, and therefore we were able to use our magnetic birefringence measurements to determine the parameters in the free energy expression Eq. (2) to be[4] $a = 0.062$ J °K^{-1}-cm^{-3}, $B = 0.47$ Jcm^{-3}, and $C = 0.79$ J cm^{-3}.

The fluctuations in order parameter may be studied by light scattering. By varying the scattering angle one may study fluctuations whose wavelength ranges from the size of the sample to half the wavelength of light. The intensity of light scattered gives the mean squared fluctuations $\langle \delta Q^2 \rangle$ and from the spectrum one may obtain the time dependence of the fluctuations. The temperature dependence which we measured for the intensity is shown in Fig. 2. According to the mean field theory this should be proportional to Eq. (5). A reasonable estimate for ξ (we take $\xi_0 = 15\text{Å}$, about the maximum dimension of a molecule) shows that $\xi^2 q^2 \ll 1$ for visible light, which is in agreement with the fact that we observed no dependence upon scattering angle. The temperature dependence of the intensity is in excellent agreement with our birefringence measurements.

In Fig. 3 I show the measured half width of the Lorentzian spectrum ($\tau = 1/\Gamma$) of light scattered by MBBA.[9] The upper curve includes the instrumental resolution (1.65 MHz) and gives the temperature dependence of A/ν. When the results are corrected for the temperature dependence of ν^4, one obtains the linear dependence of the lower curve which is also consistent with the mean field prediction.

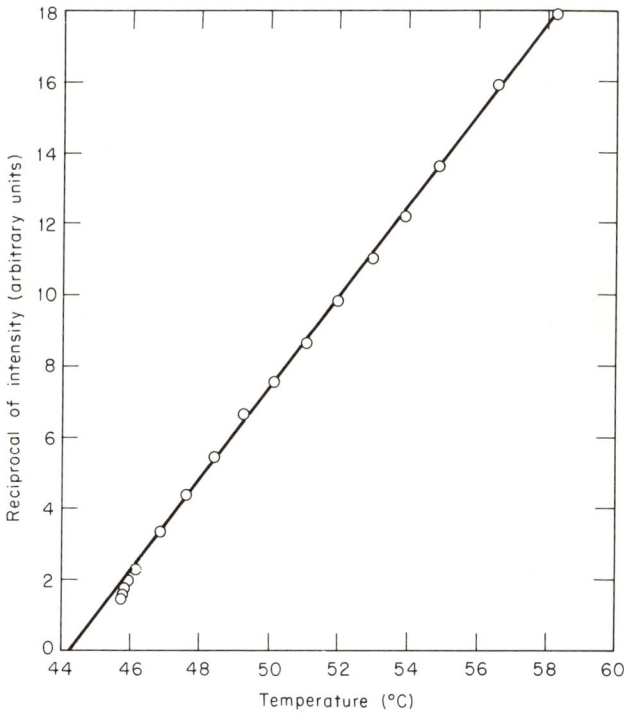

Fig. 2 *Temperature dependence of the intensity of light scattered by fluctuations in order parameter in the isotropic phase of MBBA. The solid line is a fit to Eq. (5).*

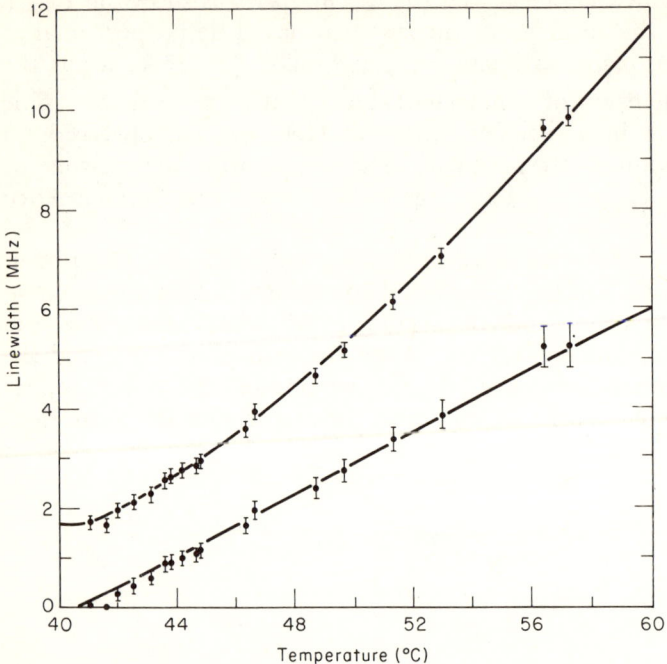

Fig. 3 *Half-width at half-height of the Lorentzian spectrum of light scattered by fluctuations in the order parameter in the isotropic phase of the nematic liquid crystal MBBA. The upper curve includes the instrumental width of 1.65 MHz and the temperature dependence of the transport coefficient ν. (See Eq. (6).) When the temperature dependence of ν is removed, in the lower plot the linear dependence is consistent with the mean field model.*

4. CRITICAL BEHAVIOR IN MBBA

In magnetic materials the mean field model provides only a qualitative description of behavior near the critical point.[1] The mean field approximation fails because it does not include the effects of fluctuations, which become increasingly important as the susceptibility diverges.[10] We have observed similar effects in MBBA. A close examination of Figs. 1 and 2 near T_K shows departures from the $(T - T_c^*)^{-1}$ dependence predicted by the mean field model. This departure is observed where the precision of our measurements is highest and is definitely a real effect.

We may use the mean field model to estimate the temperature at which departures from its predictions may be observable. We do this in the following way. Fluctuations in the liquid crystal are correlated over a finite range $\xi = \xi_0 [(T/T_c) - 1]^{-1/2}$, and so the increase in free energy associated with a fluctuation $\delta Q(\mathbf{r})$ at a point is approximately

$$\delta \Phi = v^* \left[\frac{A}{2} (\delta Q)^2 - \frac{B}{3} (\delta Q)^3 + \frac{C}{4} (\delta Q)^4 - \frac{\Delta \chi}{3} H^2 \delta Q \right] \quad (7)$$

where $v^* \simeq (4\pi/3) \xi^3$ is the correlation volume.

In Fig. 4 I show the free energy density as a function of Q for several isotherms. The curves were computed for a uniform system in zero field using our measured values of the parameters in Eq. (2).

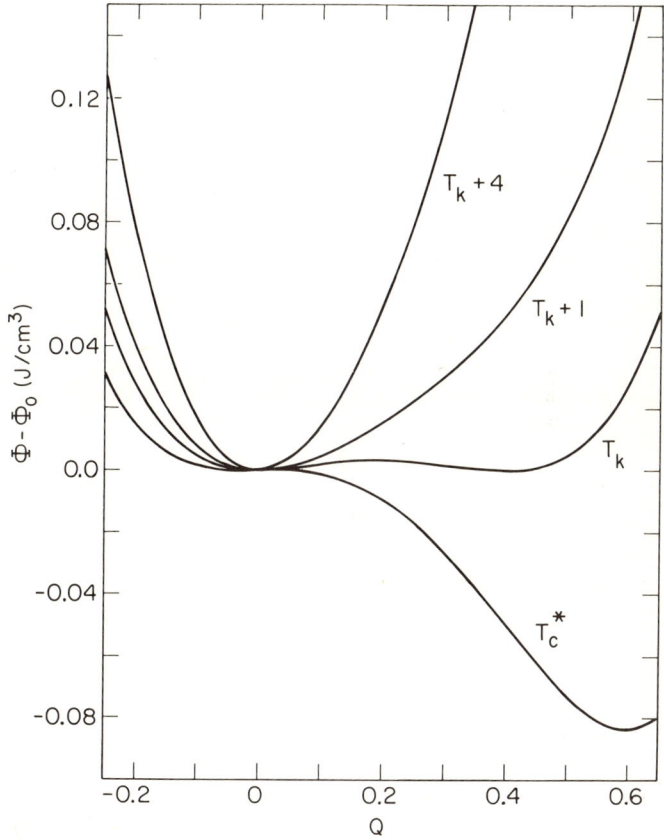

Fig. 4 *The free-energy density as a function of order parameter for several isotherms of MBBA. ($T_K = T_c^* + 1°K$; other isotherms are $T_K + 1°K$, $T_K + 4°K$.) The curves are computed from Eq. (2) with parameters obtained from Ref. 4.*

The isotherm for T_K shows two minima separated by a maximum of height $v^*(B^4/324C^3) \simeq 0.6kT$. Accordingly, we must expect increasingly large fluctuations as the phase transition is approached and the mean field approximation will fail. The equilibrium value of the order parameter (where the free energy has a minimum) remains the most probable value of Q. However because of large fluctuations and the asymmetric free energy curve, the thermodynamic average value of Q in an applied field will be larger than the most probable value. The fluctuating quantity δQ will not be a Gaussian random variable, and $\langle \delta Q^2 \rangle$ will be larger than predicted by Eq. (5). To estimate the magnitude of this effect, use Eq. (7) and take the probability of a fluctuation proportional to $\exp(-\delta\Phi/kT)$. Expanding the exponential for cubic and quartic terms gives the approximate result in zero field

$$\langle \delta Q^2(q) \rangle \simeq \frac{kT}{VA}\left[1 - \frac{3CkT}{v^* A^2} + \frac{5B^2 kT}{v^* A^3} + \cdots\right] \qquad (8)$$

The birefringence in Eq. (4) should also be multiplied by the expression in brackets in Eq. (8). Using Eq. (8), with $\xi_0 = 15$Å, we calculate the birefringence and the intensity of the scattered light should be about 6 percent greater than the mean field prediction when $T = T_K + 1°$K. This agrees well with our experimental results.

5. FLUCTUATIONS IN THE ORDERED PHASE

In the ordered phase of a nematic liquid there will not only be fluctuations in the order parameter Q, but also fluctuations in the unit vector **n**. These latter fluctuations correspond to normal modes analogous to spin waves in a ferromagnet, and give rise to intense scattering of light. The normal modes may be discussed using the Frank continuum model[11] for nematic liquid crystals. The possible distortions in **n** are: $\nabla \cdot \mathbf{n}$ [splay], $\mathbf{n} \cdot (\nabla \times \mathbf{n})$ [twist], and $\mathbf{n} \times (\nabla \times \mathbf{n})$ [bend]. The lowest energy state is a uniform vector field **n**, and fluctuations may be treated by adding elastic terms to the free energy expression (Eq. (2)). The additional terms are

$$\Phi_{el} = K_{11}[\nabla \cdot \mathbf{n}]^2 + K_{22}[\mathbf{n} \cdot (\nabla \times \mathbf{n})]^2 + K_{33}[\mathbf{n} \times (\nabla \times \mathbf{n})]^2 \qquad (9)$$

Additional terms included by Frank contribute only to surface energies and are omitted from Eq. (9). The normal modes of nematic liquid crystals have been discussed in detail by the Orsay group.[12]

For simplicity we take all three elastic constants equal to K and consider Q to be constant. When fluctuations $\delta\mathbf{n}(\mathbf{r})$ are expressed in terms of their Fourier components $\delta\mathbf{n}(\mathbf{q})$, the fluctuations in $\mathbf{n}$ are found to contribute a term to the free energy whose integral over the sample volume V is

$$\frac{V}{2} \sum_q (Kq^2 + \Delta\chi QH^2) \delta n^2(\mathbf{q})$$

These fluctuations cause large fluctuations in the off-diagonal components of the dielectric constant tensor and intense scattering of light. If $\mathbf{n}$ is taken along the z axis, we obtain from Eq. (1)

$$\langle \delta\epsilon_{zx}^2(\mathbf{q}) \rangle = \langle \delta\epsilon_{zy}^2(\mathbf{q}) \rangle = \frac{Q^2 \Delta\epsilon^2 kT}{V} \frac{1}{Kq^2 + \Delta\chi QH^2} \qquad (10)$$

It is interesting to compare this expression with that derived by Ornstein and Zernike for critical opalescence in a fluid[13]

$$\langle \delta\epsilon^2(\mathbf{q}) \rangle = \frac{kT}{V} \left(\frac{\partial\epsilon}{\partial\rho}\right)^2 \frac{1}{(1/\kappa_T) + \rho_n kT\xi_0^2 q^2} \qquad (11)$$

As the critical point of a fluid is approached, the isothermal compressibility (κ_T) diverges, leading to the well-known critical opalescence of fluids. Equation (10) suggests that a nematic liquid crystal in zero field is "critically opalescent" at all temperatures in the ordered phase. Calculations using Eqs. (10) and (11) show that MBBA in the ordered phase scatters light with the same intensity as Xe at the critical density 1 mdeg above the critical temperature.

Although Eqs. (10) and (11) are mathematically similar, the physical analogy is much closer between fluctuations in $\mathbf{n}$ and spin waves in an isotropic ferromagnet. In the isotropic ferromagnet there is no gap in the spin wave spectrum and the spin wave energies are proportional to q^2 in zero field. By measuring the intensity and spectrum of light scattered by the fluctuations in $\mathbf{n}$, one may determine the mean squared amplitude and time dependence of the normal modes of the liquid crystal. Such studies have been carried out for p-azoxyanisole at one temperature by the Orsay group[14] and throughout the nematic phase of MBBA by Haller and Litster.[15] The normal modes are overdamped and measurement of the relaxation times provides fundamental information on the hydrodynamics of anisotropic fluids as well as response times essential to the design of optical devices using liquid crystals.

6. SUMMARY AND CONCLUSIONS

We have seen that the order-disorder phase transition in nematic liquid crystals has much in common with magnetic and other critical points, but that the phase transition is ultimately a first-order one because of the quadrupolar symmetry of the ordered phase.

There are other structural classes (smectic and cholesteric) of liquid crystals to be studied in addition to nematics, and there remains a good deal of exciting physics to be done with these interesting materials. Studies of the ordered phase can provide insight into the hydrodynamics of anisotropic fluids, and experiments on phase transition will elucidate the role of symmetry as well as other interesting features of order-disorder phase transitions. Many biological materials, such as tobacco mosaic virus and transfer ribonucleic acid, exhibit liquid crystalline properties. It would be interesting to learn how this relates to their biological function.

REFERENCES

1. Heller, P.: *Rep. Prog. Phys.*, **30**:731 (1967).
2. Landau, L. D., and E. M. Lifshitz: "Statistical Physics," chap. 14, Addison-Wesley Publishing Co., Reading, Mass., 1958.
3. A useful review article is: I. G. Chistyakov: *Usp. Fiz. Nauk.*, **89**:563 (1966) *(Sov. Phys.—Uspekhi,* **9**:551 (1967)).
4. Stinson, T. W., III, and J. D. Litster: *Phys. Rev. Lett.*, **25**:503 (1970).
5. Maier, W., and A. Saupe: *Z. Naturforsch.*, **14A**:882 (1959) and *ibid.*, **15A**:287 (1960).
6. Landau, L. D., and E. M. Lifshitz: *Ibid.*, para. 116.
7. De Gennes, P. G.: *Phys. Lett.*, **30A**:454 (1969).
8. The transition temperature often varies from sample to sample of liquid crystal, probably as a result of hydrolysis of the MBBA into its amine and aldehyde constituents. Although T_C^* shows considerable variation from sample to sample, the other parameters in Eq. (2) remain constant within the sensitivity of our measurements.
9. Litster, J. D., and T. W. Stinson III: *J. Appl. Phys.*, **41**:996 (1970).
10. Kadanoff, L. P. et al.: *Rev. Mod. Phys.*, **39**:395 (1967).
11. Frank, F. C.: *Disc. Faraday Soc.*, **25**:19 (1958).
12. Groupe d'Etude des Cristaux Liquides (Orsay): *J. Chem. Phys.*, **51**:816 (1969).
13. Ornstein, L. S., and F. Zernike: *Proc. Acad. Sci. Amsterdam*, **17**:793 (1914); *Z. Physik*, **19**:134 (1918).
14. Orsay Liquid Crystal Group: *Phys. Rev. Lett.*, **22**:1361 (1969).
15. Haller, Ivan, and J. D. Litster: *Mol. Cryst. and Liq. Cryst.*, **12**:277 (1971).

DISCUSSION on *Paper* by *J. D. Litster*

C. DOMB: Am I right in understanding the interaction you are assuming as quadrupolar between molecules? If so since this depends on distance and the molecules are random in position, the magnetic analog is a random ferromagnet such as discussed by McCoy. More refined experimental results near the critical temperature should therefore be of great interest.

J. D. LITSTER: The origin of the forces in liquid crystal is presumably van der Waals forces, so they are extremely short ranged and my guess is, that unless the molecules fit right next to each other, they don't interact too strongly. But the details of what happened in a liquid crystal I don't know. The *only* theory (of Maier and Saupe) developed so far uses quadrupolar forces.

B. McCOY: The mean field theory presented to discuss the experimental results depends only on the symmetry of the order parameter and does not refer to the fact that the position of the center of masses of the molecules are randomly distributed. It is known from the studies of the rigid Ising model and the Syozi Ising model that spatially ordered and spatially mobile spin systems have different critical behavior of their spin properties. Therefore, the usefulness of mean field theory is somewhat obscure. It might be helpful to construct a lattice model which incorporates the symmetry properties of the liquid crystal order parameter and study its critical properties by means of series expansions.

J. D. LITSTER: I don't know whether the mean field model accounts for what really goes on in a liquid crystal, but in our case, the mean field theory simply works.

P. C. CLAPP: I just wish to comment that contrary to the expectation you mentioned that a first-order transition will not be anticipated in the high-temperature phase, all of the binary alloy order-disorder phase transitions that I am familiar with have very clear precursors well above T_c. Your first-order transitions also show considerable precursor behavior and one wonders if in fact this is not generally true for all first-order phase transitions.

J. D. LITSTER: My intuition would be that the reason that the two minima are aware of each other is because of thermodynamical fluctuations. How much they know about each other depends on how much it costs in free energy to create the fluctuations.

L. ONSAGER: I hate to criticize such an interesting talk; but the evidence concerning the critical behavior is a bit ambiguous, because the extrapolated critical point lies about one degree below the critical point. This is about the shift one expects from variation in the concentration of impurities. That means you may be dealing with freezing point depressions. As you freeze you will be concentrating the solute in the isotropic phase very likely and you actually get a transition at a lower temperature. My remark is only a cautionary warning.

J. D. LITSTER: It is possible. The only thing I can say is that the transition is quite sharp (with a few mdeg) and shows no sign of superheating or supercooling.

G. ALEFELD: First, I was pleased to see that you used in your description for liquid crystals an order parameter, which was as far as I could see is identical to the one I used for the two-phase transitions for hydrogen in metals. Is there any indication, that you may have a similar case as I have shown yesterday, namely that the triple point between liquid anisotropic, liquid isotropic, and gas moves towards the critical point and makes it disappear?

Second, do you have any information on the transition solid–liquid crystal? One can imagine, that this transition is less abrupt as usually melting is. Therefore, there may be some forewarning to this melting transition. The Orsay group has apparently some results on this question.

J. D. LITSTER: The first is not known. There is even no knowledge about the pressure dependence of the phase-transition temperature. The problem with the liquid crystals is that if you try to approach the critical temperature the molecules come apart.

Regarding the second, I am not familiar with these results. The transition from the solid to the liquid crystal state is much more first order compared to the case you are considering.

L. GUTTMAN: It might be interesting to add some neutral diluent, whose solubility should be as nearly the same as possible in both phases, so that the transition might become more like a critical point.

J. D. LITSTER: It would be more second order, I suspect. I would like to put impurities into the liquid crystals.

M. L. GLASSER: The Mössbauer effect as already observed in some liquid crystals also allows one to study the behavior of the coupling during the crystal–(nematic) liquid crystal transition.

J. D. LITSTER: Yes, I think so, but so far in nematic crystals, there are no Mössbauer measurements.

P. C. MARTIN: I wanted to remark on the best analogies to these type of transitions in view of Professor Domb's analogy. Fixed random impurities are not a very good analogy. Two better analogies are a ferromagnetic liquid with strong coupling between position and orientation or even more simple a gas of hard rods which tend to order as a function of pressure. In fact, one can even demonstrate the transition, in the Bernal fashion. But Meyers has shown that by shaking "rice" into a box and then pressing, one can produce the ordering.

SCALED EQUATION OF STATE AND DERIVED PROPERTIES

M. Vicentini-Missoni

Instituto di Fisica
Università di Roma, Italy

ABSTRACT

The "scaling law" hypothesis for thermodynamic functions at the critical point is discussed in relation with experimental data on magnets. Empirical forms of the asymptotic equation of state are reviewed. Properties of the exact equation of state in the parametric formulation are analyzed in terms of known models. A form of the equation of state which gives a good approximation to the Ising model and real substances while reducing to the exact equation for classical models is proposed.

1. INTRODUCTION

The asymptotic equation of state of known models has the property that it can be written in scaled form. The assumption that the same property holds for real substances leads to the existence of relations between critical exponents the validity of which can be taken as a test of "scaling." A more complete test is given by checking the form of the equation of state itself, that is, testing that in the appropriate variables all data in the critical region fall on a universal curve. Empirical forms for the scaled equation of state will be discussed in Sec. 2. In Sec. 3 the use of the parametric formulation and properties of the parametric equation of state will be illustrated. A correction to the "linear model" parametric equation of state which has the property that the form of the equation depends only on the values of the critical exponents, is shown to give a good approximation for the two-dimensional Ising model as well as the three-dimensional and real cases.

2. SCALED EQUATION OF STATE AND REAL SUBSTANCES

Only in two models is the complete form of the equation of state known in the vicinity of a critical point. They are the mean field model and the Bose gas model.[1] The equations are

$$H = M^3\left(\frac{t}{M^2} + x_0\right) + M^5\left(\frac{t}{3M^2} + \frac{1}{5}\right) + \cdots \qquad (1)$$

$$H = \frac{4}{9} M^5\left(\frac{t}{M^2} + x_0\right)^2$$
$$+ M^7\left[a_1\left(1 + \frac{3t}{2M^2}\right)^3 - 3\left(1 + \frac{3t}{2M^2}\right)\left(1 + \frac{25t}{8M^2}\right)\frac{t}{M^2}\right] + \cdots \qquad (2)$$

Both these models are classical in the sense that they can be derived from a free energy which has a series expansion in both variables at the critical point. It has since been recognized that real systems do contradict these models as the observed values of the critical exponents are different from the classical values; this fact leads to abandoning the hypothesis of analyticity of the thermodynamic functions at the critical point. The thermodynamic hypothesis which can still be assumed to be valid in the critical region has been very clearly exposed by Griffiths.[2]

The scaling hypothesis, from a thermodynamic point of view, corresponds to the assumption that the asymptotic form of the equation of state resembles the classical case and can be written

$$H = M|M|^{\delta-1} h(x) \qquad (3)$$

$$x = \frac{t}{|M|^{1/\beta}} \qquad t = \frac{(T - T_c)}{T_c} \qquad (4)$$

$h(x)$ is an analytical function of x in its range of definition: $-x_0/\infty$.

The equation reproduces exactly the phase boundary and the critical isotherm and implies the relations between critical exponents

$$\gamma = \gamma' = \beta(\delta - 1) \qquad 2 - \alpha = 2 - \alpha' = \beta(\delta + 1) \qquad (5)$$

These relations can be taken as a test of the validity of scaling. The form of the series expansion of $h(x)$ for large x is given by

$$h(x) = \sum_0^\infty \eta_{n+1} x^{\gamma - 2\beta n}$$

while around any other finite point $\bar{x}$ a power series expansion is valid. It is clear that, if $2\beta \neq 1$, it is difficult to match the form of these expansions and therefore to find a closed form expression for $h(x)$ in the field of real functions. On the other hand scaling, if valid, shows a number of problems connected with the determination of the exponents. An example is given by the fact that if all isotherms behave like the critical for large values of M, they do show discrepancies for low M that are such as to give lower H if $T < 0$ but larger H if $t > 0$. Then if one is not exactly at the critical temperature depending on whether the temperature is slightly above or below T_c, the value of δ can be biased by defect or excess. If scaling is valid, one can use the data on any isotherm provided the value of M is large enough. A closed form expression for $h(x)$ could give a representation of the data to the critical point and could also enable us to determine the values of the exponents using all data in the critical region instead of just the data on preferred curves.

Three closed form expressions have been proposed[3-5]

$$h_{an}(x) = K(x + x_0)^\gamma \qquad (6)$$

$$h_{hl}(x) = A(x + x_0)(x + c)^{\gamma - 1} \qquad (7)$$

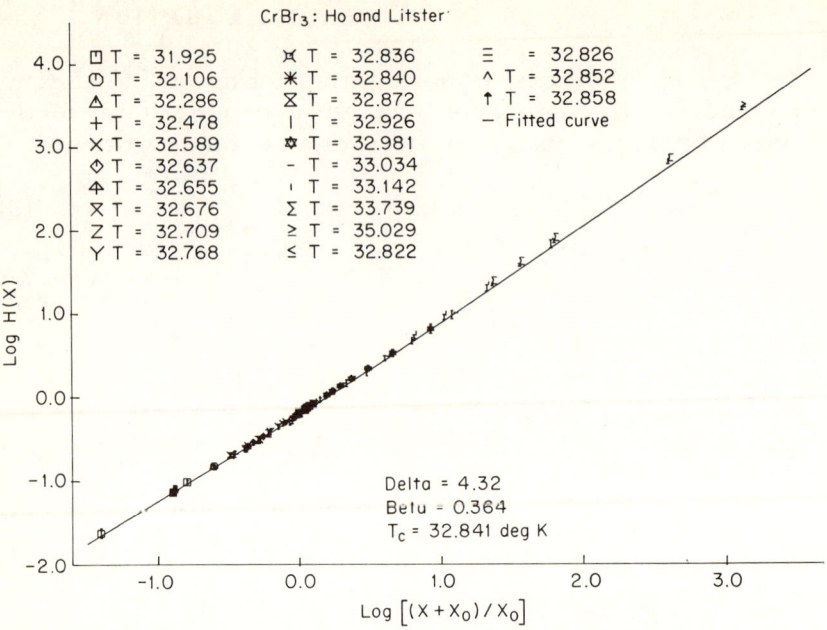

Fig. 1 *Agreement between $h_{msg}(x)$ and the experimental data on $CrBr_3$.*

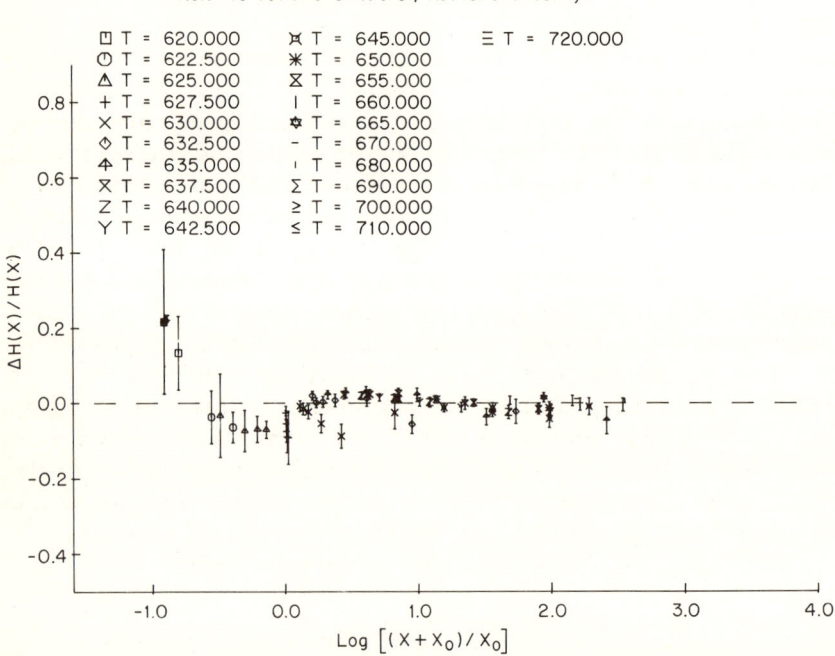

Fig. 2 *Relative deviation $[h_{msg}(x) - h_{exp}(x)]/h_{exp}(x)$ versus $\log[(x+x_o)/x_o]$ for Ni.*

$$h_{msg}(x) = E_1 \left(\frac{x+x_0}{x_0}\right)\left[1 + E_2\left(\frac{x+x_0}{x_0}\right)^{2\beta}\right]^{(\gamma-1)/2\beta} \tag{8}$$

All three satisfy some of the predicted properties of $h(x)$. The first and the third are nonanalytical at the phase boundary (but in the first one has an infinite susceptibility all along it while in the third the singularity is at higher order and could be removed by introducing another parameter). The third is the only one which has two terms of the correct form in the expansion at large x.

Only the third[5,6] has been used on a variety of systems (magnets and fluids) to determine by a nonlinear least-squares method from all the data in the critical region two of the critical exponents. An example of the agreement between the form $h_{msg}(x)$ and the experimental data is given in Fig. 1 for the data on $CrBr_3$ and in Fig. 2 for the data on Ni^7 in deviation form. From the independent analysis of the data on $CrBr_3$ it has been possible to verify that the function correctly reproduces for real cases the coefficients x_0, h_0, h_1, η_1.

Unfortunately, in the recent past few substances have been investigated with accuracy in the whole critical region. Attention has been focused mainly on the determination of the exponents. Some old data like the Weiss data on Ni^8 or the Michels data[9] on CO_2 for fluids are available.

The values of the exponents determined for magnets are given in Table 1. The agreement with the determination made by analysis of data on preferred curves is in the error for $CrBr_3$ and the Ni data of Ref. 7. For Gd^{10} and the Ni data of Ref. 8 the discrepancies are explained by the different value of the critical temperature.

New data are, of course, needed also on the specific heat since the behavior of the specific heat can be derived from the $h(x)$. For fluids a comparison between calculated and measured values has shown

TABLE 1. Critical Exponents for Magnets as Derived by Least-Squares Fit of $h_{msg}(x)$ to Experimental Data

	$T_c, °K$	x_0	β	δ	γ
$CrBr_3$	32.841 ± 0.008	0.596 ± 0.010	0.364 ± 0.005	4.32 ± 0.10	1.21
Gd	292.05 ± 0.15	0.670 ± 0.050	0.370 ± 0.010	4.39 ± 0.10	1.25
Ni	627.4 ± 0.3	0.376 ± 0.047	0.373 ± 0.016	4.44 ± 0.18	1.28
Ni	626.5 ± 0.2	0.363 ± 0.030	0.375 ± 0.013	4.48 ± 0.14	1.31

reasonable agreement. For magnets the problem is still open due to experimental difficulties in the determination of the specific heat in a magnetic field.

Another problem which is important both for the equation-of-state data and the specific heat is the fact that all theoretical hypotheses and the scaling property between the data refer to the asymptotic behavior and there are no indications on the range in which such behavior should be observed.

Therefore, on one hand, more data are needed as close as possible to the critical point, theoretical suggestions are also needed on the form and importance of correction terms.

The three closed form expressions have recently been compared with the equation of state of the Ising model [11] showing that while $h_{an}(x)$ gives a bad representation for both the two-dimensional and the three-dimensional model, $h_{hl}(x)$ and $h_{msg}(x)$ are both bad representations for the two-dimensional model but can give a good representation of the three-dimensional, the form $h_{msg}(x)$ being better in the sense that it gives a maximum discrepancy of 3.3 percent for any value of x while $h_{hl}(x)$ gives a maximum discrepancy of 7.8 percent. However for x close to $-x_0$, $h_{hl}(x)$ should be better because of the absence of singularity at the phase boundary.

3. PARAMETRIC REPRESENTATION

In searching a closed form expression for the scaled function $h(x)$ one encounters a number of difficulties due both to the fact that the exponents β and δ are not respectively the inverse of an even integer and an odd integer, and to the infinite range of the variable x. It is impossible in the field of real functions to match the form of the series expansions of $h(x)$ around $x = 0$ and for $x \to \infty$ and one should try with the real part of some complex function.

The parametric representation, introduced independently by Josephson [12] and Schofield [13] avoids these problems while preserving the scaling property.

Very briefly, the state of a system in the vicinity of a critical point, usually represented by the variables H, M, $T - T_c$, can be represented introducing a "radial" parameter R and an "angular" parameter θ. The radial parameter measures the distance from the critical point, and in consequence of this the critical point will be reached only if $R = 0$ for all values of θ. For any other value of R the system will be in the critical region away from the critical point; therefore, according to the thermodynamic assumptions of scaling, all thermodynamic functions will behave analytically in θ in the range corresponding to the one-phase region or to the stability region.

Scaling will be obeyed if one assumes that the free energy dependence on R and θ is of the form

$$f(R, \theta) = R^{\beta(\delta+1)} f(\theta) \tag{9}$$

while for H, M, t, we have the relations

$$H = R^{\beta\delta} h(\theta) \quad M = R^{\beta} m(\theta) \quad t = R t(\theta) \tag{10}$$

Assuming that one knows the transformation equations for either two of the three variables, the third relation will give the equation of state in the parametric formulation.

A scale on the θ coordinate is chosen by imposing the values of θ corresponding to the critical isochore, critical isotherm, and the phase boundary. A choice is:

$\theta = 0$ critical isochore $\quad h(0) = 0 = m(0)$
$\theta^2 = 1/a$ critical isotherm $\quad t[(1/a)^{1/2}] = 0$
$\theta^2 = 1/c$ phase boundary $\quad h[(1/c)^{1/2}] = 0, m[(1/c)^{1/2}]$ finite

The existence of a spinodal and the requirement of analyticity of the equation of state in the whole range up to the stability limit is translated in analyticity in the range from $\theta = 0$ to $\theta_s = (1/s)^{1/2}$ where s is determined by $\chi_T^{-1}(\theta_s) = 0$.

Symmetry properties when taken into consideration suggest an odd function of θ for $h(\theta)$ and $m(\theta)$ and an even function for $t(\theta)$. Schofield, assuming the simple transformation equations

$$\begin{aligned} h(\theta) &= b\theta(1 - c\theta^2) \\ t(\theta) &= 1 - a\theta^2 \end{aligned} \tag{11}$$

observes that $m(\theta)$ for real substances looks linear as it is in the mean field model. There is only a slight difference between the two cases. In the mean field theory $m(\theta)$ is linear for all values of a/c; in the real substances the approximation to a linear $m(\theta)$ is best only for one value of a/c,[14] precisely

$$\frac{a}{c} = \frac{\delta - 3}{(\delta - 1)(1 - 2\beta)} \tag{12}$$

There has been divergence of opinion [14,15] on the question, "is $m(\theta)$ a linear function of θ or is it only approximated by a linear

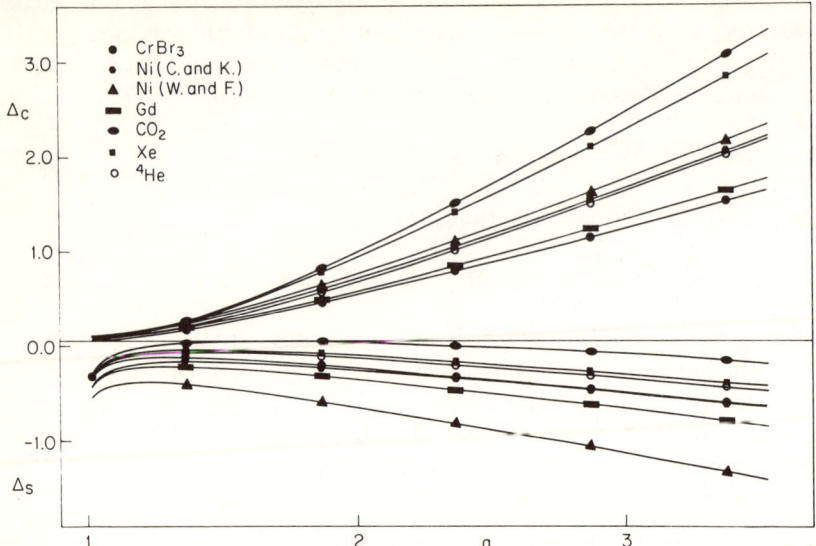

Fig. 3 *Deviations from the experimental values of the coexistence curve* (Δ_c) *and of the slope at the critical isotherm* (Δ_s) *calculated by the "linear model" as functions of the parameter "a" for c = 1.*

function for real substances?" The indications that it may be considered only as an approximation are clearly illustrated in Fig. 3 where the deviations from linearity as represented by the difference between the observed values and calculated values on the phase boundary (Δ_c) and on the slope at the critical isotherm (Δ_s) are plotted as a function of a for $c = 1$. Indeed, both deviations are close to zero around a value given by Eq. (12). However for all substances systematically the value for which $\Delta_c = 0$ is different from the value for which $\Delta_s = 0$. The systematic deviations from the "linear model" have since been conceded and the model may be considered now as a satisfactory description of the experimental data. But it is excluded that it represents the exact equation of state.

Moreover a study of the series expansion for $h(x)$ for the Ising model definitely rules out the linearity for the two-dimensional model while for the three-dimensional one again it is only an approximation.

A nice feature of the linear model is the fact that the ratio a/c depends only on critical exponents (it is undetermined for $\beta = 1/2$, $\delta = 3$ reflecting the linearity for all a/c in the mean field model). Another feature is that the specific heat c_M is θ independent. This property could be checked experimentally if one had data on the specific heat as a function of temperature and magnetization. These data are not presently available. However scaling and the parametric formulation can be translated to fluid language where data on the specific heat varying density and temperature have been obtained by Moldover[16] on He_4.

In the "linear model" $[m(\theta) = g\theta]$ the specific heat is constant along the curves

$$t = R - \left(\frac{a}{g^2}\right) R^{1-2\beta} |\Delta\rho|^2 \qquad (13)$$

Different R values are evaluated starting from the specific heat along the critical isochore $(t = R)$; one can then construct straight lines t, $|\Delta\rho|^2$ and compare the slope with the calculated value. The data are scarce but one obtains a value $g^2 = 1.8$ to be compared with the calculated value $g^2 = 1$. Also these data then seem to indicate a discrepancy with the predictions of the "linear model."

Another problem for the linear model to be valid for all values of the critical exponents arises with the Bose gas model. In this case, if one assumes the same transformation for the temperature as assumed by Schofield, and a linear function for $m(\theta)$, one derives from the known form of the equation of state in the $h(x)$ representation $[h(x) = A(x + x_0)^2]$ to the equation of state in the $h(\theta)$ representation

$$h(\theta) = b\theta(1 - c\theta^2)^2 \qquad (14)$$

which is definitely different from the $h(\theta)$ of the "linear model."

Moreover if one tries a function of the form Eq. (14) for general values of β and δ following the same procedures used by Litster, Schofield, and Ho with their transformations, one finds that a good approximation to the real $h(\theta)$ will be found only if

$$\frac{a}{c} = \frac{\delta - 5}{(1 - 2\beta)(\delta - 1)} \qquad (15)$$

(which again being undetermined for $\beta = 1/2$, $\delta = 5$ reflects the fact that for these values of the exponents the value a/c should be unimportant, $h(\theta)$ in Eq. (14) is the exact solution for any a/c).

The best way to understand the properties of a parametric transformation is to derive from the known properties of the $h(x)$ function the series expansion of the equation of state in the θ language.

We will follow the procedure of assuming the transformations for M and t in the simple form: $M = R^\beta g\theta$; $t = R(1 - a\theta^2)$ and derive the properties of $h(\theta)$. Indicate by $\theta^2 = 1/c$ the position of the phase boundary.

$h(\theta)$ must be analytic in the whole range of θ extending from the critical isochore to the phase boundary and eventually to a spinodal $\theta = \theta_s$; it is zero at $\theta = 0$ and $\theta^2 = 1/c$ and its series expansion around $\theta = 0, \theta^2 = 1/a, \theta^2 = 1/c$ are given by

$\theta \simeq 0$

$$h(\theta) = g\theta \left\{ \eta_1 + \theta^2(-\gamma a \eta_1 + \eta_2 g^2) + \theta^4 \left[\frac{1}{2}(\gamma - 1)\gamma \eta_1 a^2 - \eta_2 g^2 (\gamma - 2\beta) a + \eta_3 g^4 \right] + \cdots \right\} \quad (16)$$

$\theta^2 \simeq \dfrac{1}{a}$

$$h(\theta) = \theta \left\{ h_0 g^\delta a^{(1-\delta)/2} + (1 - a\theta^2) \left[-\frac{1}{2}(\gamma - 1) h_0 g^\delta a^{(1-\delta)/2} + h_1 g^{\delta - 1/\beta} a^{1 - \delta/2 + 1/2\beta} \right] + \cdots \right\} \quad (17)$$

$\theta^2 \simeq \dfrac{1}{c}$; $x_0 + \left(\dfrac{c-a}{c} \right) \left(\dfrac{c}{g^2} \right)^{1/2\beta} = 0$

$$h(\theta) = \theta g^{\delta - 1/\beta} c^{(1-\delta+1/\beta)/2 - 1} (c - a + 2\beta a) \left(\frac{1}{2\beta} \right)(1 - c\theta^2)$$

$$\left[f_1 + (1 - c\theta^2) \left[f_2 \left(\frac{c}{g^2} \right)^{1/2\beta} \frac{(c - a\, 2\beta a)}{2\beta c} \right. \right.$$

$$\left. \left. + f_1 \left(\frac{1}{2}(1 - \delta) + \frac{(c - a + 2\beta a + 2\beta c)}{4\beta (c - a + 2\beta a)} \right) \right] + \cdots \right] \quad (18)$$

The coefficients h_i, η_i, f_i are respectively the coefficients in the series expansions of $h(x)$ near $x = 0$, $x \to \infty$, $x = -x_0$. The constant g is unessential and may be assumed unity.

Comparison between the three series expansions shows a number of relations which must be satisfied between the coefficients; the first of these is

$$\eta_1 = h_0 a^{(1-\delta)/2} \left[1 - \frac{\delta - 1}{2} + \frac{(\delta - 1)(\delta - 3)}{8} - \frac{(\delta - 1)(\delta - 3)(\delta - 5)}{48} \right.$$

$$\left. + \cdots \right] + h_1 a^{(1-\delta+1/\beta)/2} \left[1 - \frac{\gamma - 1}{2\beta} + \frac{\frac{1}{2}(\gamma - 1)(\gamma - 2\beta - 1)}{4\beta^2} \right.$$

$$\left. + \cdots \right] + \cdots \quad (19)$$

This relation will be independent of "a" only if δ is an odd integer, $\beta = 1/2$ and $h_{\delta-\gamma} = h_{\delta-\gamma+1} = \cdots = 0$, in particular for:

$\delta = 3$, $\beta = 1/2$, $h_2 = h_3 = \cdots = 0$ (mean field model); $\delta = 5$, $\beta = 1/2$, $h_3 = h_4 = \cdots = 0$ (Bose gas model), it gives the correct relations between the coefficients η_i and h_i appropriate to the two models: $\eta_1 = h_1$ in the mean field case and $\eta_1 = h_2$ in the Bose gas case.

In all other cases therefore the relations between the η_i and the h_i will depend on the value of "a": only one particular value of "a" will give agreement for the series expansions if it satisfies simultaneously all the relations. On the other hand if we want a polynomial in θ to be the exact $h(\theta)$, "a" should satisfy another set of relations given by equating to zero all the coefficients of the terms in θ at a power larger than the degree of polynomial.

It seems more reasonable to suppose that $h(\theta)$ has an infinite series near $\theta = 0$ and thus to have only one set of relations to be satisfied.

A generalization of the "linear model" equation of state $h(\theta) = b\theta(1 - c\theta^2)$ which includes both the classical models and has an infinite series expansion near $\theta = 0$, is found in the expression

$$h(\theta) = b\theta(1 - c\theta^2)(1 - d\theta^2)^p \quad (20)$$

with p determined by the values of the exponents and such as to reduce to zero if $\beta = 1/2$, $\delta = 3$ and to unity if $\beta = 1/2$, $\delta = 5$.

Equation (20) is nonanalytical; if p is not an integer at $\theta^2 = 1/d$, one must then require that this point be out of the stability range.

Between the constants a, c, d, and the constant s characterizing the position of the spinodal the following relations must be satisfied

TABLE 2. Series Expansion Coefficients of the Ising Model

	p	η_1	η_2	η_3	η_4	h_0	h_1	h_2	f_1	f_2	X_s/X_0
II	0.375	1.0387	0.8479	0.7495 ± 0.02	0.6801 ± 0.07	0.4255 ± 0.005	2.628 ± 0.04	1.978*	1.4588 ± 0.03	5.1662	1.08
	0	/	/	0.773	0.625	0.436	2.57	1.965	1.60	3.638	1.31
	0	/	1.13	0.527	0.617	0.598	2.49	1.73	2.34	2.69	/
III	0	1.0097	0.5758	0.2419 ± 0.01	0.0637	0.345	1.384	0.252	1.1528	0.4714	1.8
	0.0625	/	/	0.218	0.030	0.341	1.35	0.228	1.18	0.378	1.79
	0†	/	0.590	0.205	0.029	0.345	1.34	0.222	1.20	0.349	/

*Onsager exact result.
†The case $p=0$ corresponds to the "linear model."

$$a > c > s > d$$
$$s > a(1 - 2\beta) \qquad (21)$$

The last relation is required because in correspondence to $\theta^2 = 1/a(1 - 2\beta)$ there is a pole in the inverse susceptibility.

Comparison of the series expansions of $h(\theta)$ as given by Eq. (20) again around $\theta = 0$, $\theta^2 = 1/a$, $\theta^2 = 1/c$ and the expansions (16), (17), and (18) gives a set of equations from which, in principle, it should be possible to determine the constants a, c, d, b, p, and to check the consistency (the number of equations is larger than the number of constants to be determined).

A numerical solution can be found for the two-dimensional Ising model using the values of x_0, η_i, h_i, f_i, calculated by Domb and Gaunt,[11] which satisfy simultaneously all the equations involving η_1, η_2, η_3, h_0, h_1, h_2, f_1, that is seven equations in five constants. The value of p turns out to be of the order of 0.3.

In order to find a solution for real cases one would need to know at least five of the coefficients η_i, h_i, or f_i. On the other hand, if one had an a priori choice for p and the ratio a/c in terms of the exponents, one could have an equation which, with the knowledge of η_1, x_0, η_2 (or h_0) could give a representation on the whole range of θ from the critical isochore to the spinodal.

The combination of exponents $p = \beta\delta - 3/2$ gives the correct value of p for the mean field model and the Bose gas model and is of the right order of magnitude for the two-dimensional Ising model.

A generalization of the expression for the best value of the ratio a/c derived by using the $h(\theta)$ appropriate to the classical models, will give as a first guess $a/c = (\delta - 3 - 4\beta p)/(\delta - 1)(1 - 2\beta)$.

The comparison between the coefficients η_i, h_i, f_i as calculated by Eq. (20) with said expressions of p and a/c in terms of the exponents, and the values in Ref. 11 for the Ising model are shown in Table 2. The values calculated using the "linear model" are also shown.

The improvement for the two-dimensional Ising model is clear. For the three-dimensional case the improvement is very slight, except for the correct position of the spinodal. Higher order coefficients are in both cases still outside of the errors in the exact values of Ref. 11.

For real substances Eq. (20) has been compared with the experimental data on $CrBr_3$. The deviation plots both for the linear model and the present correction to it are shown superimposed in Fig. 4. The systematicity of the deviations has not changed form but its entity has been reduced: the correction is at least in the right direction.

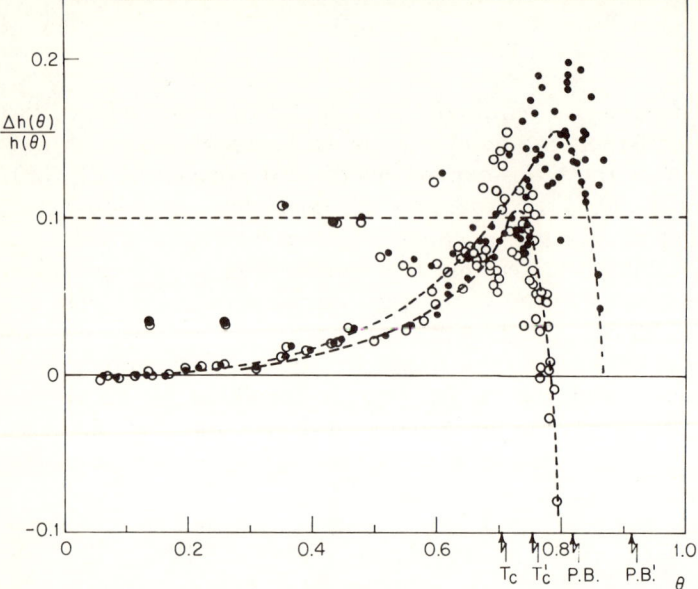

Fig. 4 *Deviations from the experimental values of $h(\vartheta)$ of the values calculated using the "linear model" (closed symbols) and the suggested correction (open symbols).*

In conclusion we have shown that a generalization of the "linear model" parametric equation of state to a form determined by the values of the critical exponents can give a good representation of the two-dimensional Ising model equation of state while reducing to the exact form for the classical models. The generalization preserves the feature that the relative position of the critical isotherm and the phase boundary is also determined by the exponents, with just another unknown parameter. The correction gives also a better approximation to the true equation of state of real substances, and predicts in all cases a spinodal as a limit of stability.

ACKNOWLEDGMENT

This work has been done in collaboration with Dr. F. Ferro Luzi of the University of Rome.

REFERENCES

1. Cooper, M. J., and M. S. Green: *Phys. Rev.*, **176**:302 (1968).
2. Griffiths, R. B.: *Ibid.*, **158**:176 (1967).

3. Arrott, A., and J. E. Noakes: *Phys. Rev. Lett.*, **19**:786 (1967).
4. Ho, J. T., and J. D. Litster: *Ibid.*, **22**:603 (1969).
5. Vicentini-Missoni, M., J. M. H. Levelt Sengers, and M. S. Green: *Ibid.*, **22**:309 (1969); *J. Res. Nat. Bureau Standards*, **73A**:563 (1969).
6. ——, ——, ——, and R. I. Joseph: *Phys. Rev.*, **B1**:2312 (1970).
7. Kouvel, J. S., and J. B. Comly: *Phys. Rev. Lett.*, **20**:1237 (1968).
8. Weiss, P., and R. Forrer: *Ann. Phys. (Paris)*, **5**:153 (1926).
9. Michels, A., B. Blaisse, and C. Michels: *Proc. Roy. Soc. (London)*, **A160**:358 (1937).
10. Graham, C. D., Jr.: *J. Appl. Phys.*, **36**:1135 (1965).
11. Gaunt, D. S., and C. Domb: *J. Phys.*, **C3**:1442 (1970).
12. Josephson, B. D.: *Ibid.*, **C2**:1113 (1969).
13. Schofield, P.: *Phys. Rev. Lett.*, **22**:606 (1969).
14. Ho, J. T., J. D. Litster, and P. Schofield: *Ibid.*, **23**:1098 (1969).
15. Cooper, M. J., M. Vicentini-Missoni, and R. I. Joseph: *Ibid.*, **23**:70 (1969).
16. Moldover, M. R.: *Phys. Rev.*, **182**:342 (1969).

DISCUSSION *on Paper by M. Vicentini-Missoni*

M. J. COOPER: I would like to comment on the first part of Dr. Missoni's paper which we believe is of real importance. Namely, that while scaling has been qualitatively demonstrated for real physical systems, there is a *real* need for the form and magnitude of correction terms. Not only are they important to determine the range of scaling, but also to understand the basic symmetries of the system. This is particularly important in the analysis of experimental data. (See paper of Cooper in this conference.)

For the real gas data, Dr. Missoni et al. (NBS reference) found scaling over the wide range of 30 percent in reduced density and also several percent in reduced temperature.

G. BENEDEK: Could you please explain how your equation of state is used to locate the position of the spinodal curve?

M. VICENTINI-MISSONI: I define the spinodal as the line where the inverse susceptibility goes to zero. The parameters were chosen in such a way that the spinodal curve is approximately located where Baker and Essam located it from their numerical study of the three-dimensional Ising model.

CRITICAL RELAXATION IN ISOTROPIC FERRO- AND ANTIFERROMAGNETS BELOW THE TRANSITION POINT

J. Villain
Institut Laue-Langevin
Grenoble, France

ABSTRACT

The goal of this paper is to give a description of relaxation processes in *isotropic* ferro- and antiferromagnets close below T_c for very small q (hydrodynamic region:[1] $q \ll \kappa = \xi^{-1}$).

Two kinds of similar descriptions have been made so far: (1) hydrodynamic description[1] based on perhaps two strong assumptions, and (2) microscopic descriptions[2,3] which are doubly disadvantageous in that they are both extremely complicated and incompatible with a finite value of the critical parameter η.

The semimacroscopic description given hereafter uses linear response theory, the mode-mode theory, and the fact that in isotropic systems there are soft, weakly damped spin waves even near T_c.[1-3] Neglecting the damping and using a R.P.A.-type decoupling, the following equation for the Fourier transform of the magnetization in ferromagnets is derived

$$M_q^z = \rho_q - \frac{1}{2S\sqrt{N}} \sum_k M_{q+k}^+ M_k^-$$

with $\langle M_{q+k}^- M_k^+ \rho_q \rangle = 0$. In antiferromagnets a similar equation holds for the staggered magnetization $\vec{P}_q$. In the above equation, the summation over $\vec{k}$ is limited to some q-dependent cut-off k_0, much larger than q but vanishing for vanishing q. From this equation it is deduced that the static longitudinal susceptibility (or staggered susceptibility in antiferromagnets) is proportional to $1/(a\kappa)^{1-\eta} aq$ where a is the side of the unit cell and κ the inverse correlation range. The formula of Ref. 2 about the longitudinal dynamical susceptibility is extended to the case $\eta \neq 0$ (with a simpler derivation): the limit of the longitudinal dynamic susceptibility for q going to zero shows a logarithmic divergence for a frequency equal to the spin wave frequency ω_q. A graphical construction to account for the damping for finite q is proposed.

A consequence of the logarithmic singularity is that the magnon damping rate in ferromagnets

$$\frac{1}{\tau_q} \sim q^4 \kappa^{-(3+\eta)/2} \log^2 \frac{q}{\kappa}$$

differs from the hydrodynamical prediction by a logarithmic factor.

In antiferromagnets, however, our result agrees with hydrodynamic[1] and microscopic[3] predictions

$$\tau_q \approx \frac{\hbar}{Js} \frac{\sqrt{a\kappa}}{a^2 q^2}$$

However, the relaxation time for the uniform longitudinal magnetization in antiferromagnets is found to disagree with the prediction of hydrodynamics

$$\tau_q^{M,Z} \approx \frac{\hbar}{Js} (aq)^{-3/2}$$

All results agree with dynamic scaling laws.

1. FERROMAGNETS

1.1 Longitudinal Magnetization

Halperin and Hohenberg[1] have shown that, even near T_c, there are soft (i.e., long wavelength) spin waves with weak damping; this will also be derived below (Eq. (17)). Because of this weak damping, the equation of motion for free spin waves[4]

$$\dot{M}_q^+ = \frac{iS}{\hbar \chi_q^\perp} M_q^+ \tag{1}$$

is nearly satisfied. Here $M_q^+ = (M_q^-)^* = M_q^x + iM_q^y$, $\vec{M}_q$ is the Fourier transform of the magnetization; $S = \langle M_0^z \rangle / N$ is the order parameter or magnetization per unit cell and

$$\chi_q^\perp \equiv \beta \langle M_{-q}^x M_q^x \rangle = \beta \frac{\langle M_q^+ M_q^- \rangle}{2} = \beta \frac{\langle M_q^- M_q^+ \rangle}{2} \approx \frac{(a\kappa)^\eta}{Ja^2 q^2} \tag{2}$$

where J is a typical exchange interaction energy and a is the side of the unit cell.

In the identity

$$\langle M_{q+k}^- M_k^+ \dot{M}_q^z \rangle = -\langle (\dot{M}_{q+k}^- M_k^+ + M_{q+k}^- \dot{M}_k^+) M_q^z \rangle \tag{3}$$

the right-hand side can be expressed by means of Eq. (1) and the left-hand side by means of the well-known identity[5]

$$\hbar \beta \langle AB \rangle = i \langle [A, B] \rangle \tag{4}$$

which holds in the classical limit, i.e., for operators with small characteristic frequencies. Now Eq. (3) reads

$$\frac{2i}{\beta^2 \hbar \sqrt{N}} (\chi_{q+k}^\perp - \chi_k^\perp) = \frac{iS}{\hbar} \left(\frac{1}{\chi_{k+q}^\perp} - \frac{1}{\chi_k^\perp} \right) \langle M_{q+k}^- M_k^+ M_q^z \rangle$$

or

$$\langle M_{q+k}^- M_k^+ M_q^z \rangle = -\frac{1}{2S\sqrt{N}} \langle M_{q+k}^- M_{q+k}^+ \rangle \langle M_k^+ M_k^- \rangle$$

$$= -\frac{1}{2S\sqrt{N}} \langle M_{q+k}^- M_k^+ M_{q+k}^+ M_k^- \rangle$$

(N equals number of unit cells.) Therefore

$$M_q^z = \rho_q^k - \frac{1}{2S\sqrt{N}} M_{q+k}^+ M_k^- \qquad \langle M_{q+k}^- M_k^+ \rho_q^k \rangle = 0 \qquad (5)$$

Now the following decoupling assumption will be made, namely*

$$\langle M_{q+k}^- M_k^+ M_{q+k'}^+ M_{k'}^- \rangle \sim 4(k_B T)^2 \chi_{k+q}^\perp \chi_k^\perp (\delta_{kk'} + \delta_{oq}) \qquad (5a)$$

Then a consequence of Eq. (5) is

$$M_q^z = \rho_q - \frac{1}{2S\sqrt{N}} \sum_k M_{q+k}^+ M_k^- \qquad (6)$$

with $\langle M_{q+k}^- M_k^+ \rho_q \rangle = 0$.

Without discussing the validity of Eq. (5a) it can be seen at once that the Eq. (5) formula, and therefore Eq. (6), is only proved if Eq. (1) is accurate enough to pass from Eq. (3) to Eq. (5), namely if

$$q c_k \tau_k \gg 1 \qquad (7)$$

where

$$\vec{c}_k = \vec{\nabla}_k \omega_k \qquad (8)$$

is the group of velocity of magnons. This means that the summation in Eq. (6) must be limited to some q-dependent cut-off.†

A consequence of the second term in Eq. (6) is that M_q^z is singular. Indeed, *assuming* that the first term produces no or weaker singularity

$$\chi_q'' \equiv \beta \langle M_{-q}^z M_q^z \rangle = \frac{1}{\beta S^2 N} \sum_k \chi_k^\perp \chi_{q+k}^\perp \approx \frac{K_B T}{J^2 s^2 (a\kappa)^{1-\eta}} \frac{1}{aq} \qquad (9)$$

where the symbol "$\approx$" means "of the order of magnitude of."

To investigate the variation of $M_q^z(t)$ with t, we write

$$M_k^+(t) = M_k^+ e^{i\omega_k t - t/\tau_k} \qquad (10)$$

*A more correct statement is: $\langle M_{q+k}^- M_k^+ M_{q+k'}^+ M_{k'}^- \rangle$ is generally negligible if situated after a summation sign when the summation is limited to a convenient cut-off.

†We believe the cut-off to be smaller than deduced from this argument, because of the approximation of Eq. (5a). See appendix.

where τ_q is the lifetime of magnons and ω_q is the frequency, given by Eq. (1)

$$\omega_q = \frac{S}{\hbar \chi_q^{\perp}} \quad (11)$$

The surfaces $\omega_k = \omega_{k+q}$ are planes, so that the Fourier transform of the time-dependent correlation function has a simple form; integrating on k in the plane, Eqs. (6), (10), and (11) yield

$$G_q''(\omega) \equiv \int_{-\infty}^{\infty} dt \, e^{i\omega t} \langle M_{-q}^z M_q^z(t) \rangle$$

$$\approx \left(\frac{K_B T}{Js}\right)^2 \frac{1}{a^{2-\eta} \kappa^{1-\eta} q \omega_q} \log \left| \frac{\omega_q + \omega}{\omega_q - \omega} \right| \quad (12)$$

This result, which agrees with scaling laws, generalizes the result of Ref. 2 which is correct for $\eta = 0$.

1.2. Spin Wave Damping

We postulate, in the familiar fashion,[6] a bilinear equation of motion

$$\dot{M}_q^+ = \frac{2i}{\hbar \sqrt{N}} \sum_k A_{k,q-k}^{\perp} M_{q-k}^z M_k^+ \quad (13)$$

The coefficients are determined by multiplying both sides of M_q^- (resp. $M_{k-q}^z M_k^+$), taking the mean value and using Eq. (4)

$$\left. \begin{array}{l} 2A_{qo}^{\perp} \chi_q^{\perp} + \dfrac{\beta}{S\sqrt{N}} \sum_{k'} A_{k',q-k'}^{\perp} \langle M_{q-k'}^z M_q^- M_{k'}^+ \rangle = 1 \\[2ex] 2A_{k,q-k}^{\perp} \chi_{q-k}'' \chi_k^{\perp} + \beta^2 \sqrt{N} A_{qo}^{\perp} S \langle M_{k-q}^z M_k^- M_q^+ \rangle \\[2ex] + \beta^2 \sum_{k' \neq q,k} A_{k',q-k'}^{\perp} \langle M_{k-q}^z M_{q-k'}^z M_k^- M_{k'}^+ \rangle = \chi_{q-k}'' - \chi_k^{\perp} \end{array} \right\} \quad (14)$$

CRITICAL PHENOMENA

In a similar way it is possible to write an equation of motion for M_q^z

$$\dot{M}_q^z = \frac{i}{\hbar\sqrt{N}} \sum_k A_{k,q+k} M_{q+k}^+ M_k^-$$

Multiplying $M_{q+k}^- M_k^+$ and taking the mean value

$$A_{k,q+k} = \frac{1}{2\chi_h^\perp} - \frac{1}{2\chi_{q+k}^\perp} \qquad (15)$$

where Eq. (5a) has been used. It will be now assumed that, for an isotropic medium

$$A_{kk'}^\perp \approx A_{kk'} \qquad (16)$$

in analogy with the microscopic equations of motion and in agreement with the first equation of Eq. (14). It is of course preferable, but much harder, to derive Eq. (16) from Eqs. (14). (See appendix.)

The magnon lifetime τ_q is now obtained by means of a Kubo formula,[5] using Eqs. (2) and (6)

$$\frac{1}{\tau_q} = \frac{\beta}{2\chi_q^\perp} \operatorname{Re} \int_0^\infty \langle \dot{M}_q^+ \dot{M}_q^-(s) \rangle e^{i\omega_q s}\, ds$$

$$\approx \frac{1}{\beta\hbar^2 N \chi_q^\perp} \sum_k \left(\frac{1}{\chi_k^\perp} - \frac{1}{\chi_{q-k}^\perp} \right)^2 \operatorname{Re} \int_0^\infty \langle M_{q-k}^+ M_{q-k}^-(s) \rangle$$

$$\langle M_{-k}^z M_k^z(s) \rangle e^{i\omega_q s}\, ds$$

This decoupling is not really correct. One should use Eq. (6) first, then decouple, thus obtaining the above term, plus another of the same order of magnitude. Using Eq. (12), the above equation yields

$$\frac{1}{\tau_q} \approx \frac{1}{\beta\hbar^2 N \chi_q^\perp} \sum_k \left(\frac{1}{\chi_q^\perp} - \frac{1}{\chi_{-k}^\perp} \right)^2 \chi_{q-k}^\perp G_k''(\omega_q - \omega_k)$$

$$\approx \frac{q^2 a^2}{\hbar^2 N} \sum_k \left(\frac{1}{\chi_k^\perp} - \frac{1}{\chi_{q-k}^\perp} \right)^2 \frac{1}{a^2(\vec{q}-\vec{k})^2} \frac{(K_B T/Js)^2}{a^{2-\eta} \kappa^{1-\eta} k\omega_k} \log \frac{\omega_k}{|\omega_k - |\omega_q - \omega_{q-k}||}$$

where $s = S(T = 0)$. The result is

$$\frac{1}{\tau_q} \approx \hbar^{-1} J s a^4 q^4 (a\kappa)^{-(3+\eta)/2} \log^2 \frac{\kappa}{q} \tag{17}$$

Although this result is right, the derivation is not correct, because the damping must be taken into account. As a result, $\log \omega_k/|\omega_k - |\omega_q - \omega_{q-k}|$ must be replaced by $\log(\omega_k \tau_k)$ for q greater than some value Q (see Fig. 1). Q is given by $Q/q = \omega_Q \tau_Q$. Let us assume Eq. (17) to be correct; then $Q \sim q g(q/\kappa)$ where $x^{\epsilon-2/3} < g(x) < x^{-2/3}$ (18) with ϵ infinitely small for small x. Limiting the integration to $q < Q$, the Kubo formula yields

$$\frac{1}{\tau_q} \sim q^4 \kappa^{-(3+\eta)/2} \int_1^{Q/q} \frac{\log x}{x} dx \sim q^4 \kappa^{-(3+\eta)/2} \log^2 g\left(\frac{q}{\kappa}\right)$$

which, by use of Eq. (18), yields Eq. (17), whose consistency is therefore proved.

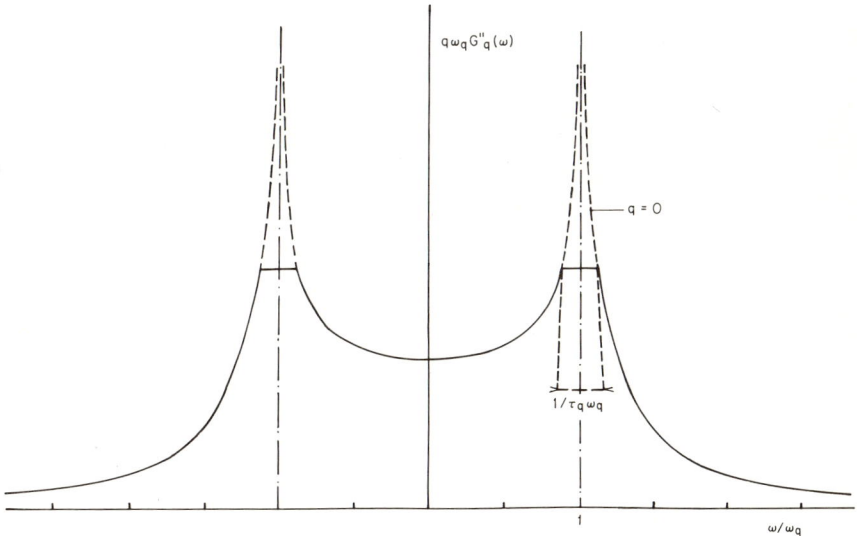

Fig. 1 *Fourier transform of the time-dependent self-correlation function of the longitudinal component of the magnetization in a ferromagnet. The curve for an isotropic antiferromagnet (staggered magnetization) differs only quantitatively. The full curve describes an approximate construction of $G_q(\omega)$ from its asymptotic form for $\vec{q}$ going to zero.*

2. ANTIFERROMAGNETS

Beside the magnetization $\vec{M}_q$, we have to consider the Fourier transform $\vec{P}_q$ of the staggered magnetization, with components P_q^z, P_q^+, P_{-q}^-. Here again we use the fact that spin waves are weakly damped (Ref. 1 and Eq. (36) below), so that the equations of motion for free magnons

$$\dot{P}_q^+ = \frac{iS}{\hbar \chi_M^\perp} M_q^+ \qquad (18)$$

$$\dot{M}_q^+ = \frac{iS}{\hbar \chi_q^\perp} P_q^+ \qquad (19)$$

are nearly satisfied. We used

$$\chi_M^\perp \equiv \frac{1}{2} \beta \langle M_q^+ M_q^- \rangle \approx \frac{1}{J} \qquad (20)$$

$$\chi_q^\perp \equiv \frac{1}{2} \beta \langle P_q^+ P_q^- \rangle \approx \frac{(a\kappa)^\eta}{Ja^2 q^2} \qquad (21)$$

$S = \langle P_0^z \rangle / N$ is the order parameter. The coefficients in Eqs. (18) and (19) can be checked by multiplying both sides by M_q^- (resp. P_q^-), taking the mean value and using Eq. (4).

In the identity

$$\langle M_{q+k}^- P_k^+ \dot{P}_q^z \rangle = - \langle (\dot{M}_{q+k}^- P_k^+ + M_{q+k}^- \dot{P}_k^+) P_q^z \rangle$$

the left-hand side can be evaluated by Eq. (4) and the right-hand side by Eq. (19), yielding after neglecting two small terms

$$\frac{2i}{\beta^2 \hbar \sqrt{N}} \chi_k^\perp = - \frac{iS}{\hbar \chi_{q+k}^\perp} \langle P_{q+k}^- P_k^+ P_q^z \rangle$$

Similarly to Eq. (6), we deduce

$$P_q^z = \rho_q - \frac{1}{2S\sqrt{N}} \sum_k P_{q+k}^+ P_k^- \qquad (22)$$

with

$$\langle P_k^- P_k^+ P_q \rangle = 0$$

Similarly to Eq. (9), we derive

$$\chi_q'' \equiv \beta \langle P_q^{z*} P_q^z \rangle \simeq \frac{K_B T}{J^2 s^2 (a\kappa)^{1-\eta}} \frac{1}{aq} \qquad (23)$$

The Fourier transform $G_q''(\omega)$ of $\langle P_q^{z*} P_q^z(t) \rangle$ is more complicated than in the ferromagnetic case because of the replacement of Eq. (11) by

$$\omega_q = \frac{S}{\hbar \sqrt{\chi_M^\perp \chi_q^\perp}} \qquad (24)$$

which is easily deduced from Eq. (18) and (19). However, the Eq. (12) formula is still acceptable as an asymptotic expression near the singularities.

2.1. Longitudinal Uniform Magnetization[7]

From the identity

$$\langle P_{q+k}^- P_k^+ \dot{M}_q^z \rangle = - \langle (\dot{P}_{q+k}^- P_k^+ + P_{q+k}^- \dot{P}_k^+) M_q^z \rangle$$

and the similar identity for $\langle M_{q+k}^- M_k^+ \dot{M}_q^z \rangle$, an equation similar to Eq. (22) can be derived

$$M_q^z = \sigma_q - \frac{1}{2S\sqrt{N}} \sum_k (M_{q+k}^+ P_k^- + M_k^- P_{q+k}^+) \qquad (25)$$

with a q-dependent cut-off in the summation, as in Eqs. (6) and (22), so that the second term is vanishingly small for small q. The main part of M_q^z is σ_q, which is expected to decay exponentially with a relaxation time $\tau_q^{M,z}$ given by a Kubo formula

$$\frac{1}{\tau_q^{M,z}} \simeq \frac{\beta}{\chi_M''} \int_0^\infty \langle \dot{M}_{-q}^z \dot{M}_q^z(t) \rangle dt \qquad (26)$$

with
$$\chi''_M = \beta \langle M^z_{-q} M^z_q \rangle$$

For $\dot{M}^z_q$ we assume

$$M^z_q = \frac{i}{\hbar \sqrt{N}} \sum_k A_{k,q+k} P^+_{q+k} P^-_k \qquad (27)$$

where $A_{kk'}$ is given by Eq. (15) as easily checked.

It is convenient to introduce the eigenmodes of Eqs. (18) and (19)

$$X^+_q = P^+_q + M^+_q \sqrt{\chi^\perp_q / \chi^\perp_M} \qquad (28)$$

$$Y^+_q = P^+_q - M^+_q \sqrt{\chi^\perp_q / \chi^\perp_M} \qquad (29)$$

Equation (26) contains two dominant contributions, one of which is

$$\frac{1}{\tau^{M,z}_q} \approx \frac{\beta}{\hbar^2 N \chi''_M} \sum_{kk'} A_{k,q+k} A_{k',q+k'} \int_0^\infty dt \langle X^-_{q+k'} X^+_{k'} X^+_{q+k}(t) X^-_k(t) \rangle$$

and the other is similar with Y instead of X. The dominant contribution is due to the k's such that $qc_k \tau_k < 1$, and the result is[8]

$$\frac{1}{\tau^{M,z}_q} \approx \frac{\beta}{\hbar^2 N \chi''_M} \sum_k \langle X^-_{q+k} X^+_k X^+_{q+k} X^-_k \rangle \tau_k A^2_{k,q+k}$$

$$\approx \frac{1}{\hbar^2 \beta N \chi''_M} \sum_k \chi^\perp_k \chi^\perp_{q+k} \tau_k \left(\frac{1}{\chi^\perp_k} - \frac{1}{\chi^\perp_{q+k}} \right)^2$$

Insertion of Eqs. (35) and (2) yields[7]

$$\tau^{M,z}_q \approx \frac{\hbar}{Js} (aq)^{-3/2} \qquad (30)$$

so that $\tau^{M,z}_q$ is found to be independent of temperature, except of course at very low temperatures. This result disagrees with hydrodynamics,[1] although it agrees with dynamic scaling laws.

2.2. Magnon Lifetime

As in the ferromagnetic case, we start with a generalization of the microscopic equations of motion[7]

$$\dot{P}_q^+ = \frac{2i}{\hbar\sqrt{N}} \sum_k (\alpha P_{q-k}^z M_k^+ - \gamma M_{q-k}^z P_k^+) \tag{31}$$

$$\dot{M}_q^+ = \frac{2i}{\hbar\sqrt{N}} \sum_k A_{k,q-k}^\perp P_{q-k}^z P_k^+ \tag{32}$$

The same argument as in the ferromagnetic case yields $A_{kk'}^\perp = A_{kk'}$ given by Eq. (15). α and γ are in principle obtained by multiplying by $M_{k-q}^z P_k$ (resp. $P_{k-q}^z M_k^-$) and taking the mean value. Near T_N, it is only possible to have the order of magnitude of α and γ

$$\alpha \approx \gamma \approx J \tag{33}$$

In the following, α is taken as equal to γ. Introducing $A_{kk'} = \mu(k^2 - k'^2)$ and $X_q^+, Y_q^+ = P_q^+ \pm q^{-1} M_q^+ \sqrt{\alpha/\mu}$, the equation of motion for X_q^+ is

$$\dot{X}_q^+ = i\omega_q X_q^+ + \frac{i}{\hbar\sqrt{N}} \sum_k \left[\sqrt{\alpha\mu}\, k P_{q-k}^z (X_k^+ - Y_k^+) \right.$$

$$\left. - \alpha M_{q-k}^z (X_k^+ + Y_k^+) + \sqrt{\alpha\mu}\left(2\frac{\vec{q}}{q}\cdot\vec{k} - q\right) P_{q-k}^z (X_k^+ + Y_k^+) \right]$$

To avoid complicated calculations, we just write a typical term of the Kubo formula which yields lifetime τ_q of the phonons

$$\frac{1}{\tau_q} \approx \frac{\beta\alpha\mu}{\hbar^2 \chi_q^\perp N} \sum_k k^2 \int_0^\infty dt\, e^{-i\omega_q t} \langle P_{k-q}^z P_{q-k}^z (t) X_k^- X_k^+(t) \rangle$$

$$\approx \frac{\alpha\mu}{\hbar^2 \chi_q^\perp N} \sum_k k^2 \chi_k^\perp G''_{q-k}(\omega_k - \omega_q)$$

A decoupling has been used. The main contribution is due to $k \gg q$, which allows the use of Eq. (12)

$$\frac{1}{\tau_q} \approx \frac{a^2 q^2}{\hbar^2 N} \left(\frac{K_B T}{s}\right)^2 \frac{1}{a^2 \kappa} \sum_k \frac{1}{k \omega_k} \log \left| \frac{|\vec{k} - \vec{q}| + k - q}{|\vec{k} - \vec{q}| - k + q} \right|$$

where we have expressed α and μ by Eqs. (33), (15), and (2), and used $S = s(a\kappa)^{1+\eta/2}$. Then using Eq. (24), the relaxation time reads

$$\frac{1}{\tau_q} \approx \frac{a^3 q^2 (K_B T)^2}{\hbar s^3 J(a\kappa)^{3/2}} \int_0^\kappa dk \, \log \frac{k}{q} \tag{34}$$

where the contribution of $k > \kappa$ has been neglected. It is easily found to be of the same order of magnitude as the contribution calculated here. Similarly, in the ferromagnetic case, the formula (Eq. (34)) is wrong because it neglects the damping. For k large enough, the log must be replaced by $\log \omega_k \tau_k$. This effect is more striking here than for ferromagnets, because the dominant contribution to the integral is due to $k \approx \kappa$. Then $\omega_k \tau_k$ is of order of magnitude one, so that the relaxation time is just

$$\tau_q \approx \frac{\hbar}{Js} \frac{\sqrt{a\kappa}}{a^2 q^2} \tag{35}$$

in agreement with the microscopic result[3] and the hydrodynamic theory.

APPENDIX: ON THE DAMPING OF SPIN WAVES

1. Justification of Eq. (16)

Using Eqs. (6) and (5a), as well as a similar factorization for the six-spin correlation, Eq. (14) writes after some calculation

$$\left. \begin{array}{l} 2A_{qo}^\perp \chi_q^\perp - \dfrac{2}{BS^2 N} \chi_q^\perp \sum_{k'} A_{k',q-k'}^\perp \chi_{k'}^\perp \approx 1 \\[2ex] 2A_{k,q-k}^\perp \chi_{q-k}'' \chi_k^\perp + \dfrac{2\chi_k^\perp}{\beta S^2 N} \sum_{k'} A_{k',q-k'}^\perp \chi_{k'}^\perp \chi_{k+k'-q}^\perp - \chi_{q-k}'' \end{array} \right\} \tag{36}$$

$$-\chi_k^\perp \left(2A_{qo}^\perp \chi_q^\perp - \frac{2\chi_q^\perp}{\beta S^2 N} \sum_{k'} A_{k',q-k'} \chi_{k'}^\perp - 1\right) \simeq 0 \qquad (37)$$

The summation on k' must be limited to some cut-off (for instance, $k' < \kappa$). This limitation of the number of thermodynamic variables taken into account is necessary to justify the decoupling procedure in the Kubo formula. The rigorous (and hypothetic) solution of Eq. (14) without cut-off would, of course, yield the rigorous equation of motion of Ref. 7.

Using Eq. (2), it is easily checked that

$$A_{kk'} = \frac{1}{2\chi_k^\perp} - \frac{1}{2\chi_{k'}^\perp} \qquad (38)$$

verifies Eq. (36), whereas, using Eq. (6), the left-hand side of Eq. (37) is found to be equal to

$$-\frac{1}{2} \chi_k^\perp \chi_{q-k}^\perp \left(\frac{1}{\chi_q^\perp} + \frac{1}{\chi_k^\perp} + \frac{1}{\chi_{q-k}^\perp}\right)$$

The condition that Eq. (37) can be considered as satisfied is that this is much smaller than $(\chi_k^\perp + \chi_{q-k}'')$, namely

$$\chi''_{q-k} \ll \chi_k^\perp \qquad (39)$$

This condition probably defines the correct cut-off in Eq. (6). It is satisfied by the scattering processes relevant to obtain Eqs. (17) and (35). Notice that the third term in the left-hand side of the second Eq. (14) (even with a cut-off) is only negligible if Eq. (39) is satisfied.

2. Discussion

Equation (13) is a very restrictive assumption. More consistent with our description would be the Ansatz

$$\dot{M}_q^+ = iN^{-1} \sum_{kk'} B_q^{kk'} M_k^+ M_{k'}^+ M_{k+k'-q}^- + iB_q M_q^+ \qquad (40)$$

with

$$\langle M_k^- M_{k'}^- M_{k+k'-q}^+ M_q^+ \rangle = - \frac{4i}{\hbar \beta^3 NS} \chi_{k+k'-q}^\perp (\chi_k^\perp + \chi_{k'}^\perp)$$

$$= \frac{8i}{\beta^3 N} B_q^{kk'} \chi_k^\perp \chi_{k'}^\perp \chi_{k+k'-q}^\perp + \frac{iB_q}{\beta} S \langle M_k^- M_{k'}^- M_{k+k'-q}^+ M_q^+ \rangle \quad (41)$$

The use of Eq. (5a) yields $B_q^{kk'} = (1/\chi_k^\perp + 1/\chi_{k'}^\perp)/2S$, which does not vanish for vanishing q; therefore the decoupling Eq. (5a) must be wrong *here*; however, its use in Eq. (14) is reasonable, because the correlation function neglected is summed over k with a cut-off.

For small q (the dominant process for the lifetime), Eq. (39) yields $B_q^{kk'}$ as a difference of two nearly equal terms, which requires a very precise evaluation of $\langle M_k^- M_{k'}^- M_q^+ M_{k+k'-q}^+ \rangle$. Therefore, a program based on Eq. (40) seems very difficult to carry out.

REFERENCES

1. Halperin, B. I., and P. C. Hohenberg: *Phys. Rev.*, **188**:898 (1969).
2. Vaks, V. G., A. I. Larkin, and S. A. Pikin: *Zh. Eksp. Teor. Fiz.*, **53**:1089 (1967); *Sov. Phys. JETP*, **26**:647 (1968).
3. Solyom, J.: *Zh. Eksp. Teor. Fiz.*, **55**:2355 (1968); *Sov. Phys. JETP*, **28**:1251 (1969).
4. This is easily derived from Eq. (4). See H. Mori and K. Kawasaki: *Prog. Theor. Phys.*, **27**:529 (1962).
5. Kubo, R.: "Lectures in Theoretical Physics," vol. 1, p. 120, Interscience Publishers, New York, 1959.
6. For instance, L. P. Kadanoff and J. W. Swift: *Phys. Rev.*, **166**:89 (1968); K. Kawasaki: To be published; J. Villain: *Phys. Stat. Sol.*, **26**:501 (1968).
7. Villain, J.: *Sol. State Com.*, **8**:31 (1970).
8. This involves the use of Eqs. (10) and (5a). One can also argue that $X_{q+k}^+ X_k^-$ is proportional to the Fourier transform $n_k(q)$ of the density $n_k(r)$ of magnons with wave vector k which satisfies a Boltzmann equation

$$\frac{\partial}{\partial t} n_k = -\vec{c}_k \cdot \vec{\nabla} n_k - \frac{1}{\tau_k} n_k$$

DISCUSSION on Paper by J. Villain

R. A. FERRELL: You have a new logarithmic factor in the width for ferromagnets, which would seem to cause difficulty with dynamical scaling.

J. VILLAIN: All formulas agree with dynamical scaling: in particular, Eq. (17) can be read

$$\frac{1}{\tau_q} \sim q^{(5-\eta)/2} f\left(\frac{\kappa}{q}\right)$$

CRITICAL MAGNETIC BEHAVIOR OF PALLADIUM-IRON ALLOYS

J. S. Kouvel

Department of Physics
University of Illinois
Chicago, Illinois

J. B. Comly

General Electric R & D Center
Schenectady, New York

ABSTRACT

Detailed magnetization-field-temperature data have been obtained on ferromagnetic Pd-Fe alloys near their Curie points. For the random f.c.c. alloy of composition Pd_3Fe, the critical exponents ($\beta \approx 0.36$, $\gamma \approx 1.32$, $\delta \approx 4.6$) closely resemble those determined earlier for nickel and those deduced for the Heisenberg model. The same alloy, when atomically ordered (Cu_3Au-type), has approximately the same Curie point and low-temperature spontaneous moment, but all the critical exponents have shifted consistently towards the molecular-field values, suggesting an increased effective range of exchange interaction. When the Fe concentration in the random alloy is reduced, β rises towards one-half, but γ does not drop correspondingly towards unity but gradually increases. Despite this apparent dichotomy between the ferromagnetic and paramagnetic behavior of the dilute **PdFe** alloys, the magnetization data for these (as well as the more concentrated) alloys obey the homogeneous Domb-Hunter-Widom equation of state. Quantitatively, moreover, the data for the dilute **PdFe** alloys confirm the existence of the "giant moments" previously revealed by neutron diffraction.

1. INTRODUCTION

Prominent among the many recent developments in the area of critical phenomena are the experimental results showing that the equilibrium critical-point properties of various ferromagnets and fluids conform very closely to the homogeneous equation of state proposed by Domb and Hunter[1] and by Widom.[2] In providing a highly simplifying description of continuous phase transitions, this equation of state is not only of general theoretical importance but is also extremely useful in any systematic comparative study of different materials. The ferromagnets investigated to date, whose properties near the Curie point appear to obey the Domb-Hunter-Widom (DHW) state equation, have been either pure metals (nickel,[3,4] cobalt,[5] gadolinium,[6,7]) or ionic compounds ($CrBr_3$,[8] $La_{1/2}Sr_{1/2}CoO_3$,[7,9] $CdCr_2Se_4$[10]), with the exception of CrO_2[11] which is an electrical conductor. Thus, the palladium-iron alloy system, which we will be reporting on in this paper, represents a new class of ferromagnetic materials for this type of detailed investigation.

There were several reasons why we were specifically attracted by the Pd-Fe system. By suitable thermal treatment, the alloys of composition near Pd_3Fe can be made to have almost perfect atomic order (of the Cu_3Au type) or they can be retained in an atomically disordered f.c.c. structure. In these two states, Pd_3Fe has closely the same zero-temperature saturation magnetization (σ_0) and Curie point (T_c), with the latter conveniently well below the atomic order-disorder temperature. Furthermore, the critical-point properties can be followed within the same crystal structure (f.c.c.) to compositions of very low Fe concentration, since T_c goes to 0°K below 0.1 at.% Fe. The ferromagnetic alloys of dilute **PdFe** are well known to have saturation magnetizations corresponding to a "giant moment" of $\sim 10\mu_B$ per Fe atom,[12,13] of which only $\sim 3\mu_B$ is located on the Fe atom itself while the rest is distributed among the Pd atoms in a large region centered about each Fe atom, as determined from neutron diffraction data.[14] At higher Fe concentrations, there is increasing overlap between the induced magnetic polarizations surrounding adjacent Fe atoms,[14] and the Pd atomic moments approach a uniform limiting value ($\sim 0.4\mu_B$ in Pd_3Fe[15]). For this alloy system there have already been several measurements of spontaneous moment vs. temperature just below T_c.[16,17] However, the magnetic order-disorder transition in these (or any other) materials can only be reliably described and illuminated by a detailed study of all the critical-point properties, especially if they are captured within an equation of state and then compared to various model predictions.

2. EXPERIMENTAL RESULTS AND ANALYSES

For this study, ingots of nominal composition $Pd_{0.75}Fe_{0.25}$, $Pd_{0.969}Fe_{0.031}$, and $Pd_{0.986}Fe_{0.014}$, were prepared by induction-melting and chill-casting. From each was machined a cylindrical sample, whose edges were beveled off in order to approximate an ellipsoid with a well-defined demagnetization factor. Each sample was given a homogenizing anneal at 1250°K for one day, followed by a water quench, for which x-ray diffraction analysis showed no superlattice evidence for atomic ordering of the f.c.c. structure. When the magnetic measurements on the disordered Pd_3Fe sample were completed, it was annealed for four days each at 1025, 975, 925, and 875°K, successively. X-ray diffraction analysis then revealed a very high degree of Cu_3Au-type order, and the magnetic measurements were repeated.

The magnetization (σ) of each alloy sample was measured with a quasi-static moving-sample magnetometer in several uniform applied fields (H_a) from 25 kOe down to 500 Oe and at small temperature intervals over a range spanning T_c. Each set of σ vs. H_a data was corrected for any thermal drift by standard interpolation methods, and the resulting isotherms were then converted to the form of σ^2 vs. H_a/σ, where a simple subtraction of the demagnetization factor (times the density) from H_a/σ gave H/σ in terms of the true internal field (H). Plotting σ^2 vs. H/σ, we obtained in each case a set of smooth isothermal curves which extended unambiguously into the σ^2 or H/σ axis, thus yielding quite reliable values for σ_s^2, the spontaneous magnetization squared, and for χ_0^{-1}, the inverse initial susceptibility, at temperatures below and above T_c, respectively.

Our results for χ_0^{-1} vs. T were analyzed by the $\gamma^*(T^*)$ method,[18] and the values deduced for T_c and the critical exponent γ in the relation

$$\chi_0^{-1} = C_+(T - T_c)^\gamma \quad \text{for} \quad T \to T_c^+ \quad (1)$$

are listed in Table 1. Using these values of T_c, we then tested our σ_s vs. T results against the power relation

$$\sigma_s = C_-(T_c - T)^\beta \quad \text{for} \quad T \to T_c^- \quad (2)$$

and got excellent agreement with the values of β shown in Table 1. We next interpolated our $\sigma(H,T)$ data to T_c for each alloy and determined the critical isotherm, which we compared with the power relation

TABLE 1. Critical Exponents and Parameters

	Experimental					Theoretical (f.c.c., $S = \frac{1}{2}$)	
	Pd_3Fe ordered	Pd_3Fe disord.	$Pd_{0.969}Fe_{0.031}$	$Pd_{0.986}Fe_{0.014}$	Nickel	Molec. Field	Ising
$T_c(°K)$	530.7	491.3	114.6	50.25	627.4		
σ_0 (emu/g)	65.2	63.6	14.45	6.85	58.6		
$\mu_0(\mu_B)$	1.10	1.07	0.275	0.130	0.616		
β	0.444	0.364	0.428	0.464	0.378	0.5	0.312
γ	1.19	1.32	1.37	1.40	1.34	1.0	1.25
δ	3.64	4.61	4.24	4.06	4.58	3.0	5.0
δ^*	3.68	4.62	4.20	4.02	4.54	3.0	5.0
$\alpha\dagger$	−0.08	−0.05	−0.23	−0.33	−0.10	0.0	0.125
m_0(emu/g)	91.9	76.3	18.2	8.25	83.3		
$h_0(kOe)$	7168	5303	538	132	15720		
m_0/σ_0	1.41	1.20	1.26	1.20	1.42	1.73	1.49
$\mu_0 h_0/kT_c$	1.00	0.78	0.086	0.023	1.04	1.73	1.52
D_-	9.1	20.4	195	472	32.8		
D_+	11.4	22.6	216	637	35.6		
$D_- m_0^3/h_0$	0.99	1.71	2.19	2.00	1.21	1.0	
$D_+ m_0^3/h_0$	1.23	1.89	2.43	2.71	1.31	1.0	1.27
A_0	0.510	0.0167	3.96	45.2	0.0331		
$A_0 m_0^\delta/h_0$	1.00	1.50	1.63	1.80	1.32	1.0	1.81

*Calculated from Eq. (4)
†Calculated from Eq. (6)

$$\sigma = C_0 H^{1/\delta} \quad \text{for} \quad T = T_c \tag{3}$$

This comparison also gave good agreement, except for the dilute-Fe alloys at our highest fields, where the measured σ reached a sizeable fraction ($\sim 1/3$) of σ_0 and began to fall consistently below the curve drawn through our lower field data following Eq. (3). Disregarding the data that exhibited this approach to saturation, we obtained the values of δ listed in Table 1. Also listed in this table are our results for σ_0 obtained by extrapolation of our low-temperature data down to $0°K$; these results are also shown in the form of the average moment per atom μ_0.

Reserving discussion of our critical exponent results for later, we simply note at this point that the numerical values in each case obey the relationship

$$\delta = 1 + \frac{\gamma}{\beta} \tag{4}$$

within experimental error ($\pm 2\%$). This relationship, which corresponds to one of Griffiths' rigorous inequalities[19] in the allowed case of an equality, is a required condition for the DHW equation of state. Hence, we proceeded to transform our data (except for that showing the saturation effects mentioned above) to isotherms of m^2 vs. h/m, where

$$m \equiv \frac{\sigma}{|1 - T/T_c|^\beta} \quad \frac{h}{m} \equiv \frac{H}{\sigma|1 - T/T_c|^\gamma} \tag{5}$$

Our transformed data points are plotted in Fig. 1a–d, and it is very clear that for each alloy they fall on two universal curves, one for $T < T_c$ and the other for $T > T_c$. Furthermore, as required by Eqs. (3) to (5), each pair of curves converges on the measured critical isotherm (shown dashed in the figures) at large values of m, and thus represents two branches of a single curve. Hence, despite the diversity in the critical exponent values for the different alloys, there is in each case a universal dependence of m^2 on h/m (and thus, of m on h) near T_c, which is a statement of the DHW equation of state.

The validity of this equation of state requires that the critical exponents obey not only Eq. (4) but also

$$\alpha = 2 - 2\beta - \gamma \tag{6}$$

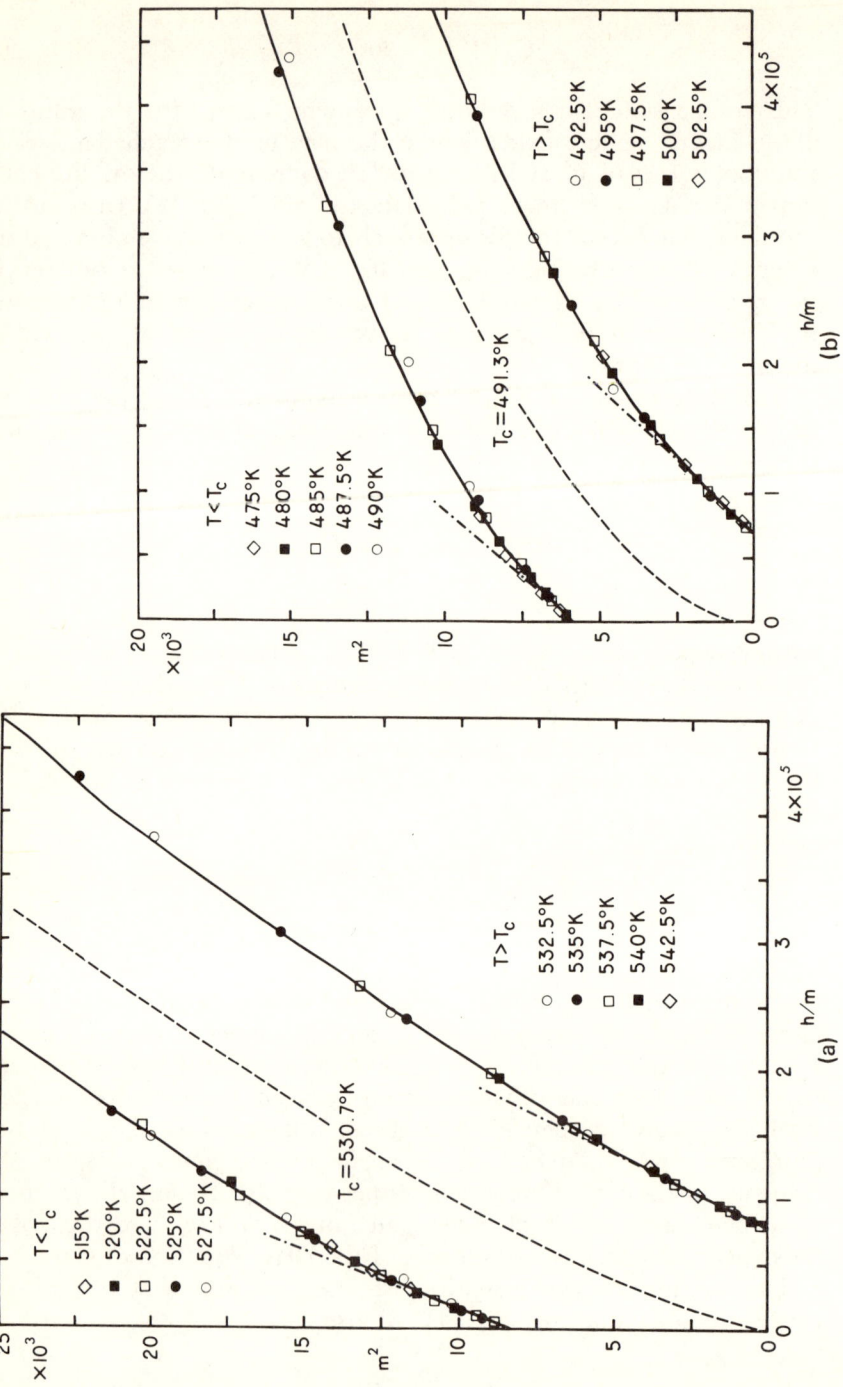

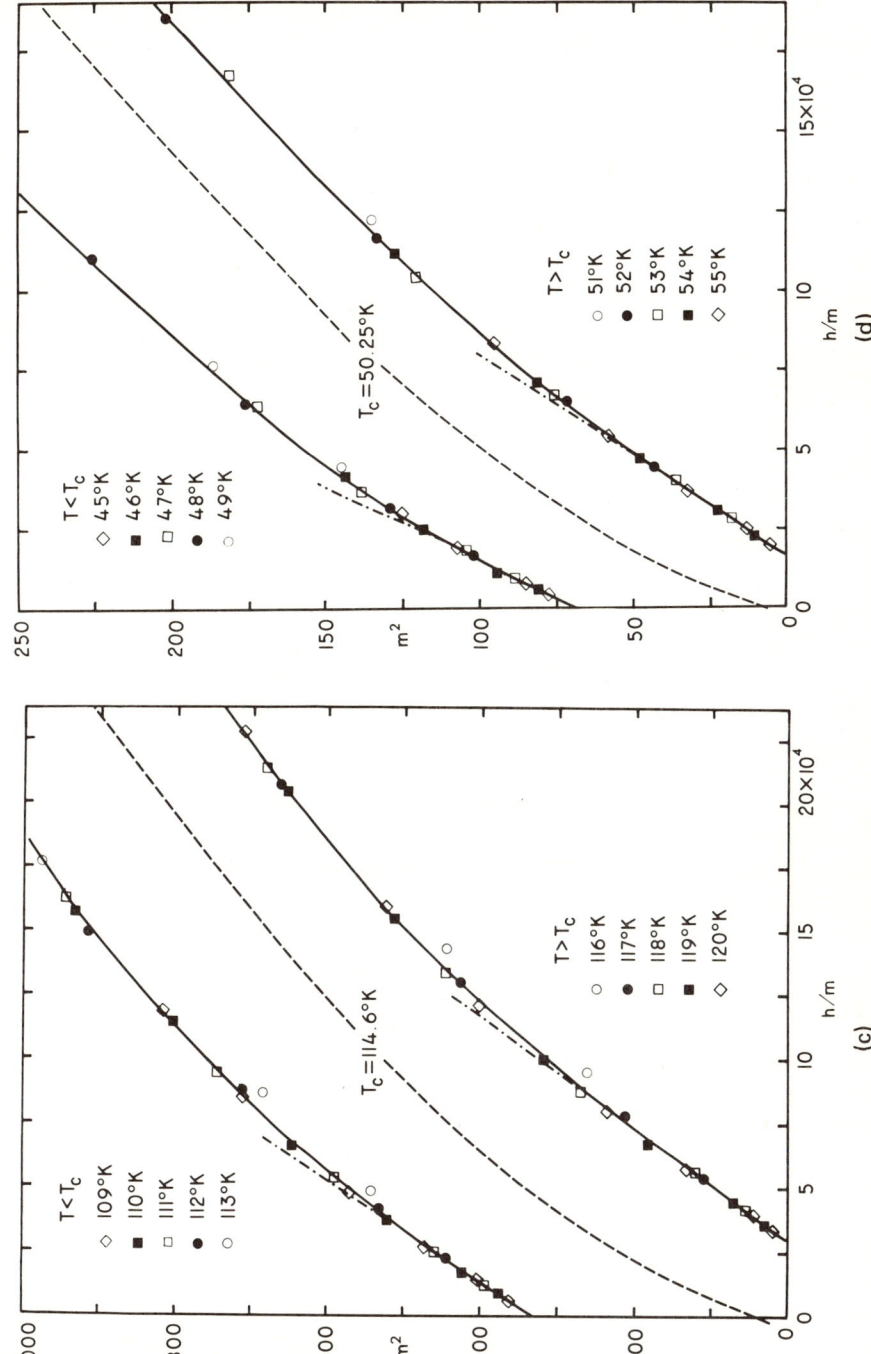

Fig. 1 Plots of m^2 vs. h/m, derived from isothermal magnetization vs. field data by means of Eq. (5), for (a) ordered Pd_3Fe, (b) disordered Pd_3Fe, (c) $Pd_{0.969}Fe_{0.031}$, and (d) $Pd_{0.986}Fe_{0.014}$. Experimental critical isotherms are shown dashed.

which is Rushbrooke's rigorous inequality [20] taken as an equality, where α in $C_v \propto T - T_c^{-\alpha}$ describes the temperature dependence of the specific heat (C_v) close to T_c. Thus, from our experimental γ and β values we can predict the corresponding values for α, and these are listed in Table 1.

Besides the critical exponents, there are several other important parameters associated with the curves in Fig. 1a–d. Defining the intercepts of these curves with the m^2 and h/m axes as m_0^2 and h_0/m_0, respectively, we determine the values of m_0 and h_0 listed in Table 1, where they are followed by the dimensionless quantities m_0/σ_0 and $\mu_0 h_0/kT_c$. We then define the inverse initial slopes of these curves for $T < T_c$ and $T > T_c$ as D_- and D_+, respectively, and convert them to the dimensionless form $D_\pm m_0^3/h_0$; their values are all given in Table 1. We also give here our results for $A_0 = H/\sigma^\delta$, taken from the critical isotherms, and for the dimensionless parameter $A_0 m_0^\delta/h_0$.

For all the experimental quantities listed in Table 1, the estimated error is ±1–2%, except for D_+ and D_- where the error is ±5%.

3. DISCUSSION

Since our critical exponent results for the Pd-Fe samples obey the scaling relation of Eq. (4) and give close agreement with the DHW equation of state, their numerical values would appear to be truly representative of the critical region and can therefore be compared meaningfully with existent theory. For this comparison, we show in Table 1 the values for the critical exponents and dimensionless quantities, such as m_0/σ_0, derived from the calculated results of Domb[21] for the three-dimensional Ising model (for $S = \frac{1}{2}$, f.c.c. lattice), as well as those readily deduced for the molecular field model (again for $S = \frac{1}{2}$). The only analogous calculations to date for the Heisenberg model ($S = \frac{1}{2}$) have been limited to a few of the critical exponents. Baker et al.[22] have deduced that $\gamma \approx 1.43$ and the gap parameter $\Delta \approx 1.82$, whereas Bowers and Woolf[23] have reported more recently that $\gamma \approx 1.38$ and $-1/8 \leq \alpha \leq -1/16$. Invoking the scaling relations of Eqs. (4) and (6) and the related identity $\Delta = \beta + \gamma$, we arrive at a compromise set of exponent values: $\alpha = -0.14, \beta = 0.37, \gamma = 1.40, \delta = 4.78$, which we will use in reference to the Heisenberg model in later discussion. We will also compare our Pd-Fe results with our previous results for nickel, which as listed in Table 1 are more complete than those reported earlier.

From this table, we immediately note two striking facts about the critical exponents for Pd_3Fe. First, all the measured quantities (β, γ, and δ) for the atomically *disordered* state of this alloy have nearly the same values as those for nickel, and both these sets of results (including the negative α values deduced from Eq. (6)) are remarkably close to the critical exponent values given above for the Heisenberg model. Secondly, for the atomically *ordered* state of this alloy, all the measured exponents have shifted consistently towards the values for the molecular field model. Our results for β differ from those of Longworth,[17] who deduced the values of 0.394 and 0.371 from Mössbauer data on disordered and ordered Pd_3Fe respectively, but the detailed consistency of all our results above and below T_c gives them a much stronger claim to reliability.

The agreement between our critical exponent values for disordered Pd_3Fe and nickel and those predicted for a Heisenberg ferromagnet might well appear to be fortuitous. Both these materials are metals, whereas a basic simplifying assumption for the model calculations is that the magnetic moments are completely localized on each atom and interact solely with their nearest neighbors. However, next nearest neighbor interactions have been included in an Ising model calculation by Domb and Dalton,[24] who found that the exponent γ is unaffected near T_c, although its effective value decreases more rapidly (toward unity) with rising temperature. Presumably, this result is qualitatively valid for an interaction that drops off sharply at any range and would also apply for a Heisenberg ferromagnet. Concerning the localization of the moment, recent calculations by Fisher and Scesney[25] for a "mobile-electron" Ising ferromagnet have shown that the values for α, β, and γ very near T_c undergo a "renormalization" which brings them remarkably close to the above values for the Heisenberg model. But if Eq. (4) is still applicable, the value for δ remains the same (i.e., at 5), which lies well outside the error limits of our results for either disordered Pd_3Fe or nickel. It is difficult, moreover, to believe that the exchange interactions in these materials have an Ising-like anisotropic form. Much more likely, they are Heisenberg-like, and our results for the critical exponents simply reflect this isotropic form of the interactions (as well as other primary factors like dimensionality).

Regarding the shift in the critical exponents towards the molecular field values when Pd_3Fe becomes ordered, this result can be understood in terms of the interactions becoming very long-ranged with no sharp cut-off. Such an effect has been calculated by Joyce[26] for the spherical and two-dimensional Ising models, where the interactions were assumed to decrease gradually with distance. This rationale is physically very reasonable for Pd_3Fe, where the exchange coupling probably arises from an indirect process involving the conduction electrons whose mean free path between collisions in the atomically ordered lattice would be much longer than in the disordered lattice at the same temperature. The fact that ordered Pd_3Fe is also different from nickel suggests that in the latter the dominant exchange interactions are those directly between the d electrons associated with neighboring atoms.

Turning to the results for the dilute **PdFe** alloys, we see in Table 1 that with decreasing Fe concentration each of the critical exponents departs monotonically from its value in disordered Pd_3Fe. However, there is no consistent change from one set of known theoretical values towards another, as there is between disordered and ordered Pd_3Fe. While the value for β increases rapidly towards the molecular field value of ½ as the alloy becomes more dilute (in agreement with recent Mössbauer results [16]), suggesting an increase in the interaction range, the value for γ does not decrease correspondingly towards unity. Instead, γ rises slowly to a value near 1.4 and thus improves its agreement with the Heisenberg model. It is indeed remarkable that despite this apparent dichotomy between them, the ferromagnetic and paramagnetic properties of dilute **PdFe** continue to obey the same equation of state.

With regard to the other experimental parameters listed in Table 1, we note that while the values of m_0 extend over an order of magnitude for the various materials, their normalized values (m_0/σ_0) all lie within a narrow range just below the Ising and molecular field values. The situation for $\mu_0 h_0/kT_c$, however, is radically different. Although in ordered and disordered Pd_3Fe (and in nickel) this quantity lies fairly close to the theoretical values near unity, it drops to conspicuously lower values in the dilute **PdFe** alloys. Since h_0 is presumably an effective exchange interaction field, the product of h_0 and an average effective moment (μ_{eff}) can be expected to equal kT_c. Such is approximately the case for ordered and disordered Pd_3Fe (and nickel) when μ_{eff} is identified with μ_0. However, this is clearly not the case for the dilute **PdFe** alloys unless μ_{eff} is taken to be very much larger than μ_0. Specifically, in order that $\mu_{eff} h_0 = kT_c$ in the 3.1 and 1.4 at.% Fe alloys, we must take μ_{eff} to equal $11.6\mu_0$ (that is, $3.2\mu_B$) and $43.5\mu_0$ ($5.7\mu_B$), respectively. Thus, from our experimental values for h_0, we are led to conclude that in the dilute **PdFe** alloys the effective moment involved in the magnetic order-disorder process is many times the average moment per atom. Indeed, its magnitude approaches the "giant moment" value of $\sim 10\mu_B$ shown to be associated with each Fe impurity atom in the highly dilute alloys, as discussed earlier.

Our results shown in Table 1 for the dimensionless quantities, $D_\pm m_0^3/h_0$ and $A_0 m_0^\delta/h_0$, are clearly not very sensitive to the chemical composition of the material, though certain definite trends are evident. In all cases, these results are in rough agreement with the available theoretical values indicated.

Although the present theoretical understanding of critical phenomena in ferromagnets is by no means complete, the variety of recent model calculations is a very valuable development. Indeed, we have shown that by using these calculations as a reference frame for our critical-point measurements we were able to extract some vital information about the physics of the various metals we have been studying.

REFERENCES

1. Domb, C., and D. L. Hunter: *Proc. Phys. Soc. (London),* **86**:1147 (1965).
2. Widom, B.: *J. Chem. Phys.,* **43**:3898 (1965).
3. Arrott, A., and J. E. Noakes: *Phys. Rev. Lett.,* **19**:786 (1967). The results here were fitted to a somewhat different equation of state.
4. Kouvel, J. S., and J. B. Comly: *Ibid.,* **20**:1237 (1968).
5. Rocker, W., and R. Kohlhaas: *Z. Angew. Phys.,* **25**:343 (1968).
6. Graham, C. D.: *J. Appl. Phys.,* **38**:979 (1967). Equation-of-state analysis of these results is given in Ref. 7.
7. Vicentini-Missoni, M., R. I. Joseph, M. S. Green, and J. M. H. Levelt Sengers: *Phys. Rev.,* **B, 1**:2312 (1970).
8. Ho, J. T., and J. D. Litster: *Phys. Rev. Lett.,* **22**:603 (1969); *J. Appl. Phys.,* **40**:1270 (1969).
9. Menyuk, N., P. M. Raccah, and K. Dwight: *Phys. Rev.,* **166**:510 (1968). Equation-of-state analysis of these results is given in Ref. 7.
10. Miyatani, K.: *J. Phys. Soc. Japan,* **28**:259 (1970).
11. Kouvel, J. S., and D. S. Rodbell: *Phys. Rev. Lett.,* **18**:215 (1967); *J. Appl. Phys.,* **38**:979 (1967).
12. Crangle, J.: *Phil. Mag.,* **5**:335 (1960).
13. Bozorth, R. M., P. A. Wolff, D. D. Davis, V. B. Compton, and J. H. Wernick: *Phys. Rev.,* **122**:1157 (1961).
14. Low, G. G., and T. M. Holden: *Proc. Phys. Soc. (London),* **89**:119 (1966).
15. Cable, J. W., E. O. Wollan, and W. C. Koehler: *Phys. Rev.,* **138**:A755 (1965).
16. Trousdale, W. L., G. Longworth, and T. A. Kitchens: *J. Appl. Phys.,* **38**:922 (1967).
17. Longworth, G.: *Phys. Rev.,* **172**:572 (1968).
18. Kouvel, J. S., and M. E. Fisher: *Phys. Rev.,* **136**:A1626 (1964).
19. Griffiths, R. B.: *Phys. Rev. Lett.,* **14**:623 (1965); *J. Chem. Phys.,* **43**:1958 (1965).
20. Rushbrooke, G. S.: *J. Chem. Phys.,* **39**:842 (1963).
21. Domb, C.: *J. Appl. Phys.,* **39**:620 (1968). Our m_0/σ_0, $\mu_0 h_0/kT_c$, $D_+ m_0^3/h_0$, and $A_0 m_0^\delta/h_0$ correspond, in Domb's notation, to w_0^{-1}, $b_1 w_0^{-1}$, $b_3/b_1 w_0^2$, and $c_0/b_1 w_0^{\delta-1}$, respectively.
22. Baker, G. A., H. E. Gilbert, J. Eve, and G. S. Rushbrooke: *Phys. Rev.,* **164**:800 (1967).
23. Bowers, R. G., and M. E. Woolf: *Phys. Rev.,* **177**:917 (1969).
24. Domb, C., and N. W. Dalton: *Proc. Phys. Soc. (London),* **89**:859 (1966).
25. Fisher, M. E., and P. E. Scesney: *J. Appl. Phys.,* **40**:1534 (1969); P. E. Scesney: *Phys. Rev.,* **B, 1**:2274 (1970).
26. Joyce, G. S.: *Phys. Rev.,* **146**:349 (1966).

DISCUSSION on Paper Presented by J. S. Kouvel

C. DOMB: The exponents for dilute alloys of iron in palladium are particularly interesting. It seems that γ becomes increasingly large while β stays near its mean field value. It is curious that there is a theoretical model with precisely this behavior—the one or two-dimensional spherical model with long-range forces. For these models with long-range forces, β is equal to ½ but as the force becomes short-ranged γ increases indefinitely but β remains unchanged. It would be interesting to determine for even more dilute alloys to see if it increases. Naturally there can be no direct association between such alloys and this model, but at least it shows that a theoretical model exists with this type of behavior of exponents.

J. S. KOUVEL: Are you saying that this behavior reflects the range of the interaction? This is in some way different then from what we observe in the chemical ordering of Pd_3Fe where everything moves toward the mean field values

B. WIDOM: I would like to make two remarks.

First, in principle such a system should show renormalization of exponents, which would have given a γ about that which was observed, had the α in the reference system been positive (though it seems not to be). Of course, β should then be simultaneously renormalized but not to so large a value as that which was observed.

Second, the exponent relations $\alpha + 2\beta + \gamma = 2$ and $\delta = 1 + \gamma/\beta$ should be satisfied by any mutually consistent set of data even if the exponents are not the final asymptotic values. The reason is that these exponent relations are almost direct consequences of thermodynamic identities alone.

J. S. KOUVEL: Since the other numbers which correspond to coefficients as opposed to exponents are coming into register with those numbers which are predicted, I am a little more confident of the asymptotic meaning of our results.

G. A. BAKER: It may be that various other effects, such as compressibility cause the apparent indices to vary, in both directions from those computed for appropriate theoretical models on rigid lattices (see for example the work of Baker and Essam).

S. C. MOSS: A couple of related points about the long-range potential. If the potential is indeed long-range it would certainly be oscillatory in nature (like RKKY) and not monotonically falling off. Then, in the disordered state, because of scattering, the range of the oscillatory potential will be severely limited or damped as de Gennes showed, and this could yield a major distinction between the behavior in ordered and disordered Pd_3Fe. But why should an oscillatory potential influence the exponents near T_c where the correlation range is much greater than $2K_F^{-1}$ (the reciprocal of twice the Fermi wave vector)?

J. S. KOUVEL: As you come closer and closer to the critical point the correlation length gets larger, sampling whatever the potential is in effect doing. Now what an oscillatory potential is doing one doesn't know precisely.

R. B. GRIFFITHS: Contrary to the remark by Professor Widom, I do not believe that the behavior of the critical indices observed in very dilute samples can be ascribed to "renormalization." I would like to emphasize that systems with random "impurities" which are more or less fixed in position have a very different mathematical structure from the situation where renormalization is appropriate. It seems to me interesting that Prof. Rushbrooke's investigation of random dilute systems indicates that γ increases as the concentration of magnetic atoms decreases, and I think it would be very interesting to determine β for such models, though I understand this is a very difficult kind of calculation.

FLUCTUATIONS NEAR THE PHASE TRANSITION IN "ONE-DIMENSIONAL" SUPERCONDUCTORS

R. J. Warburton and W. W. Webb
Department of Applied Physics
Cornell University
Ithaca, New York

ABSTRACT

The effects of thermodynamic fluctuations on the superconducting order parameter can be observed readily near the critical temperature as the fluctuation volume is restricted by reducing the physical dimensions of the system. A comparison is made between several competing theories which predict the effect of fluctuations on the temperature dependence of the electrical resistivity near the transition temperature. Measurements of the complete resistive transition of "one-dimensional" single crystals of tin have been extended through the critical region down to 10^{-5} ohms and the current-voltage characteristics have been extended to a sensitivity of 10^{-15} volts in order to thoroughly test these theories.

1. THEORETICAL BACKGROUND

Observable effects of thermodynamic fluctuations on electric currents in pure bulk superconductors are ordinarily limited to an extremely narrow temperature range, $\Delta t = |T_c - T|/T_c < 10^{-12}$, near to the critical temperature T_c by the long-range coherence characteristic of the superconducting wave function. However, if the fluctuation volume is restricted by reducing the physical dimensions of the systems, as in "one-dimensional" (wire) or "two-dimensional" (film) geometries, the fluctuation-sensitive range is vastly expanded so that fluctuation effects can be observed quite readily.[1] In the case of a "one-dimensional" wire it is assumed that the diameter of the wire is small compared to the temperature-dependent coherence length $\xi(T)$. Therefore the complex order parameter can be treated as a constant across the diameter of the wire and as a variable only along the axis.

The intense interest in the superconducting-normal phase transition and the effects of fluctuations on the phase transition as the sample geometry is reduced can be gauged by the existing number of theories describing the electrical resistance in the region near to T_c in superconductors. Figure 1 attempts to summarize the predictions of five of the existing theories for a "one-dimensional" wire, which will be briefly reviewed below.

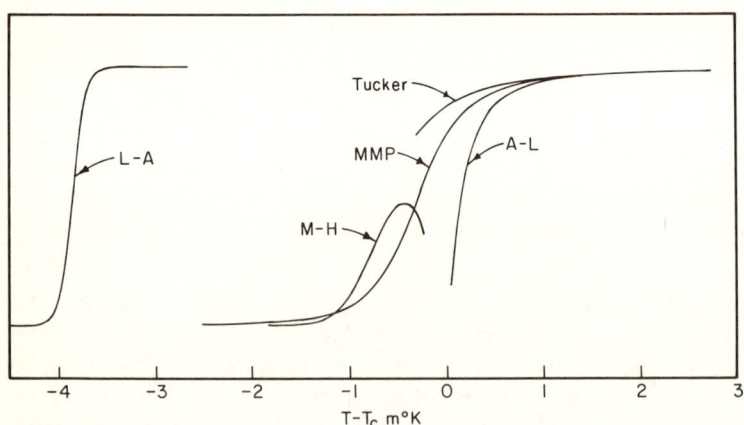

Fig. 1 *A composite of the present theories which are applicable to "one-dimensional" superconductors. See text for discussion and identification of the theories. The material parameters used for the plots are $A = 2.3 \times 10^{-9}$ cm^2, $\xi(0) = 0.54 \times 10^{-5}$ cm, $l = 0.21 \times 10^{-5}$ cm, and normal conductivity $\sigma_0 = 0.39 \times 10^6$ (ohm-cm)$^{-1}$.*

FLUCTUATIONS NEAR THE PHASE TRANSITION

At present there are two disparate theoretical viewpoints as to the nature of fluctuations in "one-dimensional" superconductors.[2] For $T < T_c$ Langer and Ambegaokar[3] propose that resistance is produced by thermal fluctuations in which the phase difference, of the superconducting order parameter, between the two ends of the wire changes by discrete "phase slips" of 2π. They consider this process to be one of thermal activation over an energy barrier $\Delta F(T)$ separating the initial state from the final state between which the difference in the phase changes by 2π radians. The barrier energy is calculated from mean field Ginzburg-Landau theory using constant voltage conditions. For $\Delta F(T) \gg k_B T$ the average rate of transitions between states is

$$\text{Rate} = N \exp \frac{-\Delta F(T)}{k_B T}$$

where the prefactor N is the attempt frequency and k_B is the Boltzmann constant.

Langer and Ambegaokar equate the rate at which the voltage across the wire increases the phase difference with the average rate at which fluctuations decrease the phase difference along the wire in order to calculate the voltage and resistance produced by a steady current. They find that

$$V = 2N\phi_0 \sinh\left(\frac{I}{2I_1}\right) \exp\left(-\frac{\Delta F(T)}{k_B T}\right) \tag{1}$$

where $I_1 = k_B T/\phi_0 = 2.5 \times 10^{-8}$ amps and ϕ_0 is the flux quantum. Langer and Ambegaokar have taken the prefactor N to be a constant equal to the frequency of electron scattering in the entire crystal in the normal state, about 10^{27}/sec.

In the limit $I \to 0$, Eq. (1) becomes

$$R_s = \frac{\phi_0}{I_1} N \exp\left(-\frac{\Delta F(T, I = 0)}{k_B T}\right) \tag{2}$$

where $\phi_0/I_1 = 8.5 \times 10^{-8} \, \Omega$ sec, $\Delta F(T, I = 0)/k_B T = 1/\pi \sqrt{3/2} I_c(\Delta t)/I_1$, and $I_c(\Delta t) = I_{co}(\Delta t)^{3/2}$ is the mean field critical current with I_{co} depending only on material parameters. R_s is the effective resistance associated with the supercurrent. The total resistance, R, is assumed to be equal to the parallel combination of R_s and the normal state resistance R_0.

McCumber and Halperin[4] have studied the theory of the prefactor in much greater detail and find it to be both current and temperature dependent. N is expressed as the product of the number of statistically independent subsystems in the crystal divided by a characteristic relaxation rate. They take the number of independent subsystems to be of order $L/\xi(t)$; i.e., crystal length/coherence length, which leads to a decrease of a factor of about 10^{10} from the value of N suggested by Langer and Ambegaokar. The time dependent Ginzburg-Landau equations are used to introduce a characteristic time $\tau = \pi \hbar / 8 k_B T_c \Delta t$ and to calculate small corrections due to the overlap of the fluctuations along the wire. This leads to a further decrease of approximately 10^5 from the value of N suggested by Langer and Ambegaokar. The current dependence of order I^2 and higher has not been calculated, and there currently are no estimates of its magnitude.[5] McCumber and Halperin stress that although general statistics support their 10^{11} reduction of the prefactor, recent results suggest that the time-dependent Ginzburg-Landau theory is not exactly correct. Therefore their prefactor may still not be exactly correct.[4]

The final form of the McCumber-Halperin prefactor in the limit of $I \to 0$ is

$$N = N_0 \Delta t^{9/4} \quad \text{where} \quad N_0 = \frac{L}{\xi(0)} \frac{I_1}{e} \frac{4}{\pi} \sqrt{\frac{3}{\pi}} \left[\frac{1}{\pi} \sqrt{\frac{3}{2}} \frac{I_{co}}{I_1} \right]^{1/2} \quad (3)$$

is a function only of material parameters. For our crystals at a temperature $T_c - T \simeq 1 m^\circ K$, $N = 10^{10}$/sec and the prefactor on the right-hand side of Eq. (2) is approximately 10^3 ohms. For a typical normal state resistivity of one ohm, this means that fluctuations with energies as large as 20 $k_B T$ give rise to observable resistances of 10^{-3} ohms. This fact, that the resistivity is governed by large and improbable fluctuations justifies the calculation of $\Delta F(T)$ using mean field theory. McCumber and Halperin is used to refer to the Langer-Ambegaokar theory (Eq. (2)) when the McCumber and Halperin prefactor is used.

Figure 1 shows the sharp resistive transition predicted by Langer-Ambegaokar and the effects of the reduced McCumber-Halperin prefactor. The effect is to broaden the resistive transition and to decrease the magnitude of the temperature shift $T_c - T$ at the resistance onset by at least a factor of four. These predictions are consistent with recent experiments on "one-dimensional" superconductors.[6,7] McCumber and Halperin is not expected to be valid close to T_c where $\Delta F(T)/k_B T \lesssim 1$.

The basic idea of the calculations above the critical temperature $T > T_c$ is that fluctuations produce regions of partial superconductivity in which the order parameter is not zero. In one dimension the lengths of the regions are determined by the coherence length $\xi(T)$ and hence get larger as $T \to T_c$. The order parameter of these regions approaches its equilibrium value zero with the microscopic diffusion time $\tau = \pi\hbar/8k_B T\Delta t$ which becomes larger as T_c is approached.[2] This was first calculated from the BCS microscopic theory by Aslamazov and Larkin[8] for 1, 2, and 3 dimensional specimens. The same result has also been calculated by Abrahams and Woo[9] using time-dependent Ginzburg-Landau theory.

The fractional change in the resistance R from the normal state value R_0 is predicted to be

One dimensional: $\dfrac{R_0 - R}{R} = \dfrac{\pi}{16} \dfrac{e^2}{\hbar} \dfrac{R_0 \xi(0)}{A} \Delta t^{-3/2}$

(4)

Two dimensional: $\dfrac{R_0 - R}{R} = \dfrac{e^2 R_0}{16\hbar} \Delta t^{-1}$

Aslamazov and Larkin predict a long tail (see Fig. 1) in the resistance at high temperatures in contrast to the sharp onset of resistance predicted by Langer and Ambegaokar and McCumber and Halperin. This is due to the basic difference between the effect of fluctuations below and above T_c. Below T_c, Langer and Ambegaokar and McCumber and Halperin theories describe a thermally activated process where only fluctuations which cause the order parameter to momentarily go to zero contribute to the resistance. However, above T_c there is no such restriction on the fluctuations. The normal electrons carry most of the current and fluctuations which produce only infinitesimal deviation of the order parameter from zero decrease the resistance proportional to a power of Δt as in Eq. (4) instead of an exponential as for $T < T_c$.

Although Glover[10] found excellent agreement with Aslamazov and Larkin in very dirty "two-dimensional" bismuth films, Strongin et al.[11] and Masker and Parks[12] have found in aluminum that the excess conductivity could be up to ten times larger than predicted by

Aslamazov and Larkin for films with large mean free paths. Crow et al.[13] have shown that this large excess conductivity is consistent with the excess conductivity predicted by Maki[14] due to the interaction of the normal electrons with the superconducting pairs. As a magnetic field is applied they find that the conductivity decreases and approaches that of Aslamazov and Larkin. In fact these are some of the experiments that show that the time-dependent Ginzburg-Landau theory is not exactly correct. Masker, Marčelja, and Parks[15] have extended the Aslamazov and Larkin theory from above T_c by including the effect of the fourth-order term in the Ginzburg-Landau theory, which represents the interaction between superfluid excitations. For $T > T_c$ this theory simplifies to Aslamazov and Larkin and for $T < T_c$ predicts a cubic dependence of the excess conductivity σ' in one dimension and an exponential dependence in two dimensions

One dimensional: $\quad \sigma'_{T<T_c} \propto (\Delta t)^3$

Two dimensional: $\quad \sigma'_{T<T_c} \propto \exp(-c\Delta t)$

(5)

where c depends only on material parameters. In two-dimensional aluminum films, Masker, Marčelja, and Parks[15] have obtained good agreement with Eq. (5) for $T < T_c$ even when the sample mean free paths were varied over two decades.

Tucker and Halperin[16] have recalculated the theory presented by Masker, Marčelja, and Parks by using a Hartree-Fock approximation for the superconducting wave function instead of the Hartree approximation used by Masker, Marčelja, and Parks. The effect is a more gradual decrease in R/R_0 as $T \to T_c$ than that predicted by Marčelja, and Parks. Tucker and Halperin are less optimistic about the validity of their calculation for $T < T_c$ and believe that deviations should begin to occur at $T \lesssim T_c$.

Figure 1 also summarizes Aslamazov and Larkin, Masker, Marčelja, and Parks, and Tucker. It is apparent that either Masker, Marčelja, and Parks or Tucker could smoothly extrapolate to McCumber and Halperin through the region $T \simeq T_c$. It is also clear how drastically different the McCumber and Halperin and Masker, Marčelja, and Parks prediction for the onset of resistance are.

2. EXPERIMENTAL

We have found continued measurements[6,7] on the entire resistive transition through the critical region on single tin whisker crystals.

FLUCTUATIONS NEAR THE PHASE TRANSITION

These "one-dimensional" crystals have diameters small compared to the temperature-dependent coherence length in the temperature range of the measurements. Under these conditions it is expected that the transition shape is determined by thermodynamic fluctuations, and the theories reviewed above can be experimentally tested.

The whisker crystals, which were prepared by conventional techniques,[17] were carefully soldered to superconducting electrodes with Wood's metal solder. A major difficulty of this experiment was production of these contacts without damage to the crystal. Such damage produced readily identifiable structure in the resistive transition. The mounted crystals were screened by examining the transition using a conventional nanovoltmeter and accepting only those which showed no obvious structure in the transition. Only about 5 percent of the crystals mounted were acceptable by this criterion.

Figure 2 shows a typical transition at a constant current of 2.2×10^{-8} amps for a Sn-2% In alloy whisker crystal which has identical parameters to those used in Fig. 1. Since the experimental critical temperature is unknown, the theory curves (and T_c) can be moved along the horizontal temperature axis in order to obtain a fit. In Fig. 2, T_c is shifted such that the theory of McCumber and Halperin

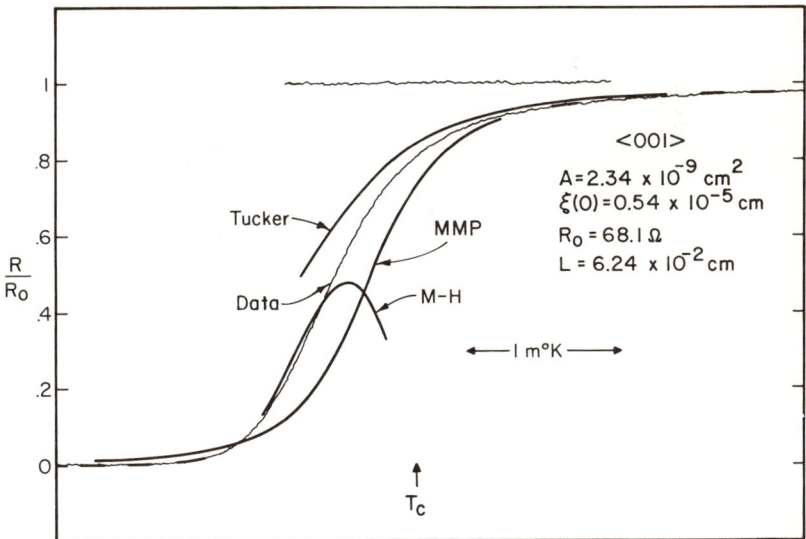

Fig. 2 *The resistive transition of an alloy Sn-2% In whisker crystal with material parameters as in Fig. 1. An approximate fit to this data is provided by McCumber and Halperin for $T < T_c$ and Tucker for $T \gtrsim T_c$. Measurement is made at a constant current of 2.2×10^{-8} amps.*

overlaps the onset of resistance. However, the McCumber and Halperin theory predicts a resistance that increases too quickly which indicates that the prefactor is still slightly too large.

In Figs. 3 and 4 a closer look is taken at the onset of the resistive transition for an alloy crystal and a pure tin crystal. The measurements are made at 2.2×10^{-8} and 10^{-7} amps respectively. Although 10^{-7} amps is four times larger than I_1, since the transition shift due to this current is less than 0.02 m°K, the results presented here are effectively in the zero current limit.

The solid line is the best fit to Eq. (2) taking the prefactor N_0 and T_c as variable parameters but using the experimentally determined

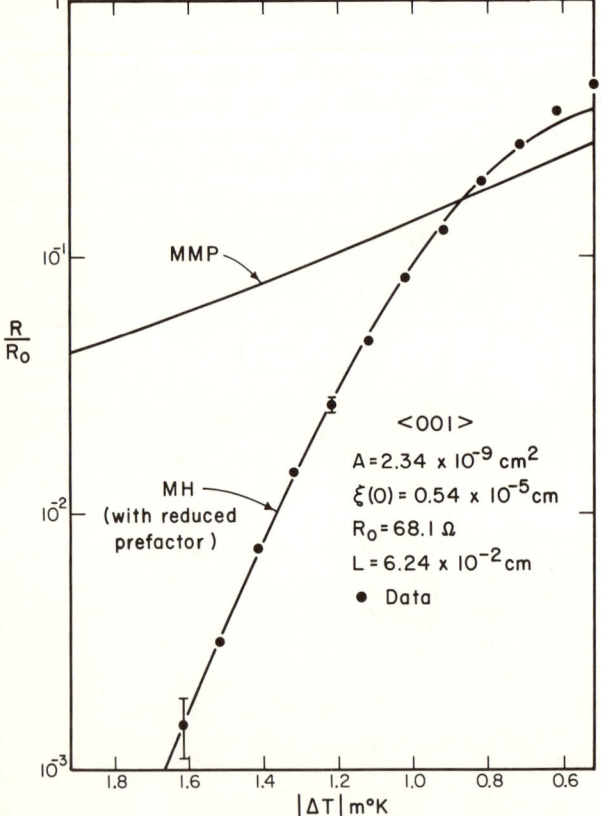

Fig. 3 *Onset of the total resistance of the Sn-2% In crystal of Fig. 2 with temperature. Points ● are data. Line through data is a two-parameter fit to McCumber and Halperin theory with N_0 reduced by a factor of 1.3. Masker, Marčelja, and Parks theory is shown to be inconsistent with the data for $T < T_c$.*

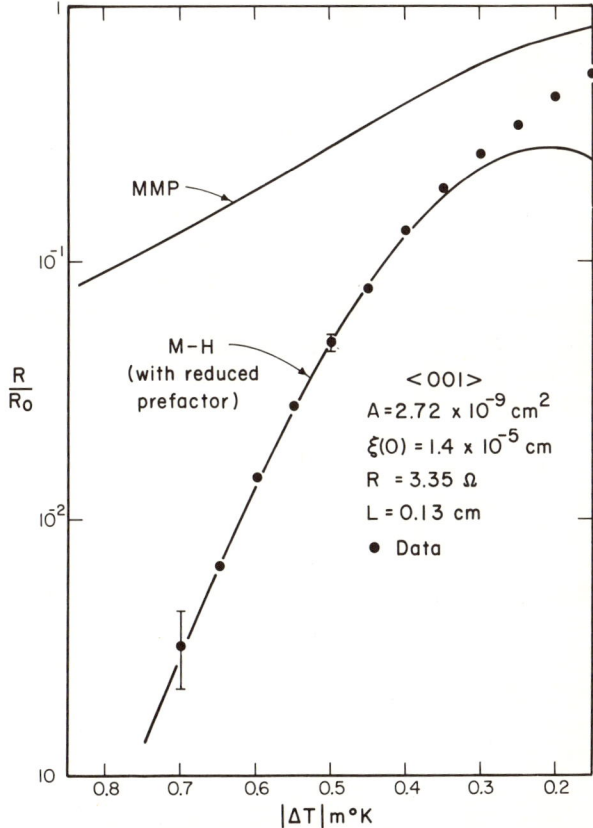

Fig. 4 *Onset of the total resistance of a pure tin whisker crystal with temperature. Line through data is a two-parameter fit to McCumber and Halperin theory with N_0 reduced by a factor of 14. The mean free path $l = 10 \times 10^{-5}$ cm. Masker, Marčelja, and Parks theory is also shown.*

value of I_{co} where $I_c = I_{co} \Delta t^{3/2}$. The best fit values of N_0 obtained in this way are less than the McCumber and Halperin value by factors of $10^0 - 10^2$ for a number of samples with areas ranging from 10^{-9} to 10^{-8} cm² and electron mean free paths of 0.5×10^{-5} to 30×10^{-5} cm. This is to be contrasted with a deviation of about 10^{15} from the attempt frequency originally suggested by Langer and Ambegaokar. Crystals with short mean free paths and hence reduced coherence length generally have values of N_0 closer to the theoretical value. The suggestion of Masker, Marčelja, and Parks that $R \propto \Delta t^{-3}$ for small resistances is definitely excluded by the exponential decrease observed.[18]

The best fit value of N_0 is determined primarily by the curvature of the data on $\log R$ vs. T. This curvature is greatest for high resistance—just where one expects corrections to the theory to become important. The basic criterion for the applicability of the theory is that the ratio of the energy barrier height ΔF to the thermal energy $k_B T$ be large. In Figs. 3 and 4 the data and theory began to deviate where $\Delta F/k_B T$ is about 3. Thus, the precision with which the various parameters can be specified depends on where corrections to Eq. (2) become important. An independent value for T_c would define the regime where the corrections are serious and thus permit more certain determination of the experimental parameters and allow a test of the temperature dependence of the prefactor. We are currently making measurements of the temperature-dependent penetration depth to obtain such a value.

I_{co} was determined in two ways. First, the temperature dependence of the current necessary to produce a given resistance in the crystal (chosen to be about 1% R_0) was measured between about 10 and 30 μa. A plot of $I_c^{2/3}$ versus temperature gave a straight line of slope $I_{co}^{2/3}$. This slope could be determined within a few percent. Although data close to T_c was not used it is possible, though we feel it is unlikely, that an additional error of several percent could have been introduced by the effect of fluctuations on the mean field slope. Next we checked the value of I_{co} by comparing it to the value calculated from the measured cross sectional area and coherence length[19]

$$I_{co} = \frac{16}{3\sqrt{3}} \frac{e}{\hbar} \frac{(1.82 H_0)^2}{8\pi} \xi(0) A \qquad (6)$$

where the critical magnetic intensity $H_0 = 306$ gauss for tin. The area was determined from the room temperature resistance R_{RT} using the standard relationship $A_{<>} = \rho_{<>} L/R_{RT}$ where orientations are obtained from electron diffraction patterns and $\rho_{<>} = 9.9, 11.1,$ and 15.3 $\mu\Omega$-cm for the $<100>$, $<101>$, and $<001>$ orientations respectively.[20] The mean free path at 4.2°K, $l_{4.2}$, is determined from the relation $l_{4.2} = (R_{RT} - R_{4.2}/R_{4.2}) l_{RT}$ where $l_{RT} = 6.4, 9.0,$ and 4.3×10^{-7} cm for the $<100>$, $<101>$, and $<001>$ orientations.[21] The coherence length is then determined using the standard formula[22]

$$\xi(0) = .74 \xi_0 \left[1 + .752 \left(\frac{\xi_0}{l}\right)\right]^{-1/2} \qquad (7)$$

where $\xi_0 \simeq 2.1$ to 2.3×10^{-5} cm for tin.[23] The uncertainty in this calculation is about 10 percent which is comparable with the uncertainty of ξ_0. The calculated and measured values of I_{co} agree within this error. The best fit value of N_0 varies by about a factor of 2 when I_{co} is varied within its experimental uncertainty.

Referring back to Fig. 2, for $T \gtrsim T_c$ it is clear that the decrease in resistance from the normal resistance is the same type behavior as that predicted by Masker, Marčelja, and Parks and Tucker. However, Masker, Marčelja, and Parks predicts that the resistance decreases too quickly from R_0. An extremely good fit can be made for $R/R_0 > 0.8$, to the theory of Tucker, by varying only T_c. This is shown in Figs. 5 and 6. Such a fit, however, necessitates a higher value of T_c than that used to fit McCumber and Halperin. This discrepancy between the values of T_c necessary to fit the theories of McCumber and Halperin and Tucker, ranges from 0.5 to 0.1 m°K and appears to decrease as the mean free path decreases. However, this observation has been made on only a few crystals so that no exact dependence has been determined. An independent measurement of T_c would certainly clarify the situation.

We have not attempted to investigate the Aslamazov and Larkin effect or the extra conductivity predicted by Maki in these experiments. These effects are observable in approximately the top few percent of the R/R_0 transition and for $T \gg T_c$.

In order to further test the theory of Langer and Ambegaokar and McCumber and Halperin we have extended measurements of the onset of the resistive transition and of the current-voltage characteristics below the critical temperature on single crystal tin whiskers to voltages of 10^{-15} volts (an improvement of six orders of magnitude over previous measurements) and power levels of 10^{-22} watts. These measurements of the onset of the transition employed the circuits shown in Fig. 7 where a superconducting magnetometer serves as a femto-volt amplifier[24] arranged alternatively to measure voltage across a crystal specimen or the current circulating in a superconducting loop containing the crystal. A comparison of the resolution of the nanovoltmeter and femtovoltmeter is made in Fig. 8 where traces of the resistive transition are shown for the same crystal.

The crystals were placed in helium exchange gas inside a copper can which could be temperature regulated to ±1 microKelvin. The ambient magnetic field level was about 10^{-5} gauss. The crystals were surrounded by several layers of shielding including copper and superconducting cans. The final copper can which enclosed the crystal had only high impedence current leads entering it. These leads were well filtered. Additional filters were added at helium

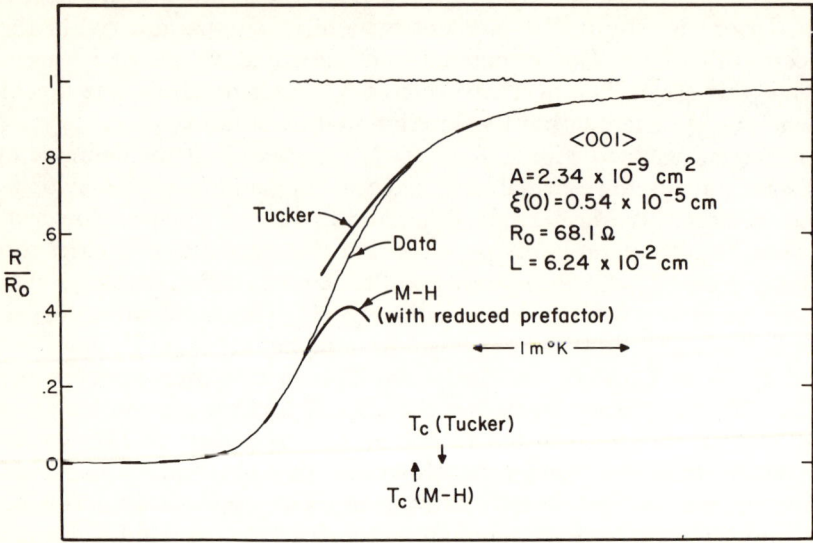

Fig. 5 *Independent fits of McCumber and Halperin and Tucker to the resistive transition for the Sn-2% In crystal of Figs. 2 and 3 are displayed. Where theory and data overlap the theory is indicated by dashed lines. The discrepancy between fitted values of T_c is 0.13 m°K.*

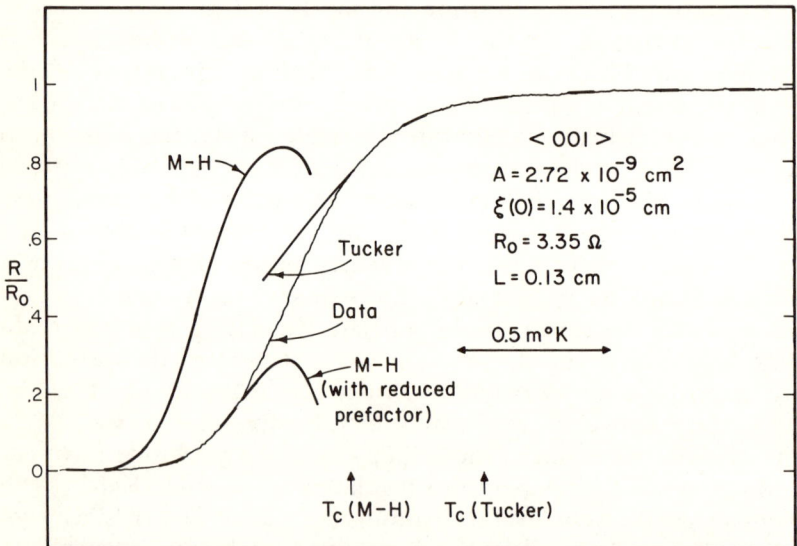

Fig. 6 *Independent fits of McCumber and Halperin and Tucker to the data for the Sn crystal of Fig. 4 are displayed. The effect of reducing the McCumber and Halperin prefactor by a factor of 14 is shown. The discrepancy between the fitted values of T_c is 0.45 m°K.*

FLUCTUATIONS NEAR THE PHASE TRANSITION

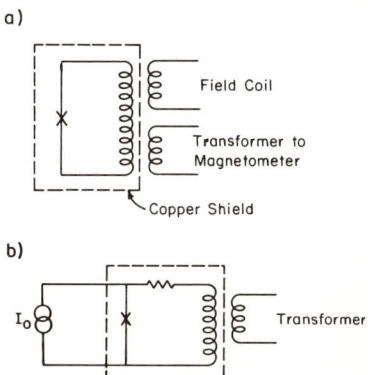

Fig. 7 *Measuring circuits of the superconducting femtovolt amplifier. The tin crystal is indicated by X. (a) Ammeter configuration using closed superconducting loop with magnetically induced current. (b) Voltmeter configuration with external current leads.*

temperature, but no change in the results was found. The most convincing evidence that external noise had no effect on the measurements was obtained in the configuration of Fig. 7a where current was magnetically induced in the crystal. In this case no wires of any kind penetrated the shield can. Using several cans with cutoff frequencies ranging from 45 Hz to about 1kHz there was no change from the results obtained with circuit 7b. Any significant magnetic noise below the cutoff frequency would have been directly observed with the magnetometer, which was part of the superconducting voltmeter.

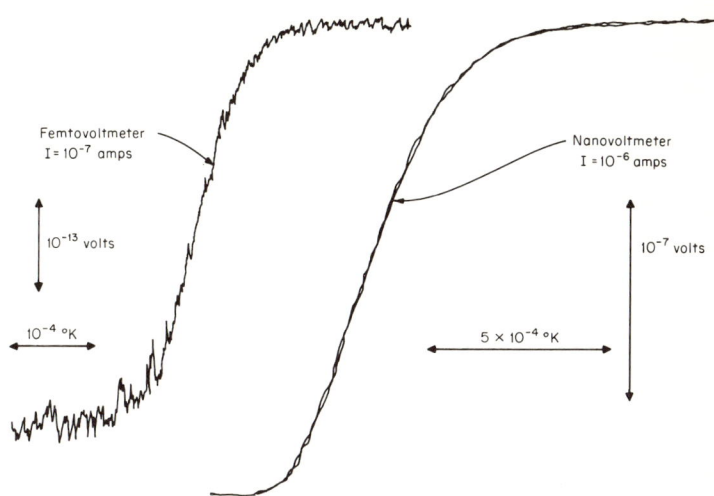

Fig. 8 *A comparison of the data from a conventional nanovoltmeter and the superconducting femtovoltmeter.*

The most direct evidence for the validity of the Langer and Ambegaokar model of voltage arising from the unique phase slips of magnitude 2π is provided by the current dependence of the voltage V at constant temperature. At constant temperature, the theory predicts to first order, $V \propto \sinh I/2I_1$. This relationship follows directly from the model assumed and does not depend on detailed calculations of the energy barriers, on sample parameters, or on the boundary conditions.[25] Figure 9 shows a measured $I - V$ curve compared to the sinh relation. For $I \gg 2I_1$, the theory gives $(\ln V)/I_1 = 20.4 (\mu a)^{-1}$. The measured slope is $S = \Delta \ln V/\Delta I = 17.2$ $(\mu a)^{-1} \pm 2\%$ at $0.15\,\mu a$. This value of S shows some variation, usually in the range of $15-18$ $(\mu a)^{-1}$ depending on the sample measured and the temperature, $\Delta t = 1 - T/T_c$, of measurement. S tends to

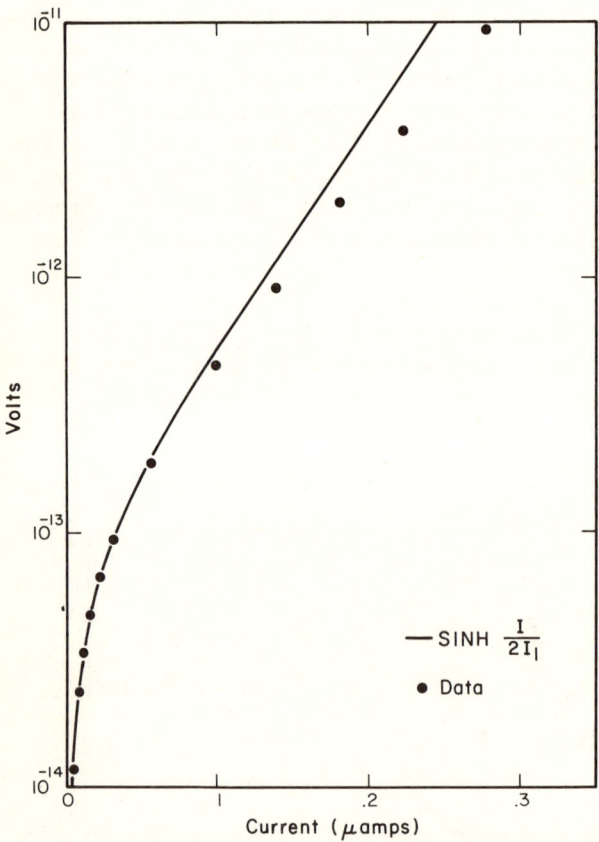

Fig. 9 *Current-voltage characteristic at fixed temperature. Solid line, $V = \sinh I/2I_1$, • are data points.*

increase with Δt. Inclusion of higher order terms in I that appear in the Langer and Ambegaokar calculation of the energy barrier gives slightly better agreement between theory and experiment. The still unknown higher order current dependence of the prefactor, presumably depending both on sample parameters and temperature, may explain the small remaining discrepancy.

The data in Fig. 9 was obtained with the crystal connected in a closed superconducting loop with an inductance of 10^{-6} H as shown in Fig. 7a. Current was induced in the loop by ramping an external magnetic field and was measured with a superconducting magnetometer. The voltage was determined from the measured rate of change of flux linking the loop. The value of the inductance was then changed to 2×10^{-8} H to permit observation of flux changes through the loop of less than ϕ_0. As the induced current in the loop was increased toward the critical current we observed that flux entered in discrete units of ϕ_0 corresponding to the 2π phase slips of the Langer and Ambegaokar model. Furthermore, at constant ambient conditions near T_c, with no induced current the loop current spontaneously fluctuated among several discrete values corresponding to the low level quantum states. This behavior is shown in Fig. 10. This is the same type of behavior previously observed in low inductance loops containing thin film weak links.[26] The value of S that was measured in the low inductance loop remained 17.2 $(\mu a)^{-1}$.

The crystal was then connected in the voltmeter configuration shown in Fig. 7b. Here the loop contains a resistance so that fluxoid quantization is not required. However, the measured value of S again remained unchanged indicating that the voltage was still produced by discrete 2π phase slips as assumed by Langer and Ambegaokar and

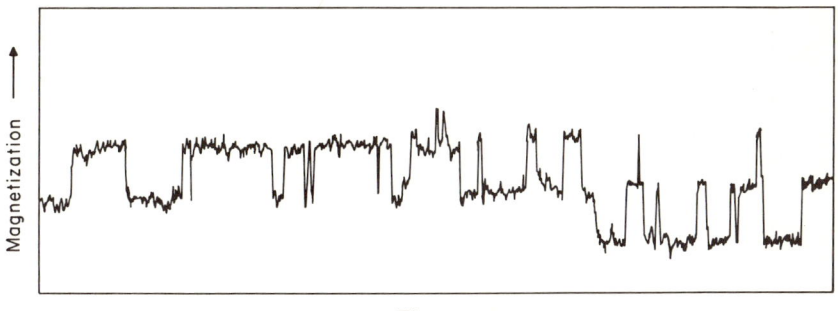

Fig. 10 *Spontaneous fluctuations of the loop current, for the circuit of Fig. 7a, among several discrete quantum states. The temperature is adjusted near to T_c and the applied magnetic field is close to a multiple of ϕ_0.*

directly observed in the closed loop. This data also tends to confirm McCumber's calculation that the boundary conditions on the crystals are relatively unimportant.[27]

In the configuration of Fig. 7b the resistive transition is also measured. Figure 11 shows such a resistive transition measured in the zero current limit extended down to 10^{-5} ohms. Data on the lower part of the transition was obtained with the superconducting voltmeter. Data for $R > 0.1$ was also obtained with a Keithley nanovoltmeter. The consistency of the two measuring techniques is shown by the region of overlapping data. The same fitting procedures used for Figs. 3 and 4 are used here. In this case the prefactor must be reduced by a factor of 30 to obtain a good fit. Deviations from theory occur where $\Delta F/k_B T$ is approximately 3 which is in agreement with Figs. 3 and 4.

If all three parameters, N_0, T_c, and I_{co} are varied, the fit obtained above cannot be improved. It becomes significantly worse for values of I_{co} deviating by more than 20 percent from the measured value.

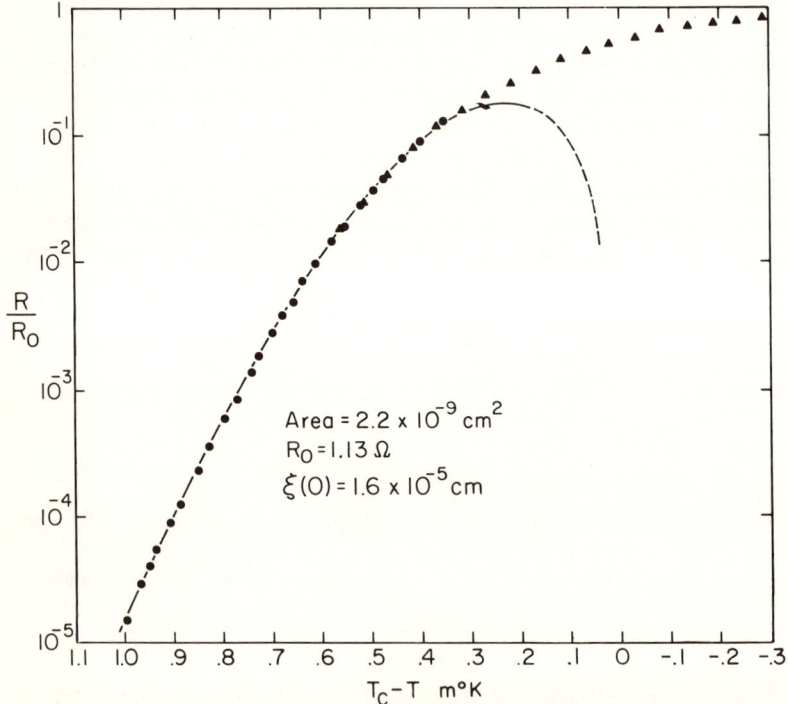

Fig. 11 *Total resistance of tin crystal as a function of temperature.* ● *measured with superconducting femtovoltmeter;* ▲ *measured with nanovoltmeter. Line is a two-parameter fit to McCumber and Halperin theory.*

3. CONCLUSIONS

We conclude that the voltage appearing near the onset of the transition in uniform "one-dimensional" crystals is due to thermally activated quantum phase slips as proposed by Langer and Ambegaokar. The Langer and Ambegaokar calculation of the zero current activation energy barrier also agrees with the value that we observe. The basic current-voltage dependence is also confirmed; however, significant deviations from theory are observed at higher currents where terms of order $(I/I_c)^2$ are not negligible. This discrepancy may well arise from the uncalculated current dependence of the attempt frequency, although our measurement does not distinguish this possibility from that of additional current dependence in the barrier height. The generally close agreement of the measured zero-current attempt frequency to that predicted by McCumber and Halperin appears to verify their statistical arguments. However, the prefactor appears to be more strongly dependent on the sample mean free path than predicted by McCumber and Halperin, and approaches the value predicted by McCumber and Halperin as the mean free path is decreased.

We also observe that for $T \gtrsim T_c$ the theory of Tucker describes the resistive transition for $R/R_0 > 0.8$, where only the critical temperature T_c is adjusted. This, however, leads to an unexplained discrepancy with the value of T_c found by fitting McCumber and Halperin to the transition onset. Fitting the data to Tucker indicates values of T_c from 0.1 to 0.5 m°K higher than that found from McCumber and Halperin. The discrepancy between the two values of T_c decreases as the mean free path is decreased.

As a result of the two mean free path dependencies indicated above, we find that in our Sn-2% In alloyed crystals a smooth extrapolation between the theories of McCumber and Halperin and Tucker provide a good description of the complete experimental transition region.

ACKNOWLEDGMENTS

We owe special thanks to J. E. Lukens with whom we have collaborated in doing these experiments. We are also pleased to acknowledge helpful discussions with B. I. Halperin, J. W. Wilkins, and J. Tucker, the gift of a flux detector by R. A. Buhrman, the support of the U.S. Atomic Energy Commission and the use of facilities at Cornell provided by the Advanced Research Projects Agency.

REFERENCES

1. See for a review, P. C. Hohenberg: *The 11th Int. Conf. on Low Temperatrue Physics*, St. Andrews, Scotland, August 1968.
2. See for a review, V. Ambegaokar, NATO Advanced Study Institute on Superconductivity, McGill University, Montreal, June 1968.
3. Langer, J. S., and V. Ambegaokar: *Phys. Rev.*, **164**:498 (1967).
4. McCumber, D. E., and B. I. Halperin: *Phys. Rev.*, **B1**:1054 (1970).
5. B. I. Halperin: Private communication.
6. Webb, W. W., and R. J. Warburton: *Phys. Rev. Lett.*, **20**:461 (1968).
7. Warburton, R. J., B. R. Patton, W. W. Webb, and J. W. Wilkins: *Proc. Conf. on the Science of Superconductivity*, Stanford University, to be published.
8. Aslamazov, L. G., and A. J. Larkin: Soviet Physics—Solid State, **10**:875 (1968); *Phys. Rev. Lett.*, **26A**:238 (1968).
9. Abrahams, E., and J. W. F. Woo: *Phys. Rev. Lett.*, **27A**:117 (1968).
10. Glover, R. E.: *Ibid.*, **25A**:542 (1967); Naugle, D. G., and R. E. Glover: *Ibid.*, **28A**:110 (1968).
11. Strongin, M., O. F. Kammerer, J. Crow, R. S. Thompson, and H. L. Fine: *Ibid.*, **20**:922 (1968).
12. Masker, W. E., and R. D. Parks: *Phys. Rev.*, **B1**:2164 (1970).
13. Crow, J. E., R. S. Thompson, M. A. Klenin, and A. K. Bhatnager: *Phys. Rev. Lett.*, **24**:371 (1970).
14. Maki, K.: *Progr. Theoret. Phys. (Kyoto)*, **39**:897 (1968).
15. Masker, W. E., S. Marčelja, and R. D. Parks: *Phys. Rev.*, **188**:745 (1969).
16. Tucker, J., and B. I. Halperin: Private communication.
17. Ellis, W. C., D. F. Gibbons, and R. G. Treiling: "Growth and Perfection of Crystals," p. 102, John Wiley & Sons, N. Y., 1958.
18. Preliminary data of Ref. 7 supported the theory of Masker, Marčelji, and Parks for Sn-2% In alloyed whisker crystals. (See Fig. 6 of Ref. 7.) However, this spurious result was caused by fitting the theory to the data by using a clean coherence length of $\xi(0) = 0.74\xi_0$ and by varying the cross sectional area A by 10 percent. Using the more accurately determined parameters contained in this work, this no longer provides a good description of the experimental resistive transition.
19. De Gennes, P. G.: "Superconductivity of Metals and Alloys," p. 184, W. A. Benjamin, New York, 1966.
20. Burckbuchler, F. V., D. Markowitz, and C. A. Reynolds: *Phys. Rev.*, **175**:543 (1968).
21. Cook, J. W., Jr., M. J. Skove and E. P. Stillwell: Preprint.
22. Goodman, B. B.: *Rep. Prog. Phys.*, **29**:445 (1966).
23. The standard value of ξ_0 of 2.3×10^{-5} cm is presumably based on early measurement of the anomalous skin effect. However, $\xi_0 = 2.1 \times 10^{-1}$ cm is obtained by J. Fedor and D. S. McLachlan, *Phys. Rev.*, **177**:763 (1969).
24. Lukens, J. E., R. J. Warburton, and W. W. Webb: *J. Appl. Phys.*, to be published.
25. McCumber, D. E.: *Phys. Rev.*, **172**:427 (1968).
26. Lukens, J. E., and J. M. Goodkind: *Phys. Rev. Lett.*, **20**:1363 (1968).
27. McCumber, D. E.: *Phys. Rev.*, **181**:716 (1969).

DISCUSSION on Paper Presented by R. J. Warburton

R. A. FERRELL: I want to make some additional remarks. In our problem we have a free energy which has not only a double minimum but a continuous degeneracy of minima. In other words we have an order parameter which has a certain definite size but a continuous phase. We are dealing with a system, leaving one minimum and going over a barrier into a second minimum. Somewhere along the wire the order parameter has to vanish and the phase is changed. That means we have a certain free energy difference and this appears in the exponent. This is the dominant factor, I think, in fitting your data and if it is dominant it's hard to see why there is so much a difference between the pure and the disordered material because it is really a question of thermodynamics.

R. J. WARBURTON: I see no reason why there should be any difference between the pure and the dirty limit. But the same difference is seen for temperature above T_c. In effect the basic voltage relation I showed you is not affected by the mean free path.

R. B. GRIFFITHS: But in all expressions the $(T - T_c)$ term is important and as the bulk transition temperature is not known, how do you know that the agreement is worth more in the pure than in the impure case?

R. J. WARBURTON: In both cases we have a two-parameter fit using the prefactor and the critical temperature and a fit is obtained. However, the fit in the alloy-crystal, with a much smaller mean free path, is much closer to that which you expect from the theory of McCumber and Halperin. That means the prefactor is closer to the theoretical value for the impure samples.

R. B. GRIFFITHS: But if the prefactor is not controlled by thermodynamics it could depend in some way on the impurity content of this material.

L. P. KADANOFF: That factor describes how fast you are leaving one minimum of the free energy and go up the free energy barrier. It depends on a lot of detailed properties of the crystal.

Agenda Discussion:
MAGNETS AND SUPERCONDUCTORS

R. A. Ferrell[*]
Department of Physics and Astronomy
University of Maryland
College Park, Maryland

H. Büttner[†]
Battelle Institute
Frankfurt/Main
West Germany

In reviewing the agenda discussion on superconductors and magnets, it is useful first to emphasize that a wealth of fluctuation effects has now been uncovered in superconductors. This conference has concentrated on the effect of fluctuations on a one-dimensional superconductor. Similar resistive effects are manifested in two- and

[*]Chairman
[†]Secretary

three-dimensional superconductors. Especially clear is the two-dimensional case, where thanks to the highly developed art of preparing evaporated films, parameters such as thickness and electron mean free path can be carefully controlled. Here again, as reported by Dr. Warburton in the one-dimensional case, the "clean," or long-mean-free-path samples, seem to show a significant deviation from the behavior of the short-mean-free-path samples. The microscopic theory predicts some deviation, but some basic difficulties remain in the theory and it seems likely that the subject has not yet reached its final state. This certainly is true on the experimental side, where the fluctuation contribution to the onset of the Meissner effect promises to exhibit many interesting facets, such as orientation dependence ensuing from crystalline anisotropy. Further effects of fluctuations on the onset of the Josephson effect, on the specific heat, on nuclear spin relaxation, etc., have hardly begun to be explored. One can safely predict further vigorous theoretical and experimental activity in the study of superconductors near the phase transition.

An interesting aspect of the fluctuations in superconductors is that their very weakness, which delayed the detection of the fluctuation effects, may actually be regarded as an asset. The effects set in so close to the transition temperature that other noncritical sources of temperature variation can be neglected compared to the critical variation coming from the fluctuations. This permits the clean separation of the critical effects.

The situation in magnets is, of course, much different. Here the fluctuations are much stronger and the self-consistent treatment of their interaction is essential. This is a difficult computational problem which has been under serious mathematical attack by Résibois and coworkers. They necessarily have to introduce approximations. The crudest of these, which can be regarded as a "decoupling" or factoring procedure, leads to remarkably accurate comparison with experiment. To explain why this simple procedure works so well remains one of the principal tasks of the theory. One can expect that more and more sophisticated mathematical techniques will be brought to bear to obtain self-consistent solutions from the relevant integral equations. At the same time, it is likely that the neutron scattering experiments will continue to fill in the dynamical details of the magnets. This is especially to be expected in the antiferromagnets, where the interplay of the two different kinds of fluctuations (staggered and uniform) leads to a great richness of phenomena. It is especially desirable in this case that one particular system be studied with many different experimental techniques. For example, the neutron experiments need to be supplemented by nuclear resonance frequency and line-width measurements, as well as thermodynamic specific heat and magnetization studies, *all on the same material*. Only in this way can we expect to arrive at a complete and consistent picture of the critical dynamical properties of magnet materials.

1. SUPERCONDUCTORS

The first part of the discussion was centered around the problem of fluctuation in superconductors. In continuation of Dr. Warburton's paper, Dr. Büttner gave the following short introduction to the theoretical background and the problems associated with this field.

The main problem is the calculation of the transition rates of the system activated by thermal fluctuations. The rate constant R is written as

$$R = \phi e^{-\beta \Delta F}$$

where ΔF is the free energy barrier for the process and ϕ is the prefactor discussed this morning.

1. The determination of ΔF is a thermodynamical problem. One starts with the usual mean field approximation by expanding the free energy $F[\psi(x)]$ as a functional of the order parameter $\psi(x)$. The equilibrium condition then leads to a differential equation for $\psi(x)$, the well-known Ginzburg-Landau equation. In the case of a one-dimensional system the minimum solution of the linearized equation is

$$\psi(x) = C \cdot e^{ikx}$$

where C is a constant and $k = 2\pi h/l$ ($l = L/\xi$, L is the length of the wire and ξ the coherence length of the system). So there are a number of minima for which the order parameter differs only by a phase factor. Between the minima there are other extremal solutions located which correspond to saddle points in the free energy. In order to make a transition from one minimum to another the system has to overcome this saddlepoint. A nice discussion of the problem is given by Langer and Ambegoakar[1] (see also McCumber and Halperin[2]).

2. The next step, the calculation of the prefactor ϕ, is much more problematic. As we heard this morning the results of McCumber and

Halperin[2] differ by twelve orders of magnitude from the value assumed by Langer and Ambegoakar. I would like to make a comment on some of the difficulties involved. As we now are interested in the dynamics of the system, we have to discuss the time dependence of the order parameter. This is done by using a time-dependent Ginzburg-Landau equation, derived by several authors (one of the latest discussions is given by Eilenberger,[3] which also contains earlier references). But there are still problems connected with the derivation from the microscopic Gorkov equation. As McCumber and Halperin[2] pointed out, they are only interested in the right order of magnitude of ϕ, and this should be independent of the detailed structure of the time-dependent G. L. equation.

In order to incorporate the influence of fluctuations one usually adds an additional random force. By carefully discussing the resulting Fokker-Planck equation, McCumber and Halperin[2] were able to determine a quite accurate prefactor, which fits very nicely the experiments for whiskers with alloy structure (the so-called dirty limit) as Dr. Warburton reported. The agreement with the measurements in clean materials is not so good and off by an order of magnitude. The problem in these clean limits is that it is important to look for the coupling between the Cooper-pairs and the single electrons, because they also can carry a current in these materials (see Eilenberger[3]).

Finally, I should mention that the difficulty in describing fluctuations in a clean superconductor also exists above T_c.

2. LIQUID CRYSTALS

The discussion on liquid crystals started with some remarks of Prof. M. J. Stephen on the possibility of second-order transition in these materials. The question is: Under what circumstances does one expect a second-order transition? The proposed answer was: If the positive and negative values of the order parameter correspond to equivalent situations, then the transition can be of second order. In a reply to Prof. Domb's suggestion, that a liquid crystal may behave like an Ising system with random impurities, Prof. Stephen questioned this analogy, because the translational motion, which is quite rapid in these crystals, tends to average out the interaction.

After these remarks Prof. J. D. Litster recalled Uhlenbeck's work on the validity of the mean field theory (MFT) for weak long-range forces. From his work on liquid crystals, where presumably short-range van der Waals forces are dominant, he found that MFT seems to work well in these subtances. He suggested that MFT is also valid for short-range forces if fluctuations are not too large.

Dr. G. Alefeld wondered whether there are really short-range forces or whether the torque between different molecules produces long-range forces.

Prof. B. Widom argued that one has to distinguish carefully between long-range forces and long-range correlation. The propagation of a correlation effect can go well beyond the range of the force itself, as is well known in fluids near their critical points.

A way to decide whether the range of correlations or of forces is involved was proposed by Dr. P. C. Clapp. One twists one molecule and holds all others fixed to their positions, except one far away which can move freely. If it responds to the motion of the first molecule then there is an interaction; otherwise we have only a propagation of order.

Prof. L. P. Kadanoff responded to the original question of Prof. Litster by saying that the experiments seem to agree with both the MFT and *also* with the standard theory (Ginzburg) for the limits of validity of MFT. The experiments, therefore, do not bring into question the "conventional wisdom" about MFT. The next question by Dr. H. E. Cook was about the elastic constants of liquid crystals and whether they are anisotropic. Furthermore he raised the problem of the range of the elastic interaction. If the crystal is elastically isotropic then there is a theorem which says that there is no long-range interaction. Only in anisotropic crystals can the interaction be long-ranged.

Prof. J. D. Litster confirmed that there are normal elastic constants for compression in liquid crystals. But Dr. G. Alefeld in his reply to the second problem, said that one has to be careful in discussing elastic interactions because the interaction of two centers of dilatation is different in finite or infinite elastic continua. In the infinite isotropic medium the centers interact only if they are on top of each other, while in finite media they can interact via image forces. In anisotropic systems there is always an interaction.

Finally, Dr. D. de Fontaine made some comments on the distinction between long-range forces and long-range propagation, which seems to be not apparent in the mean field approximation. For example, in the microscopic theory of Dr. Cook and Dr. de Fontaine for elastic effects they start with short-range coupling parameters and after elimination of displacement variables end up with long-range effective forces. In a sense, the mean field approximation replaces the propagation of short-range forces by an effective long-range force.

Regarding the question raised by Dr. Cook concerning the liquid crystal work, one must bear in mind that Crum's theorem for isotopic systems applies only to centers of dilatation. The molecules of the liquid crystals act as elastic dipoles, however, and can generate long-range interactions even in elastically isotropic systems.

3. MAGNETIC SYSTEMS

In this part of the session two topics were discussed intensively: The problem of parametric representation and the effect of impurities on the critical indices.

In a first remark Prof. J. D. Litster took a somewhat more enthusiastic attitude than Prof. Vincentini-Missoni toward the linear model proposed by Schofield, Ho, and himself. The force for the equation of state is in this model

$$H = \alpha\theta(1 - \theta^2)r\beta\delta \qquad T - T_c = (1 - b^2\theta^2)r$$
$$M = m(\theta)r\beta = K\theta r\beta$$

The virtues of this equation of state are as follows: only five parameters, T_c, β, δ, a, and k, are required to specify the entire equation of state. The value of $b^2 = (\delta - 3)/[(\delta - 1)(1 - 2\beta)]$ is chosen because the parameter r which measures the distance from the critical point then does so in a physically significant sense. This because the specific heat C_M is proportional to $r^{-\alpha}$ independent of the parameter θ. Slides of the data of Roach and Douglas for He^4, Wallace and Meyer for He^3, and Ho and Litster for $CrBr_3$, showed that the linear model is good to about 1 percent for these materials. One therefore has a simple equation of state that provides an accurate representation of thermodynamic properties in the critical region. This is well suited for engineering calculations of the properties of critical fluids—an example was given by Dr. Barmatz in the morning session.

In another comment Prof. Litster pointed out that it is possible to choose a slightly different form for $m(\theta)$ and to obtain an equation in which lines of constant r are paths of constant susceptibility. As the slide showed, the fit for $CrBr_3$ to this equation of state was not so good and he had no explanation for this. It would be interesting to know why choosing r to measure the magnetization correlation function instead of the energy-density correlation function makes this difference.

Prof. C. Domb agreed with Prof. Litster that the linear parametric formula is useful as long as one considers it as an engineering type approximation. But he would not attribute theoretical significance to the formula or look for deep-seated explanations. He tried to fit the Ising calculations to the formula, and since, as he stated, the true Ising solution is infinitely complex there was no possibility of a genuine fit. However, in three dimensions he was able to get quite a good fit by suitable choice of parameters but in two dimensions this was not possible. In a reply Prof. Litster said that Dr. Schofield certainly would agree with the statement, that there is no deeper philosophical significance in the particular fit obtained.

Prof. M. Vincentini-Missoni discussed a way of checking the validity of the "linear model." According to the model, $r \propto C_M^{1/\alpha}$. Therefore, measurements of the specific heat as a function of field would allow to check the constancy of C_M along curves at constant r. As for the use of the "linear model" equation of state for engineering purposes she pointed out that the $h(x)$ function derived from it gives a worse agreement with experimental data than the $h_{MSQ}(x)$ function. She agreed on the other hand that the linear model provides an easier way to evaluate the free energy, and therefore the specific heat (one doesn't need numerical integration as in the $X, h(x)$ language).

Prof. L. P. Kadanoff argued in favor of the linear model, because the concept of smoothness would imply the universality of the equation of state for fluids: That is to say that different gases should be fit by the very same equation of state, with only some multiplication constant in front of $(\mu - \mu_c)$ $(T - T_c)$, and $(P - P_c)$ varied. The analysis of Litster agreed very well in his words with the idea since the equations of state for He^3 and He^4 are precisely the same for these multiplicative constants. Since He^3 and He^4 have very different Λ^*, this is an impressive confirmation of universality.

At this point another topic from the morning session came into the discussion and Prof. Domb made a comment on a possible association between Kouvel's data and the model discussed by Rushbrooke. He argued that there is in fact little connection; the concentrations used by Kouvel are well below the percolation limit in Rushbrooke's system, and the only possibility of accounting for Kouvel's data is by means of an itinerant electron model (this is equivalent to a very long-range force). His point of view was that a change in concentration changes the effective interaction between the itinerant electrons and hence changes the exponents. He also said that the apparent increase of γ in Prof. Rushbrooke's calculations as the concentration decreases may be spurious and due to inadequate convergence.

Prof. G. S. Rushbrooke didn't agree with the last statement but supported the first point, concerning the connection with Kouvel's data.

In a short remark Prof. R. B. Griffiths mentioned that we have for years been applying the results of Heisenberg model calculations to understand the critical phenomena in a metal like nickel where the magnetism is due to itinerant electrons. He then expressed his hope that Rushbrooke's calculations for a dilute system would give us some hint as to what is happening in the case of alloys in which magnetic atoms are randomly distributed even though the physical effects giving rise to ferromagnetism and critical indices are quite complicated.

The effect of a possible change of the critical indices due to impurities was also discussed by Prof. L. P. Kadanoff. He pointed out that Griffiths in his paper has described how critical indices do not change as parameters change unless there is some kind of disaster like a change of symmetry. This idea of smoothness or "universality" provides some explanation of why an itinerant ferromagnet like Ni might have the same critical indices as the Heisenberg ferromagnet. But from model calculations presented at this conference it seems to be quite clear that impurities do produce a "disaster"—they can change critical indices. Hence Prof. Kouvel's result that small amounts of Pd do change the apparent critical indices of Fe seems to be consistent with those models.

On the other hand, Dr. G. A. Baker argued that these effects may not be a complicated phenomena related to the critical aspects themselves, but rather a modification of the apparent exponents due to the changes in the compressibility of the underlying lattice with concentration and temperature. In order to investigate this possibility accurate accompaning measurements of the lattice parameter might be very helpful.

In addition to this, Dr. H. Stenschke pointed out that the interaction between magnetic moments in the ordered state might be effectively larger than in a random arrangement since, in the latter case, all anisotropic parts of the interaction are averaged out. In his opinion the situation seems to be quite similar to the problem of reduction of the line width of NMR spectra when a substance is investigated in its liquid state as compared with the spectra in the solid state. The line width is mainly determined by perturbations of a particular spin due to the dipole-dipole interactions with neighboring spins. As far as ferromagnetism is concerned, a longer range of the interactions being present in an ordered arrangement of magnetic moments would favor the mean-field approximation.

In the final part of the session Prof. R. A. Ferrell discussed some theoretical aspects of dynamical effects in magnetic materials, especially isotopic ferromagnets.

We start with a static property, the wave number (q) dependent susceptibility at a certain temperature. The well known Ornstein-Zernike formula is

$$\chi(q,\kappa) \propto \frac{1}{q^2 + \kappa^2(T)}$$

where $\kappa(T)$ is the temperature-dependent reciprocal correlation length. The static properties show a temperature continuity which means that the temperature variation is mild if κ is comparable to q. We therefore can replace κ by q as we go from large values of κ to smaller ones.

In dynamical scaling theory we study wave number dependent fluctuations and we characterize the spectrum by a width $\Gamma(q,\kappa)$ and again use the rule, that crossing the boundary $q = \kappa$, κ is replaced by q.

Starting in the hydrodynamic region we have

$$\Gamma \propto q^2 D(T)$$

where the temperature dependence has to disappear in the limit of small κ values. But we know that in this region

$$\Gamma \propto q^{5/2}$$

so that we get

$$D(T) \propto \kappa^{1/2} \propto |T - T_c|^{1/3}$$

where the last proportionality comes from static considerations.

More recent advances have to do with getting away from the proportionality signs by introducing numerical coefficients and there seems to be good agreement with experiment.

As an interpolating formula for two regions in the q, κ plane we write

$$\Gamma = Cq^2(q^2 + \kappa^2)^{1/4} \sigma\left(\frac{q}{\kappa}\right)$$

where the constant C is introduced in order to normalize the function σ to $\sigma(0) = 1$.

In the limits discussed above we have

$$\Gamma = C \cdot \begin{cases} q^{5/2} \\ q^2 \kappa^{1/2} \sigma(\infty) \end{cases}$$

The function $\sigma(x)$ was calculated by Prof. Kawasaki for a classical gas, while Résibois has determined it for spin systems. Instead of discussing these results in greater detail, I would like to point out another method for calculating the diffusion constant. This method uses the Kubo formula. One can relate the diffusion problem to a conductivity problem by writing

$$D = \frac{\lambda}{\chi}$$

where the Onsager coefficient λ describes the flow of spins and is given by

$$\lambda = \frac{1}{6T} \int d^4 12 \langle \mathbf{J}_x(2) \cdot \mathbf{J}_x(1) \rangle$$

with the magnetization current component $\mathbf{J}_x$.

The main assumptions in calculating λ are:

1. The factorization of the four-spin correlation into two-spin correlation functions. This is an approximation which seems to work quite well.

2. The use of the Ornstein-Zernike expression for the equal-time correlation.

In summary, the main theoretical task remaining seems to be to understand why the first assumption above works so well.

REFERENCES

1. Langer, J. S. and V. Ambegaokar: *Phys. Rev.*, **164**: 498 (1967).
2. McCumber, D. E., and B. I. Halperin: *Ibid.*, **B1**: 1054 (1970).
3. Eilenberger, G.: *Z. Physik*, **236**: 1 (1970).

———————————————— *Part Five*

TRANSPORT PROPERTIES

NMR AND NEUTRON SCATTERING STUDIES OF CRITICAL FLUCTUATIONS IN UNIAXIAL ANTIFERROMAGNETS

Peter Heller

Physics Department
Brandeis University
Waltham, Massachusetts

Detailed measurements of the F^{19} nuclear magnetic resonance (NMR) linewidth in the paramagnetic state of the anisotropic antiferromagnet FeF_2 have been made. The results exhibit the dependence on the direction of a weak externally applied field to be expected on the basis of the Kubo-Tomita[1] expression for the linewidth. According to this expression, the linewidth is due to the fluctuation in the local hyperfine field due to electron-spin motion. By examining the data for two different orientations of the applied field, the contribution $\delta\nu_\parallel$ to the linewidth from the longitudinal (c axis) spin fluctuations may be deduced. In the critical region above

T_N, the results may be written

$$\delta\nu_{\|} = A\left[\frac{T - T_N}{T_N}\right]^{-n} \qquad (1)$$

with

$$A = 0.0361 \pm 0.0015 \; MHz$$

$$n = 0.675 \pm 0.01$$

This power law extends over almost three decades in $T - T_N$. The exponent n may be compared with the theoretical prediction $n = 2/3$ deduced from the application of dynamical scaling to the anisotropic antiferromagnet by Riedel and Wegner.[2]

It is possible to relate the NMR linewidth to the inelastic neutron scattering[3] by writing the Kubo-Tomita expression in terms of the Van Hove scattering function $S(\mathbf{K},\omega)$. For the F^{19} resonance in MnF_2 or FeF_2 we have

$$\delta\nu_{\|} = \frac{1}{\sqrt{3}\hbar^2}\frac{1}{V_K}\int_{V_K} d^3K |A_{cc}(K)|^2 S_{\|}(\mathbf{K},0) \qquad (2)$$

Here $\delta\nu_{\|}$ is the longitudinal fluctuation contribution to the NMR linewidth (between derivative extrema). The integral is carried out over the first Brillouin zone of the Mn^{2+} ion reciprocal lattice of volume

$$V_K = \frac{2(2\pi)^3}{a^2 c}$$

where a and c are the dimensions of the tetragonal unit cell containing two magnetic ions. The function $A(K)$ can be computed from the known[4] hyperfine couplings between the F^{19} nucleus and the neighboring magnetic spins.

Assuming $S_{\|}(\mathbf{K},\omega)$ to be Lorentzian in ω of half width at half maximum $\Gamma_{\|}(\mathbf{K},T)$, it follows that Eq. (2) may be written

$$\delta\nu_{\|} = \frac{S(S+1)}{3\pi\sqrt{3}\hbar^2}\frac{1}{V_K}\int d^3K |A_{cc}(K)|^2 \frac{\hat{\chi}_{\|}(K)}{\Gamma_{\|}(K)} \qquad (3)$$

Here

$$\hat{\chi}_\parallel(K, T) = \frac{\chi_\parallel(K, T)}{\chi_{\text{Curie}}(T)}$$

where $\chi_\parallel(K, T)$ is the wavelength dependent static susceptibility.

Equation (3) invites a direct comparison of NMR and neutron scattering data. This has been accomplished for the case of MnF_2 using the NMR data of Heller and Benedek[5] and the neutron scattering data of Schulhof, Heller, Nathans, and Linz.[6] In order to apply Eq. (3), the following requirements had to be met.

1. The susceptibilities $\chi_\parallel(K)$, which are provided on a *relative* basis by neutron scattering, have to be given absolutely.

2. The neutron scattering data are available only near an antiferromagnetic point in the zone. A general expression for the susceptibilities $\chi_\parallel(K)$ compatible with this data, and also with the experimental value[7] of the ordinary uniform susceptibility at $K = 0$ must be given.

3. An expression must be given for the decay rates $\Gamma_\parallel(K)$ over the entire zone. This expression must be compatible with the neutron scattering data near the antiferromagnetic wave vector. It must also be compatible with the theoretical requirement that $\Gamma_\parallel$ be small near $K = 0$, i.e., with "kinematical slowing down."

Requirement 1 was first met by comparing the neutron scattering data[6] for the transverse staggered susceptibility with Moriya's prediction[8]

$$\chi_\perp(0°K) = \frac{M_0(0)}{H_A(0)} = \frac{C}{1.36°K}$$

where C is the Curie constant. This then gave $\hat{\chi}_\parallel(T_N + 1°K) \simeq 225$.

For requirement 2, we may expect that since corner to body-centered couplings are dominant, we have for $T > T_N$

$$\frac{1}{\hat{\chi}_\parallel(K)} = A(T) + B(T) \cos\frac{K_x a}{2} \cos\frac{K_y a}{2} \cos\frac{K_z c}{2} \qquad (4)$$

A somewhat more elaborate expression, to be used near the critical point, has been given by Fisher and Buford;[9] it is not appreciably different from Eq. (4). The parameters $A(T)$ and $B(T)$ can be determined by requiring Eq. (4) to be compatible with the normalization found in requirement 1 and with the observed value of the longitudinal inverse correlation length $\kappa_\parallel$ found in the neutron scattering experiments[6] above T_N. The resulting calculated value of the uniform susceptibility differs from the experimental value by about 10 percent. Alternatively, the parameters A and B can be chosen for compatibility with $\kappa_\parallel$ and the experimental uniform susceptibility. This gives a normalization differing by 10 percent from that found in requirement 1.

Requirement 3 was met by using an expression [10] given by Mori and Kawasaki

$$\Gamma_\parallel(\mathbf{K}, T) = [F(T) + G(T)\hat{J}(\mathbf{K})][1 - \hat{J}(\mathbf{K})] \tag{5}$$

where

$$\hat{J}(\mathbf{K}) = \cos\frac{K_x a}{2} \cos\frac{K_y a}{2} \cos\frac{K_z c}{2}$$

with F and G adjusted to fit the neutron scattering data[6] near the antiferromagnetic wave vector at each temperature $T > T_N$. The second factor on the right in Eq. (5) satisfies the requirement of kinematical slowing down.

Using Eqs. (4) and (5), the NMR linewidth integral (Eq. (3)) was calculated numerically on a large digital computer at a number of temperatures. At $T = T_N + 7.5°K$, where about 20 percent of the integral comes from the region of $\mathbf{K}$ space covered by the neutron data, the computed and measured linewidths are 60kHz and 58kHz, respectively. This agreement may be primarily regarded as showing that the extrapolated curves of $\Gamma_\parallel$ vs. $\mathbf{K}$ provided by Eq. (5) are satisfactory for computing the linewidth integral. Closer to T_N, where a much larger proportion of the integral comes from the region measured by neutrons, the agreement was also quite satisfactory.

The absolute calibration of the neutron scattering data was also of interest for another reason. A physical argument [11] shows that for temperatures below and not too close to T_N, we may expect that the longitudinal staggered susceptibility (per spin) observed in the neutron scattering experiments should equal the quantity

$$\chi_{th} = \frac{T}{C_H^*}\left(\frac{dM_0}{dT}\right) \tag{6}$$

where C_H^* is the measured [12] magnetic contribution to the specific heat per spin, and $M_0(T)$ is the measured[4,5] thermal average magnetic moment per spin. The susceptibility χ_{th} corresponds to the fluctuation in the local energy density. Using the calibration found in requirements 1 and 2 above, the data[6] for the staggered susceptibility for $T < 0.95 T_N$ are shown to be consistent with this expectation. Closer to T_N, there is an indication that the observed staggered susceptibility is larger than χ_{th} by a numerical factor between 1 and 2.5. This result is consistent with available theories [13] of the ratio of the specific heat at constant staggered magnetization to that at constant staggered field. The argument leading to Eq. (6) also shows that the spectral diffusivity of the scattering observed by Schulhof et al. below T_N, should be interpreted as a manifestation of thermal diffusion taking place within the spin system. A closely related theory has also been advanced by Halperin and Hohenberg.[14]

The NMR linewidth measurements reported here, together with the comparison of the NMR and neutron scattering data in MnF_2 has been carried out in collaboration with Dr. A. M. Gottlieb of Brandeis University. We are grateful to Dr. Gottlieb for permission to reproduce this data prior to publication. This work[15,16] and also the details[11,17] of the absolute calibration and interpretation of the data of Schulhof et al.[6] on antiferromagnetic MnF_2 will be published elsewhere.

This research was supported by the U.S. Air Force Office of Scientific Research, Grant No. AF68-1480.

REFERENCES

1. Kubo, R., and K. Tomita: *J. Phys. Soc. Japan*, 9:888 (1954).
2. Riedel, E., and F. Wegner: *Phys. Rev. Lett.*, 24:730 (1970).
3. Heller, P.: *Rep. Prog. Phys.*, 30:731 (1967).
4. Clogston, A. M., et al.: *Phys. Rev.*, 117:1222 (1960); Stout, L. W., and R. G. Shulman: *Ibid.*, 118:1136 (1960).
5. Heller, P., and G. B. Benedek: *Phys. Rev. Lett.*, 8:428 (1962).
6. Schulhof, M. P., P. Heller, R. Nathans, and A. Linz: *Ibid.*, 24:1184 (1970); *Phys. Rev.*, B1:2304 (1970).
7. Trapp, C.: Doctoral dissertation, Univ. of Chicago, 1963 (unpublished).
8. Moriya, T.: *Prog. Theor. Phys.*, 28:371 (1962).
9. Fisher, M. E., and R. J. Burford: *Phys. Rev.*, 156:583 (1967).
10. Mori, H., and K. Kawasaki: *Prog. Theor. Phys.*, 27:529 (1962).
11. Heller, P.: *Proc. Conf. Critical Phenomena*, Fordham University, New York, June 1970. *Int. J. Magnetism*, 1:53 (1970).
12. Teaney, D. T.: *Phys. Rev. Lett.*, 14:898 (1965).
13. Baker, G. A., and D. S. Gaunt: *Phys. Rev.*, 155:545 (1967); Schofield, P., J. D. Litster, and J. T. Ho: *Phys. Rev. Lett.*, 23:1098 (1969).
14. Halperin, B. I., and P. C. Hohenberg: *Phys. Rev.*, 188:898 (1969).
15. Gottlieb, A. M.: Thesis, Brandeis University, 1970 (unpublished).

16. —— and P. Heller: *Phys. Rev.*, **B3**:3615 (1971).
17. Heller, P., M. P. Schulhof, R. Nathans, and A. Linz: *J. Appl. Phys.*, **42**:1258 (1970); Schulhof, M. P., R. Nathans, P. Heller, and A. Linz: *Phys. Rev.*, **B**:Sept. 1971 (To be published).

DISCUSSION on *Paper* by *P. Heller*

L. KADANOFF: You assume in your analysis that the coupling between the lattice and the spin system is so weak that energy is not lost from the spin system within the observed relaxation time for heat flow within the spin system. Is the coupling really that weak?

P. HELLER: I do not know. We need an estimate of a spin lattice time. My analysis requires q to be large enough so that the flow of heat within the spin system is much faster than the flow of heat to the lattice. At small q this condition will fail.

L. KADANOFF: Can we learn about the spin lattice coupling from ultrasonic attenuation in MnF_2?

K. KAWASAKI: If ultrasonic attenuation in MnF_2 is dominated by a process in which sound wave couples to the energy density of spin system and if the time variation of energy density proceeds predominantly through the energy transfer between the spin system and the lattice, the sound attenuation becomes proportional to the spin lattice relaxation time.

DYNAMICAL THEORY OF CRITICAL FLUCTUATIONS WITH APPLICATIONS TO TRANSPORT PROPERTIES IN THE CRITICAL REGION

Kyozi Kawasaki

Department of Physics
Temple University
Philadelphia, Pennsylvania

ABSTRACT

A general kinetic equation is derived for the probability distribution function of critical fluctuations which allows a general discussion of dynamical scaling. A method is developed to obtain time correlation functions of critical fluctuations which enables us to study transport properties in hydrodynamical as well as in nonhydrodynamical regimes. We illustrate the theory for binary critical solutions where we discuss the decay rate of concentration fluctuations and a possible anomaly in the shear viscosity.

1. INTRODUCTION

A distinctive feature common to all systems in the neighborhood of their critical points is the existence of long-lived fluctuations of the order parameters that extend over large semimacroscopic spatial regions.[1] Thus the system possesses a kind of turbidity which manifests itself most spectacularly in the phenomena called the critical opalescence. This is quite analogous to a turbulent fluid flow where one observes large-scale velocity fluctuations. Full microscopic treatment of dynamics of such critical fluctuations may still be premature in view of lack of a satisfactory microscopic equilibrium theory of critical phenomena. Thus here we aim at developing a quasimicroscopic kinetic theory of critical fluctuations where all the equilibrium properties known so far are taken as input to the theory. Specifically we consider in Part 2 a probability distribution function of critical fluctuations and construct a kinetic equation for the distribution function. The equation is very useful for discussing the dynamical scaling (Part 3). Then in Part 4 we develop a method to calculate time correlation functions of fluctuations for a simplified version of the general kinetic equation. In Part 5 we illustrate the method for binary critical solutions.

2. KINETIC EQUATION

The first task in constructing a kinetic equation is to choose a proper set of gross variables that enter the kinetic equation. Here an important criterion is the slow time variation. The order parameter meets this criterion because of the critical slowing down.[2] Then the only conceivable remaining gross variables are the ones that are connected to the conservation laws, i.e., the hydrodynamic variables. Thus, in the most complete version of the kinetic theory the gross variables consist of the local hydrodynamic variables plus the local order parameter when the latter is not included in the former. We denote these variables by a_j, and a denotes the whole set $\{a_j\}$. We restrict ourselves to the cases where classical mechanics applies, and the phase functions corresponding to a_j and a are denoted by $A_j(x)$ and $A(x)$, respectively. An example of such a set for one component fluids is the Fourier components of local mass density, local velocity, and local entropy density where wave numbers are restricted to be much smaller than the inverse of an average intermolecular distance.

Having chosen the gross variables, the gross state of the system is given by specifying the values of these gross variables. More generally, we may consider a probability distribution function for the entire set of the gross variables $g(a, t)$. Such distribution function was introduced in 1952 by M. S. Green[3] and later by R. W. Zwanzig,[4] who have demonstrated that $g(a, t)$ obeys the following generalized Fokker-Planck equation

$$\left[\frac{\partial}{\partial t} + \mathcal{H}(a)\right] g(a, t) = 0 \tag{1}$$

$$\mathcal{H}(a) \equiv -\sum_j L_j \frac{\partial^2}{\partial a_j \partial a_j^*} + \sum_j \frac{\partial}{\partial a_j} [v_j(a) + L_j F_j^*(a)] \tag{2}$$

where L_j is the Onsager kinetic coefficient for the irreversible processes involving a_j in which we assume the absence of interference between irreversible processes involving different a's. $F_j(a)$ is the thermodynamic driving force which is obtained from the thermodynamic potential $\Phi(a)$ by

$$F_j(a) = -\frac{\partial \Phi(a)}{\partial a_j} \tag{3}$$

and $a_{j*} = a_j^*$ defines j^*. $v_j(a)$ is the average of $\dot{A}_j$ in the constrained equilibrium state where $A(x) = a$, which is denoted as

$$v_j(a) = \langle \dot{A}_j(x) ; a \rangle \tag{4}$$

We choose a's so that the equilibrium average $\langle a \rangle$ vanishes.

The Equation (1) describes a time evolution of $g(a, t)$ over the time scales much greater than the times that characterize rapid molecular random processes, and thus it constitutes the basic kinetic equation in this theory.

In the absence of $v_j(a)$, we recover the following macroscopic law of irreversible thermodynamics for the averages $\overline{a_j(t)} = \int a_j g(a, t) da$

$$\frac{d}{dt} \overline{a_j(t)} = L_j F_{j*}[\bar{a}(t)] \tag{5}$$

where we have ignored fluctuations a's in F_{j*}. Since L's are expressed in terms of time-correlation functions of "random forces" which are statistically independent of a's, it is natural to assume that L's remain finite at the critical point. Then all the critical anomaly in Eq.

(5) is contained in the thermodynamic driving force $F_{j^*}(\bar{a})$. This is nothing but the thermodynamic theory of critical slowing down (or, also referred to as the conventional theory).[2] Since the thermodynamic theory is known to fail in many instances,[2] we must consider the deviations from the thermodynamic theory which are contained in $v_j(a)$.

Let us now quote without proof the two expressions for $v_j(a)$ which are more convenient to handle than Eq. (4).[5] The first expression is

$$v_j(a) = -k_B T \sum_l \left\{ \frac{\partial}{\partial a_l} \langle\langle A_j, A_l \rangle; a\rangle + \langle\langle A_j, A_l \rangle; a\rangle F_l(a) \right\} \quad (6)$$

where (X, Y) denotes the Poisson bracket of two phase functions X and Y. The second expression is expressed in terms of an infinite set of variables composed of various products of a's, which is arranged in a column matrix $\mathcal{C}$ defined by

$$\mathcal{C} = \begin{pmatrix} \{a_j\} \\ \{a_j a_l\} - \langle A_j A_l \rangle \\ \vdots \end{pmatrix} \quad (7)$$

where $\{a_j\}$ denotes the set $a_1, a_2, \cdots$ and $\{A_j A_l - \langle A_j A_l \rangle\}$ the set $a_1 a_2 - \langle A_1 A_2 \rangle, a_1 a_3 - \langle A_1 A_3 \rangle$, and so on. Then we have in matrix notation

$$v_j(a) = k_B T \langle (a_j, \mathcal{C}^\dagger) \rangle \cdot \langle \mathcal{C} \mathcal{C}^\dagger \rangle^{-1} \cdot \mathcal{C} \quad (8)$$

where $\mathcal{C}$ is obtained from $\mathcal{C}$ by replacing all the a's by A's, and $\mathcal{C}^\dagger$ is the Hermitian adjoint row matrix of $\mathcal{C}$.

3. DYNAMICAL SCALING

In order to illustrate the usefulness of the kinetic equation described in the preceding section, we examine the dynamical scaling[6] from this point of view. As a first example we consider a classical isotropic Heisenberg ferromagnet in the disordered phase. Here we choose A_j to be $\hat{S}_k^\alpha$, $(\alpha = x, y, z)$, defined by

$$\hat{S}_k^\alpha = V^{-1/2} \sum_r \hat{S}_r^\alpha e^{-i\mathbf{k}\cdot\mathbf{r}} \quad (V \text{ is the volume}) \tag{9}$$

where $\hat{S}_r^\alpha$ is the spin variable on the rth lattice site, and $|\mathbf{k}|$ is restricted to be much smaller compared to the inverse lattice spacing. Then $(\hat{S}_k^\alpha, \hat{S}_{k'}^\beta) = V^{-1/2} \Sigma_\gamma \hat{\epsilon}_{\alpha\beta\gamma} \hat{S}_{k+k'}^\gamma$, where $\hat{\epsilon}_{\alpha\beta\gamma}$ is the antisymmetric tensor with $\hat{\epsilon}_{xyz} = 1$, and by (6) we have

$$v_k^\alpha(S) = -\frac{k_B T}{V^{1/2}} \sum_{k'} \sum_{\beta\gamma} \hat{\epsilon}_{\alpha\beta\gamma} S_{k+k'}^\gamma F_{\beta k'}(S) \tag{10}$$

where S_k^α is the variable a corresponding to $\hat{S}_k^\alpha$ and $\{S\}$ is the whole set of S's.

We now write the kinetic equation as

$$\left(\frac{\partial}{\partial t} + \mathcal{H}_d + \mathcal{H}_s\right) g(a, t) = 0 \tag{11}$$

where

$$\mathcal{H}_d \equiv \sum_k \sum_\alpha \frac{\partial}{\partial S_k^\alpha} v_{k\alpha}(\{S\}) \tag{12}$$

$$\mathcal{H}_s \equiv -\sum_k \sum_\alpha L_{k\alpha} \left[\frac{\partial^2}{\partial S_k^\alpha \partial S_k^{\alpha\,*}} - \frac{\partial}{\partial S_k^\alpha} F_{-k\alpha}(\{S\})\right] \tag{13}$$

Note all the critical anomalies are now contained in the thermodynamic driving force $F_{k\alpha}(\{S\})$, which in turn is derived from the thermodynamic potential $\Phi(\{S\})$ by Eq. (3). Thus for studying *dynamic* scaling we use the *static* scaling form for Φ which is, when only a uniform magnetization density S is present,

$$\Phi(\epsilon, S) = V\epsilon^{2-\alpha} K(\{S/\epsilon^\beta\}) \tag{14}$$

where $\epsilon \equiv |T - T_c|/T_c$ is the dimensionless temperature distance from the critical point.[1,7]* In the presence of a nonuniform

---

*Here and after we employ the customary notation for critical exponents.[1]

magnetization $\{S_k^\alpha\}$, Eq. (14) is readily generalized to

$$\Phi(\{S\}) = V\epsilon^{2-\alpha} K(\{S/V^{1/2}\epsilon^\beta\}, \{k/\kappa\})$$

Here we have introduced the inverse correlation range of critical fluctuations $\kappa = \epsilon^\nu$ expressed in dimensionless units. This can also be written as

$$\Phi(\{S\}) = V\epsilon^{2-\alpha} K(\{(V\kappa^d)^{-1/2}\epsilon^{\gamma/2} S\}, \{k/\kappa\}) \qquad (15)$$

where we have used $\beta = (d\nu - \gamma)/2$ (d is the dimensionality of space) which follows from the scaling law.[1,7]

Then, $F_{\mathbf{q}\alpha}(\{S\})$ has the following scaling law form

$$F_{\mathbf{q}\alpha}(\{S\}) = \epsilon^{\gamma/2}(V\kappa^d)^{1/2} \hat{F}_\alpha(\{V\kappa^d)^{-1/2}\epsilon^{\gamma/2} S\}, \mathbf{q}/\kappa, \{k/\kappa\}) \qquad (16)$$

where we have used $2 - \alpha = d\nu$ which follows from the scaling law relations.[1,7]

With these preparations, we see that $\mathcal{H}_d$ and $\mathcal{H}_s$ take the following forms

$$\mathcal{H}_d = \epsilon^{x\nu} \hat{\mathcal{H}}_d(\{\epsilon^{\gamma/2} S\}, V\kappa^d, \{k/\kappa\}) \qquad (17)$$

$$\mathcal{H}_s = \epsilon^{2\nu+\gamma} \hat{\mathcal{H}}_s(\{\epsilon^{\gamma/2} S\}, V\kappa^d, \{k/\kappa\}) \qquad (18)$$

where $x \equiv (1/2)d + (1/2)\gamma/\nu = (d + 2 - \eta)/2$ using $\gamma = (2 - \eta)\nu$. Since $x\nu < 2\nu + \gamma$ for $2 \leq d \leq 5$, here $\mathcal{H}_d$ dominates the critical dynamics. Then, the scaling transformation $S = \epsilon^{-\gamma/2}\tilde{S}, \mathbf{k} = \kappa \mathbf{y}$, and $t = \epsilon^{-x\nu}\tau$ eliminates the critical anomaly except in the combination $V\kappa^d$. Now, let us choose V so that $V\kappa^d$ remains a large fixed number (which eventually goes to infinity in the thermodynamic limit). Then the dependence upon $V\kappa^d$ does not affect the critical anomaly. Hence we can now conclude that the characteristic frequency spectrum $\Omega_\mathbf{q}$ of a critical fluctuation with wave number q behaves as

$$\Omega_\mathbf{q} = \kappa^x f(q/\kappa) \qquad (19)$$

which coincides with the prediction of the dynamical scaling.[5]

Similar analyses have been made for isotropic antiferromagnets,[5] planar ferromagnets, and the λ transition in liquid helium* where $\mathcal{H}_d$

---

*Although this is purely quantum mechanical phenomenon, we make use of the fact that the system may be described by classical field variables. See, for instance, L. P. Pitaevskii, *Soviet Phys. JETP*, 8:282 (1959).

scales as $\epsilon^{d\nu/2}$ and $\mathcal{H}_s$ as ϵ^γ. Thus for $d \leq 3$, $d\nu/2 < \gamma$ and $\mathcal{H}_d$ dominates, and the characteristic frequency spectrum of critical fluctuations has the form $\kappa^{d/2} f(q/\kappa)$ in agreement with the predictions of the dynamical scaling.[6]

Another kind of example is typified by critical binary solutions. Here analysis is somewhat more involved, and we only give a very brief account.[5] Here the variables a_j are the local concentration c_k and transverse components of the local velocity u_k^α ($\alpha = x, y, z$ and $\sum_\alpha k^\alpha u_k^\alpha = 0$). First we divide $\mathcal{H}$ into $\mathcal{H}_d$ and $\mathcal{H}_s$ in the manner analogous to Eqs. (11) to (13). Then it is found that dominant parts of $\mathcal{H}_d$ and $\mathcal{H}_s$ scale as $\epsilon^{(1+d/2)\nu}$ and $\epsilon^{2\nu}$, respectively. Since $1 + 1/2d \geq 2$ for $d \geq 2$ we are not allowed to ignore $\mathcal{H}_s$ in contrast to earlier examples, and the dynamical scaling is here far less founded. Later we shall see that the dynamical scaling is recovered here also by calculating the decay rate of concentration fluctuation arising from a specific type of processes.

4. TIME CORRELATION FUNCTIONS

Although the kinetic equation (Eq. (1)) introduced in Part 2 is rather useful in studying some general aspects of critical dynamics such as the dynamical scaling, it is still too complicated for detailed study of some specific aspects of the problem like time correlation functions of A's or transport coefficients, since $v_j(a)$ involves infinite power series in a's. Thus a useful nontrivial simplification of the kinetic equation is obtained (1) by retaining in the expansion (Eq. (8)) only the linear and quadratic terms in a's, and (2) by retaining only the linear terms in $F_j^*(a)$ which appears in Eq. (2).* If we choose a's such that $\langle A_j A_l^* \rangle = \langle (A_j, A_l^*) \rangle = 0$ when $j \neq l$, the kinetic equation (Eq. (1)) reduces to

$$\left\{ \frac{\partial}{\partial t} + \mathcal{H}_0(a) + \mathcal{H}_1(a) \right\} g(a, t) = 0 \qquad (20)$$

$$\mathcal{H}_0(a) \equiv \sum_j \left\{ -L_j \frac{\partial^2}{\partial a_j \partial a_j^*} + \frac{\partial}{\partial a_j} \left(i\omega_j - \frac{L_j}{\chi_j} \right) a_j \right\} \qquad (21)$$

*This also implies that $\Phi(a)$ is bilinear in a's and hence the equilibrium distribution function of a, $P(a)$ is Gaussian.

$$\mathcal{H}_1(a) = -i \sum_{jlm} \mathcal{O}_{jlm} \frac{\partial}{\partial a_j} (a_l a_m - \langle a_l a_m \rangle) \qquad (22)$$

where

$$\chi_j \equiv \langle |a_j|^2 \rangle = \langle |A_j|^2 \rangle \qquad (23)$$

$$\omega_j \equiv \frac{i k_B T \langle (A_j, A_j^*) \rangle}{\chi_j} \qquad (24)$$

$$\mathcal{O}_{jlm} \equiv i \frac{k_B T}{2} \frac{\langle (A_j, A_l^* A_m^*) \rangle}{(\chi_l \chi_m)} \qquad (25)$$

ω_j gives the frequency of possible oscillations in hydrodynamic motions and $\mathcal{O}_{jlm}$ gives the coupling of modes described by the variables a_j, a_l, and a_m.

In the absence of $\mathcal{H}_1$, Eq. (20) simply represents a Gaussian random process familiar in the Brownian motion theory and the equation can be solved easily in the form of a product of $g_j^0(a_j, a_j^*; t)$, the explicit form of which is not needed in the following. Taking this to be an unperturbed solution, we can develop a perturbation theory with respect to $\mathcal{H}_1$. A formal and compact way of carrying this out is first to put the solution of Eq. (20) with the initial condition that $a = a'$ at $t = 0$ in the form

$$g(aa', t) = e^{-t\mathcal{H}(a)} \delta(a - a') = e^{-t\widetilde{\mathcal{H}}(a')} \delta(a - a') \qquad (26)$$

where $\widetilde{\mathcal{H}}(a)$ is the adjoint operator of $\mathcal{H}(a)$.* Then use the ordered exponential form for the perturbation series of the operator $e^{-t\widetilde{\mathcal{H}}(a')}$.

Each term in the perturbation expansion allows a graphical representation and a use of renormalization technique familiar in quantum field theory allows us to resum a suitable subset of infinite number of graphs. We do not give here details of this technique since the principle is the same and fine details are of not much interest here.[5]

Using this method we have studied the propagator of gross variables defined by

$$G_j(t) = \frac{\langle a_j(t) a_j^*(0) \rangle}{\chi_j} \equiv \frac{1}{\chi_j} \iint da\, da'\, g(aa'; t) a_j a_j'^* P(a'), \quad (t > 0) \qquad (27)$$

*Here and after a stands for the set $\{a_j\}$ which includes its complex conjugates.

Then we obtain a self-consistent set of equations for determining $G_j(t)$ which still contains an infinite number of graphs. The simplest nontrivial approximation to the above equation is obtained by ignoring all the vertex corrections and we find

$$\left(\frac{\partial}{\partial t} - i\omega_j + \gamma_j\right) G_j(t)$$

$$= -2 \sum_{lm} \frac{\chi_l \chi_m}{\chi_j} |\mathcal{O}_{jlm}|^2 \int_0^t ds\, G_l(t-s) G_m(t-s) G_j(s) \quad (28)$$

with $G_j(0) = 1$. This is also expressed graphically as

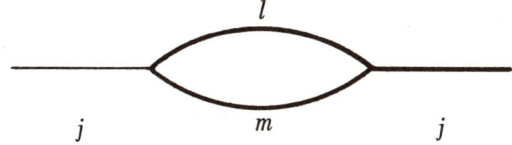

where thin and thick straight lines represent the unrenormalized (i.e., without $\mathcal{H}_1$) and the renormalized propagators, respectively. It should be noted that in deriving Eq. (28) no restriction was imposed which would limit the applicability of the equation to certain regions near the critical point except that the equation deal with phenomena occurring over space-time regions much greater than microscopic sizes. Thus the theory provides a sound basis for studying nonhydrodynamical transport phenomena near the critical points as well as ordinary transport coefficients.

5. APPLICATIONS

The simplest application of our general theory is to the isotropic ferromagnet discussed in Part 3. The resulting self-consistent equation for the time-displaced spin correlation function of the Eq. (28) form agrees with that which was found recently by many authors and was discussed extensively,[1,8] and we do not discuss it here again. The same is more or less true for other magnetic models.[1,8,9]

Thus here we wish to discuss binary critical solutions. We have two kinds of propagators, one for the concentration fluctuation denoted by $G_{k_c}(t)$ and another for the transverse component of local velocity $G_{k_u}(t)$ defined by

$$\langle u_k^\alpha(t) u_k^{\beta *}(0) \rangle = G_{ku}(t) \langle u_k^\alpha u_k^{\beta *} \rangle \quad (29)$$

The self-consistent set of equations for the propagators then reads

$$\left(\frac{\partial}{\partial t} + q^2 \frac{\eta^0}{\rho}\right) G_{qu}(t) = -\frac{k_B T}{2\rho V} \sum_{k}' [k^2 - (\mathbf{q}\cdot\mathbf{k}/q)^2] \chi_k \chi_{\mathbf{q}-\mathbf{k}}$$

$$\times \left(\frac{1}{\chi_k} - \frac{1}{\chi_{\mathbf{q}-\mathbf{k}}}\right)^2 \int G_{kc}(t-s) G_{|\mathbf{q}-\mathbf{k}|c}(t-s) G_{qu}(s)\, ds \quad (30)$$

$$\left(\frac{\partial}{\partial t} + q^2 D_q^0\right) G_{qc}(t) = \frac{k_B T}{2\rho V} \sum_{k}' (q^2 - (\mathbf{q}\cdot\mathbf{k}/k)^2) \frac{\chi_{\mathbf{q}-\mathbf{k}}}{\chi_q}$$

$$\times \int_0^t G_{ku}(t-s) G_{|\mathbf{q}-\mathbf{k}|c}(t-s) G_{qc}(s)\, ds \quad (31)$$

with the initial conditions $G_{qc}(0) = G_{qu}(0) = 1$. Here ρ is the mass density, $\chi_k = \langle |c_k - \langle c_k \rangle|^2 \rangle$, and η^0 and D_q^0 are the shear viscosity and the diffusion constant of the thermodynamic theory, respectively, and $\sum_k'$ limits k to be much smaller than inverse microscopic distance. Thus η^0 is finite and D_q^0 is proportional to $1/\chi_q$. In Eq. (30) we have also ignored the processes in which a transverse velocity breaks up into two other transverse velocities since contributions from such processes turn out to be negligible.

Now, in Part 3 we have seen that η^0 and D_q^0 cannot be ignored completely. This conclusion also follows from Eqs. (30) and (31) by assuming that η^0 and D_q^0 can be ignored and by making the scaling transformation $q = \kappa x$, $k = \kappa y$, $t = \kappa^{-(1+d/2)}\tau$, $s = \kappa^{-1+1/2d}\sigma$, $G_{qu}(t) = G_u(x,\tau)$, and $G_{qc}(t) = G_c(\kappa,\tau)$. Then all the critical anomaly disappears from Eqs. (30) and (31) and the characteristic frequency scales as $\epsilon^{(1+d/2)\nu}$, which leads to a contradiction. This means that within the framework of the present theory the shear viscosity cannot exhibit a true divergence as long as η^0 stays finite.[2] * This then allows us to set $G_{ku}(t-s) = e^{-(k^2\eta/\rho)(t-s)}$ and $G_{|\mathbf{q}-\mathbf{k}|c}(t-s) \simeq 1$ in Eq. (31) and we find $G_{qc}(t) = e^{-t\Gamma_q}$ with

$$\Gamma_q = \frac{k_B T}{6\pi\eta} \kappa^3 K\left(\frac{q}{\kappa}\right) \quad (32)$$

where the function $K(q/\kappa)$ is expressed in terms of χ_k for which the

---

*If η^0 were to be ignored, the shear viscosity must be dominated by long wave length fluctuations which requires the convergence of the k integral in Eq. (30) at large k, which turns out to be not the case.

strong scaling form[7] is assumed. In particular, if we use the Ornstein-Zernike form for χ_k, $K(x)$ is obtained in the following closed form[2,10]

$$K(x) = (3/4)[1 + x^2 + (x^3 - x^{-1})\tan^{-1} x] \qquad (33)$$

which was tested experimentally by Volochine and coworkers.[11]

Let us now return to the question of a possible anomaly in the shear viscosity. Using the expression for $G_{kc}(t)$ just obtained and taking $q \ll \kappa$ in Eq. (30) we immediately find an anomalous contribution to the shear viscosity $\Delta\eta = \eta - \eta^0$ as follows

$$\Delta\eta = \frac{k_B T}{15(2\pi)^2} \int_0^{k_m} dk\, k^4 \left(\frac{\partial \ln(\chi_k)}{\partial k}\right)^2 \Gamma_k^{-1} \qquad (34)$$

where k_m is some cut-off wave number which is much smaller than the inverse microscopic distance. Assuming again the Ornstein-Zernike form for χ_k, the integrand of Eq. (34) behaves as $(64\eta/k_B T)k^{-1}$ for $k \gg \kappa$ and as $(24\pi\eta/k_B T)k^4 \kappa^{-5}$ for $k \ll \kappa$. Thus we may approximate Eq. (34) by

$$\Delta\eta \simeq \frac{16\eta}{15\pi^2} \int_\kappa^{k_m} \frac{dk}{k} = \frac{16\eta}{15\pi^2} \ln \frac{k_m}{\kappa} \simeq - \frac{16\eta\nu}{15\pi^2} \ln \epsilon \qquad (35)$$

This indicates an apparent logarithmic divergence for $\Delta\eta$. However, this behavior cannot continue indefinitely close to the critical point because Eq. (35) together with $\Delta\eta = \eta - \eta^0$ yields

$$\eta = \left(1 + \frac{16\nu}{15\pi^2} \ln \epsilon\right)^{-1} \eta^0 \qquad (36)$$

which becomes negative for $\epsilon < \epsilon_0$ where $\epsilon_0 = \exp(-15\pi^2/16\nu) \simeq 10^{-6}$ with $\nu = 2/3$. As ϵ approaches ϵ_0 a more detailed study of the self-consistent set of Eqs. (30) and (31) is necessary. Without such a detailed study we can only say that η has to be finite at $\epsilon = 0$ if the argument for finiteness of η given before is indeed correct and the value of ϵ_0 will be further reduced because η in Eq. (35) will be reduced.* The apparent divergence of η for $\epsilon > \epsilon_0$ found here may explain sharp increases found experimentally for the shear viscosities of binary critical solutions.[12]

6. CONCLUDING REMARKS

In the preceding sections we have given a general kinetic equation for gross variables together with applications. This is a Fokker-Planck version of our kinetic equation based on the generalized Langevin equation.[2] There still remains a number of remaining problems,[5] the most important of which is concerned with the choice of gross variables. Although a general prescription is given in Part 2 we have often used a subset of complete gross variables in the actual example discussed, and only qualitative arguments for justifying such choices have been given elsewhere.[2] We hope to improve our approach by clarifying this problem.

REFERENCES

1. Fisher, M. E.: *Reps. Prog. Phys.*, **30**:616 (1967), and the reference cited therein.
2. Kawasaki, K.: *Ann. Phys.*, **61**:1 (1970), and the earlier references cited therein.
3. Green, M. S.: *J. Chem. Phys.*, **20**:1281 (1952); *ibid.*, **22**:398 (1954).
4. Zwanzig, R. W.: *Phys. Rev.*, **124**:983 (1961).
5. Details of these will appear in the *Proc. 1970 Varenna Summer School on Critical Phenomena*, to be published.
6. Ferrell, R. A., et al.: *Ann. Phys.*, **47**:565 (1968); Halperin, B. I., and P. C. Hohenberg: *Phys. Rev.*, **177**:952 (1969).
7. Kadanoff, L. P., et al: *Rev. Mod. Phys.*, **39**:395 (1967).
8. Kawasaki, K.: *J. Appl. Phys.*, **41**:1311 (1970) and the references cited therein; Hubbard, J.: *AERE Rep.* TP 400 (1970).
9. Riedel, E.: Paper presented at Fordham Conference on Dynamical Aspects of Critical Phenomena, June 1970 (to be published).
10. Ferrell, R. A.: *Phys. Rev. Lett.* **24**:1169 (1970).
11. Berge, P., et al: *Ibid.*, **24**:1223 (1970).
12. Alocvito, G., et al.: *Ibid.*, **22**:1040 (1969); Allen, G.: Private communication.

DISCUSSION on *Paper by* K. Kawasaki

H. E. COOK: At high temperatures (small fluctuations) does your equation reduce to a linear mean field equation that includes Brownian motion?

*Note that this η should in fact be η^* of L. P. Kadanoff and Swift (*Phys. Rev.*, **166**:89 (1968)) which is smaller than η.

K. KAWASAKI: In the high temperature limit the mode coupling terms become very small and the equations reduce to the mean field theory.

R. FERRELL: Concerning the term "mode coupling," it should be noted that an equivalent formalism can be built on a "decoupled mode" approximation. This approach is based on applying the Kubo formula to the fluctuations of the relevant current involved. If this current is bilinear in the relevant variables, its autocorrelation function can be evaluated by a factoring, or "decoupling," approximation.

P. MARTIN: If the decoupling works so that one can take the first term in the series expansions, one expects that this rapid convergence must be due to the existence of a small parameter.

These are terribly complicated problems, and many assumptions were made. Why does it work so well? Is there a small parameter and what is it?

L. KADANOFF: It is important to notice that, in most cases, the calculations described here by Kawasaki involve a perturbation expansion which can be described as the coupling among many modes. Since Kawasaki simplifies to include only a few terms in the many-mode expansion, in general we expect that the corrections to the divergent transport coefficients are about as large as the calculated values. However, a special simplification occurs in the particle diffusivity of a critical mixture or equivalently the thermal diffusivity of a liquid-gas phase transition. Because the viscous flow causes no order parameter variations, there is an effective decoupling of viscous diffusion from order parameter diffusion. This decoupling causes the perturbation theory to converge especially rapidly so that the term calculated by Kawasaki might be the only term which need be considered. On the other hand, in other calculations—for example, calculations of the bulk viscosity or of spin-wave damping—higher order terms are important. These terms include three or more modes in intermediate states and can be expected to be of the same order of magnitude as the calculated two-mode terms.

P. MARTIN: Can you say what the small parameter might be?

L. KADANOFF: I cannot say what the small parameter might be. But the essential physical idea which eliminates many terms in the perturbation expansion is that the viscous diffusion produces little or no order parameter fluctuations.

P. MARTIN: Do you then believe that the answer is rigorous rather than an expansion in some parameter?

L. KADANOFF: I wonder if we might eventually push the formalism to see whether the answer is in fact rigorous. But the formalism is not yet good enough. All we can see is that the next order terms vanish for this transport coefficient and not for others.

K. KAWASAKI: If Professor Ferrell's decoupling is right this result *is* rigorous.

Note added in proof:
The question raised in the discussion of this paper, namely, convergence of the perturbation series, has been examined recently by S. M. Lo and the author. We considered the contributions of the simplest vertex corrections to the decay rate of concentration fluctuation Eq. (32). The correction turned out to be 2.4 percent for $q \ll \kappa$, and 0.4 percent for $q \gg \kappa$. On the other hand, for magnets, vertex corrections appear to be much more important (P. Résibois, private communication). These results in fact confirm the point made by Kadanoff here.

ON THE SPECTRUM AND THE INTENSITY OF LIGHT SCATTERED FROM A FLUID NEAR ITS CRITICAL POINT

George B. Benedek
Department of Physics
Massachusetts Institute of Technology
Cambridge, Massachusetts

Joseph B. Lastovka
Bell Telephone Laboratory
Murray Hill, New Jersey

Marzio Giglio
University of Milan
Milan, Italy

David Cannell
Department of Physics,
University of California
Santa Barbara, California

ABSTRACT

We present the information contained in recent measurements of the intensity and the spectrum of light scattered from a pure fluid near its critical point. The magnitude and angular dependence of the scattered and reflected light permits a determination of the compressibility, the density, and the correlation range along various paths in the pressure-temperature plane. On the other hand, measurements of the spectral distribution of the scattered light measures a combination of the transport coefficients and the equilibrium susceptibilities. By combining the spectral measurements with the intensity measurements one can determine separately the nature of the divergence of the equilibrium properties, the transport coefficients and the correlation range. Experimental results are presented for xenon and sulfur hexafluoride, and the theoretical significance of these data is discussed.

1. INTRODUCTION

It has long been understood that the magnitude and angular dependence of the intensity of light scattered from a fluid near its critical point provides a means of measuring its compressibility and the correlation range. The theoretical work of von Smoluchowski,[1] Einstein,[2] and Ornstein and Zernike[3] has shown that if P_s is the power scattered per unit solid angle around some angle θ in a plane perpendicular to the polarization of the incident light, then P_s is related to the incident power P_I by the formula

$$\left(\frac{P_s}{P_I}\right) = \left\{\frac{\pi^2 L}{\lambda_0^4} k_B T \rho^2 \left(\frac{\partial \epsilon}{\partial \rho}\right)_T^2\right\} \kappa_T \frac{1}{(1 + K^2 \xi^2)} \tag{1}$$

Here L is the length of the illuminated region, ρ the density, ϵ the dielectric constant, κ_T the isothermal compressibility, K the scattering vector, and ξ is the correlation range.

$$K = \left[\frac{4\pi}{\lambda/n}\right] \sin\left(\frac{\theta}{2}\right) \tag{2}$$

Furthermore it is also possible to measure the density of a fluid near its critical point by measuring its index of refraction, since the refractive index and density are connected by the Lorentz-Lorenz relation. By combining such optical measurements of density and compressibility with a determination of the state of the fluid, i.e., pressure and temperature, it is in fact, possible to accurately measure the equation of state near its critical point. By making such measurements along a series of paths near the critical point it is possible to obtain sufficiently accurate data to establish all the equilibrium properties of the fluid (except for C_v) needed, e.g., in the recent parametric representations of the equation of state.[4]

Despite the information potentially available, it is only recently that experiments have been performed which are beginning to provide extensive optical measurements of density and compressibility and correlation range along a variety of paths near the critical point.

In addition to the information contained in the intensity of the scattered light, the spectrum of light provides a measurement of the transport properties of the fluid. The temporal fluctuation in the amplitude of the scattered electric field is a measure of the temporal fluctuations in the density. The characteristic "slowing down" in the fluctuation of the order parameter is reflected in the spectrum of the scattered light by the appearance of a strong Rayleigh component whose spectral width measures the correlation time for the density fluctuations. The correlation time τ which describes the dying away of the major portion of the density fluctuations is related to the thermal conductivity Λ and the specific heat per unit mass C_p by the formula

$$\frac{1}{\tau} = 2\pi(\Delta\nu)_{1/2} = \left(\frac{\Lambda}{\rho C_p}\right) K^2 \qquad (3)$$

when $K\xi \ll 1$. In this formula $\Delta\nu_{1/2}$ is the half width of the quasielastic or "Rayleigh" component in the scattered light, and K is the scattering vector. When $K\xi$ is not small compared to unity, Kawasaki[5] has shown that Eq. (3) has a form reminiscent of the Ornstein-Zernike effect, namely

$$2\pi(\Delta\nu)_{1/2} = \left(\frac{\Lambda}{\rho C_p}\right) K^2 \left[1 + \left(\frac{3}{5}\right)(K\xi)^2\right] \qquad (4)$$

when $K\xi < 1$. On the other hand when $K\xi \gg 1$ dynamical scaling arguments show[5] that $2\pi(\Delta\nu)_{1/2} \propto K^3$ and is independent of the correlation range ξ. In the intervening region between $K\xi \ll 1$ and $K\xi \gg 1$, Kawasaki[5] used a mode-mode coupling calculation to predict that the line width should vary with scattering angle in accordance with the formula

$$2\pi(\Delta\nu)_{1/2} = \frac{k_B T}{8\pi\eta\xi^3}\left[1 + K^2\xi^2 + \left(K^3\xi^3 - \frac{1}{K\xi}\right)\tan^{-1}K\xi\right] \quad (5)$$

Here η is the shear viscosity. R. Ferrell[6] has also obtained this result using the Kubo formula and a "decoupling approximation."

Apart from its importance as an interpolation formula between the region of large and small $K\xi$ regions, Eq. (5) contains in essence an important physical interpretation of the physical significance of the thermal diffusivity $(\Lambda/\rho C_p)$ near the critical point. Let us examine this interpretation. In the limit $K\xi \ll 1$, Eq. (5) takes on the form

$$2\pi(\Delta\nu)_{1/2} = \frac{k_B T}{6\pi\eta\xi} K^2 \quad (6)$$

Thus according to the mode-mode coupling schemes[7] the singular part of the thermal diffusivity should be related to the correlation range ξ and the shear viscosity η by the formula

$$\left(\frac{\Lambda}{\rho C_p}\right) = \left(\frac{k_B T}{6\pi\eta\xi}\right) \quad (7)$$

This result has a very simple physical interpretation. The left-hand side is the diffusion coefficient for the transfer of heat from one region of space to another. If one regards the flow of heat as taking place by the diffusion of spheres whose characteristic size is ξ, then from the Stokes-Einstein relation we know that the diffusion coefficient of such particles is given by

$$D = \frac{k_B T}{6\pi\eta\xi} \quad (8)$$

This is identical to the right-hand side of Eq. (7). Thus the mode-mode coupling schemes essentially regard the diffusion of heat as occurring by the spatial diffusion of regions of size ξ. This interpretation of the diffusion coefficient is quite natural when the wavelength of the scattering vector is much larger than the correlation range ξ. However, when this wavelength is comparable to or even smaller than ξ it is physically clear that the actual structure of the diffusing heat units will play an important role. It is tempting to speculate that the Kawasaki scaling function shown in Eq. (5) in effect takes this structure into account. We believe that it is very important to determine the essential physical content of the scaling function introduced by Kawasaki. Even in the absence of this physical understanding it is clearly of great importance to verify experimentally this connection between the thermal diffusivity and the correlation range.

It is interesting to observe that according to the mode-mode coupling scheme the behavior of the transport coefficient Λ does not provide a new independent measure of the critical behavior. Λ diverges in these theories in a way determined by the correlation range ξ. This result which at first may appear surprising is really a reflection of the dynamical scaling law hypothesis that there is only one fundamental scale of divergence in the critical region and this is the divergence of the correlation range. According to the theorists then, the behavior of the transport coefficients near the critical point are determined by the behavior of the equilibrium properties as characterized by the correlation range ξ. Present studies of the dynamical behavior of fluids, ferromagnets, and critical mixtures near critical points are essentially efforts to determine whether these general dynamical scaling ideas apply to transport coefficients in the same way that the static scaling laws apply to the equilibrium properties.

The discussion above shows that the dynamical and the equilibrium properties are closely interwoven. A beautiful example of this intermingling can be seen by considering some of the information which can be obtained from the Brillouin portion of the spectrum of light scattered from a fluid near its critical point. Till now we have discussed the Rayleigh or quasielastic component of the spectrum. In addition to the scattering of light from these entropy fluctuations at constant pressure, the fluid can scatter light off pressure fluctuations at constant entropy. These thermally generated sound waves are one of the mechanisms for the spontaneous relaxation of the density fluctuations. The splitting of the Brillouin doublets measures the sound speed of waves whose wave vector is equal to that of the light scattering vector $\mathbf{K}$. The width of the Brillouin doublets measures the attenuation of these hypersonic sound waves. The relative intensity of the Rayleigh and Brillouin components in a so-called normal fluid measures the specific heat ratio C_p/C_v. By normal fluid is meant one whose transport coefficients show no frequency or wave vector dependence. In the critical region, however, even low frequency sound propagation experiments show that there is considerable relaxation in the sound speed and sound attenuation.

In a relaxing fluid the hydrodynamic equations predict that the relative intensity of the Brillouin and Rayleigh components is not given simply by the ratio C_p/C_v. If the fluid has a single relaxation process and if the Brillouin splittings are much greater than the relaxation frequency, then David Cannell[8] has observed that the experimentally measured ratio of intensities denoted by $C_p(K,\omega)/C_v(K,\omega)$ is related to the zero frequency infinite wavelength C_p/C_v ratio by

$$\left(\frac{C_p}{C_v}\right) = \left(\frac{C_p(K,\omega)}{C_v(K,\omega)}\right)\left(\frac{C_{00}^2}{C_\infty^2}\right)[1 + (K\xi)^2] \tag{9}$$

The factor C_∞^2/C_{00}^2 is the square of the ratio of the infinite frequency sound speed to the zero frequency sound speed. This ratio in effect removes the frequency dependence of $C_p(K,\omega)/C_v(K,\omega)$. The factor $1 + (K\xi)^2$ corrects for the wave vector dependence C_p by the Ornstein-Zernike theory. Eq. (9) can be used as the basis of an experimental determination of the correlation range as follows. According to thermodynamics we have

$$\left(\frac{C_p}{C_v}\right) = \left(\frac{\kappa_T}{\kappa_S}\right) \tag{10}$$

Furthermore the zero frequency sound speed is related to the adiabatic compressibility κ_S by

$$\rho_0 \kappa_S = \frac{1}{C_{00}^2} \tag{11}$$

According to the Ornstein-Zernike theory, as confirmed in the measurements of M. Giglio and G. Benedek,[9] the correlation range is related to κ_T by

$$\frac{\xi^2}{R^2} = \frac{\kappa_T}{\kappa_I} \tag{12}$$

where κ_I is the compressibility of an ideal gas at density ρ and temperature T, that is, $\kappa_I = 1/nk_B T$ where n = number of molecules/cc. The "direct correlation range" R in the Ornstein-Zernike theory is a constant independent of temperature and density whose size is comparable to the scale of the intermolecular forces. On replacing Eq. (10) for C_p/C_v in Eq. (9) and by eliminating κ_T, and κ_S in the resulting equation using Eq. (11) and (12), we find that the factor $C_{00}{}^2$ is eliminated from Eq. (9) giving the following equation for the correlation range ξ in terms of experimentally measured quantities

$$\frac{\xi^2}{R^2} = \left(\frac{C_p(K,\omega)}{C_v(K,\omega)}\right) \frac{(1 + K^2 \xi^2)}{\rho_0 \kappa_I C_\infty{}^2} \qquad (13)$$

Since $C_\infty{}^2$ is effectively measured from the Brillouin splittings, Eq. (13) shows how the structural features (magnitudes and splittings) of the Rayleigh-Brillouin spectrum can be used to determine the correlation range ξ. In essence then a measurement of the dynamical properties (the spectrum) can be used to determine equilibrium properties. It should be mentioned that previous workers had observed that the ratio $C_p(K,\omega)/C_v(K,\omega)$ did not diverge as rapidly as $T \to T_c$ as one would have expected from P-V-T data. The analysis of Dr. Cannell shows why this occurs.

The use of spectral measurements near the critical point of single component systems and critical mixtures began in 1965.[10,11] The discovery of the c-w helium neon laser in 1961 made these measurements possible. The conventional light sources available before then had such a large spectral width that the natural features of the spectra could not be resolved. Moreover, even with extremely monochromatic lasers, the line widths of the Rayleigh line are so narrow (less than ~ 5 kHz) that it was necessary to develop the new techniques of optical mixing spectroscopy[12-14] to provide the resolution needed to measure the exceptionally narrow spectral width of the scattered light. Thus, while in principle measurements of the intensity of the scattered light could have been made long ago it is only recently that the dynamical information available from the spectrum could be obtained. It is clear then that both the intensity and the spectrum of the light scattered from a fluid near its critical point can provide extensive information on the dynamical and the equilibrium properties of that system. In fact the measurements which have been made using this experimental approach during the past year alone have given us so much information that it is important to condense and summarize it simply. With this purpose in view I should like to confine my attention to two fluids, xenon and sulfur hexafluoride. First I discuss xenon.

2. INTENSITY AND SPECTRAL MEASUREMENTS ON XENON

By making measurements of the intensity and the angular dependence of the light scattered from xenon, M. Giglio[9] was able to determine the divergence of the compressibility κ_T along both the gas and the liquid sides of the coexistence curve. He found that the quantity that scales properly is not κ_T but $(\partial \rho/\partial \mu)_T$

$$\left(\frac{\partial \rho}{\partial \mu}\right)_T = \rho^2 \kappa_T \tag{14}$$

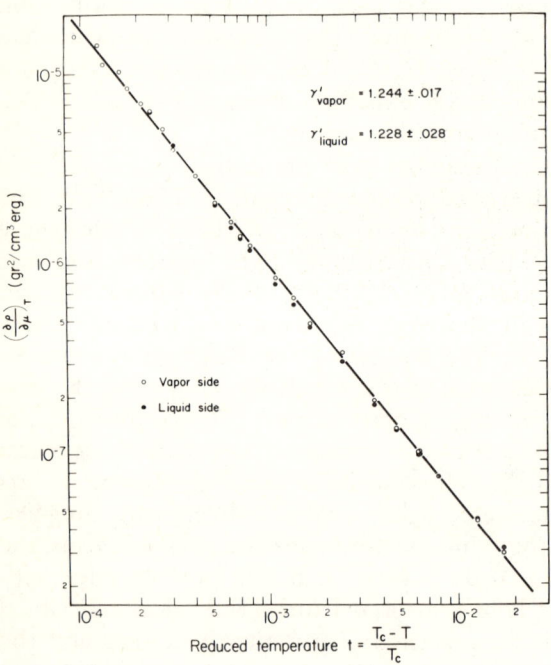

Fig. 1 $(\partial \rho/\partial \mu)_T$ vs. temperature along both sides of the coexistence curve for xenon. The value of the exponent γ' for both the liquid and the gas sides of the coexistence curve is $\gamma' = 1.235 \pm 0.03$.

where μ is the chemical potential per unit mass. In Fig. 1 we show Dr. Giglio's data on $(\partial \rho/\partial \mu)_T$ on both sides of the coexistence curve. It can accurately be represented over a temperature range $10^{-4} < (1 - T/T_c) < 10^{-2}$ by

$$\left(\frac{\partial \rho}{\partial \mu}\right)_{\text{liquid}} = (1.90 \pm 0.15) \times 10^{-10} \epsilon^{-1.23 \pm 0.03} \text{ g}^2/\text{erg cm}^3$$

$$\left(\frac{\partial \rho}{\partial \mu}\right)_{\text{gas}} = (1.79 \pm 0.07) \times 10^{-10} \epsilon^{-1.24 \pm 0.02} \text{ g}^2/\text{erg cm}^3$$

(15)

It is important to realize that while $(\partial \rho/\partial \mu)_T$ is essentially identical on both sides of the coexistence curve, the compressibility κ_T is not. For example, at $\epsilon = (T_c - T)/T_c = 0.0173$, $\kappa_{T_{\text{vapor}}} = 6\kappa_{T_{\text{liquid}}}$! So there is an enormous difference in κ_T on both sides of the coexistence curve.

In Fig. 2 we show Dr. Giglio's data on the correlation range along the vapor side of the coexistence curve. This data can be represented by

$$\xi = \xi_0 \epsilon^{-\nu}$$

where

$$\left.\begin{array}{l} \nu = 0.57 \pm 0.05 \\ \xi_0 = 1.84 \pm 0.2 \text{Å} \end{array}\right\} T < T_c \qquad (16)$$

Dr. Giglio also was able to compare his measurements of ξ and his measurement of κ_T and was able to show that the Ornstein-Zernike relation

$$\frac{\xi^2}{R^2} = \frac{\kappa_T}{\kappa_I} \qquad (17)$$

is correct when K_I is $1/nk_B T$ and $R \sim 6.0$Å both above and below T_c. In the data above T_c Dr. Giglio did not follow an isochore but followed a curving path. Nevertheless at each point in the $\rho - T$ plane he studied, he found that Eq. (17) applied.

To determine how ξ varies along the critical isochore above T_c we must use Dr. Cannell's deduction of the correlation range. This is

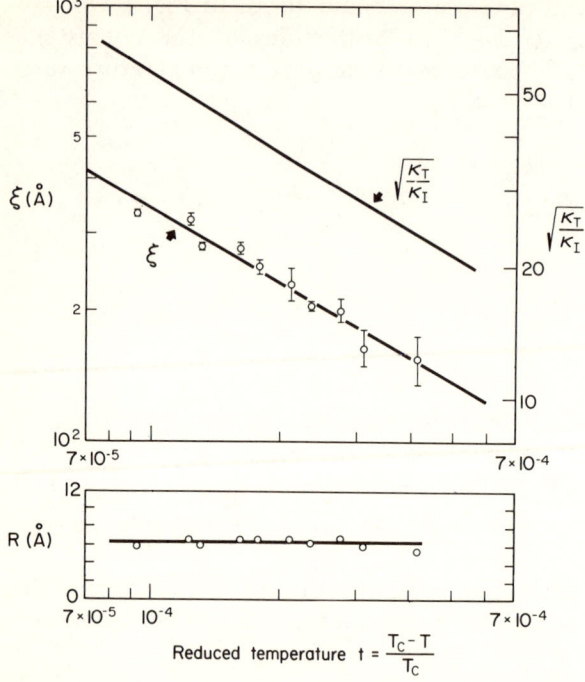

Fig. 2 The correlation range ξ; the ratio of the square root of the compressibility of xenon to the compressibility of an ideal gas of the same density, $(\kappa_T/\kappa_I)^{1/2}$ and the "direct correlation range" R vs. temperature along the vapor side of the coexistence curve for xenon.

shown in Fig. 3. Dr. Cannell's results along the critical isochore can be fit by an equation of the form

$$\xi = 3.02\epsilon^{-0.60} \text{ Å} \qquad (18)$$

In Fig. 4 we compare Dr. Cannell's values of ξ for $T > T_c$ with Dr. Giglio's data along the coexistence curve.

In Fig. 5 we see Dr. Cannell's results[8] for $(\partial\rho/\partial\mu)_T$ as measured along the critical isochore using the relative intensities of the Rayleigh and Brillouin components, as part of the analysis described above. This data is consistent with

$$\left(\frac{\partial\rho}{\partial\mu}\right)_T = 16.3 \times 10^{-10} \left(\frac{T - T_c}{T_c}\right)^{-1.21 \pm 0.05} \text{gm}^2/\text{erg cc} \qquad (19)$$

ON THE SPECTRUM AND INTENSITY OF LIGHT

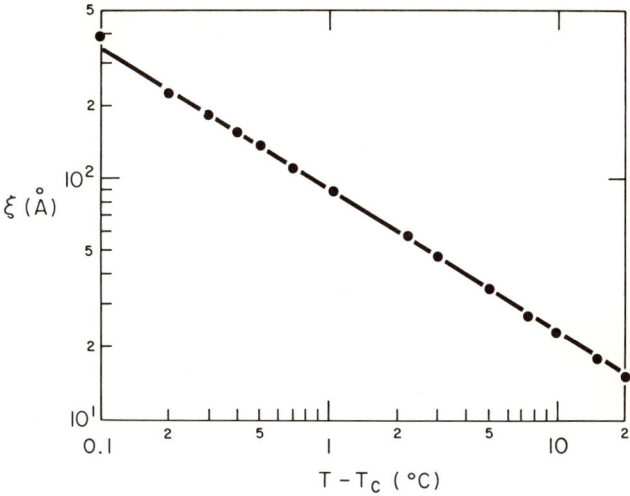

Fig. 3 *The correlation range ξ for xenon along the critical isochore (from measurements of the Rayleigh-Brillouin spectra).*

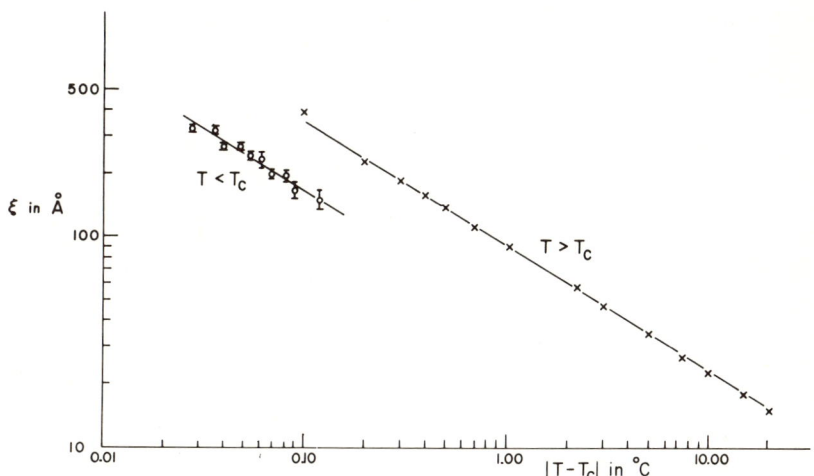

Fig. 4 *Temperature dependence of ξ along the critical isochore $T > T_c$ and along the vapor side of the coexistence curve $(T < T_c)$.*

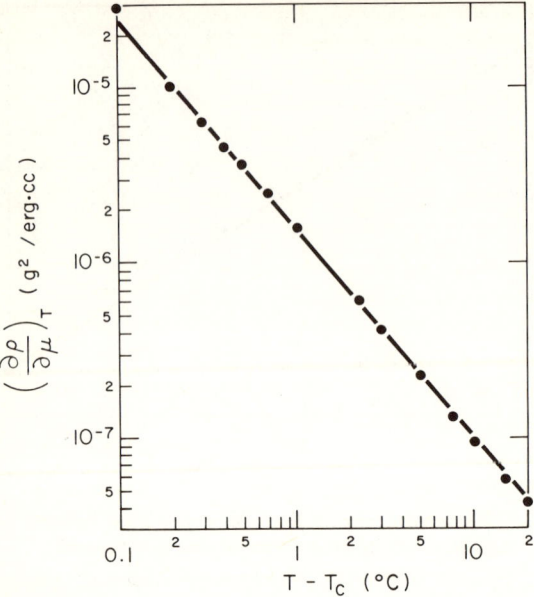

Fig. 5 $(\partial\rho/\partial\mu)_T$ as a function of temperature for xenon along the critical isochore.

Let us now examine the behavior of the thermal diffusivity of xenon as determined from the spectral width of the scattered light. The thermal diffusivity $\Lambda/\rho C_p$ has been measured by D. L. Henry, H. L. Swinney, and H. Z. Cummins along the critical isochore;[15] they find:

$$\left(\frac{\Lambda}{\rho C_p}\right) = (6.94 \pm 0.21) \times 10^{-21} \epsilon^{0.751} \text{cm}^2/\text{sec} \qquad (20)$$

We may examine the significance of this result in the following way: Since $C_p \propto \epsilon^{-\gamma}$ and if we assume that the thermal conductivity diverges like $\Lambda \propto \epsilon^{-\psi}$ we have that $0.751 = \gamma - \psi$. Since $\gamma \simeq 1.24$ from the measurements of Giglio and Benedek and the scaling law predicts that $\gamma = \gamma'$, we find that the thermal conductivity in xenon must diverge like $\Lambda \propto \epsilon^{-0.49}$, that is, $\psi = 0.49$. On the other hand, the Kawasaki-Kadanoff-Swift theory predicts that $\gamma - \psi = \nu$, where ν is the exponent describing the divergence of the correlation range. Since ν is certainly no greater than 0.64, the K-K-S theory predicts $\psi \leq 0.60$. This theoretical prediction is clearly a stronger divergence than is experimentally observed. In considering the reason for this disagreement, it is well to point out that experimental data of Sengers suggests that the thermal conductivity has a "background" or nondivergent part. The existence of such a component will have the effect of reducing the apparent divergence in Λ.

D. Henry, H. Swinney, and H. Z. Cummins also examined the validity of Kawasaki's formula (Eq. (5)) for the dependence of line width on $K\xi$. They found that the formula fit their data provided that they assumed $\xi = 0.58\epsilon^{-0.751}$ Å. This assumed form for the temperature dependence of ξ is not in agreement with the experimental measurements here reported on ξ for $T > T_c$. Furthermore it implies a value of ν much greater than can be expected from Ising-model calculations or scaling law estimates. Clearly it would be most appropriate to check the Kawasaki theory for $(\Delta\nu)_{1/2}$ by measuring ξ independently and then comparing the theoretical prediction for the right-hand side of Eq. 5 with the experimental values of the left-hand side. Despite this criticism, it should be mentioned that the value of $(k_B T/6\pi\eta)$ that these authors needed to fit to the Kawasaki theory was 4.0×10^{-12} cm^3/sec. This compares satisfactorily with a direct computation of this number using $\eta = 5.0 \times 10^{-4}$ poise which Dr. John Zollweg[16] has measured at M.I.T. using the spectrum of light scattered from thermally excited surface waves in xenon. The direct computation gives $(k_B T/6\pi\eta) = 4.3 \times 10^{-12}$ cm^3/sec.

3. INTENSITY AND SPECTRAL MEASUREMENTS ON SULFUR HEXAFLUORIDE

Let me now turn to the presently available data on the equilibrium and transport properties of SF_6. I shall describe in the main experiments conducted by Dr. A. Saxman, Dr. J. B. Lastovka, and G. Feke. I shall also discuss the results found by Professor N. C. Ford and V. G. Puglielli[17] at the University of Massachusetts. The work of Dr. Lastovka is particularly ambitious in that he has undertaken to measure on the same sample the equilibrium equation of state and the transport properties over a wide temperature and density range around the critical point. Dr. Lastovka and Mr. Feke are now engaged in these measurements. In what follows I will present the results they have so far obtained. Since all the measurements are not yet complete we must regard the results quoted here as preliminary. In Fig. 6 we see $(\partial\rho/\partial\mu)_T$ along the liquid side of the coexistence curve. The data is represented analytically by

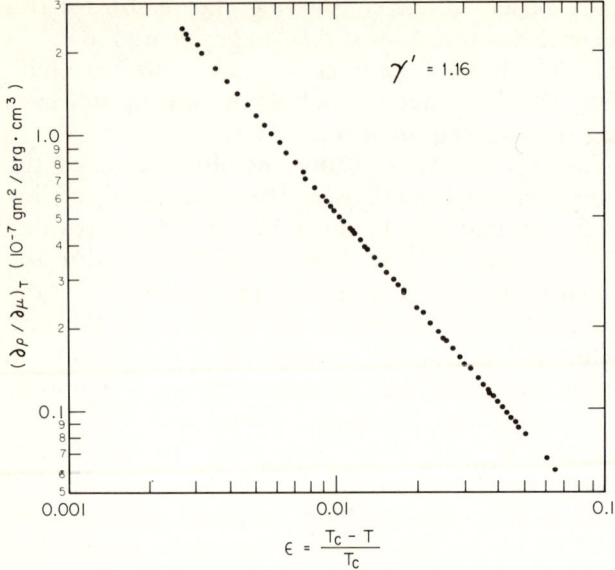

Fig. 6 $(\partial\rho/\partial\mu)_T$ *as a function of temperature along the liquid side of the coexistence curve for sulfur hexafluoride.*

$$\left(\frac{\partial\rho}{\partial\mu}\right)_T = 2.5 \times 10^{-10}\left(\frac{T_c - T}{T_c}\right)^{-1.16} \frac{\text{gm}^2}{\text{erg cc}} \quad (21)$$

In Fig. 7 we see a graph which shows both Dr. Saxman's and Dr. Lastovka's measurement of the shape of the coexistence curve. This data has the form

$$\left(\frac{\rho_l - \rho_g}{2\rho_c}\right) = 1.81\left(\frac{T_c - T}{T_c}\right)^{0.333} \quad (22)$$

that is

$$\beta = 0.333$$

Along the critical isotherm, Dr. Lastovka has measured both $(\partial\rho/\partial\mu)_T$ and the pressure versus density relation. Figures 8 and 9 show this data. $(\partial\rho/\partial\mu)$ and the p vs. ρ data should be related to one another in such a way that the exponent for the latter is δ while that

ON THE SPECTRUM AND INTENSITY OF LIGHT

for the former is $-\delta + 1$. The data now available starts at $\Delta\rho \simeq 15\%$ and extends to $\Delta\rho \sim 40\%$. It can be represented by

$$\left(\frac{\partial \rho}{\partial \mu}\right)_T = 1.08 \times 10^{-9} \left(\frac{\rho - \rho_c}{\rho_c}\right)^{-3.25} \frac{\text{gm}^2}{\text{erg cc}} \quad (23)$$

$$\frac{P - P_c}{\delta \cdot \rho + \rho_c} = 387 \left(\frac{\rho - \rho_c}{\rho_c}\right)^{4.25} \frac{\text{psi cm}^3}{\text{gm}} \quad (24)$$

Here we see $\delta \simeq 4.25$.

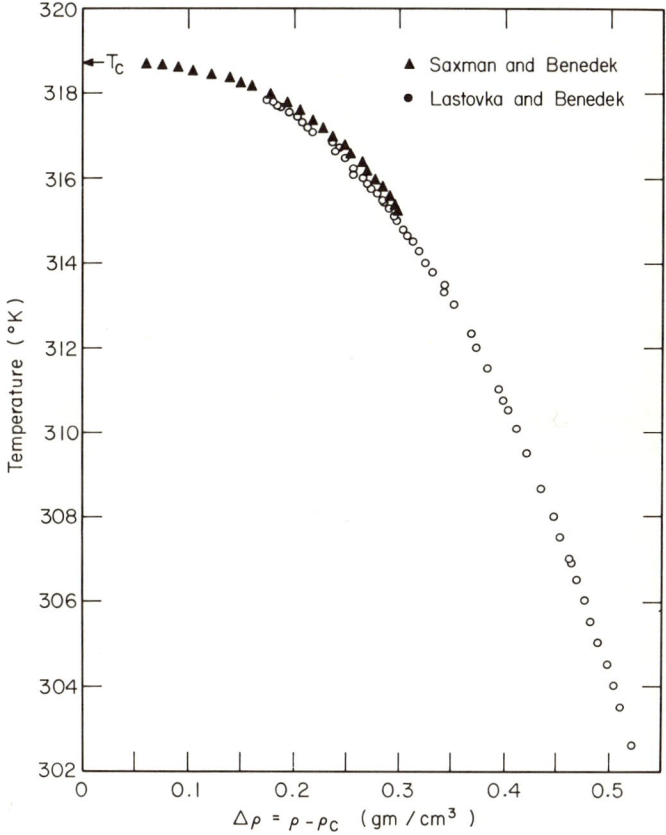

Fig. 7 *Shape of the coexistence curve for sulfur hexafluoride. The open circles are the data of J. B. Lastovka. The triangles are the data of Dr. Saxman taken close to the critical density.*

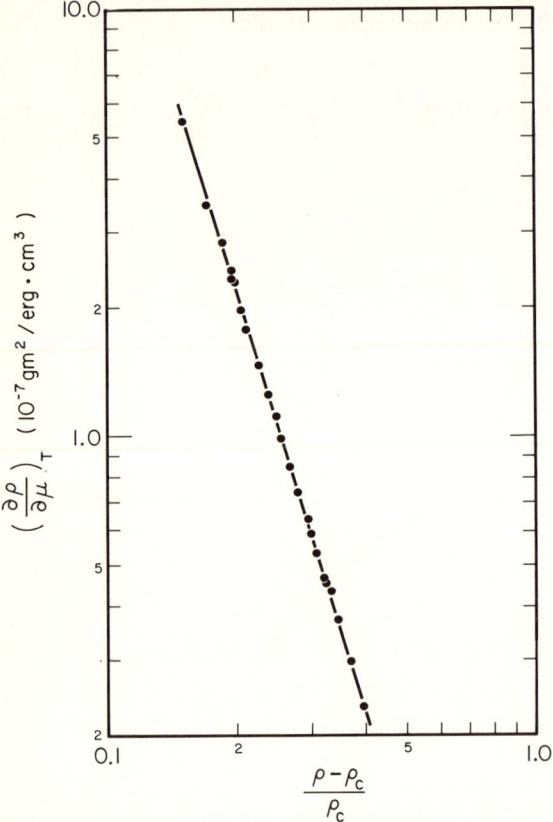

Fig. 8 $(\partial \rho / \partial \mu)_T$ along the critical isotherm for sulfur hexafluoride.

For $T > T_c$ we have the data of Ford and Puglielli.[17] They have measured the compressibility or essentially $(\partial \rho / \partial \mu)_T$ and the correlation range ξ along the critical isochore by measuring the loss in intensity of the *transmitted* beam. This of course is equal to the integral over all angles of the scattered intensity per unit solid angle. From data taken for $T \gg T_c$ they can deduce γ. And from the departure of the intensity from the power law variation with temperature they can deduce the correlation range when T is sufficiently close to T_c. This procedure indicates that the compressibility varies with temperature as

ON THE SPECTRUM AND INTENSITY OF LIGHT

$$\kappa_T = (1.18 \pm 0.16) \times 10^{-3} \left(\frac{T - T_c}{T_c} \right)^{-1.22 \pm 0.02} \text{atm}^{-1} \quad (25)$$

They also deduce the following temperature variation of ξ along the critical isochore

$$\xi = (1.5 \pm 0.23) \left(\frac{T_c - T}{T_c} \right)^{-0.67 \pm 0.07} \text{Å} \quad (26)$$

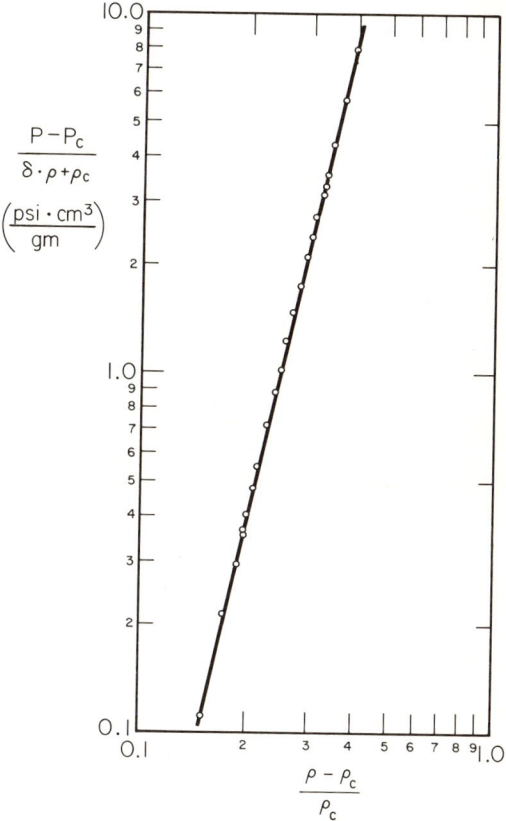

Fig. 9 *The pressure-density relation along the critical isotherm in sulfur hexafluoride. According to the scaling law theories the shape of the critical isotherm is to be analyzed in terms of the equation* $(P - P_c)/(\rho\delta + \rho_c) = \text{const.} \, (\rho - \rho_c)^\delta$.

Let us now discuss the transport properties of SF_6 as determined from the spectrum of the scattered light. In Fig. 10 we see Dr. Saxman's 1967 measurement[12] of the line width of the quasielastic component in the scattered light. The data above T_c was measured along the critical isochore. The data below T_c was measured along both the gas and the liquid sides of the coexistence curve. As indicated previously this line width measures $\Lambda/\rho C_p$ where Λ is the thermal conductivity and ρ is the mass density and C_p the specific heat per unit mass. The most striking feature of this graph is that as we approach T_c from above the line width approaches zero with an exponent greater than unity. In fact, Dr. Saxman's data can be fit to an equation of the form

$$\frac{\Lambda}{\rho C_p} = (1.26 \pm 0.04) \times 10^{-2} \left(\frac{T - T_c}{T_c}\right)^{1.26 \pm 0.02} \frac{cm^2}{sec} \quad (27)$$

for $T > T_c$. On the other hand, for T below T_c the thermal diffusivity varies as

$$\frac{\Lambda}{\rho C_p} \simeq (1.78 \pm 0.03) \times 10^{-3} \left(\frac{T_c - T}{T_c}\right)^{0.633 \pm 0.003} \frac{cm^2}{sec} \quad (28)$$

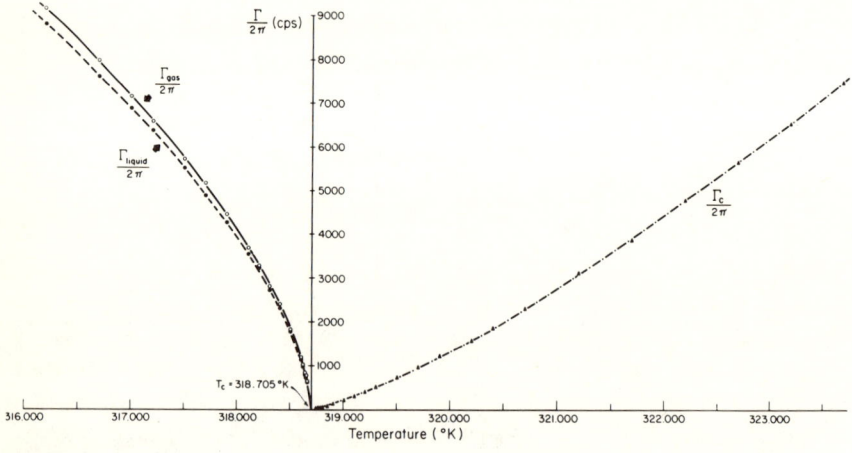

Fig. 10 *Line width of the Rayleigh component of light scattered from sulfur hexafluoride. The data above T_c is along the critical isochore. The data below T_c is along both the gas and liquid sides of the coexistence curve. (Data of A. Saxman.[12])*

The fact that the line width appears to approach T_c with an infinite slope from below T_c and a zero slope from above T_c was quite perplexing for the following reason. If the quantity ρC_p diverges the same way as $T \to T_c$ from below as from above, then the data on the line width implies that Λ has quite a different temperature dependence as one approaches T_c from either above or below T_c. This conclusion seemed very much at variance with what one expected about the symmetry of exponents describing divergences below and above T_c.

This dilemma has recently been resolved by Dr. Lastovka and Mr. G. Feke who have pointed out that ρC_p does not in fact diverge with a single exponent as the critical point is approached along the coexistence curve. Their data on the equilibrium properties along the coexistence curve shows dramatically that ρC_p does not vary like a simple exponent function along the coexistence curve. Their analysis further shows that in order to deduce the temperature dependence of Λ one must actually measure ρC_p along the same path as one measures $\Lambda/\rho C_p$. Then one should compute the product to determine the temperature dependence of Λ. Upon carrying out this procedure we should not be surprised to find that Λ does not vary like $(T - T_c)$ raised to a single exponent. This should be evident from the very outset since the thermal conductivity has a divergent part and a non-divergent part. While the divergent part may be characterized with single exponents the sum of the two parts will not in general be describable by a single exponent. The data on xenon mentioned previously also suggested this situation exists there.

Let us see how in fact ρC_p varies along the coexistence curve. To determine this let us remind ourselves how C_p can be determined. Thermodynamics tells us that

$$(C_p - C_v) = \frac{T}{\rho^3}\left(\frac{\partial P}{\partial T}\right)_v^2 \left(\frac{\partial \rho}{\partial \mu}\right)_T \tag{29}$$

When using Eq. (29) along the coexistence curve it is important to realize that $(\partial P/\partial T)_v$ is *not* a slowly varying function of the temperature. As a result $C_p - C_v$ does *not* vary with temperature in the same way as $(\partial \rho/\partial \mu)_T$. The quantity that varies slowly with temperature along the coexistence curve is the slope of the vapor pressure curve $(\partial P/\partial T)_{l.v.}$. This derivative represents the change in vapor pressure when the temperature of the system consisting of both gas and liquid together changes. In such a temperature change the density of both liquid and gas change. The derivative required by

Eq. (29) is the pressure temperature derivative of a single phase under the condition that the density of that phase is held constant. These two derivatives $(\partial P/\partial T)_{l.v.}$ and $(\partial P/\partial T)_v$ are related as follows

$$\left(\frac{\partial P}{\partial T}\right)_{l.v.} = \left(\frac{\partial P}{\partial T}\right)_v + \left(\frac{\partial P}{\partial V}\right)_T \left(\frac{\partial V}{\partial T}\right)_{l.v.} \qquad (30)$$

where $(\partial V/\partial T)_{l.v.}$ is essentially the derivative of the shape of the coexistence curve: that is, $(\partial V/\partial T)$ is evaluated along the coexistence curve. On putting Eq. (30) into Eq. (29) and using $\partial \rho/\rho = -\partial V/V$ we find the following expression for ρC_p

$$\rho C_p = \rho C_v + \frac{T}{\rho^2}\left[\left(\frac{\partial P}{\partial T}\right)_{l.v.} + \left(\frac{\partial P}{\partial \rho}\right)_T \left(\frac{\partial \rho}{\partial T}\right)_{l.v.}\right]^2 \left(\frac{\partial \rho}{\partial \mu}\right)_T$$

Now Dr. Lastovka and Mr. Feke have measured $(\partial \rho/\partial \mu)_T$ and $(\partial P/\partial \rho)_T$ along the coexistence curve. They and Dr. Saxman have

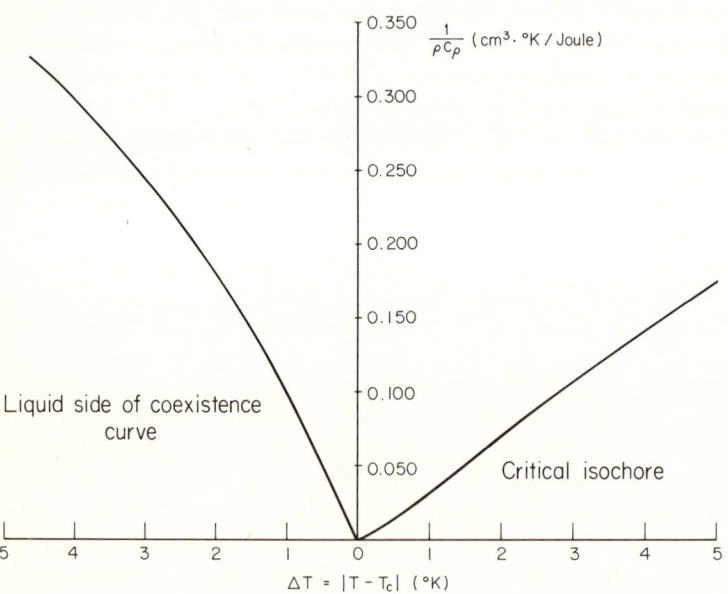

Fig. 11 *A plot of $1/\rho C_p$ for sulfur hexafluoride along the coexistence curve and the critical isochore.*

measured the shape of the coexistence curve and therefore know $(\partial \rho/\partial T)_{l \cdot v} \cdot (\partial P/\partial T)_{l \cdot v}$ is also known from their P, ρ, T data. Thus all the quantities in the second term on the right-hand side of Eq. (31) are known experimentally. It is important to observe that even though $(\partial \rho/\partial \mu)_T$ varies like a simple exponent, the quantity $(\partial P/\partial \rho)_T (\partial \rho/\partial T)$ goes to zero as $T \to T_c$ so that $(\partial P/\partial T)_v$ undergoes a very marked variation along the coexistence curve. Using data on the low frequency sound speed they also were able to determine the variation of ρC_v along the coexistence curve and along the critical isochore. They then plotted the experimentally determined quantity $(1/\rho C_p)$ vs. temperature along the coexistence curve and the critical isochore and obtained the results shown in Fig. 11. This figure shows dramatically that $1/\rho C_p$ *appears* to approach zero with an infinite slope as $T \to T_c$ from below along the coexistence curve. While it is true that asymptotically $1/\rho C_p$ must approach zero in the same way as $(\partial \rho/\partial \mu)_T$, the temperature range in which this asymptotic region applies is too small to be apparent except for data taken extremely close to T_c. Certainly in the experimental region of Dr. Saxman's data the asymptotic expressions do not apply. This graph shows very clearly the great dangers of approximating the actual temperature variation of the equilibrium properties along the coexistence curve with asymptotic approximations. In Fig. 11 we also see how $1/\rho C_p$ approaches zero as $T \to T_c$ along the critical isochore. We see from this graph how much this looks like a linear approach to zero. In computing $1/\rho C_p$ along this path the Ford-Puglielli data on κ_T was used.

Now that we see that $(1/\rho C_p)$ can, in fact, appear to have quite different temperature dependences above and below T_c let us continue by including in the discussion recent measurements of the line width made by Braun, Hammer, Tscharnuter, and Weinzeirl[18] who studied the line width for $T > T_c$, and measurements of Lastovka and Feke for $T < T_c$ along the liquid side of the coexistence curve. For purposes of comparison, Lastovka and Feke have used thermal conductivity data of Lis and Kellard[19] to determine the nondivergent part of the thermal conductivity in SF_6. If one computes from the Lis and Kellard data[19] the ratio $\Lambda_{background}/\rho C_p$ along the coexistence curve and the critical isochore one obtains the smooth curves drawn in Fig. 12.

Below T_c the triangles represent the most recent measurements of $\Lambda/\rho C_p$ taken by Dr. Lastovka. These data have so far been taken from $T - T_c = 5°C$ to $(T - T_c) = 1°C$. The open circles are the data of Braun, et al. The solid circles are those of Saxman. It seems clear that it will be necessary to extend the triangle points into the region

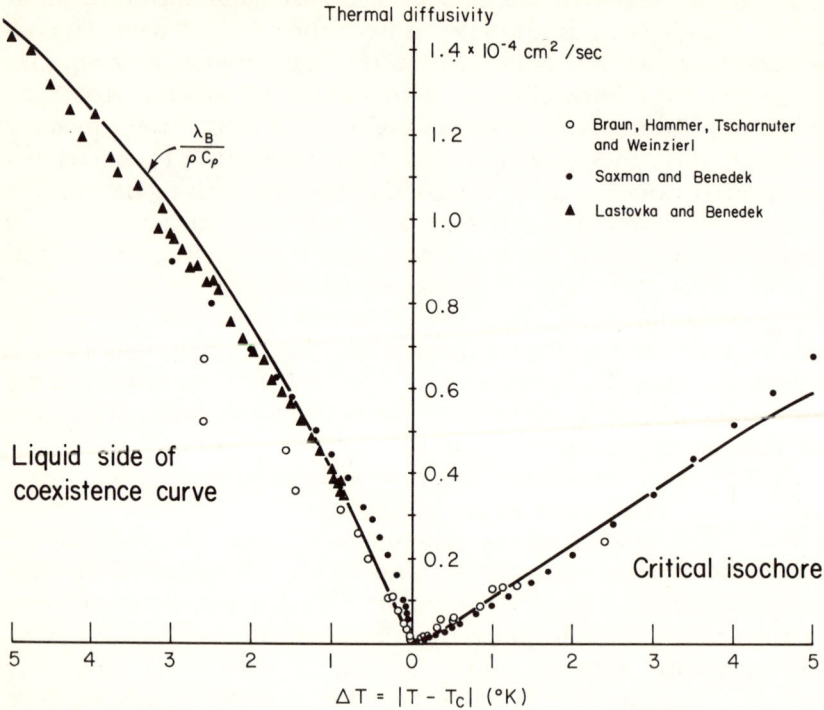

Fig. 12 *Summary of present data on the thermal diffusivity of sulfur hexafluoride both above and below T_c. The smooth line is the calculated value of $\Lambda_{background}/\rho C_p$.*

below $\Delta T = 1°C$ to establish the size of the divergent part of the thermal conductivity. It is interesting to note that the close match between the smooth curve and the data of Braun, et al. means that their data suggests very little divergence in the thermal conductivity below T_c.

Above T_c there is only the original Saxman data and the Braun-Hammer-Weinzeirl data. On the scale shown there does not seem to be much difference between the two measurements, and it appears quite difficult to deduce convincingly the divergent part of the thermal conductivity from the data now available. In Fig. 13 we present a magnified comparison of the data of Saxman and of Braun, et al. in the region $0 < \Delta T < 2.5°C$. Here we see a clear difference between the two experiments. However the scatter in data of Braun, et al. do not appear to permit a reliable deduction of the temperature dependence of Λ.

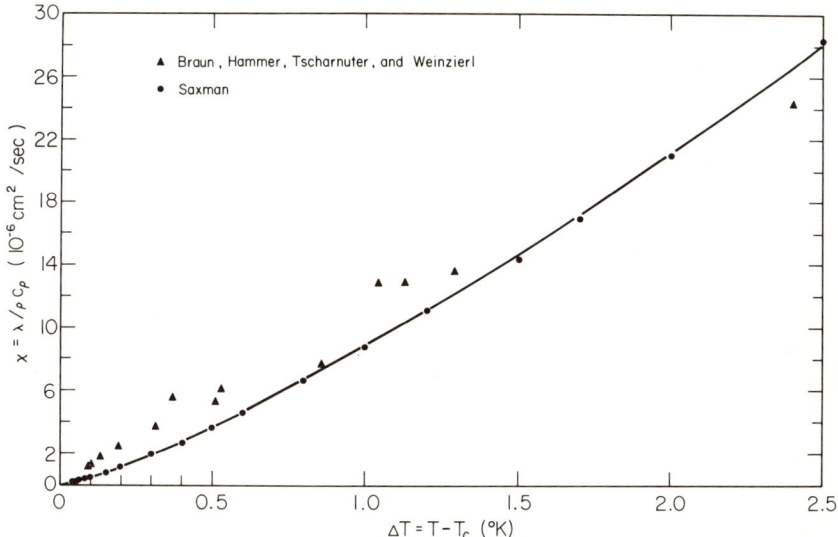

Fig. 13. *A magnified comparison of the data of Saxman and of Braun, et al. above T_c for sulfur hexafluoride.*

To summarize the present position on the behavior of the thermal conductivity in SF_6 near its critical point, it seems fair to say that both above and below T_c, the diverging part of the thermal conductivity makes itself felt only quite close to T_c. It is clearly necessary to get very accurate data in the region between $\Delta T = 1°C$ and $\Delta T \sim 0.030°C$ to reliably permit the deduction of Λ from spectroscopic measurements of $\Lambda/\rho C_p$ and intensity and equation of state measurements of ρC_p.

4. CONCLUSIONS

We have presented the results of recent optical measurements of the divergences of the equilibrium and transport properties of xenon and sulfur hexafluoride near their critical points. In particular $(\partial \rho/\partial \mu)_T$ and the correlation range ξ are now established along a variety of paths which pass through the critical point.

We have shown that it is necessary to combine these actual measurements of equilibrium properties with the spectral measurements of the thermal diffusivity to deduce the variation of the thermal conductivity Λ. We have in particular criticized the assumption that ρC_p varies along the coexistence curve in a way that can be characterized by a single exponent γ'. We have shown that the asymptotic region in which such a characterization applies is very small indeed. These observations as to the temperature dependence of ρC_p along the coexistence curve will require a reinterpretation of the measurements of thermal diffusivity of CO_2.[20] In the case of SF_6, the experimental data on ρC_p and $\Lambda/\rho C_p$, while not complete, suggest that the diverging part of the thermal conductivity is small and does not make itself felt until within $\sim 1°K$ of the critical point.

We have also discussed the experimental evidence[15] as to the correctness of the Kawasaki dynamical scaling function in xenon. Present tests of this function[15] in xenon have the flaw that the values of correlation range were unknown and the authors calculated the temperature variation of ξ that was needed to bring the Kawasaki function into agreement with the measurements of line width. Unfortunately, however, their calculated temperature variation of ξ is in disagreement with the measurements made by Dr. D. Cannell. On the other hand in the extreme limit $K\xi \gg 1$ the Kawasaki formula predicts a value for the quasielastic line width which is independent of correlation range. The theoretical magnitude of this limiting line width however is in reasonable agreement with the experimental result.

Despite these criticisms, both neutron and light scattering data at this point generally substantiate the dynamical scaling theories. This agreement focuses attention on the correlation range as being the fundamental parameter in characterizing both the equilibrium and transport properties near the critical point.

ACKNOWLEDGMENTS

The authors acknowledge with thanks the help of Professor Klaus Fritsch who kindly measured for us the sound velocity in SF_6 along the critical isochore. This data enabled us to accurately determine C_p along this path. This research was supported by funds from the Army Research Office—Durham and the Advanced Research Projects Agency, Department of Defense, under contract SD 90 with the Massachusetts Institute of Technology.

Note added in proof:
A redetermination of the geometrical factors involved in the absolute determination[9] of the light intensity scattered from xenon shows that the prefactors in Eq. (15) should be replaced by the number $(3.0 \pm 0.1) \times 10^{-10}$ g²/erg-cm³. Also we find $(\partial \rho/\partial \mu)_T =$

$(1.38 \pm 0.05)\epsilon^{-1.21} \times 10^{-9}$ g²/erg-cm³ along the critical isochore for $T > T_c$. This direct measurement differs by ~15 percent from Cannell's deduced values [Eq. (19)]. The redetermination of $(\partial\rho/\partial\mu)_T$ has the following effect: (a) The value of the direct correlation length R is (4.9 ± 0.1) Å for $10^{-4} \lesssim \epsilon \lesssim 10^{-3}$; (b) Cannell's deduced values of ξ, as shown in Figs. 3 and 4 and Eq. (18), are reduced at all temperatures by ~15 percent. These changes do not alter the conclusions given above.

REFERENCES

1. von Smolouchowski, M.: *Ann. Physik*, 25:205 (1908).
2. Einstein, A.: *Ibid.*, 33:1275 (1910).
3. Ornstein, L., and F. Zernike: *Proc. Acad. Sci. Amst.*, 17:793 (1914), and *Physik. Z.*, 19:134 (1918).
4. Schofield, P., J. D. Litster, and J. Ho: *Phys. Rev. Lett.*, 23:1098 (1969).
5. Kawasaki, K.: *Phys. Lett.*, 30A:325 (1969); *Phys. Rev.*, A1:1750 (1970).
6. Ferrell, R.: *Phys. Rev. Lett.*, 24:1169 (1970).
7. Kadanoff, L., and J. Swift: *Phys. Rev.*, 166:89 (1968).
8. Cannell, D., and G. B. Benedek: *Phys. Rev. Lett.* (in press).
9. Giglio, M., and G. B. Benedek: *Ibid.*, 22:1145 (1969).
10. Ford, N. C., Jr., and G. B. Benedek: *Ibid.*, 15:649 (1965).
11. Alpert, S., Y. Yeh, E. Lipworth: *Ibid.*, 14:486 (1965).
12. Benedek, G. B.: Optical Mixing Spectroscopy with Applications to Problems in Physics, Chemistry, Biology, and Engineering, in Jubilee Volume in honor of Prof. A. Kastler, pp. 49-84, Presses Universitaires, Paris, 1969.
13. Swinney, H., and H. E. Cummins: in E. Wolf (ed.) "Progress in Optics," vol. 8, North Holland Publishing Co., 1969.
14. Forrester, H. T.: *J.O.S.A.*, 51:253 (1961).
15. Henry, D. L., H. C. Swinney, and H. Z. Cummins: *Phys. Rev. Lett.* (to be published).
16. J. Zollweg: Private communication.
17. Puglielli, V. G., and N. C. Ford, Jr.: *Phys. Rev. Lett.*, 25:143 (1970).
18. Braun, P., D. Hammer, W. Tscharnuter, and P. Weinzeirl: *Physics Lett.*, 32A:390 (1970).
19. Lis, J., and P. O. Kellard: *Central Electricity Res. Lab. Note* #RD/L/N133/64, Leatherhead, England.
20. Swinney, H. L., and H. Z. Cummins: *Phys. Rev.*, 171:152 (1968).

DISCUSSION *on Paper by G. Benedek*

The discussion of this paper was combined with the discussion on the next two papers. (See p. 559.)

EXPERIMENTAL STUDY OF THE DYNAMICS OF CRITICAL FLUCTUATIONS IN A BINARY MIXTURE

P. Berge and B. Volochine

Service de Physique du Solide et de Résonance Magnétique
Centre d'Etudes Nucléaires de Saclay
Gif-sur-Yvette, France

All the results discussed here are relative to a binary mixture of cyclohexane-aniline at critical concentration (47% aniline-53% cyclohexane in weight). The experiments were performed by means of the photon-beating spectroscopic method.

In optical spectroscopy, usually one chooses the line width of the optical spectrum as a parameter. But this implies that this parameter can be defined always in the same way throughout the whole study.

A theoretical approach shows that, far from T_c, the shape of the spectral line is always the same, but when one gets closer to T_c, it *might* change for two reasons as it was shown first by G. B. Benedek,[1] and later by Cummins, et al.[2,3] and by F. T. Arecchi, et al.[4]

1. The relaxation of concentration fluctuations might have a new dependence in time.

2. The statistics of the scattered field might introduce an alteration of the spectral line, if this is defined by self-beating technique, independently of the problem which is studied.

So, the first question we will discuss now is the determination of the shape of the spectral line.

1. SHAPE OF THE SPECTRAL LINE

If we solve with the usual approximations the well-known diffusion equation, we find that the relaxation of concentration fluctuations is exponential, with a characteristic decay time depending on K and $(T - T_c)$. In these conditions, far from T_c, when the fluctuations are not strongly correlated, the spectrum of the scattered light corresponding to such time dependence will be Lorentzian, as well in self-beating technique as in heterodyne technique, and we can define, without any ambiguity, a line width Γ.

Furthermore we know that, the spectrum obtained by the heterodyne[5,6] technique gives directly the shape of the real spectrum of the scattered light whatever the statistical distribution of the scattered field might be; whereas the self-beating spectrum is the convolution of the real optical spectrum by itself, only if the statistical distribution of the scattered field is Gaussian.[1]

So, by means of the heterodyne technique, one can study directly the relaxation of concentration fluctuations whatever the reduced temperature $(T - T_c)/T_c$ might be.

We performed such experiments very close to T_c and we found that the spectra which we obtained in that way were Lorentzian as shown in Fig. 1. In fact on this diagram we fitted our spectra to a square root of a Lorentzian line, because we do not use a squarer in our experimental device. We can see that the agreement between the experimental spectrum and the theoretical square root of a Lorentzian line is very good. This result means that *the relaxation of concentration fluctuations is exponential up to 1 or 2 millidegrees from T_c*, and confirms the assumption made by Kawasaki on this subject in a recent paper.[7]

Furthermore, if we study the shape of the self-beating spectra, obtained in the same range of temperature and for the same scattering angle ($\theta = 170°$) we find that it is Lorentzian as shown in Fig. 2. The average line width of these spectra is $\Gamma = 1395 \pm 60$ Hz. For corresponding heterodyne spectra, we found an average line width $\Gamma = 649 \pm 50$ Hz. So we can see that the self-beating spectra

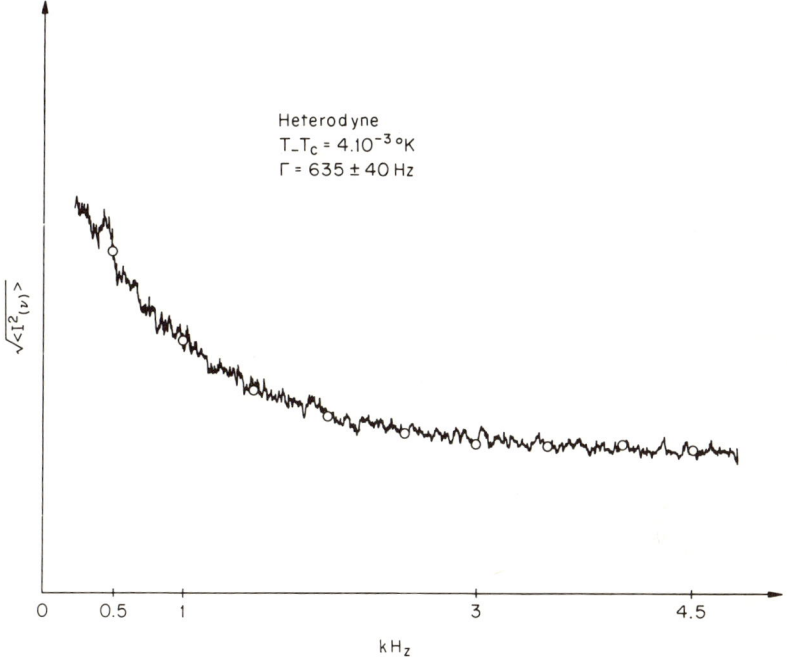

Fig. 1 *Heterodyne spectrum obtained at $4 \times 10^{-3}\,°K$ o: points corresponding to the computed Lorentz line.*

are Lorentzian and are, within our experimental precision, systematically twice broader than those obtained by heterodyne technique (Fig. 3). The conclusion of this result is that the self-beating spectra *are the convolution* of the corresponding heterodyne spectra by themselves. That means that the statistical distribution of the scattered field *seems to be Gaussian*.

This last point could be criticized because perhaps the photon-beating technique is not the proper experimental way to study the statistical distribution of the scattered field.

Anyway, the main point is that *the shape of the self-beating spectra is always Lorentzian even at 1 or 2 millidegrees from T_c*.

2. DYNAMICS OF CONCENTRATION FLUCTUATIONS

One knows that (cf. H. Z. Cummins and H. L. Swinney[8]) the thermal fluctuations in fluids near a critical point will give an optical spectrum which is Lorentzian and the line width Γ of which is given by

$$\Gamma = DK^2(1 + \xi^2 K^2)^{1-\eta/2} \tag{1}$$

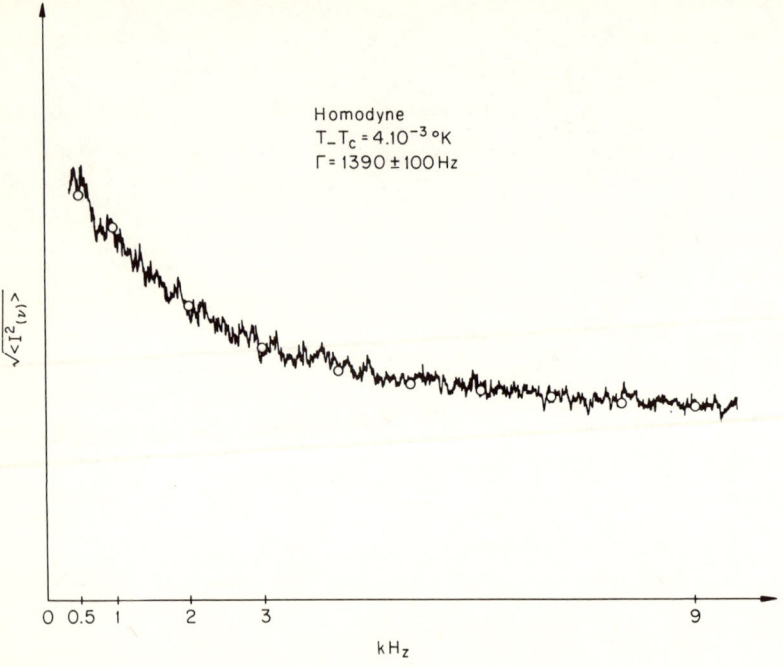

Fig. 2 *Self-beating spectrum obtained at 4×10^{-3} °K o: points corresponding to the computed Lorentz line.*

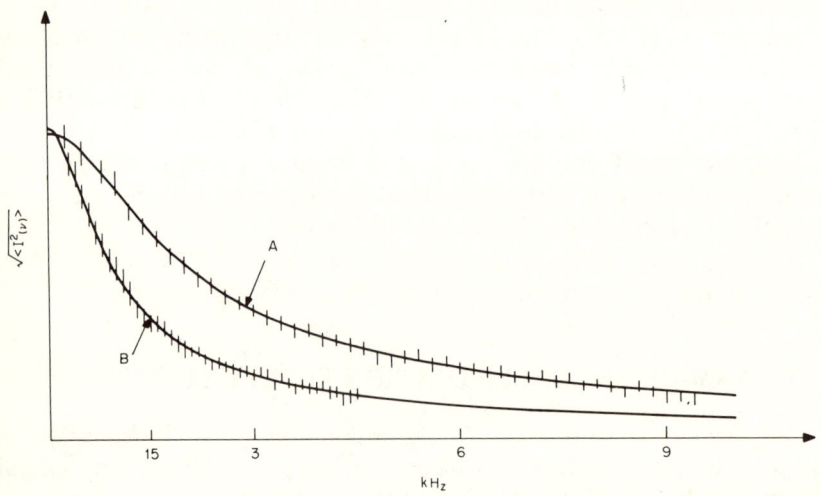

Fig. 3 *(a) Mean value of all self-beating spectra. (b) Mean value of all heterodyne spectra with the same scale in frequency.*

where D is a diffusion constant, K the scattering vector defined by $|K| = (4\pi/\lambda_0)n \sin\theta/2$, and ξ is the correlation length defined by the asymptotic form $(r \to \infty)$ of the modified Ornstein-Zernike formula

$$g(r) = \frac{e^{-r/\xi}}{r^{1+\eta/2}}$$

η being small compared to unity.

We have checked this formula as a function of temperature very carefully in the case of a binary mixture of cyclohexane-aniline.

For small values of vector K ($\theta = 10°12'$) we find a reasonable straight line with a slope equal to 0.6 as shown in Fig. 4; in this case, Eq. (1) reduces itself to $\Gamma = DK^2$ which is the classical hydrodynamical equation. This study allowed us to determine the dependence of the diffusion coefficient as a function of temperature. This dependence is given by

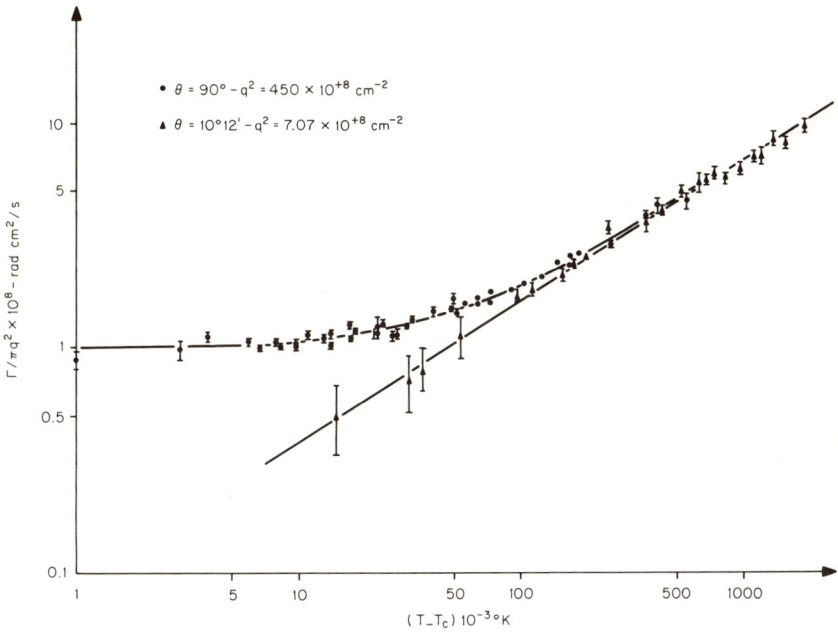

Fig. 4 *Dependence of Γ/K^2 vs. $T-T_c$ (log-log plot); the filled circles correspond to a scattering angle of $\theta = 90°$, the triangles correspond to a scattering angle of $\theta = 10°$. The limits of errors on temperature (not represented here) are equal to 3/1000 of a degree for almost all points.*

$$D = D_0 \left(\frac{T - T_c}{T_c}\right)^{0.6}$$

with $D_0 = 7.7 \times 10^{-6}$ cm^{2}s^{-1}.

For greater values of vector $K(\theta = 90°)$ Fig. 4 shows a great divergence from the former law, which corresponds to the second term of Eq. (1).

If we look closer at Fig. 4, we notice immediately that the line width Γ seems to be independent of temperature as soon as we get closer to T_c. This peculiar behavior is not predicted by Eq. (1) and will be examined.

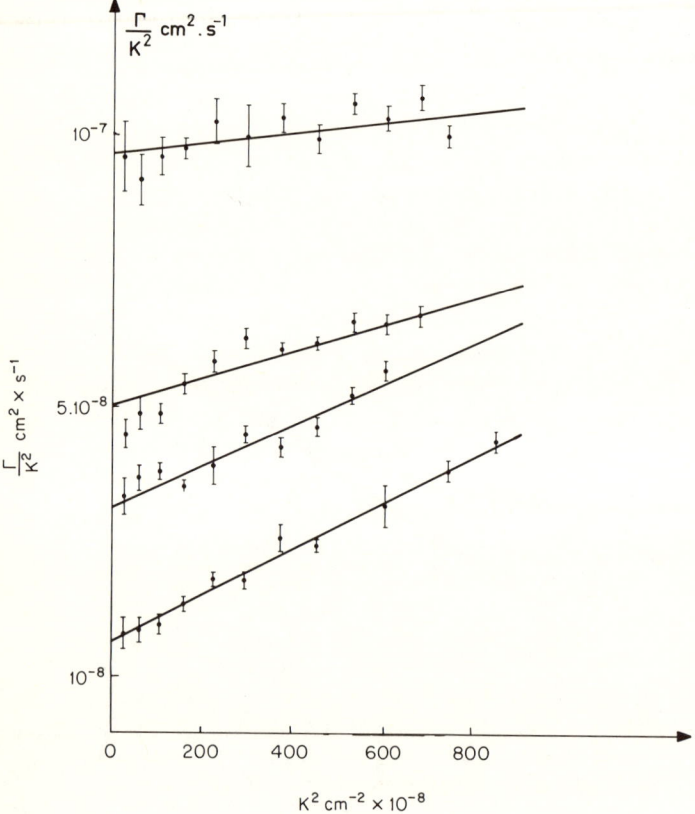

Fig. 5 *Dependence of Γ/K^2 vs. K^2 for different temperatures. The values of $T-T_c$ are (from the lower curve upward): 4×10^{-3}, 69×10^{-3}, 121×10^{-3}, 282×10^{-3} °K.*

The next step was a systematic study of the dependence of Γ as a function of the scattering angle, i.e., the scattering vector $\mathbf{K}$. Figure 5 shows some results of this study. The system of straight lines obtained gave, by comparison with Eq. (1) for D and ξ^2, under the condition that $K\xi < 1$, the following results

$$D = (7.7 \pm 0.7) \, 10^{-6} \left(\frac{T - T_c}{T_c}\right)^{0.61 \pm 0.07} \text{cm}^2 \, \text{s}^{-1} \qquad (2)$$

and

$$\xi^2 = \xi_0^2 \left(\frac{T - T_c}{T_c}\right)^{-1.21 \pm 0.05} \text{cm}^2 \qquad (3)$$

with $\xi_0 = 1.65 \pm 0.07$Å. The point-by-point computation of ξ^2 from an experiment performed at $T - T_c = 4 \times 10^{-3}$°K (lower straight line of Fig. 5), using the value of D given by Eq. (2), shows that ξ^2 *depends on K as soon as $K\xi > 1$*; only values obtained for $K\xi \leq 1$ are in a good agreement with those given by Eq. (3).

The conclusion is that Eq. (1) is no longer valid, as soon as $K\xi \simeq 1$. So it was worthwhile to study systematically the behavior of Γ for great values of $K\xi$, i.e., very close to T_c in the region of the Fig. 4 where Γ seems to be temperature independent. Figure 6 shows the results of such a study, consistent with a law such as $\Gamma = AK^3$, A being a coefficient independent of temperature. This last diagram shows also the data obtained at $T - T_c = 12 \times 10^{-3}$°K. At this temperature the value $K\xi = 1$ is attained for a scattering angle equal to 70 degrees; thus, this line represents the transition between the "critical" and the "hydrodynamical" regions.

The last diagram allowed us to determine the constant A as being equal to

$$A = 1.52 \times 10^{-13} \, \text{cm}^3 \, \text{s}^{-1}$$

This "critical" behavior is justified by a theoretical study developed by Halperin and Hohenberg on dynamical scaling laws.[9]

Recently Kawasaki[7] has developed a mode-mode coupling theory according to which the spectral line width Γ is given by

$$\Gamma = AK^3 F\left(\frac{1}{K\xi}\right)$$

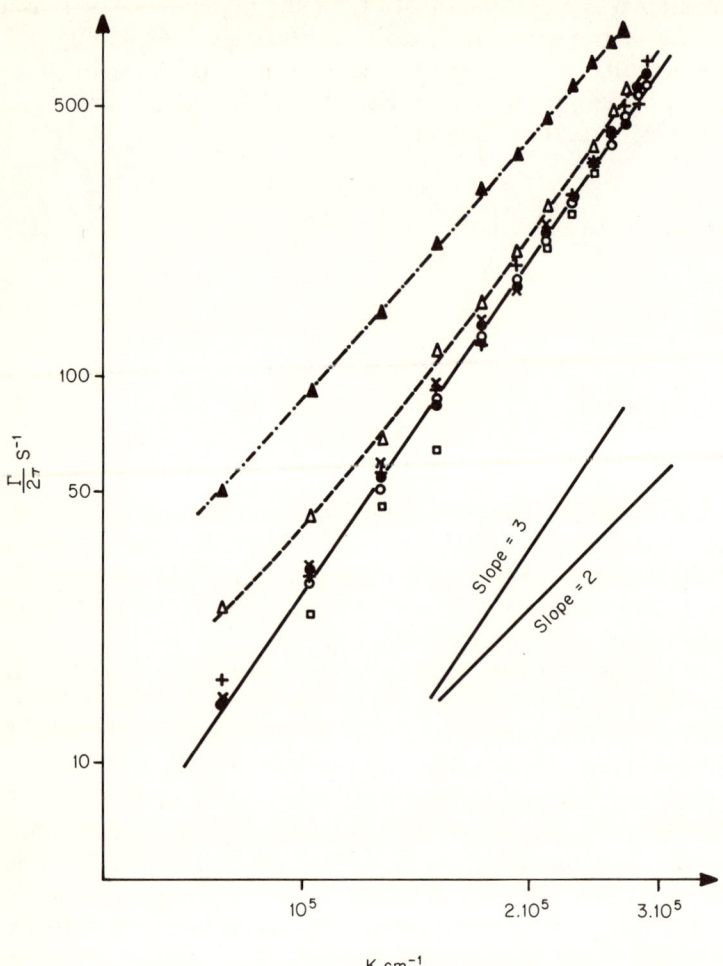

Fig. 6 *Angular dependence of the spectral width Γ vs. $T-T_c$. Closed triangles, 121×10^{-3} °K; open triangles, 12×10^{-3} °K; closed circles, 6×10^{-3} °K; crosses, 5.5×10^{-3} °K; plusses, 4×10^{-3} °K; open squares, 3.5×10^{-3} °K; open circles, 1.5×10^{-3} °K.*

with $A = (3\pi/8) D\xi = k_B T/16\eta^*$ (η^* being the high frequency part of the viscosity) and $F(1/K\xi)$ is an analytical function of the form

$$F(x) = \frac{2}{\pi}\left[x + x^3 + (1 + x^4)\,\mathrm{arctg}\,\frac{1}{x}\right]$$

In the hydrodynamical region $K\xi \ll 1$ so that

$$F(x) = \frac{8x}{3\pi}$$

therefore

$$\Gamma = \frac{8A}{3\pi\xi} K^2 \equiv DK^2$$

In the nonlocal hydrodynamical region $K\xi < 1$ and one obtains

$$\Gamma = DK^2 \left(1 + \frac{3}{5} K^2 \xi^2\right)$$

This last formula differs from Fixman's equation because of the 3/5 factor and of the exponent $1 - \eta/2$.

Finally, in the critical region $K\xi \geq 1$, $F(x) \to 1$ and one gets

$$\Gamma = AK^3$$

Thus, Kawasaki's mode-mode coupling theory describes *continuously* the behavior of a binary mixture whatever the reduced temperature might be.

Fisher's critical exponent η is taken equal to zero and the high frequency part η^* of the viscosity is supposed temperature independent.

Quite recently Ferrell[10] obtained the same formula for the K dependent diffusion coefficient by calculating the critical slowing down of the diffusion in a binary mixture from the fluctuation-dissipation theorem.

So it was worthwhile to compare our experimental data obtained on the cyclohexane-aniline mixture for different temperatures and scattering angles to Kawasaki's equation giving Γ.

This fit was made using a statistical refining program developed by M. Tournarie[11] and giving a plot of Γ/AK^3 as a function of $[K\xi_0 (T - T_c/T_c)^{-\nu}]^{-1}$ where Γ, K, and $(T - T_c)/T_c$ were known experimentally, and A, ξ_0, and ν were determined by the computer in order to get the best fit to Kawasaki's function $F(1/K\xi)$.

Figure 7 shows the obtained result. The 262 experimental points reported on this diagram were obtained using different scattering

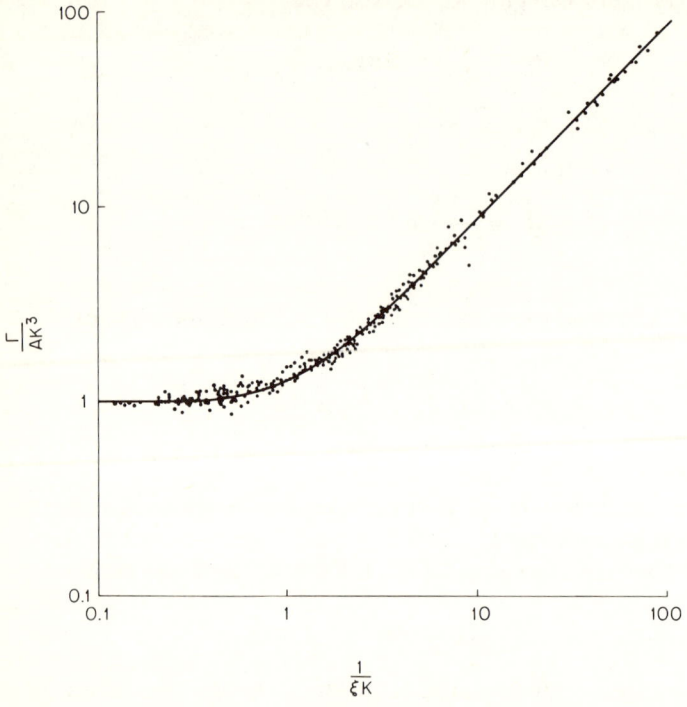

Fig. 7 Plot of Γ/AK^3 vs. $1/\xi K$. The solid line represents the best fit of a function having Kawasaki's form.

cells under different conditions. The plain line represents the best fit to these points of a function having Kawasaki's form. The best-fit curve is obtained for the following values of A, ξ_0, and ν

$$A = 1.51 \pm 0.02 \; 10^{-13} \, \text{cm}^3 \, \text{s}^{-1}$$

$$\xi_0 = 2.11 \pm 0.08 \; 10^{-8} \, \text{cm}$$

$$\nu = 0.588 \pm 0.06$$

These values, except maybe for ν, are in good agreement with those obtained directly by considering the previous three regions, using Fixman's equation, and which are respectively

$$A = 1.52 \pm 0.03 \; 10^{-13} \, \text{cm}^3 \, \text{s}^{-1}$$

$$\xi_0 = 1.65 \pm 0.08 \; 10^{-8} \, \text{cm}$$

$$\nu = 0.61 \pm 0.07$$

Notice that this last value of ξ_0 must be multiplied by $\sqrt{5/3}$ in order to be compared to that deduced from Kawasaki's formula. This gives the value $\xi_0 = 2.14 \times 10^{-8}$ cm which is in good agreement with the previous one.

Using Kawasaki's formula $A = k_B T/16\eta^*$ and our experimental value $A = 1.51 \times 10^{-3}$ cm^3 s^{-1}, we calculated η^* and obtained

$$\eta^* = 1.97 \times 10^{-2} \text{ stokes}$$

which is in excellent agreement with the static value determined by Arcovito, et al.[12] using a capillary flow viscosimeter. The found $\eta \simeq 1.9 \times 10^{-2}$ stokes.

In conclusion, Kawasaki's formula gives a complete and accurate description of the behavior of the spectrum of the scattered light of a binary mixture in the hydrodynamical and nonhydrodynamical regimes. Furthermore the assumption of an exponential relaxation of concentration fluctuations, even very close to T_c when the fluctuations are strongly correlated, is confirmed by the experimental data. However the comparison of experimental data to Kawasaki's function is perhaps not the best way to determine the critical exponent ν.

References 13 to 19 are the main publications strictly related to this experimental work, whereas references 20 to 26 represent a short list of theoretical papers which we used in the interpretation of this experimental work.

REFERENCES

1. Benedek, G. B.: Lectures at the Brandeis Summer Institute, Gordon & Breach Science Publishers, Inc.,
2. Cummins, H. Z.: Summer School of Varenna, 1967.
3. —— and H. L. Swinney: Prog. in Optics, VIII (1969).
4. Arecchi, F. T., M. Giglio, and U. Tartari: *Phys. Rev.*, 163:1, 186 (1967).
5. Forrester, A. T.: *J. Opt. Soc. Am.*, 51:253 (1961).
6. Adam, M., A. Hamelin, and P. Berge: *Optical Acta*, 16:3 (1969).
7. Kawasaki, K.: (To be published); also *Phys. Lett.*, 30A:325 (1969).
8. Cummins, H. Z., and H. L. Swinney: *J. Chem. Phys.*, 45:4438 (1966).
9. Halperin, B. I., and P. C. Hohenberg: *Phys. Rev. Lett.*, 19:700 (1967).
10. Ferrell, R. A.: *Ibid.*, 24:1169 (1970).
11. Tournarie, M.: *J. Physique*, 30:737 (1969).
12. Arcovito, G. et al.: *Phys. Rev. Lett.*, 22:1040 (1969).
13. Berge, P., and B. Volochine: *Phys. Lett.*, 26A:267 (1968).
14. ——, P. Calmettes, and B. Volochine: *Ibid.*, 27A:637 (1968).
15. ——, ——, ——, and C. Laj: *Ibid.*, 30A:7 (1969).
16. ——, ——, ——, and ——: *Phys. Rev. Lett.*, 23:13 (1969).
17. ——, ——, C. Laj, M. Tournarie, and B. Volochine: *Ibid.*, 24:1223 (1970).

18. Volochine, B. P. Berge, and I. Lagües: *Phys. Rev. Lett.*, to be published.
19. ——, and ——: *J. Physique,* to be published.
20. Chu, B., F. J. Schoenes, M. E. Fisher: *Phys. Rev.,* 185:219 (1969).
21. Fixman, M.: Pontifica Academia Scientiarum Scripta Varia 31 on the Molecular Forces, p. 329, (1966).
22. Felderhof, B. U.: *J. Chem. Phys.,* 44:602 (1966).
23. Botch, W., and M. Fixman: *J. Chem. Phys.,* 42:196 (1965).
24. Fisher, M. E.: *J. Math Phys.,* 5:944 (1964).
25. Swift, J.: *Phys. Rev.,* 173:257 (1968).
26. Kadanoff, L. P., et al.: *Rev. Mod. Phys.,* 39:395 (1967).

DISCUSSION *on Paper by B. Volochine*

The discussion of this paper was combined with the discussion of the preceding and the next papers (see p. 559).

ULTRASONIC MEASUREMENTS NEAR THE LIQUID-GAS CRITICAL POINT

M. Barmatz
Bell Telephone Laboratories
Murray Hill, New Jersey

ABSTRACT

Recent low frequency measurements of sound velocity near the He^4 critical point are presented and compared to the theoretical predictions of the linear model parametric equation of state. The effect of gravity on measurements of the sound velocity and specific heat is calculated in detail and found to be consistent with the best available experimental data. Low frequency dispersion measurements agree with the temperature dependence of a recent mode-mode coupling calculation of Kawasaki.

1. INTRODUCTION

Ultrasonic measurements near the liquid-gas critical point are a sensitive tool for investigating the nature of static and dynamic critical phenomena. The ideal behavior of the sound velocity near the transition is obscured experimentally for two reasons: (1) Dispersion effects become important for measurements at finite frequency, and (2) gravity produces a significant density gradient in a sample of finite height. Both of these effects result in a measured velocity which is higher than the zero frequency velocity in a homogeneous system. In this article, we examine in detail the influence of a gravitational field on ultrasonic measurements near the critical point.

The isentropic or zero frequency sound velocity u and the adiabatic compressibility κ_s, for a homogeneous system, are obtained from the thermodynamic expression

$$u^2 = (\rho \kappa_s)^{-1} = (\rho \kappa_T)^{-1} + \left(\frac{\partial P}{\partial T}\right)_v^2 \left(\frac{T}{\rho^2 C_v}\right) \qquad (1)$$

where κ_T is the isothermal compressibility and C_v is the specific heat at constant volume. It is assumed that C_v diverges as $|t|^{-\alpha}$, $t = (T - T_c)/T_c$. As the transition is approached along the critical isochore, $(\rho \kappa_T)^{-1}$ goes to zero, $(\partial P/\partial T)_v$ approaches a constant, and therefore $u \propto C_v^{-1/2} \to 0$. In Sec. 2, we present velocity measurements[1] for He^4 along several isochores above the transition, which were performed in a specially designed cylindrical resonator of small vertical height. Except very near the transition, these low frequency measurements yield the isentropic sound velocity.

Considerable progress has been made in our understanding of static critical phenomena with the advent of scaling laws which relate the various critical indices. Recently, a linear model parametric equation of state[2] was proposed which requires a small number of parameters and has convenient expressions for thermodynamic properties near the critical point. In Sec. 3, this model is tested in the region where gravity effects are negligible, using the critical exponent α as the only adjustable parameter. The jump in the specific heat $C_v(-t) - C_v(t)$ calculated from the model is used to fix this parameter. We perform such a fit on He^4 and Xe, and find the values $\alpha = 0.07$ and $\alpha = 0.05$ respectively. We also calculate the velocity in the whole (ρ, T) plane in He^4, using $\alpha = 0.07$, and find agreement with low frequency measurements for temperatures not too close to T_c.

Quantitative calculations of the gravity effect on measurements of u and C_v are carried out in Sec. 4, using this linear model equation of state.[3] Except very near the transition the average specific heat $\tilde{C}_v$ is greater than its gravity free value for $T > T_c$ and less for $T < T_c$. This fact may account for the apparent inequality $\alpha > \alpha'$ obtained in many previous experimental analyses using power law fits. Numerical calculations of $\tilde{C}_v$ in Xe correctly predicts a maximum at a temperature *below* T_c and accurately reproduces the data of Edwards, Lipa, and Buckingham[4] within experimental error. The average sound velocity measured in a resonance experiment is determined from a calculation of the normal modes of a cylindrical resonator. We describe a method for the numerical solution of the wave equation[5] in an inhomogeneous medium and discuss the mode dependence of the gravity corrections. For measurements along the critical isochore, the minimum in the isentropic sound velocity is predicted to occur *above* the transition at a temperature which depends on the resonator mode and sample height. Calculations for the lowest order radial mode give good agreement with the 1.8 KHz velocity data in He^4.

In recent years, dynamic scaling[6] and mode-mode coupling theories[7,8] have made explicit predictions concerning the divergences of transport coefficients at the critical point. In particular, Kawasaki[9] has calculated in detail the behavior of the sound attenuation and dispersion in the hydrodynamic and nonhydrodynamic regimes. In Sec. 5, this theory is compared to ultrasonic measurements in He^4.

2. ULTRASONIC MEASUREMENTS

The singularity in the isentropic sound velocity was investigated near the He^4 critical point using a resonant technique. The sound velocity in the frequency range 1.5–50 KHz was measured in a cylindrical resonator, 0.49 cm high and 5 cm in diameter. The resonant frequencies f of a homogeneous fluid, in a cylindrical resonator of height h and radius a, are given by

$$f_{pmn} = \frac{1}{2} u \left[\left(\frac{p}{h}\right)^2 + \left(\frac{\alpha_{mn}}{a}\right)^2 \right]^{1/2} \tag{2}$$

where p is a nonnegative integer and α_{mn} is a solution of the boundary condition $d[J_m(\pi\alpha)]/d\alpha = 0$, where $J_m(\pi\alpha)$ is a Bessel function of the first kind. This relation is only valid for a homogeneous sample; in Sec. 4, we discuss solutions of the wave equation for an inhomogeneous medium. The plane wave modes ($m = n = 0$) correspond to wave motion which depends on the vertical coordinate only. For these modes, the integer p designates the harmonic. When the motion is independent of the vertical coordinate ($p = 0$) there are two types of pure modes, the symmetric or radial modes ($p = m = 0$) and the antisymmetric modes ($p = n = 0$). For these nonplane-wave resonances, the integers m and n represent the number of nodal diameters and nodal circles.

The resonator was terminated at both ends with identical electrostatic transducers. Figure 1a shows the backplate of these (specially designed) transducers. A transducer is formed by stretching a 0.00063 cm thick sheet of Mylar or Kapton across the backplate. The solid dielectric is metallized on one side and acts as a movable plate of a parallel plate capacitor. Transducers of this type have previously been used to investigate the λ transition in He^4.[10] The backplate is composed of four electrically isolated segments, each capable of functioning as a separate electrostatic drive. Various normal modes are excited depending on the polarities applied to the segments; the polarity arrangement shown in the figure will generate the lowest order radial mode which has one nodal circle. The

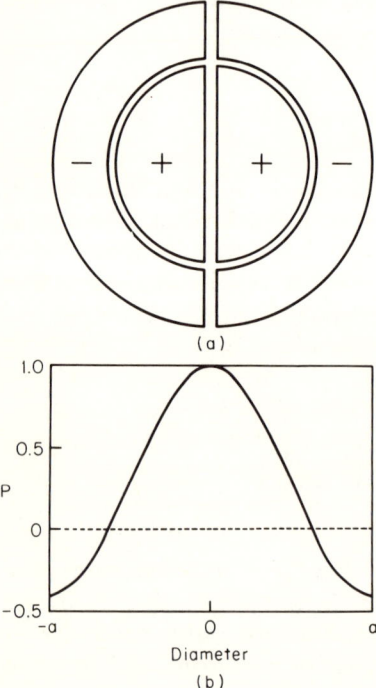

Fig. 1 *(a) Electrostatic transducer. Segments of backplate may be individually polarized. Polarity arrangement corresponds to lowest order radial mode. (b) Pressure distribution for radial mode.*

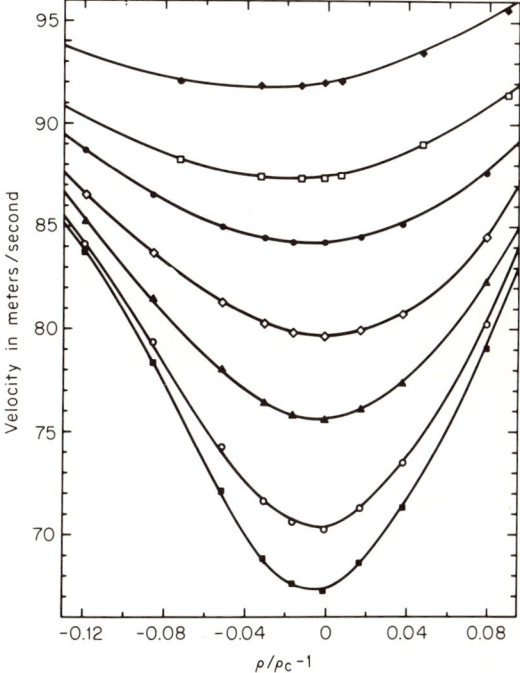

Fig. 2 *Velocity measurements along isochores in the He^4 critical region. The solid lines represent velocity profiles along isotherms. The symbols ◆, □, ●, ◇, ▲, ○, and ■ correspond to isotherms with $t =$ 1.1×10^{-2}, 8.2×10^{-3}, 3.9×10^{-3}, 1.9×10^{-3}, 9.5×10^{-4}, 3.4×10^{-4}, and 1.6×10^{-4}, respectively.*

corresponding pressure distribution along any diameter is shown in Fig. 1*b*. Near the He^4 critical point this mode had a frequency of ~ 1.8 KHz. These transducers can also generate the plane wave modes if all backplate segments have the same polarity.

Low frequency velocity measurements along several isochores in the He^4 critical region are shown in Fig. 2. These measurements have a relative accuracy better than 0.1 percent and an absolute accuracy of 0.4 percent. For all isochores, the velocity was measured at the same absolute temperatures. The solid lines drawn through the data at each temperature represent velocity profiles along isotherms. The farthest isotherms from T_c have velocity minima at densities which are less than the critical value ρ_c. As $T \to T_c$ the velocity minima asymptotically approach the same density which was taken to be ρ_c. In the next section we compare these data with the linear model parametric equation of state.

3. LINEAR MODEL EQUATION OF STATE

A comprehensive analysis of thermodynamic properties near the critical point requires a knowledge of gravity-induced gradients which are present in any experimental cell of finite height. At this time, direct measurements of density profiles are not sufficient to give a quantitative description of the temperature variation of these gradients. However, the influence of gravity can be calculated from recently proposed scaled equations of state. The expression for the chemical potential used by Vicentini-Missoni, et al.[11] is suited for least square fit analysis of P.V.T. data and has been successful in describing thermodynamic properties in fluid and magnetic systems. However, this equation of state does not satisfy all analyticity requirements[12] and does not give closed form expressions for the specific heat and velocity in the entire critical region. On the other hand, the parametric equation of state considered by Josephson[13] and Schofield[14] satisfies all analyticity requirements and has the desirable feature of having simple closed forms for thermodynamic functions near the critical point.

Calculations of the gravity effect in fluids are presented using a linear model of the parametric equation of state[2] which is entirely determined by two critical indices and a few scaling parameters. In this model the asymptotic form of thermodynamic functions is given by $g(\theta)r^{-z}$, where z is the appropriate critical index. The parameter r is a measure of the "radial distance" from the critical point and θ represents an "angle" along contours of constant r. The equation of state may be described by the transformation

$$\frac{[\mu(\rho,T) - \mu_0(T)]}{\mu_c} \equiv a\theta(1 - \theta^2)r^{2-(\alpha+\beta)} \tag{3a}$$

$$\frac{(T - T_c)}{T_c} \equiv t \equiv (1 - b^2\theta^2)r \tag{3b}$$

$$\frac{(\rho - \rho_c)}{\rho_c} \equiv k\theta r^\beta \tag{3b}$$

where μ_c, T_c, and ρ_c are respectively the critical values of the chemical potential μ, the temperature T, and the density ρ. In this paper, we use the usual notation[12] for the critical exponents; thus α and β are associated respectively with the divergence of the specific heat C_v and order parameter $(\rho - \rho_c)/\rho_c$, and other critical exponents may be obtained from the scaling relations $2 - (\alpha + \beta) = \beta\delta = \gamma + \beta$. The linear model parameters $a\mu_c$ and k are determined from the experimental coefficients[2,11] B and C given by $(\rho - \rho_c)/\rho_c = B|t|^{-\beta}$ and $K_T = Ct^{-\gamma}$. The quantity b is obtained from the "minimization condition" $b^2 = (\delta - 3)/(\delta - 1)(1 - 2\beta)$. This minimization relation has not been theoretically justified, but it is consistent with presently available experimental data and the linearity assumption between the order parameter and θ (see Eq. (3c)). In our application of the linear model, we shall take β from the experiment and leave α as an undetermined parameter.

The Helmholtz free energy per unit volume $a(\rho, T)$ is determined by integrating the relation $\mu = (\partial a/\partial \rho)_T$ and has the form

$$a(\rho, T) = F(\rho, T) + \mu_0(T)\rho + a_0(T) \tag{4}$$

where $F(\rho, T) \sim r^{2-\alpha}$ is the singular part of a and $\mu_0(T)$ and the integration constant $a_0(T)$ are analytic functions of T with $\mu_0(T_c) = \mu_c$. All thermodynamic properties can now be derived from this form of the free energy. For $\rho = \rho_c$ the singular behavior in the specific heat in the region where gravity effects are negligible is given by[2]

$$\frac{\rho C_v}{T} = -\left(\frac{\partial^2 a}{\partial T^2}\right)_\rho = A^{\gtrless}|t|^{-\alpha} + C_0, \quad t \gtrless 0 \tag{5}$$

where the constants $A^{\gtrless}$ are determined from the scaling parameters and critical exponents. Simple scaling theories assume that C_0 is a regular function of t and thus the only discontinuous term in C_v is $A^{\gtrless}|t|^{-\alpha}$, higher order terms being neglected. Using this assumption, the difference $C_v(-t) - C_v(t)$ in the gravity free region depends only on $A^{\gtrless}$ and α. This "jump," calculated from the linear model (or other scaling theory[11]) is sensitive to the value of α and can be used to fix this parameter. Specific heat measurements[15,4] in He^4 and Xe^4 which have been analyzed in this way, yield the values of $\alpha = 0.07$ and $\alpha = 0.05$ respectively.

The isentropic sound velocity was calculated in the entire He^4 critical region using Eq. (1) and $\alpha = 0.07$. Figure 3 shows a

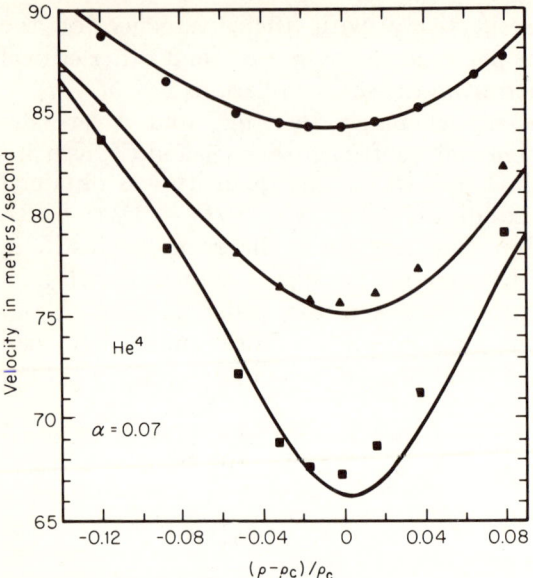

Fig. 3 *1.8 KHz sound velocity vs. reduced density in He^4. The symbols represent the same isotherms as in Fig. 2. The zero frequency velocity calculated from the linear model without gravity corrections is shown by the solid lines.*

comparison between the theory and experimental measurements at ~ 1.8 KHz. The constants $d^2 a_0/dT^2$ and $d^2 \mu_0/dT^2$ were adjusted to fit the data at $t = 3.9 \times 10^{-3}$. Agreement is satisfactory for measurements not too close to T_c. At $t = 9.3 \times 10^{-4}$, however, the small discrepancies for $\rho > \rho_c$ are probably not within experimental error, but must be investigated further before they may be attributed to a breakdown of the theory. In the region very near the transition, gravity and dispersion corrections become important and these will be discussed in more detail in Sec. 4.3 and 5 below. Using the same parameters, we also find similar agreement with the He^4 specific heat measurements of Moldover.[15]

Previous ultrasonic studies in the two-phase region indicate a rapid decrease in the velocity of the liquid and saturated vapor as the critical point is approached. In several investigations,[16] the velocity curves of the liquid and vapor were observed to intersect near the transition. The linear model predicts such a crossing; however, the crossover temperature is sensitive to many parameters of the model. It should be noted that the "jump," $u(-t) - u(t)$, between velocity measurements along the coexistence curve and critical isochore may also be used to determine the parameter a of the linear model. This approach, however, must await precision low frequency measurements in the two-phase region.

4. GRAVITY EFFECT

4.1. Theory

In a gravitational field, the divergence of the isothermal compressibility near the critical point produces a significant vertical density gradient, or equivalently a gradient in the chemical potential. The linear model equation of state was shown in the previous section to be consistent with measurements of the sound velocity and specific heat in the region where gravity effects are negligible. We now use this model to determine the local values of these thermodynamic functions in a gravitational field. For simplicity the discussion is limited to samples of uniform cross section. In calculating the gravity average of thermodynamic properties, we assume that extensive variables, such as the Helmholtz free energy, are the sum of contributions from local volume elements. We neglect nonlocal effects which are only important when the correlation length ξ becomes comparable to the height. This is a good approximation for most static experiments since $\xi \simeq 10^{-4}$ cm at $t = 10^{-6}$. The situation is quite different for transport phenomena such as sound attenuation and dispersion which is discussed in Sec. 5.

In an isothermal fluid, the variation of the chemical potential μ with vertical height z is given by the relation $d\mu = -g\,dz$, where g is the acceleration of gravity. The origin of z is chosen at the height where $\rho = \rho_c$, or equivalently $\mu = \mu_0(T)$. By integrating this relation and combining it with Eq. (3a) we obtain

$$z = -h_0\, a\theta(1 - \theta^2) r^{2-(\alpha+\beta)} \tag{6}$$

where $h_0 \equiv \mu_c/g$. The local values of any thermodynamic function can now be derived using $\theta(z, t)$, obtained from Eqs. (6) and (3b) and $r = t/(1 - b^2\theta^2)$. For the case of a cylindrical sample of height h, the exact position of the origin, $z = 0$, is obtained from the relation

$$\tilde{\rho} = \int_{-z_1}^{h-z_1} \rho(z, t)\, dz \tag{7}$$

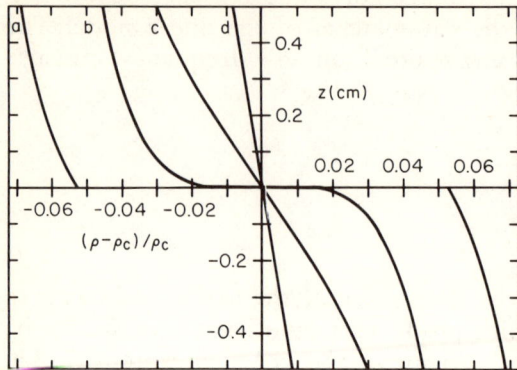

Fig. 4 *Density gradients near the critical point of He^4. The curves a, b, c, and d were calculated from the linear model and correspond to $t = -10^{-4}$, o, 3×10^{-3}, and 10^{-3}, respectively.*

where $\tilde{\rho}$ is the average density and $-z_1$ is the position of the bottom of the cell. For the most interesting case $\tilde{\rho} = \rho_c$, the symmetry inherent in the linear model requires the critical density to be at the center of the cylinder for all temperatures $t \geq 0$. The calculated density distribution for various temperatures near the He^4 critical point are shown in Fig. 4. At the critical temperature, the density gradient obeys the expression

$$\frac{(\rho - \rho_c)}{\rho_c} = -d_0 z^{1/\delta} \tag{8a}$$

where

$$d_0 = kb^{-1}[h_0 ab^{-1}(1 - b^2)]^{-1/\delta} \tag{8b}$$

Using Eqs. (6) and (3) we can estimate a characteristic temperature

TABLE 1. Values of the Constants d_o in CGS Units and t_o for Several Liquid-gas Critical Points ($h = 1$ cm).

	He^4	Ar	CO_2	Xe
d_o	5.35×10^{-2}	4.53×10^{-2}	4.54×10^{-2}	5.35×10^{-2}
t_o	3.44×10^{-4}	1.18×10^{-4}	6.39×10^{-5}	1.53×10^{-4}

range, in which gravity effects are important, as $t_0(h) \equiv (h/ah_0)^{1/\beta\delta}$. Values of d_0 and t_0 are listed in Table 1 for several liquid-gas transitions. It is interesting to note that while the maximum gradient at T_c is about the same, the temperature region affected by gravity varies over five-fold among these systems.

4.2. Specific Heat

In a specific heat measurement, the sample is heated at constant total volume; however, near the critical point a local volume element does not follow a path of constant density. Thus, as $T \to T_c$, the average specific heat $\tilde{C}_v$ is not the average of the local C_v, but is the derivative of the average entropy $\tilde{S}$.[4] Thermodynamic averages are calculated using the fact that a volume element at a height z follows a path of constant $\Delta\mu = \mu(T) - \mu_0(T)$. Thus for $\tilde{\rho} = \rho_c$ we have

$$\tilde{C}_v = T\left(\frac{d\tilde{S}(T)}{dT}\right)_{\tilde{\rho}} = T\frac{d}{dT}\int_{-h/2}^{h/2} S(z,T)\,dz \qquad (9)$$

where $S(z, T)$ is the local entropy. We can carry out the differentiation at constant z under the integral sign by expressing the entropy in the form $S[\theta(z, T), T]$. In Fig. 5, numerical calculations of $\tilde{C}_v$ from Eq. (9) are compared to the precision Xe measurements of Edwards, Lipa, and Buckingham,[4] which were obtained from heating experiments. The theoretical curve corresponds to $h = 1$ cm and $\alpha = 0.05$, and was normalized to the data at $t = 10^{-3}$. The authors quote variations of as much as 10 percent between heating and cooling curves in the region $|t| < 3 \times 10^{-4}$. We see that the theory reproduces the general shape, and correctly predicts the absolute value of $\tilde{C}_v$ at T_c and the position of the specific heat maximum, within experimental error. The maximum occurs *below* T_c at a temperature $t_M = -4.5 \times 10^{-5} \approx 0.3 t_0$. Theoretical curves of a similar shape are found in He^4, Ar, and CO_2 using appropriate parameters. For $T = T_c$, we can integrate Eq. (9) directly and find

$$\tilde{C}_v(T_c) \propto h^{-\alpha/\beta\delta} \qquad (10)$$

which is strongly dependent on α. In the Xe analysis, for example, $\alpha = 1/8$ gives $\tilde{C}_v(T_c) = 279$ Joules/mole°K, which is ~ 1.7 times larger than the experimental value.

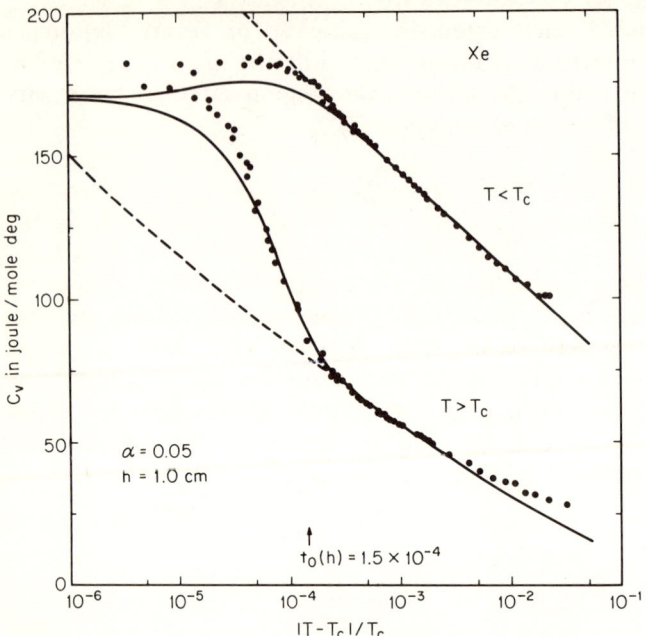

Fig. 5 *Temperature dependence of the specific heat in Xe for $\tilde{\rho} = \rho_c$. The solid lines are the gravity average $\tilde{C}_v$ and the dashed lines are the theoretical C_v in the absence of gravity, both calculated using the linear model. The points are measurements of Edwards, et al.*[4]

4.3. Sound Velocity

Low frequency velocity measurements in fluids are usually performed in cylindrical enclosures using a standing wave technique. In a gravitational field, the local velocity is a function of height and the normal modes of the resonator must be determined from the solution of the wave equation in an inhomogeneous medium. Let us consider a cylindrical coordinate system (r', θ', z') with its origin centered along the axis of a rigid cylindrical resonator of radius a and height h. The wave equation and boundary condition which the acoustic pressure P must satisfy are, respectively

$$\nabla^2 P - u^{-2}\left(\frac{\partial^2 P}{\partial t^2}\right) = 0 \qquad (11a)$$

and

$$\nabla P = 0 \tag{11b}$$

where ∇^2 is the Laplacian operator. In a homogeneous medium, where the local velocity is independent of the coordinates, the eigenvalues or normal mode frequencies are those given in Eq. (2). For the case of a gravity induced velocity gradient, we determine the local velocity $u(z')$ from the linear model equation of state and numerically calculate the resonant frequencies.

In solving the wave equation it is convenient to use the product-type solution

$$P(r', \theta', z') = \Phi(\theta') R(r') Z(z') e^{-i\omega t} \tag{12}$$

After substitution into Eq. (11a) and separation of variables, we obtain the following second order differential equation in $Z(z')$

$$\frac{d^2 Z}{dz'^2} - \left[\frac{A_{mn}^2 - 2\pi \tilde{f}_{pmn}}{u^2(z')} \right] Z = 0 \tag{13}$$

where $A_{mn} = \pi \alpha_{mn} a$ and $\tilde{f}_{pmn}$ are the normal mode frequencies in the presence of gravity. This equation may be solved numerically by reducing it to the solution of several first order differential equations using a technique recently developed by Dr. E. Wassertrom.[5] This method, which may also be used for solving the wave equation in geometries other than cylindrical, is outlined below.

Consider the general eigenvalue problem

$$L_i f_i(z') = \lambda_i f_i(z') \tag{14}$$

where L_i is a linear self-adjoint operator and λ_i and $f_i(z')$ are its real eigenvalue and eigenfunction respectively. The subscript i denotes one of the normal modes of the resonator. Let L_i belong to a one parameter family of operators $L(s)$ such that for fixed s, the s dependent eigenfunction $f(z', s)$ and eigenvalue $\lambda(s)$ satisfy the relation

$$L(s) f(z', s) = \lambda(s) f(z', s) \tag{15}$$

The transformation $L(s)$ is defined in such a way that

1. For $s = 0$, we have a trivial eigenvalue problem $L(0)f(z',0) = \lambda(0)f(z',0)$ whose eigenvalue and eigenfunction are known.
2. For $s = 1$, we have the original relation, Eq. (14), that is, $L_i \equiv L(1)$ and $\lambda_i \equiv \lambda(1)$. By taking the derivative of Eq. (15) with respect to s, and then the inner product of the resulting relation with $f(z', s)$, we arrive at the expression

$$\frac{d\lambda(s)}{ds} = G(s, \lambda) \qquad (16)$$

This is a first-order nonlinear, ordinary differential equation with an initial value $\lambda(0)$ and can be integrated from $s = 0$ to $s = 1$ using numerical integration programs. The eigenvalue $\lambda(1)$ corresponding to $s = 1$ is then the desired solution to the original expression Eq. (14). An important feature of this method is that the resulting first-order differential equation for a given mode is uncoupled with respect to the other modes. For example, the resonant frequency of the tenth plane wave mode may be calculated without previously solving for the first nine modes. Figure 6 shows the dependence of the gravity correction on the resonator mode. It is seen that the odd plane wave modes are more strongly affected by gravity than the even modes.

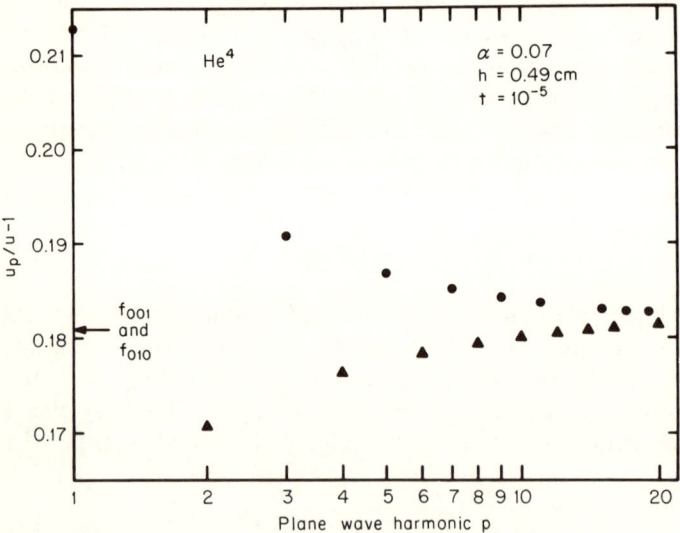

Fig. 6 *Gravity correction vs. resonator plane wave mode in He^4. The odd harmonics have a pressure node at the center of the cell and are most strongly affected by gravity. The fundamental mode has the largest gravity correction.*

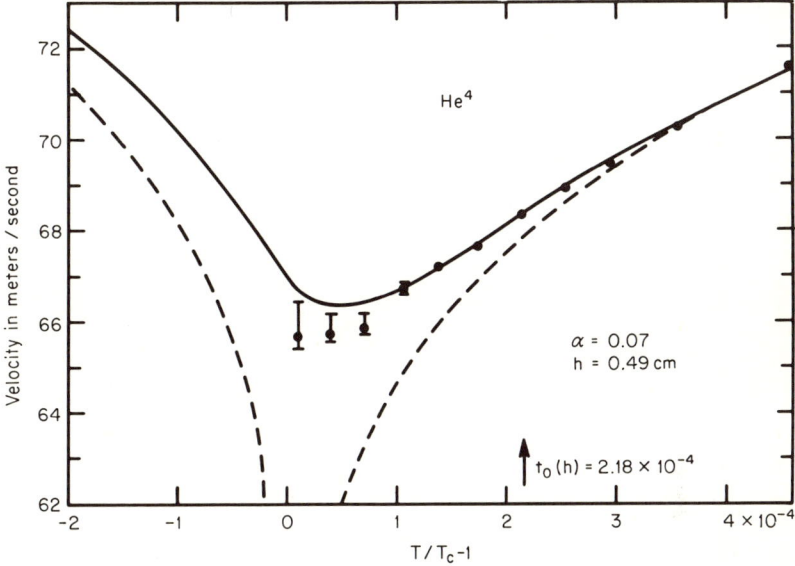

Fig. 7 *Temperature dependence of the sound velocity in He^4 for $\tilde{\rho} = \rho_c$. The solid circles are 1.8 KHz measurements. The solid curve is the gravity-averaged velocity $\tilde{u}$ and the dashed curve is the gravity-free velocity, calculated from the linear model. No dispersion corrections have been made.*

For higher harmonics ($p \gtrsim 50$), all modes of the resonator have essentially the same gravity correction. In Fig. 7, calculations for the lowest order radial mode ($\tilde{f}_{001}$) are compared with He^4 measurements for an isochore within 0.5 percent of ρ_c. Linear model parameters determined from measurements in the gravity-free region were used in the calculation. For $\alpha = 0.07$, the theory predicts a velocity minimum *above* T_c at a temperature $|t_m = 4.2 \times 10^{-5} \approx 0.2 t_0(h)$. A larger value of α decreases t_m and leads to smaller gravity corrections. In the region $t < 10^{-4}$, dispersion corrections are no longer negligible and will increase the disagreement between experiment and theory.

Sound velocity measurements along the critical isochore in He^4 were referenced to a transition temperature derived from the gravity analysis presented above. The adiabatic compressibility κ_s, determined from the radial mode over three decades in temperature, is shown in Fig. 8. A least squares fit to the relation

$$\kappa_s = \left(\frac{A}{\alpha}\right)[t^{-\alpha} - 1] + Bt + c \qquad (17)$$

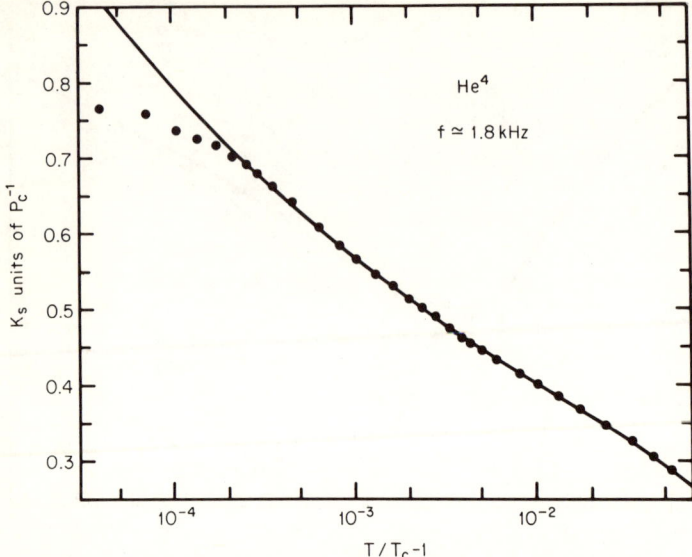

Fig. 8 *Adiabatic compressibility vs. reduced temperature for the critical isochore of He^4. The solid line represents a least squares fit to the 1.8 KHz data in the range $3 \times 10^{-4} < t < 6 \times 10^{-2}$.*

was performed on the data in the gravity-free region $3 \times 10^{-4} < t < 6 \times 10^{-2}$. The solid line is the best fit which corresponds to the parameters $A = 0.0301$, $B = -0.556$, $c = 0.211$, and $\alpha = 0.146$ with κ_s in units of P_c^{-1}. The maximum error in α is estimated to be 0.05, which includes the uncertainty in determining the temperature difference between the velocity minimum and T_c.

The CO_2 velocity measurements of Feke, Fritsch, and Carome[17] were derived from the fundamental plane wave mode of a cylindrical resonator 4.28 cm in height. Their analysis of the data was consistent with a logarithmic dependence in κ_s ($\alpha = 0$); however, gravity effects were not taken into account. We have reanalyzed these measurements in light of the present theory. A least square fit to the power law expression (Eq. (17)) in the same range as in Ref. 17 ($1.4 \times 10^{-4} < t < \times 10^{-2}$) yields $\alpha \approx 0.12$, which is similar to the He^4 results. The velocity measurements in He^4 and CO_2 lead to a critical exponent which is considerably larger than the value $\alpha = 0.07$ derived from the linear model. It must be noted, however, that different temperature intervals were used in these analyses. We find that in the range $t < 5 \times 10^{-3}$, the linear model fits the experimental velocity in He^4 and CO_2 and the specific heat in He^4 and Xe (see

Fig. 5, $T > T_c$). Outside this temperature interval, correction terms to a simple power law behavior seem to be necessary in order to explain experimental measurements.

In this section we have shown that the linear model self-consistently chooses the same exponent α in both the gravity-dominated and gravity-free regions. Further precision measurements, particularly as a function of height, are necessary to determine whether this model must be modified. If this is the case, the present method may still be applied to gravity calculations and the resulting critical exponent may differ from $\alpha = 0.06 \pm 0.02$ found in the present work.

5. DISPERSION

Near the critical point there are nonequilibrium processes which have characteristic relaxation times that diverge. The critical fluctuations associated with these processes lead to divergences in various transport coefficients. Theories based on the mode-mode coupling approach[7,8] have predicted the magnitude and form of these singularities. In particular, Kawasaki[9] has recently considered the contribution to sound attenuation and dispersion from a process in which the sound wave breaks up into two heat conduction modes. For the frequency-dependent velocity, he finds

$$u(\omega) - u(0) = \frac{k_B T^3}{2\pi^2 \rho u C_v^2} \left(\frac{\partial P}{\partial T}\right)_v^2 \kappa \left(\frac{\partial \kappa}{\partial T}\right)_s^2 I \qquad (18a)$$

where

$$I = \int_0^\infty \frac{x^2\, dx}{(1+x^2)^2} \left[1 + \left(\frac{k_B T}{3\pi\eta}\right)^2 \kappa^6 \omega^{-2} K^2(x)\right]^{-1} \qquad (18b)$$

$$K(x) = \tfrac{3}{4}[1 + x^2 + (x^3 - x^{-1})\tan^{-1} x] \qquad (18c)$$

$\kappa = \kappa_0 t^\nu$ is the inverse correlation range, and η is the shear viscosity.

Dispersion information is obtained experimentally from simultaneous measurements of the velocity for several modes of the resonator. Figure 9 shows the temperature-dependent velocity difference between the second and third plane wave modes and the

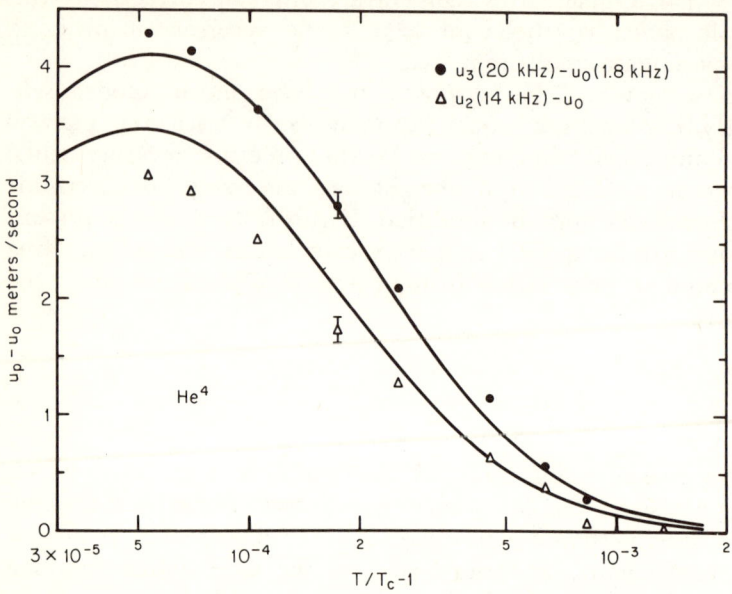

Fig. 9 *Dispersion vs. reduced temperature in He^4. The symbols ▲ and ● represent the velocity difference between the second and third plane wave modes and the lowest order radial mode, respectively. The solid lines are the theoretical prediction of Eq. (18) normalized to the higher frequency data at $t = 1.5 \times 10^{-4}$. Parameters used in the fit are $\alpha = 0.07$, $\kappa_o = 8.5 \times 10^7$ cm^{-1}, and $D = 0.38$.*

lowest order radial mode, for $\rho \approx \rho_c$. As we have seen in Fig. 6 the magnitude of the gravity correction depends on the resonator mode and thus leads to an apparent dispersion in the present case. The data in the range $t < 3 \times 10^{-4}$ were corrected for these gravity effects. Uncertainties in the difference between resonant frequencies at each temperature are indicated by the error bars. A comparable uncertainty in the background position of each set of measurements permits a vertical shift of the data which does not affect the temperature dependence. Equation (18) was compared to the higher frequency data of Fig. 9, using κ_0 and an overall multiplicative constant D as adjustable parameters, with $\alpha = 0.07$, $\nu = (2-\alpha)/3$ and experimental values for the thermodynamic derivatives. The solid lines, normalized to the higher frequency data at $t = 1.5 \times 10^{-4}$, represent the theoretical prediction for $\kappa_0 = 8.5 \times 10^7$ cm^{-1} and $D = 0.38$. The theory correctly predicts the measured temperature dependence, but appears to underestimate the magnitude of the dispersion. We have not considered gravity corrections to the thermodynamic parameters multiplying I in Eq. (18). These may significantly shift the theoretical curves for $t < 10^{-4}$.

ACKNOWLEDGMENT

The gravity analysis presented in this paper was carried out in collaboration with P. C. Hohenberg. The author is grateful to Dr. E. Wasserstrom for developing a convenient numerical solution to the wave equation and to V. Chirba for technical assistance.

REFERENCES

1. Barmatz, M; *Phys. Rev. Lett.*, 24: 651 (1970).
2. Schofield, P., D. Litster, and J. T. Ho: *Ibid.*, 23: 1098 (1969).
3. Barmatz, M., and P. C. Hohenberg: *Ibid.*, 24: 1225 (1970).
4. Edwards, C., J. A. Lipa, and M. J. Buckingham: *Ibid.*, 20: 496 (1968).
5. Wasserstrom, E.: (To be published.)
6. Ferrell, R. A., N. Ményhard, H. Schmidt, F. Schwabl, and P. Szépfalusy: *Ann. Phys.*, 47: 565 (1968); Halperin, B. I., and P. C. Hohenberg: *Phys. Rev.*, 177: 592 (1969).
7. Kadanoff, L. P., and J. Swift: *Ibid.*, 166: 89 (1968).
8. Kawasaki, K.: *Ann. Phys.*, 61: 1 (1970).
9. ———: *Phys. Rev.*, A1: 1750 (1970).
10. For example, Barmatz, M., and I. Rudnick: *Ibid.*, 170: 224 (1968).
11. Vicentini-Missoni, M., J. M. H. Levelt Sengers, and M. S. Green: *J. Res. Natl. Bur. Std.*, 73A: 563 (1969).
12. Griffiths, R. B.: *Phys. Rev.*, 158: 176 (1967).
13. Josephson, B. D.: *J. Phys. Chem. Solid State Phys.*, 2: 1113 (1969).
14. Schofield, P.: *Phys. Rev. Lett.*, 22: 606 (1969).
15. Moldover, M. R.: *Phys. Rev.*, 182: 342 (1969).
16. See, for instance, Schneider, W. G.: *Can. J. Chem.*, 29: 243 (1951).
17. Feke, G. T., K. Fritsch, and E. F. Carome: *Phys. Rev. Lett.*, 23: 1282 (1969).

DISCUSSION on *Papers by G. Benedek, B. Volochine, and M. Barmatz*

R. FERRELL (to B. Volochine): You determined A in two different ways, the first from the Kawasaki fit. What was the second?

B. VOLOCHINE: The other determination came from the critical region in which Γ is proportional to k^3. The data in this region determines A directly.

R. FERRELL: So that both ways of determining A come from the line width data. The first way being a fit from the Kawasaki formula; the second a direct fit in the critical region.

B. VOLOCHINE: Yes.

R. FERRELL (to G. Benedek): When C_p values are available, would it not be useful to regard line width values as a determination of Λ directly? So then you could present the data as Λ measurements.

G. BENEDEK: Yes that is the way I think it should be done.

J. LITSTER (to G. Benedek): What can you say about the value of η?

G. BENEDEK: Our data is consistent when $\eta = 0$.

J. LITSTER: Plus or minus what?

G. BENEDEK: η is undetectable in our measurements as it is in Volochine's.

B. VOLOCHINE: That is so.

L. KADANOFF: How much does the uncertainty in η affect the possible values of ν?

G. BENEDEK: Write

$$\gamma = (2 - \eta)\nu$$

$$\gamma = 1.23 \pm 0.01$$

$$\nu = 0.57 \pm 0.05$$

$$\eta = -0.14 \pm 0.15$$

L. KADANOFF: Therefore you have no information about η within the accuracy of this data. As I understand it, you assume $\eta = 0$ to find ν.

G. BENEDEK: No you don't assume that. You use the slope of the Ornstein-Zernike plots as a function of temperature and you find whether they go as $(k\xi)^2$.

The work of Volochine is most impressive. But in addition to finding ξ by using the Kawasaki theory, it should be most valuable to determine it directly. This would provide a substantiation of this procedure. A measurement of the angular dependence of the scattering, for example, would be most helpful in this context.

B. WIDOM: How did Ford and Poglielli find ξ from their turbidity measurement?

G. BENEDEK: Ford obtained ξ from the difference between the extrapolated temperature dependence of the turbidity and the actual measured turbidity. The difference depends on ξ in a way which can be calculated from the Ornstein-Zernike theory.

R. FERRELL: This method does not depend upon the detailed Ornstein-Zernicke picture but only upon the $(k\xi)^2$ correction to the scattering cross section.

Agenda Discussion:
TRANSPORT PROPERTIES

L. P. Kadanoff[*]
Physics Department
Brown University
Providence, Rhode Island

P. R. Sievert[†]
Department of Physics
Battelle Memorial Institute
Columbus, Ohio

The first main question posed to the group was:
> Do the currently fashionable "mode coupling" and "dynamic scaling" theories of nonequilibrium critical phenomena provide a viable description of the observed experimental data? Where do the main disparities between theory and experiment occur? What future experiments are necessary to explore these disparities?

The group began with the fact that the theory apparently fits the linewidth data in Xe and CO_2 very much better than corresponding data in SF_6.

[*]Chairman
[†]Secretary

R. GRIFFITHS (to G. Benedek): If you apply the same analysis to Xe as you used for SF_6 and subtract off the background thermal conductivity, how large a change in Λ results?

G. BENEDEK: You should follow the procedure suggested in my talk and multiply the line width by measured values of ρC_p to find Λ. This has not yet been done. Instead, Swinney, Cummins, and Henry have only used their data for $\Lambda/\rho C_p$ to find exponents.

R. GRIFFITHS: Thus in the meantime we should take the exponent for $\Lambda/\rho C_p$ with some caution.

G. BENEDEK: Surely that should be done. First of all notice that the difference of 0.1 between the exponents of Swinney, Cummins, and Henry and the theoretical prediction might be qualitatively explicable in terms of an unsubtracted background which would tend to reduce the apparent exponent. There is now sufficient data available from the very recent work of Cannell to make this analysis. In my opinion the exponential type description is a handy back of the envelope judgment as to how things are going, but it conceals a tremendous amount of what you are actually measuring.

P. MARTIN: Doesn't the theory predict $\Lambda/\rho C_p$ as well as it predicts Λ? Shouldn't you say that this experiment gives exponents but because of backgrounds—for example, from rotational levels—these exponents have large errors?

R. FERRELL: Λ has an additive background term which is not determined by theory. If that background term happens to large then it can obscure the singular part.

P. MARTIN: You can express this as a corresponding uncertainty in the exponent.

G. BENEDEK: But you cannot represent

$$A\epsilon^{-x} + B$$

by a form $C\epsilon^{-y}$. That's the point.

P. MARTIN: Then in that case the error on the exponent should be large.

G. BENEDEK: I don't understand it that way. You have two different functional forms and you shouldn't try to fit a simple power law to something which is not that.

J. LITSTER: But Paul's point is that people can fit data with power law.

P. MARTIN: When they do that they should quote an error consistent with large nonsingular terms.

L. KADANOFF: One difference between SF_6 and Xe is the larger background specific heat in the former. Another difference is that the mode-mode term in Λ does not apparently contribute appreciably within the experimental region. Does this indicate
1. a large viscosity
2. a large background conductivity or
3. an error in the theory?

Can we distinguish among these possibilities?

G. BENEDEK: You say that if the viscosity is big enough then the divergent term in Λ will be small. But we do not know the viscosity. On the other hand, there are measurements indicative of an anamoly in Λ by Lis and Keller.

L. KADANOFF: But you don't see a divergent part of Λ in your data?

G. BENEDEK: In the temperature regime of the graphs I showed no divergent term has become apparent. It only appears when you get much closer in SF_6. The data is not yet available on the isochore to see when it might begin to be significant. You have to get much closer to T_c to see the divergence in SF_6 than is required for Xe.

L. KADANOFF: As yet we have no understanding of why this numerical difference occurs.

G. BENEDEK: Exactly.

L. KADANOFF: We must then provisionally say that these SF_6 experiments neither confirm nor deny the theory.

G. BENEDEK: I think that's the position at present.

The tentative conclusion is that the SF_6 experiments no longer seem to violate the theory of Fixman, Kawasaki, Kadanoff, and Swift. Benedek's criticism of the previous methods of data analyses also promises to cast light on the smaller differences between theory and experiment in CO_2 and Xe. The discussion then turned to a more general consideration of whether the mode coupling theory is in substantial agreement or disagreement with experiment.

G. BENEDEK: The data of Cummins, Henry, and Swinney are also in substantial support of the Kawasaki formula for Γ as is Volochine's.

L. KADANOFF: Is Benedek's criticism of the analysis of SF_6 equally applicable to the work of these three authors on CO_2?

G. BENEDEK: As far as the data below T_c is concerned there is a substantial problem caused by the fact that ρC_p is not describable with a single exponent.

M. BARMATZ: There is also the work on ultrasonic attenuation in Xe by Garland and coworkers which is in very good agreement with the function proposed by Kawasaki.

G. BENEDEK: There are indications of an extra relaxation time in Xe not predicted by the theoretical formulas of Kawasaki. This work will be published soon. The trouble lies in the high frequency data rather than the low.

L. KADANOFF: There also seem to be substantial disagreements between the theory of say Laramore and Kadanoff and ultrasonic attenuation in magnetic insulators.

G. BENEDEK: I would rather focus on the question of what is the physical essence behind mode-mode coupling which enables us to get the good agreement we do find. Mode-mode coupling and dynamical scaling are working and let's try to understand why.

With the above comment, Benedek summarized the previous discussion and signaled the move to the next main question before the group:

Should the theories be regarded as mere phenomenology or have they been put on a sound basis? What improvements (or replacements) of the theories are likely to be necessary?

L. KADANOFF: The theory contains two essential statements:
   1. the idea of Ferrell, et al. that the frequencies should scale;
   2. the idea derived from the work of Fixman, Kawasaki, and later Kadanoff and Swift that the various divergences can be understood as a result of couplings among the hydrodynamic modes. These mode-couplings are handled in analogy with the quasiparticles of solid-state physics or the resonance of elementary particle theory. Recently Halpern and Kawasaki have pointed out that you should also couple to the order parameter even when it does not appear in the hydrodynamics.

Problems in the theory include
   1. attenuation in magnetic insulators;
   2. Suzuki's calculation of a vanishing Onsager coefficient—
   which is not yet understood microscopically or in physical terms.

B. McCOY: Do the pulse measurements of ultrasonic attenuation present a particular problem since the pulse may be short in comparison with the correlation length?

M. BARMATZ: But the pulses are not that short except perhaps at the very highest frequencies?

G. BENEDEK: Do we yet have a satisfying intuitive picture of the theory? In my own mind, in the region where the wavelength of fluctuation is much longer than ξ it is reasonable to look on the thermal diffusivity as the random walk of particles all of which have the size ξ. However when the wavelength becomes comparable to ξ we must take into account that heat can be transformed in units whose densities range from 0 to larger than ξ. Thus we may expect that the diffusion occurs from a distribution of sizes of heat carrying regions which could lead I believe to a dependence of Γ on K like that found by Kawasaki.

R. FERRELL: The Kubo formula calculation of Γ in terms of current fluctuations is a realization of the extension of the random walk picture.

P. MARTIN: But in your evaluation you make a simple factorization? What is the small parameter that justifies that?

L. KADANOFF: There seems to be a missing link in the formalism in that we have no way of seeing why the higher order diagrams are all unimportant.

R. FERRELL: There is a physical reason for making the factorizations in the gas. Remember that the shear modes last for a long time and carry along the concentration fluctuations with them. One assumes that the Brownian motion of the shear modes is unaffected by concentration fluctuations. This is the reason that we can decouple the modes. On the other hand, in the magnetic case all the modes are intertwined and the decoupling is far less justified.

M. BARMATZ: Recently, ultrasonic attenuation measurements near the He^4 λ point were successfully interpreted using the Landau-Khlatnikov relaxation mechanism. Is this mechanism applicable to sound attenuation near other critical points?

L. KADANOFF: Yes, this kind of mechanism is important whenever the order parameter is not conserved as in superconductors or anisotropic magnets. Then the relaxation of the order parameter must be included.

In summary, the experiments seem to suggest that the mode-coupling theory is satisfactory in describing the divergent portion of the transport coefficients. The theorists, however, do not seem to have come to a complete agreement about correction terms in the theory.

ELECTRONIC PHASE TRANSITIONS

David Adler

Department of Electrical Engineering
Massachusetts Institute of Technology
Cambridge, Massachusetts

ABSTRACT

Phase transitions that are accompanied by sharp anomalies in electrical conductivity are classified and discussed in detail. A sharp distinction is drawn between transitions that occur entirely under equilibrium conditions and those that take place in nonequilibrium situations. The former are induced by varying pressure, stress, carrier concentration, or temperature. The latter can be brought about by increasing the intensity of light impinging on a solid or the applied electric field.

Three classes for equilibrium transitions can be differentiated—those induced by band overlap, those due to shifts in the electronic states, and Mott transitions. Nonequilibrium transitions can result from a sharp increase in free-carrier concentration, screening effects, dissipation of excess electronic energy into heat, or mobility increases due to the filling of traps.

The nonequilibrium threshold and memory switching that has recently been found in thin films of covalent amorphous semiconductors is discussed in some detail. It now appears fairly certain that memory switching results from phase separation and crystallization in a nonequilibrium conducting state, but threshold switching is not yet completely understood. Several possible mechanisms are discussed in light of the present experimental data.

1. INTRODUCTION

In a sense, all phase transformations in a solid are electronic in nature. Assuming the adiabatic approximation is valid and we can indeed separate electronic motion from lattice vibrations, we find that the characteristic electronic energies are at least an order of magnitude larger than the phonon energies, and it is thus extremely unlikely that phonon energy and entropy are ever significant enough by themselves to drive a phase transition. Magnetic transitions and most crystallographic phase transformations result primarily from a crossing of the electronic free energies of two possible states, the phonons themselves generally playing only a small role. Consequently, we shall ignore the effects of phonons except for cases in which electron-phonon coupling is large.

In general, electronic phase transitions can be induced either by a repopulation of the electronic states of a material or by shifts in these states due to a change in the external parameters of the system. Such transitions are of special interest when they are accompanied by large jumps in electrical conductivity. These are usually referred to as insulator-metal transitions, regardless of the actual nature of the two states. We shall restrict our discussion to transitions of this type.

Since a large variety of different types of insulator-metal transitions have been studied in recent years, it is desirable to attempt a classification before discussing the examples in detail. A simple classification scheme is one based on thermodynamics: We can differentiate between transitions in which the system always remains in thermodynamic equilibrium and transitions in which nonequilibrium conditions prevail. Examples of equilibrium transitions are those induced by varying hydrostatic pressure, uniaxial stress, stoichiometry, doping concentration, or temperature. Nonequilibrium insulator-metal transitions can be brought about by changing the applied electric field, the magnetic field, or the intensity of light impinging on a material. We shall discuss both classes in detail, paying particular attention to the various mechanisms which can account for the conductivity anomalies.

2. EQUILIBRIUM TRANSITIONS

2.1. Experimental Data

2.1.1. Pressure-induced Transitions. Conductivity anomalies under the application of high pressure are more the rule than the exception.[1] Jumps in conductivity in excess of a factor of 10^8 have been observed. In most cases, although not all, the transitions are

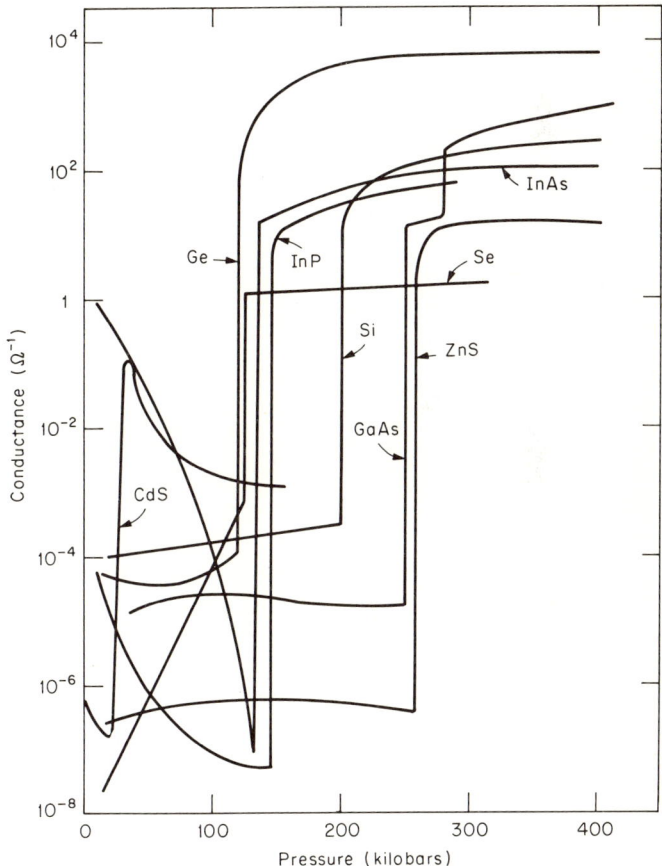

Fig. 1 *Electrical conductivity as a function of pressure for several crystalline solids.*

accompanied by a change in crystal structure. Even when no symmetry change is observed, there is always a discontinuous volume contraction at the critical pressure. Thus, the transitions always appear to be of first order. In general, no order parameter can be defined. Several examples are shown in Fig. 1.

2.1.2. Temperature-induced Transitions. Many common semiconductors transform to metals upon melting.[2] Figure 2 shows the extent of the conductivity anomalies in a number of cases. Of course, these transitions are of first order. In each case, a sharp change in the short-range order (i.e., the local coordination number) occurs at the melting temperature.

More striking is the large number of insulator-metal transitions that have been observed in transition-metal compounds.[3] To date, such transitions have been found in Ti_2O_3, Ti_3O_5, Ti_4O_7, Ti_5O_9, Ti_6O_{11}, V_2O_3, V_3O_5, V_4O_7, V_5O_9, V_6O_{11}, V_8O_{15}, VO_2, V_6O_{13}, NbO_2, Fe_3O_4, $BaVS_3$, CrS, FeS, NiS, and $FeSi_2$, among others. Several examples are shown in Fig. 3. A change in crystalline symmetry usually accompanies these transitions, although Ti_2O_3 and

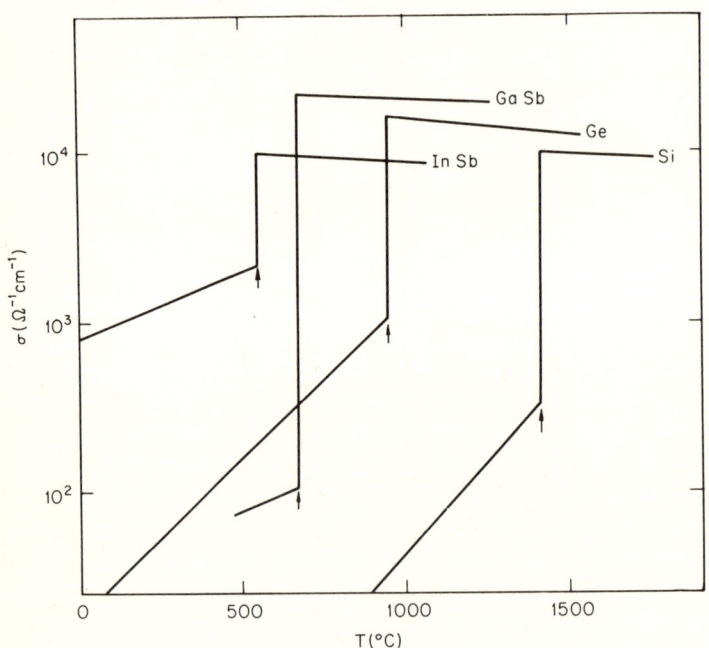

Fig. 2 *Electrical conductivity as a function of temperature for several crystalline semiconductors. The melting temperature in each case is indicated by an arrow.*

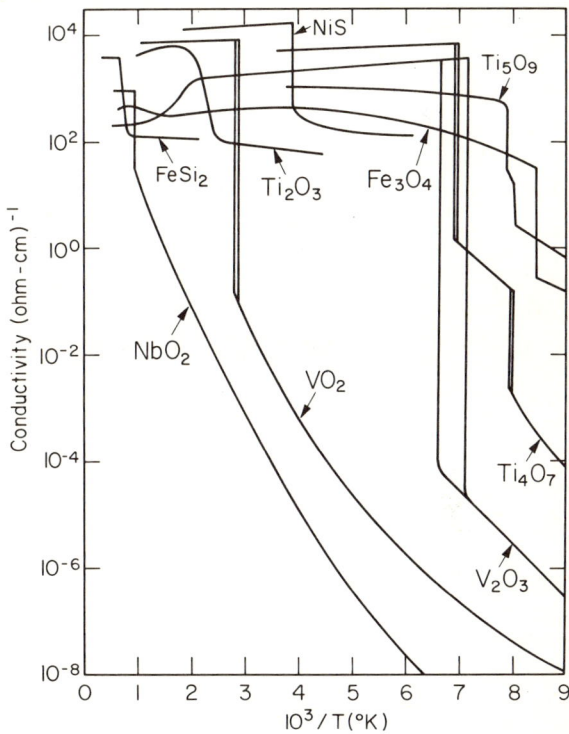

Fig. 3 *Electrical conductivity as a function of reciprocal temperature for several crystalline transition-metal compounds.*

NiS retain their symmetry through the conductivity anomaly. In most cases, the low-temperature structure is a distortion of the high-temperature structure, and thus an order parameter, the fractional distortion, can be introduced. A few of the insulator-metal transitions, in particular those of V_2O_3 and NiS, also represent a transition from an antiferromagnetically ordered state to a state with no long-range magnetic order, and the extent of the magnetic alignment provides another order parameter. Most of the transitions are of first order, although several are of second or higher order. The conductivity anomalies can be either very sharp or quite gradual. For example, V_2O_3 exhibits a conductivity jump of a factor in excess of 10^7 in a temperature range of hundredths of a degree, while the conductivity of Ti_2O_3 increases only by a factor of 100 over a $400°K$ temperature range. But even in Ti_2O_3, which shows no changes in magnetic order or crystal symmetry through the transition, there is now definite evidence of a 20 percent softening of a phonon mode.[4]

2.1.3. Composition-induced Transitions. As the composition of certain covalent alloys is changed, sharp conductivity anomalies exist. A striking example occurs in the Sb-Se system, near the stoichiometric compound, Sb_2Se_3.[5] The conductivity as a function of composition is given in Fig. 4. A sharp conductivity jump of a factor of 10^4 was found within a 0.01 percent change of stoichiometry. Although, in principle, the composition of covalent alloys could be controlled by varying the appropriate chemical potentials, the low diffusion rates at low temperatures make the attainment of equilibrium conditions take a hopelessly long time. At high

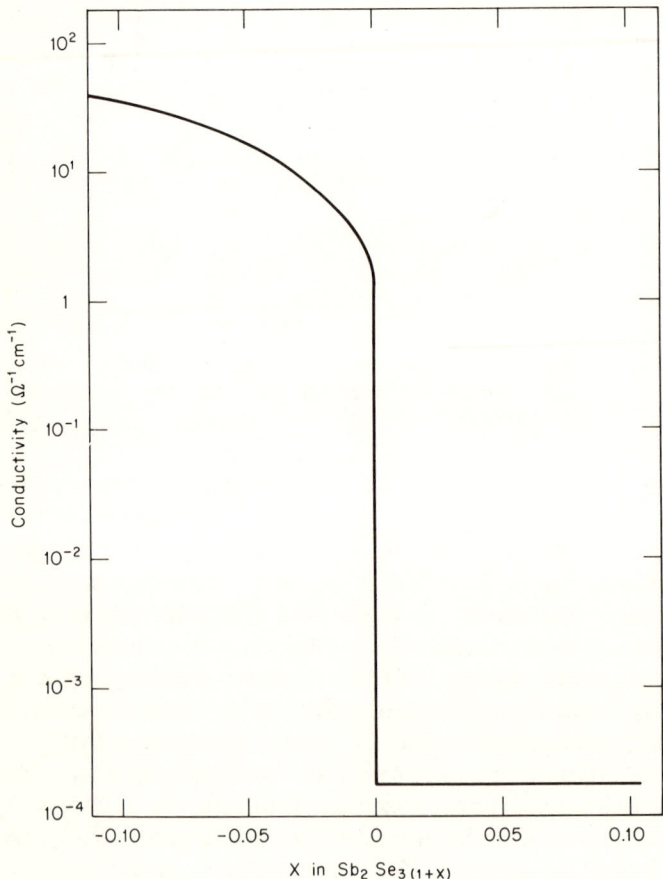

Fig. 4 *Electrical conductivity of crystalline Sb_2Se_3 as a function of composition.*

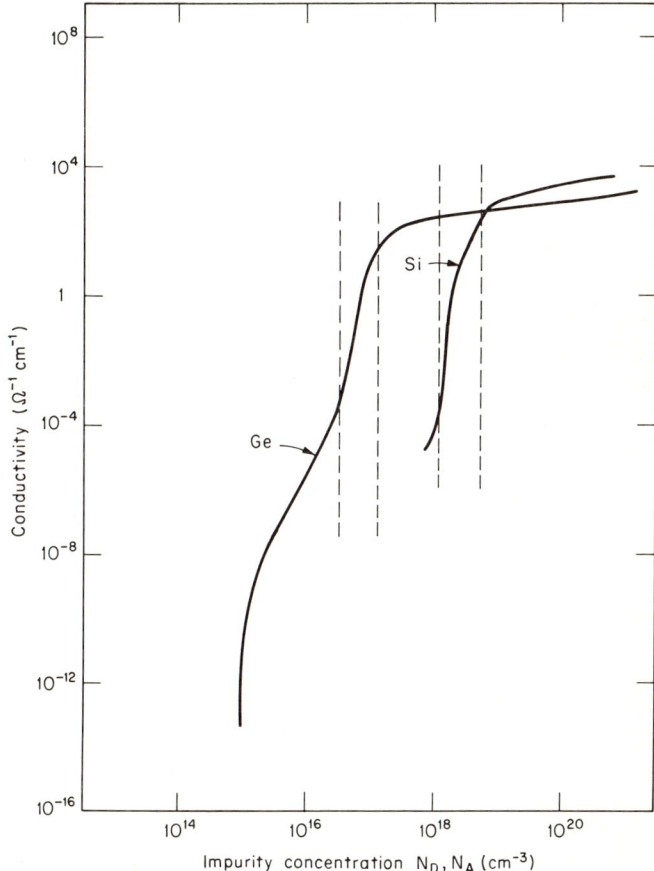

Fig. 5 *Electrical conductivity of doped, crystalline Si and Ge as a function of impurity concentration.*

temperatures, where equilibrium is attained quickly the distinction between an insulator and a metal disappears, and the transition is washed out. Thus, an experimental study of the thermodynamics of this type of transition is virtually impossible, although the theoretical analysis is generally straightforward.

Insulator-metal transitions can also be induced in a doped semiconductor by increasing the impurity concentration.[6,7] Two examples are shown in Fig. 5. This type of transition is not generally very sharp, although the conductivity changes by a factor of 10^6 within a four-fold increase in donor or acceptor concentration. For the same reason as in composition-induced transitions, experimental investigation of the thermodynamics of impurity-induced transitions is essentially impossible.

2.2. Mechanisms

An astoundingly large number of physical mechanisms which can account for the observed conductivity anomalies have been proposed,[8] and there now appears to be more models even than experimental examples. All of these models, however, can be placed into one of three classes—band-overlap transitions, electronically induced phase transformations, and Mott transitions. The basic physical ideas behind these mechanisms are discussed in this subsection.

2.2.1. Band-overlap Transitions.
An ordinary metal is a material whose outer electrons move in a partially filled band. A simple insulator is a material all of whose bands are either completely filled or completely empty, and, in addition, has an energy gap between the top of the highest filled band and the bottom of the lowest empty band. Thus, a particularly simple insulator-metal transition can be brought about by a band gap that moves through zero as any physical parameter is varied. For example, if a semiconductor with a small gap is subjected to high pressures, and the gap decreases with increasing pressure, then the gap may become negative at a critical pressure, thus inducing a semiconductor-semimetal transition. This type of transition can occur as a function of any physical variable on which the energy gap depends. Consequently, it is a very general mechanism. However, since a narrow-gap semiconductor has high conductivity for an insulator and a semimetal with small overlap has low conductivity for a metal, a simple band-overlap transition is necessarily rather gradual.

Most observed transitions are not gradual, but, in fact, are quite sharp. Several modifications of band-overlap transitions can be conjured up which will sharpen up the conductivity anomaly. The simplest is that a first-order transition occurs, in which the gap suddenly changes from positive to negative at the critical value of the parameter being varied. Since the primitive cell volume depends on electronic structure, a first-order transition can be expected as a function of pressure. But the total entropy of any state is also a function of electronic structure, and a first-order temperature-induced transition is a possibility.

A second possibility for attaining sharp band-overlap transitions is via a change in crystal symmetry. Since the electronic band structure sensitively depends on the crystal structure, it is not difficult to conceive of a band-overlap semiconductor-semimetal transition accompanying a crystallographic phase transformation. Since such transformations are commonly induced by pressure or temperature, this mechanism can explain any of the conductivity jumps observed in Figs. 1 and 3, provided a change in symmetry is observed at the transition point.

A final possibility is that another type of insulating phase forms at low temperatures—an excitonic insulator.[9] Such a state can be shown to be stable at $T = 0$ when the binding energy of an exciton (bound state of an electron-hole pair) exceeds the absolute magnitude of the energy gap. The state, which may be thought of as a Bose condensate of excitons, exhibits a suppressed conductivity relative to that of an ordinary semimetal. Above a certain critical temperature, an excitonic-insulator-to-semimetal transition occurs. In the weak-binding approximation, the transition turns out to be of second order, although a first-order transition is also possible. At a constant temperature, the transition can be induced by pressure.

2.2.2. Electronically Induced Phase Transformations. This mechanism makes use of the fact that, in narrow-band materials, a metallic state can be unstable at $T = 0$ towards a crystallographic distortion to lower symmetry.[10] This arises because, for any metal, a distortion can always be envisioned which reduces the energy of just the number of states that are filled, but raises the energy of the number of states that are empty. Such a distortion always minimizes the electronic energy of the system, and can result in an insulating state if a real energy gap in the density of states is produced. This is more likely if the bands are narrow. The distortion will occur spontaneously only if the total energy is reduced, i.e., if the electronic energy decrease outweighs the increase in strain energy. Once again, it turns out that such is the case only if the relevant energy bands are narrow.[10] As the temperature increases, the stabilization of the distorted state decreases with thermal excitation of free carriers, and this taken together with the greater electronic entropy of the metallic state results in a first-order insulator-metal transition at a critical temperature. This model was explicitly applied to the transition-metal oxides that exhibit sharp conductivity jumps, on the basis that these were indeed narrow-band materials.[11] In such a model, it is important to consider the electron-phonon interactions[12-15] and the electron-electron interactions[16] explicitly, but these introduce no qualitative changes to the theory.

Although the model as originally presented involved a pairing type of distortion, it should be remarked that no increase in the size of the primitive cell is necessary. The additional crystalline-field splittings that arise from the lowering of symmetry could easily provide the additional energy gap necessary to obtain an insulating ground state. Furthermore even a lowering of symmetry itself is not an absolute necessity, particularly when the high-temperature phase already has a complex structure. In this case, motions of the ions within a single primitive cell could, in principle, produce an energy gap by intensifying crystalline-field and structure-type splittings that already exist but are insufficient to produce a gap in the density-of-states. This type of transition is more like a semiconductor-semimetal transition. It could manifest itself structurally via sharp changes in either the lattice parameter or the basis. Ultimately, once the electron-phonon interaction is introduced in detail, the transition could conceivably be driven merely by the softening of a phonon mode.

An analogous, but theoretically more difficult, mechanism involves an antiferromagnetic ordering of spins. Slater[17] was the first to point out that antiferromagnetism is equivalent to a pairwise distortion as far as its effect on the electronic band structure is concerned. If any antiferromagnet has a half-filled band in the absence of magnetic ordering, the antiferromagnetic state could then be insulating. In such a case, the Néel temperature marks the point of an insulator-metal transition.[18-20] This type of transition is usually of second order, although certain band shapes can lead to a first-order transition.[18]

2.2.3. Mott Transitions.

Mott[21] was the first to point out that an insulating state need not have an energy gap in the one-electron density of states. When the outer energy bands of a material are extremely narrow, electronic correlations lead to an insulating ground state, in which the valence electrons remain localized on their respective ions. Mott[22] further presented arguments which indicated that, as the interatomic separation decreases, a sharp transition to a metallic state should occur. Any such change, from a state in which all electrons are localized to one in which the valence electrons delocalize into a partially filled band, can be called a Mott transition.

It is clear that Mott transitions can be induced by applied pressure, since this is a direct means of reducing the interatomic spacing. It is equally clear that such a transition can occur as a function of doping concentration, by delocalizing the electrons in the impurity band. A composition-induced Mott transition could arise from an analogous delocalization of electrons on the donor or acceptor band that stems from the presence of vacancies in a nonstoichiometric alloy. It is somewhat more difficult to conceive of a temperature-induced insulator-metal Mott transition, especially since entropy considerations militate against such a situation, although it is possible.[2] Fröhlich[23] has proposed that screening of the electronic Coulomb repulsion due to thermal excitation of free carriers could lead to a temperature-induced Mott transition, and this model has been explicitly applied to the temperature-induced transitions in the vanadium oxides.[24,25] Falicov and Kimball[26,27] have proposed a model in which a Mott transition is driven by an effective crossing of an ordinary conduction band and a correlation-split valence band, the transition arising from the attraction between thermally excited electron-hole pairs. Recently, Mott[28] suggested that the transitions can be driven with the assistance of the electron-phonon interactions, the insulating state being unstable above a critical temperature or pressure towards the formation of a degenerate gas of electron-hole pairs stabilized by their interactions with phonons.

2.3. Comparison of Theory and Experiment

The pressure-induced insulator-metal transitions, in general, appear to be examples of simple band overlap, almost always occurring simultaneously with a crystallographic phase transformation.[1,29] No evidence has yet been found for an excitonic-insulator state,[30] nor has a Mott transition under pressure been found unambiguously in a pure material.[31]

Very little detailed work has been carried out on the composition-induced transitions. They may be Mott transitions within the donor or acceptor bands, or they could be caused by overlap of these bands with the conduction or valence bands of the stoichiometric material. In the case of Sb_2Se_3 (Fig. 4), although cavalierly described as a Mott transition by the original investigators,[4,32] the fact that it occurs so near the stoichiometric composition indicates the probability of another mechanism. It has been proposed[33] that in PbTe, the vacancy states of both Pb and Te lie well inside the valence and conduction bands, respectively, thus leading to semimetallic behavior in the nonstoichiometric materials. It is not unreasonable to postulate that for Sb_2Se_3, the localized state resulting from a Se vacancy is above the conduction band edge, while the corresponding state for an Sb vacancy lies in the energy gap. This model is consistent with the results shown in Fig. 4, and is easily extended to the similar material Sb_2S_3 which shows no such transition,[32] by concluding that the states arising from S vacancies in the wider gap material are *below* the conduction band edge.

The two mechanisms of band-overlap and Mott transitions are also possible explanations for the transition induced by varying the doping concentration in ordinary semiconductors. In this case, however, a detailed review of the presently available data[6] suggests that both mechanisms occur, with the Mott transition taking place at slightly lower impurity concentrations. A complication in discussing insulator-metal transitions within impurity or vacancy bands is that there are *two* possible causes for localization—electronic correlations and disorder.[34] It is clear that both are present in impurity and vacancy bands. Although delocalization with increasing density of states in either case can be properly termed a Mott transition, recent investigations of amorphous semiconducting systems have led to a distinction between the two.[35] When the localization just at the transition point is due primarily to the effects of disorder, the transition may be called an Anderson transition. Although the theoretical distinction between Mott and Anderson transitions is clear, it is extremely difficult to distinguish between the two experimentally.

Unfortunately, the basic mechanisms underlying the large number of temperature-induced transitions in the transition-metal compounds are far from clear at the present time. The lack of any magnetic order in the low-temperature phases of nearly all of the materials that exhibit these transitions militate against a Mott transition being the driving force, since localized moments must magnetically order at sufficiently low temperatures. Any other type of electronic localization, such as into molecular orbitals of zero angular momentum, can be treated entirely within the one-electron approximation, and leads to a state which is no more a Mott insulator than is NaCl. Thus, of the more than 20 temperature-induced transitions presently known, only V_2O_3, NiS, FeS, and perhaps Fe_3O_4 can be examples of Mott transitions.

Although it is tempting to conclude that all the transitions must have a common origin, this hope does not seem to be be borne out by the presently available experimental data. It is clear from Fig. 3 that the electrical conductivity behavior is qualitatively different for different materials. In addition, the crystallographic, magnetic, and optical properties are strikingly diverse.[3,8] We shall discuss the available data for several materials, although neither the theoretical nor the experimental situation is clear at present.

Ti_2O_3 has the corundum structure at all temperatures.[36] However, between 450°K and 600°K, the lattice parameters vary sharply,[37] and a specific heat anomaly has been observed.[38] After much discrepancy, it now appears virtually certain that no magnetic ordering exists in the material.[39] The electrical properties of Ti_2O_3 indicate that at low temperatures it is a relatively wide-band semiconductor with a small gap, of the order of 0.06 eV.[40,41] Above approximately 800°K, the conductivity looks semimetallic. Thus, Ti_2O_3 looks to be an example of a band-overlap transition, although the conductivity anomaly appears to be somewhat sharper than expected from the simplest version of that model.[42] The recent discovery of a 20 percent softening of a phonon mode as the temperature increases through the transition[4] sheds light on the exact mechanism responsible for the band overlap. The transition in Ti_2O_3 may well be an example of a variation of the electronically induced phase transformation model, in which the metallic-state instability at low temperatures manifests itself as an ionic reorientation rather than a change of symmetry. However, the exact ionic positions as a function of temperature must be determined before any quantitative model can be applied.

VO_2 has tetragonal symmetry above 340°K; below 340°K, an 8 percent monoclinic distortion occurs, in which the V^{4+} ions pair up.[43] No long-range magnetic order exists.[44] Below 340°K, the conductivity is thermally activated, with a slope increasing from 0.15 eV at very low temperatures to 0.45 eV just below the transition. At 340°K, the conductivity jumps by a factor of 10^5, and at higher temperatures it acts like an ordinary metal or semimetal.[45] The mobility is not very temperature dependent, and does not change much at the transition. Optical measurements[45] show that the semiconducting phase is characterized by an energy gap that decreases from about 0.75 eV at very low temperatures to about 0.62 eV just below the transition. This gap sharply disappears at the transition temperature.[46]

Because of the large pairing-type of distortion, VO_2 appears to be an ideal example of an electronically induced phase transformation, particularly if the effects of electron-phonon coupling are considered in detail.[47] Information about the possible softening of phonon modes in anticipation of the transition is, at present, lacking, and it is consequently difficult to differentiate between phonon-induced and electronically induced phase transformations.[48] This is a somewhat nebulous question to discuss in any event, since the energy difference in any first-order phase transformation is primarily due to the electronic contribution, but the entropy discontinuity may well be dominated by the phonon contribution, primarily if the bands are relatively wide. Furthermore, the importance of the electron-phonon interaction in transition-metal oxides makes a separation of electronic and vibrational motion inappropriate. In any event, the association of the conductivity anomaly in VO_2 with the lattice distortion has received striking confirmation from recent experiments on amorphous films.[49,50] When the sputtered films were deposited on a substrate whose temperature was above 340°K during deposition, the material was metallic at all temperatures. When the substrate was initially held below 340°K, the films were semiconducting at all temperatures. In neither case did the semiconduction-metal transition take place. Crystalline films of comparable thickness all exhibited the usual conductivity anomaly. The amorphous films deposited at low temperatures presumably have the short-range order on monoclinic VO_2, while those deposited above 340°K have the short-range order of tetragonal VO_2. In both cases, the lack of any long-range order prevents the crystallographic transformation from occuring, thus suppressing the conductivity transition. Thus, we can conclude that the crystalline distortion in VO_2 is necessary to obtain the semiconductor-metal transition.

The most intriguing material that exhibits a conductivity anomaly is V_2O_3. At high temperatures, V_2O_3 has the corundum structure, but below about 150°K a monoclinic distortion occurs.[51] The transition is of first order, and a large latent heat has been observed.[52] Below the transition, V_2O_3 is antiferromagnetic, with a moment of about $1.2\mu_B$ per vanadium ion.[53] The low-temperature state is semiconducting, with an activation energy of about 0.15 eV; at high temperatures, the material is metallic or semimetallic, with an anomaly in the vicinity of 600°K.[52] At the transition, the conductivity sharply jumps by a factor of greater than 10^7. Optical experiments indicate that an energy gap of about 0.3 eV opens up in the low-temperature phase.[52] The Hall constant is independent of temperature above the transition, and indicates the presence of about one free carrier per vanadium ion, if we assume the predominance of only one type of carrier.[54]

Recent experiments[55-66] have been interpreted as proof that the transition in V_2O_3 is a Mott transition, the lattice distortion being magnetostrictive in origin and of little consequence. Although V_2O_3 is antiferromagnetic at low temperatures, and thus can be an example of a Mott transition, it is difficult to account for the reason why it should be so different from the seven other oxides of vanadium that exhibit similar conductivity transitions without the presence of magnetic order. In each of these other cases that has been investigated, a lattice distortion does occur, however. Furthermore, amorphous V_2O_3 films fail to switch,[49] additional evidence for the importance of the lattice distortion to the insulator-metal transition. It would seem more reasonable, at present, to conclude that lattice effects, not necessarily involving a change of symmetry, are of major significance in all the temperature-induced transitions. Nevertheless, we are left with the problem of whether or not an accompanying Mott transition does occur in V_2O_3, even if it is not the driving force of the conductivity anomaly. This question can be answered only by experimentally determining whether the low-temperature, semiconducting state is one in which the spins are localized or itinerant. An itinerant antiferromagnetic state that has semiconducting properties was first suggested by Slater many years ago,[17] and it is perfectly consistent with the observed moment of 1.2 μ_B, which is significantly less than the saturation value of 2 μ_B expected if the spins were purely localized. Experimentally, a localized $3d^2$ configuration could manifest itself in several ways. As already indicated, the saturation magnetization value is not approached. The crystalline-field excitations (Frenkel excitons) of the localized spins could be observable in the optical spectra, as occurs in known Mott insulators such as NiO and Cr_2O_3, but none have yet been found.[66] The absorption edge may well cut all of these off, but the question remains why the peaks are not shifted to lower energies by overlap and screening effects. The "white-line" X-ray spectra characteristic of Mott insulators is also not observed in vanadium oxides, and, in fact, no change in the X-ray emission spectrum has been found between the semiconducting and metallic states. This is difficult to reconcile with a sharp Mott transition. In addition, absolutely no change in the positron lifetime between the two states has been measured in positron-annihilation experiments,[68] although measurement of the momentum distribution would provide a more sensitive test of whether or not a Mott transition is occurring. A large mobility change at the transition or a thermally activated hopping conduction at low temperatures would indicate electronic localization in the semiconducting state, but neither has been observed. In short, there is as yet no clear-cut evidence for the Mott insulating nature of the low-temperature state in V_2O_3.

It has further been proposed that the high-temperature anomaly in the vicinity of 600°K in V_2O_3 represents a continuous (supercritical) metal-insulator Mott transition.[55-66] This is surprising in view of the facts that the resistivity appears to monotonically increase up to 900°K,[52,54] the Hall coefficient is constant from 150°K through 700°K,[54,69] and no anomalous temperature dependence between 400°K and 600°K can be determined from the NMR data.[70] The identification of this anomaly with a Mott transition was based on the general phase diagram obtained for the $(V_{1-x}Cr_x)_2O_3$ system, in which the entropy change associated with the high-temperature transition was taken to be inconsistent with a band uncrossing as the temperature increases. However, this conclusion overlooks the possibility of a change in entropy associated with the phonon spectrum, analogous to that observed in Ti_2O_3,[4] which could easily more than compensate for the negative electronic entropy change associated with a band uncrossing. Double first-order conductivity anomalies have been obtained before, as in the olivine-spinel-olivine transformations in Fe_2SiO_4 under pressure.[71] Furthermore, the additional effects of disorder in the mixed system complicate the problem.[72] Clearly, the final words on V_2O_3 have not yet been written.

3. NONEQUILIBRIUM TRANSITIONS

3.1. Experimental Data

When electric or magnetic fields are applied, a material is subjected to electromagnetic radiation, or thermal gradients are introduced, equilibrium conditions no longer maintain, and we can no longer analyze the physics in terms of the ordinary concepts of reversible thermodynamics. In particular, all insulators eventually become conducting in a sufficiently intense external electric field. This so-called "dielectric breakdown" can be either destructive or reversible, depending on the material, and often on the geometry of the experiment as well. This may be thought of as an example of a field-induced insulator-metal transition. Since the conducting path in such cases is always filamentary in nature,[73] the actual value of the conductivity is difficult to measure. Usually, just the current-voltage characteristics are plotted, as is shown in Fig. 6, which is an example of the reversible threshold-type switching that is found in thin films of certain amorphous semiconductors.[74] Threshold switching can be characterized by the fact that only one state exists under equilibrium conditions (zero field). The transition is extremely sharp in the sense that the switching time is of the order of 10^{-10} sec.

A second type of field-induced transition is shown in Fig. 7. These characteristics can be sharply distinguished from those of Fig. 6 by the fact that two different equilibrium states exist for the same material, a state of low conductivity and a state of high conductivity. This is known as memory-type switching.[74]

The above types of transition can also be induced by exposing the material to light of sufficiently high intensity and frequency.[75] Light-induced insulator-metal transitions have also been observed in crystalline materials, such as Ge.[76]

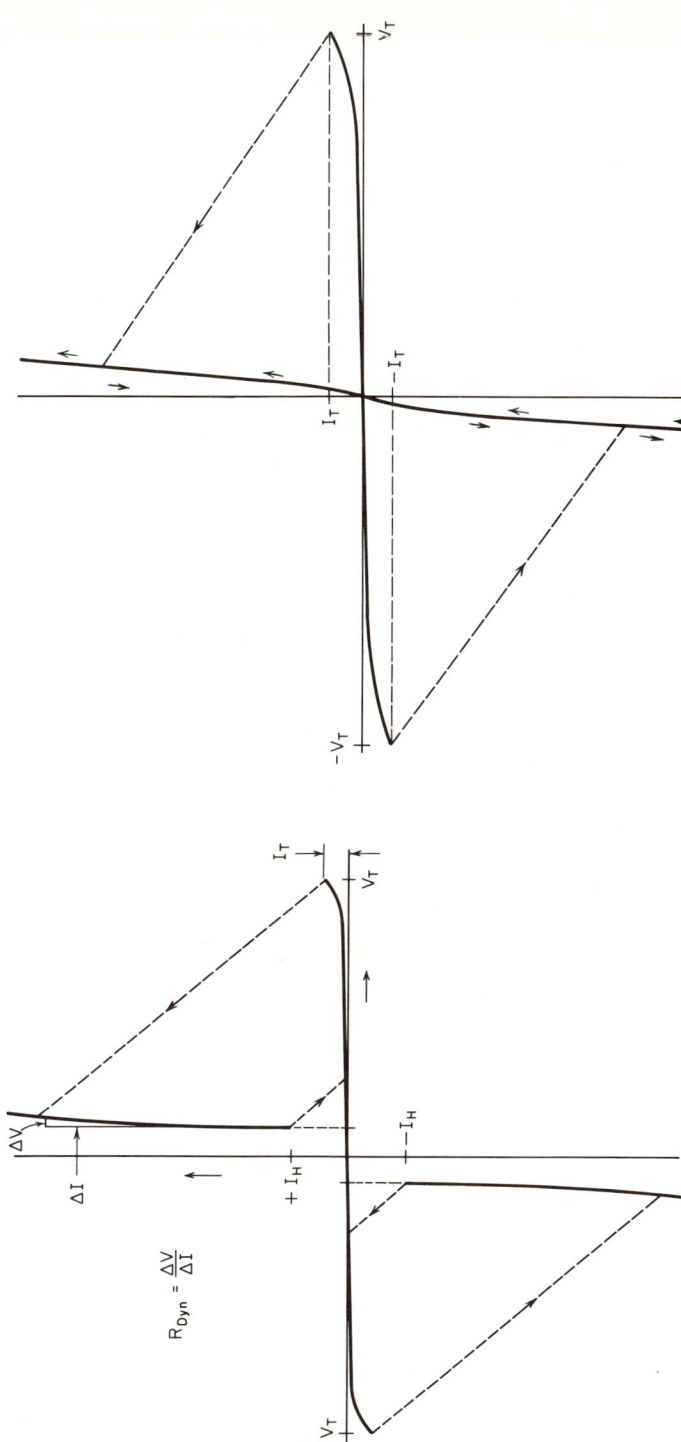

Fig. 6 Current-voltage characteristics for a thin film of a chalcogenide glass that exhibits reversible threshold-type switching.

Fig. 7 Current-voltage characteristics for a thin film of a chalcogenide glass that exhibits reversible memory-type switching.

3.2. Mechanisms

To even a greater extent than is true for equilibrium transitions, an immense number of mechanisms can be envisioned to account for nonequilibrium conductivity jumps. These might roughly be grouped into four main classes, depending on whether the conducting state is primarily due to the sudden direct excitation of a very high concentration of free carriers, screening effects brought about the nonequilibrium excitation of relatively small free-carrier concentrations, the dissipation of the excess energy obtained from the nonequilibrium process into heat, or sharp mobility increases due to the filling of traps. These mechanisms will be discussed briefly in the following subsection.

3.2.1. Sharp Increase in Free-Carrier Concentration.

Application of an electric field or light of frequency greater than the band gap creates excess free carriers, in addition to those generated thermally. Ordinarily the excess carrier concentration is not very large, and just produces some non-ohmic currents of photocurrents. In very intense fields or light beams, however, very sharp increases in free-carrier concentration become possible, for one of three reasons. The first is via the mechanism of tunneling from the valence to the conduction band, a phenomenon which increases exponentially in probability with increasing field. When the tunneling rate exceeds the recombination rate, there is suddenly a sharp increase in carrier concentration, known as Zener breakdown. Zener breakdown is not usually characterized by a negative-resistance region, since a reduction in field leads to a failure to maintain breakdown conditions, and rapid recombinations will destroy the high-conductivity state.

A second possibility for a sharp increase in free-carrier concentration is via the mechanism of impact ionization, which represents the creation of an electron-hole pair by inelastic scattering of a free electron by a bound electron. Free carriers are accelerated by the field, and can attain an energy in excess of the band gap. It then becomes possible for such a carrier to excite an electron across the gap, thus creating an electron-hole pair, and still maintain sufficient energy to remain in the conduction band itself. At sufficiently large values of the field per unit mean free path, a chain reaction completely analogous to that in a nuclear explosion can occur, a phenomenon known as avalanche breakdown. If space-charge effects are unimportant, the field remains essentially homogeneous across the material, and only a small negative-resistance region due to secondary effects should exist. However, if the effective masses or the mobilities of electrons and holes are very different, a large space-charge region can be set up, and the field will become much larger at one or both electrodes. Since the critical field need only then be exceeded in the high-field region, the holding voltage can be significantly lower than the threshold voltage, and a negative-resistance region can occur.

Finally, a strong electric field in a partially ionic solid can induce a lowering of the energy of the system by means of an ionic reorientation. Such a displacement will take place when the energy reduction due to lining up dipoles in the field exceeds the increase in energy from breaking the semipolar bonds. Once the bonds are broken, a sharp increase in the free-carrier concentration leads to a quasi-metallic state. We might call this type of transition a field-induced phase transformation. In amorphous materials, it is a possibility that the displaced phase has a lower free energy than the original phase, leading to memory-type behavior. Neither Zener nor avalanche breakdown can explain memory switching.

3.2.2. Screening Effects. Creation of a nonequilibrium concentration of free carriers always leads to a screening of the Coulomb interaction, in addition to that contributed by the polarization of the atoms of the solid and reflected in the static dielectric constant. Because of this additional screening, several types of nonequilibrium phase transitions become possible, particularly if the solid at equilibrium contains localized outer electrons. For example, if a material is a Mott insulator, but near the point of a Mott transition, a reduction of the intraionic Coulomb repulsion by nonequilibrium screening can lead to a sharp delocalization of all the outer electrons and a quasi-metallic state. We might call this a nonequilibrium Mott transition.

A second possibility exists if the solid is one whose free carriers become localized as a result of strong interactions with the phonons. In a narrow-band ionic material, the relatively massive carriers remain in the vicinity of a given set of ions sufficiently long to induce a local lattice deformation, e.g., an excess electron will attract its surrounding positive ions toward it and repel the negative ions in its environment away from it. This local deformation acts as a potential well, with which the electron can form a bound state and be self-trapped, no longer contributing to conduction without the assistance of phonons. Nonequilibrium screening of the Coulomb interaction, in this case between electrons and ion cores, can lead to a screened potential well too weak to produce any bound states, and thus a sharp increase in free-carrier concentration. This transition might be termed a nonequilibrium electron-phonon decoupling.

A third possibility exists for the case of amorphous systems with a large carrier concentration in states localized only because of the fluctuations in potential arising from the disorder. Screening of the Coulomb interaction due to excess free-carrier excitation can lead to a reversal of the criterion for Anderson localization,[34] and thus induce a nonequilibrium Anderson transition.

Note that in all of these cases, the switching is triggered by a sharp increase in carrier *mobility*. This distinguishes this group of mechanisms from those in the previous subsection, in which the transitions arise from an increase in carrier concentration.

3.2.3. Thermal Runaway. Nearly all the excess energy induced by nonequilibrium processes is eventually transferred into ionic vibrations, which leads to an increase in the temperature of the material. In particular, a current passing through any material with resistance leads to Joule heating, proportional to $I^2 R$. Since the conductivity of all insulators increases exponentially with increasing temperature as

$$\sigma = \sigma_0 e^{-E/kT} \quad (1)$$

the current will increase with Joule heating. In turn, this leads to accelerated Joule heating until a steady state is reached in which the heat dissipation through the electrodes just balances the Joule heat produced. Above a critical voltage, the heating is sufficiently intense that a very rapid temperature increase occurs, too rapid to achieve a steady state, given the thermal conductivity of the electrodes. The increasing temperature leads to an exponentially increasing Joule heating via Eq. (1), and a negative-resistance region occurs. Eventually, the temperature becomes sufficiently high so that the conductivity no longer rises much with additional increase of T, and the steady state can be achieved. This phenomenon is called thermal runaway. Since a fluctuation in temperature always leads to a localized hot spot somewhere in the material, a filamentary conducting path exists, just as in avalanche and Zener breakdown. If the material is not decomposed or otherwise destroyed by the fast heating rate and the high steady-state temperature, thermal runaway can be reversible. A major distinction between thermal runaway and the mechanisms of Sec. 3.1 is that it is assumed that currents are ohmic in the thermal theory. The conductivity is always determined only by the temperature, and is taken as independent of field. This model is also completely independent of polarity.

We have previously noted that many materials exhibit insulator-metal transitions as a function of temperature, either at the melting point or below. Such transitions clearly make thermally induced nonequilibrium transitions likely, without the necessity for attaining thermal runaway conditions. Thermally induced threshold switching in this special class of materials can be much more rapid than ordinary thermal runaway.

3.2.4. **Trap-filling Mechanisms.** These mechanisms become possible because of the peculiar nature of the highly disordered chalcogenide glasses. The currently accepted band model for these materials[77] suggests that large but equal concentrations of positive and negative traps exist at all temperatures. Because of this, a highly conducting state becomes possible if all traps are filled, since the material is then neutral. Recently, several mechanisms for threshold switching have been proposed which make explicit use of this possibility. Henisch, et al.[78] suggested a model in which the main feature is the formation of a negative space charge near the cathode, caused by the trapping of injected electrons, and similarly a positive space charge near the anode. These space-charge regions limit the current and act as a virtual electrodes, increasing the field in the bulk. Eventually, the two space-charge regions overlap in the center of the film leading to a region in which all positive and negative traps are filled. But since *equal* concentrations of these traps are present, this is a neutral region, and is thus highly conducting. Consequently, the state is unstable, and electrodes will be accelerated toward the anode, holes toward the cathode. Very rapidly, a highly conductive state is formed, and is sustained by means of double injection, provided a holding voltage greater than the mobility gap is maintained across the film.

Schmidlin[79] suggested that oxide barriers form on both electrodes, resulting in the buildup of accumulation regions that eventually produce a conducting mode sustained by double injection. This model is very similar to that of Henisch, et al.[78] except that it is the existence of poor contacts that leads to switching.

Finally, Fritzsche and Ovshinsky[80] suggested a somewhat different model, which is sketched in Fig. 8. They assume that the field-induced excess carriers lead to the formation of depletion regions near the electrodes, since the electrodes are unable to supply sufficient electrons and holes to maintain the nonequilibrium current. The resulting Schottky barriers grow with increasing voltage, but the large density of traps in chalcogenide glasses allows the entire depletion region to be accommodated within 30Å of the electrodes, and tunneling can easily take place. Trap-limited current proceeds until all traps are filled, thus triggering the switching. Once again, the conducting state is maintained if the applied voltage exceeds the mobility gap.

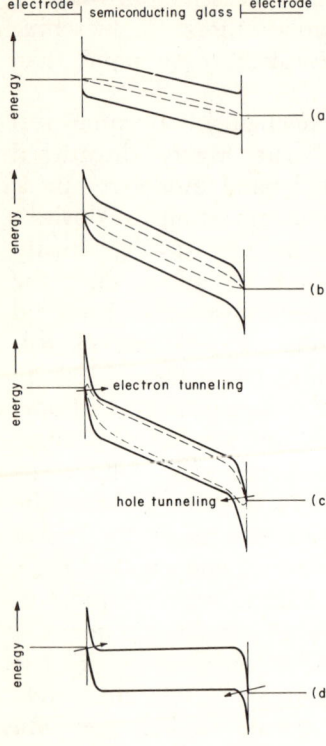

Fig. 8 *Model of Fritzsche and Ovshinsky for threshold switching in glasses: (a) Initial state of low conductivity after voltage is applied; (b) formation of Schottky barriers resulting from nonequilibrium carrier concentration in the bulk; (c) prebreakdown state with trap-limited current; (d) stable conducting state reached after all traps are filled.*

3.3. Comparison of Theory and Experiment

Almost all of the mechanisms discussed in the previous section have been observed in some class of materials. Because of the recent interest in threshold and memory switching in chalcogenide glasses, we shall restrict discussion to this particular group of amorphous solids.

Avalanche breakdown would appear to be just about the least likely mechanism for switching in chalcogenide glasses. The energy gap in these materials is of the order of 1 eV, the mean free path of carriers is about 5Å, and the threshold field is of the order of 10^5 V/cm. This means that the carrier acquires an energy of the order of 0.005 eV between collisions, and any impact ionization across a 1 eV gap would appear to be impossible. However, the possibility of a mean free path that increases sharply with increasing energy above the bottom of the conduction band makes an immediate dismissal of any impact ionization somewhat cavalier. Nevertheless, no hot-electron effects have yet been observed in chalcogenide glasses.

Zener breakdown can also be eliminated, either on the basis that fields of 10^5 V/cm will not yield a tunneling probability in a 1 eV band gap material sufficiently large to overcome the recombination rate, or because of the fact that a large negative-resistance region is incompatible with the necessity of maintaining a high tunneling rate. In addition, the recombination rate in disordered systems, which contain large densities of traps throughout the gap, should be particularly high.

We cannot eliminate the possibility of a nonequilibrium Mott transition being responsible for the switching, although it does not appear likely. In chalcogenide glasses, carriers, excited across the energy gap are already delocalized, and excitation effects appear to be unimportant. The low mobility appears to be a reflection of intense scattering arising from atomic disorder, and is not due to a localization of the carriers. Thus, a real Mott transition cannot occur. Similarly, nonequilibrium electron-phonon decoupling can be eliminated. An Anderson transition which delocalizes the states in the band tails is a possibility, but the density of localized states is probably insufficient to explain the conductivity jump on switching, and the sharpness of this type of transition is dubious. A field-induced phase transformation is unlikely because of the small degree of ionicity in the chalcogenide glasses. A recent analysis[81] has shown that the magnitude of the effective charge on a Te atom with one nearest-neighbor Ge atom in amorphous Te-Ge alloys is less than 0.2 e.

Thermal runaway has been suggested as the explanation of threshold switching on numerous occasions.[82-84] Although there are many examples of thermal runaway in thicker samples, the breakdown field appears to be too small in thinner films to be explained on a purely thermal basis. In addition, there are several other experimental indications against thermal runaway being the primary cause of the switching.[85]

The double-injection models are consistent with a large number of experimental results,[85] but the models of Henisch, et al.[78] and Schmidlin[79] appear to be eliminated by the essential independence of the preswitching delay time on polarity reversal.[86,87] This result is, however, consistent with the Fritzsche-Ovshinsky model, which has the present advantage of not yet being in major conflict with any of the observations.

Memory switching, as is evident from Fig. 7, is preceded by threshold switching, and, in fact, is obtained only after the material is held in the conducting state for a minimum time.[88] In at least one composition, $Te_{81}Ge_{15}As_4$, a consistent model for memory switching has appeared.[89,90] In this material, a nonequilibrium phase separation takes place under breakdown conditions, in which Te-rich regions form and crystallize. Polycrystalline As-doped Te is semi-metallic in nature and leads to the conducting equilibrium state. The resistive state is only reattained after application of a short, intense current pulse, which melts and quenches the conducting filament, thus reforming the homogeneous glass.

ACKNOWLEDGMENT

This research was supported by the Advanced Research Projects Agency of the Department of Defense and was monitored by U.S. Army Research Office—Durham, under Contract No. DAHCO4-70-C-0048.

REFERENCES

1. Drickamer, H. G.: *Solid State Phys.*, **17**:1 (1965).
2. Ioffe, A. F., and A. R. Regel: *Progr. Semicond.*, **4**:237 (1960).
3. Adler, D.: *Rev. Mod. Phys.*, **40**:714 (1968).
4. Raccah, P. M., and A. Mooradian: *Proc. 10th Int. Conf. Phys. Semicond., Cambridge, Mass., 1970* (in press).
5. Gildart, L., and D. F. Clifton: *AFOSR Tech. Note* 58, p. 849, 1958, unpublished.
6. Alexander, M. N., and D. F. Holcomb: *Rev. Mod. Phys.*, **40**:815 (1968).
7. Fritzsche, H.: *J. Phys. Chem. Solids*, **6**:69 (1958).
8. See Adler, D: *Essays in Physics*, **1**:33 (1970) for a review of the theoretical situation as of mid-1969.
9. Halperin, B. I., and T. M. Rice: *Solid State Phys.*, **21**:115 (1968).
10. Adler, D., and H. Brooks: *Phys. Rev.*, **155**:826 (1967).
11. Adler, D., J. Feinleib, H. Brooks, and W. Paul: *Ibid.*, **155**:851 (1967).
12. Hanamura, E.: *Rev. Mod. Phys.*, **40**:744 (1968).
13. Aronov, A. G., and E. K. Kudinov: *Sov. Phys. JETP*, **28**:704 (1969).
14. Kristoffel, N., and P. Konsin: *Phys. State. Sol.*, **28**:731 (1968).
15. Mattis, D. C., and W. D. Langer: *Phys. Rev. Lett.*, **25**:376 (1970).
16. de Graaf, A. M., and R. Luzzi: *Helv. Phys. Acta*, **41**:764 (1968).
17. Slater, J. C.: *Phys. Rev.*, **82**:538 (1951).
18. Adler, D.: *Phys. Rev. Lett.*, **17**:139 (1966).
19. Johansson, B., and K. F. Berggren: *Phys. Rev.*, **181**: 855 (1969).
20. Bulaevskii, L. N., and D. I. Khomskii: *Sov. Phys.-Sol. State*, **9**:2422 (1968).
21. Mott, N. F.: *Proc. Phys. Soc.*, **A62**:416 (1949).
22. ———: *Phil. Mag.* **6**:287 (1961).
23. Fröhlich, H.: in P. O. Lowdin (ed.), "Quantum Theory of Atoms, Molecules, and the Solid State," p. 465, Academic Press, Inc. New York, 1966.
24. Hyland, G. J.: *J. Phys.*, **C1**:189 (1968).
25. ———: *Rev. Mod. Phys.*, **40**:739 (1968).
26. Falicov, L. M., and J. C. Kimball: *Phys. Rev. Lett.*, **22**:997 (1969).
27. Ramirez, R., L. M. Falicov, and J. C. Kimball: *Phys. Rev. B* (in press).
28. Mott, N. F.: *Phil. Mag.*, **20**:1 (1969).
29. Samara, G.: *J. Geophys. Res.*, **72**:671 (1967).
30. Jerome, D., M. Rieux, and M. Achard: *Proc. Int. Conf. on Phys. of Solids under Pressure, Grenoble, France, 1969*, Editions du C.N.R.S., Paris, 1970.
31. However, as we shall discuss later, there is evidence for a possible Mott transition under pressure in the $Cr_2xV_{2(1-x)}O_3$ system (see Ref. 55).
32. Gildart, L.: *J. Non-Crystalline Solids*, **2**:240 (1970).
33. Parada, N. J., and G. W. Pratt, Jr.: *Phys. Rev. Lett.*, **22**:180 (1969).
34. Anderson, P. W.: *Phys. Rev.*, **109**:1492 (1958).
35. Mott, N. F.: *Rev. Mod. Phys.*, **40**:677 (1968).
36. Straumanis, M. E., and T. Ejima: *Acta Cryst.*, **15**:404 (1962).
37. Pearson, A. D.: *J. Phys. Chem. Sol.*, **5**:316 (1958).
38. Naylor, B. F.: *J. Am. Chem. Soc.*, **68**:1077 (1946).
39. Moon, R. M., T. Riste, W. C. Koehler, and S. C. Abrahams: *J. Appl Phys.*, **40**:1445 (1969).
40. Yahia, J., and H. P. R. Frederikse: *Phys. Rev.*, **123**:1257 (1961).
41. Honig, J. M., and T. B. Reed: *Ibid.*, **174**:1020 (1968).
42. Van Zandt, L. L., J. M. Honig, and J. B. Goodenough: *J. Appl. Phys.*, **39**:594 (1968).
43. Anderson, G.: *Acta Chem. Scand.*, **10**:623 (1956).
44. Kosuge, K.: *J. Phys. Soc. Japan*, **22**:551 (1967).
45. Ladd, L. A., and W. Paul: *Solid State Commun.*, **7**:425 (1969).
46. Verleur, H. W., A. S. Barker, and C. N. Berglund: *Phys. Rev.*, **172**:788 (1968).
47. Paul, W.: *Mat. Res. Bull.*, **5**:000 (1970).
48. Adler, D.: *Proc. Int. Conf. on Phys. of Solids under Pressure, Grenoble, France, 1969*, p. 165, Editions du C.N.R.S., Paris, 1970.
49. Kennedy, T. N., and J. Mackenzie: *J. Non-Cryastalline Solids*, **1**:326 (1969).

50. Mackenzie, J. D.: *J. Vac. Sci. Tech.*, **6**:658 (1969).
51. Dernier, P. D., and M. Marezio: To be published.
52. Feinleib, J., and W. Paul: *Phys. Rev.*, **155**:841 (1967).
53. Moon, R. M.: *Phys. Rev. Lett.*, **25**:527 (1970).
54. Austin, I. G., and C. E. Turner: *Phil. Mag.*, **19**:939 (1969).
55. McWhan, D. B., T. M. Rice, and J. D. Remeika: *Phys. Rev. Lett.*, **23**:1384 (1969).
56. Rice, T. M., and D. B. McWhan: *IBM J. Res. Dev.*, **14**:251 (1970).
57. McWhan, D. B., and J. P. Remeika: *Phys. Rev. B* (in press).
58. Jayaraman, A., D. B. McWhan, J. P. Remeika, and P. D. Dernier: *Phys. Rev. B* (in press).
59. Menth, A., and J. P. Remeika: *Phys. Rev. B* (in press).
60. Gossard, A. C., D. B. McWhan, and J. P. Remeika: *Phys. Rev. B* (in press).
61. Wertheim, G. K., J. P. Remeika, H. J. Guggenheim, and D. N. E. Buchanan: *Phys. Rev. Lett.*, **25**:94 (1970).
62. McWhan, D. B., and T. M. Rice: *Phys. Rev. Lett.*, **22**:887 (1969).
63. Rice, T. M., D. B. McWhan, and W. F. Brinkman: *Proc. 10th Int. Conf. Phys. Semicond., Cambridge, Mass., 1970* (in press).
64. Rice, T. M., and W. F. Brinkman: This volume.
65. McWhan, D. B., T. M. Rice, and J. P. Remeika: *Proc. Int. Conf. on Phys. of Solids under Pressure, Grenoble, France, 1969*, Editions du C.N.R.S., Paris, 1970.
66. Barker, A. S., H. C. Montgomery, and J. P. Remeika: *Bull. Am. Phys. Soc.*, **51**:386 (1970).
67. Fischer, D. W.: *J. Appl. Phys.*, **40**:4151 (1969).
68. Gainotti, A., C. Ghezzi, and M. Manfredi: *Nuova Cimento*, **62B**:121 (1969).
69. Zhuze, V. P., A. A. Andreev, and A. I. Shelykh: *Sov. Phys.-Solid State*, **10**:2914 (1969).
70. Jones, E. D.: *J. Phys. Soc. Japan*, **27**:1692 (1969).
71. Akimoto, S., and H. Fujisawa: *Tech. Rept. of ISSP*, Ser. A, No. 135, 1964, unpublished.
72. See discussion to Ref. 65.
73. Ridley, B. K.: *Proc. Phys. Soc. (London)*, **81**:996 (1963).
74. Ovshinsky, D. R.: *Phys. Rev. Lett.*, **21**:1450 (1968).
75. Feinleib, J.: *Int. J. Magnetism* (in press).
76. Rogachev, A. A.: *Proc. 9th Int. Conf. Phys. Semicond., Moscow, 1968*, p. 407, Publishing House, Leningrad, 1968.
77. Cohen, M. H., H. Fritzsche, and S. R. Ovshinsky: *Phys. Rev. Lett.*, **22**:1065 (1969).
78. Henisch, H. K., E. A. Fagen, and S. R. Ovshinsky: *J. Non-Crystalline Solids*, **4**:538 (1970).
79. Schmidlin, F. W.: *Phys. Rev.*, **B1**:1583 (1970).
80. Fritzsche, H., and S. R. Ovshinsky: *J. Non-Crystalline Solids*, **2**:393 (1970).
81. Adler, D., M. H. Cohen, E. A. Fagen, and J. C. Thompson: *Ibid.*, **3**:402 (1970).
82. Stocker, J. J., C. A. Barlow, Jr., and D. F. Weirauch: *Ibid.*, **4**:523 (1970).
83. Thomas, D. L., and A. C. Warren: *Electronics Lett.*, **6**:62 (1970).
84. Collins, F. M.: *J. Non-Crystalline Solids*, **2**:498 (1970).
85. Adler, D.: *Int. J. Magnetism* (in press).
86. Shanks, S. R.: *J. Non-Crystalline Solids*, **2**:504 (1970).
87. Balberg, I.: *Appl. Phys. Lett.*, **16**:491 (1970).
88. Evans, E. J., J. H. Helbers, and S. R. Ovshinsky: *J. Non-Crystalline Solids*, **2**:334 (1970).
89. Adler, D., J. M. Franz, C. R. Hewes, B. P. Kraemer, D. J. Sellmyer, and S. D. Senturia: *Ibid.*, **4**:330 (1970).
90. Bienenstock, A., F. Betts, and S. R. Ovshinsky: *Ibid.*, **2**:347 (1970).

DISCUSSION *on Paper by D. Adler*

This paper was discussed with the paper by T. M. Rice.

SOME ASPECTS OF THE
THEORY OF THE MOTT TRANSITION

T.M. Rice and W. F. Brinkman
Bell Telephone Laboratories, Inc.
Murray Hill, New Jersey

ABSTRACT

Two new results for the Hubbard model with a single half-filled band are presented. First, using the approximate variational approach of Gutzwiller, it is shown that in the strongly correlated metallic state, in which there are only a few doubly occupied sites, the screening constant is proportional to the number of doubly occupied sites and therefore goes to zero at the metal-insulator transition. Second, an approximation which gives a reasonable description of the properties of a hole (or electron) in the half-filled band in the atomic limit, which was derived recently by the present authors using a graphical technique, is formulated here using a Green's function technique. In this approximation the infinite hierarchy of Green's function equations is not truncated or decoupled but replaced by a set of recurrence relations which have a simple solution.

1. INTRODUCTION

Some 20 years ago Mott[1] raised the interesting question of how the transition between a localized magnetic insulating phase and an itinerant metallic phase would proceed. He proposed that this transition, which is now generally referred to as a Mott transition, should be first order, arguing that a phase with arbitrarily small numbers of carriers would not be stable because of the long-range nature of the Coulomb interaction. Since that time there has been considerable interest both in the search for such a transition experimentally and in attempting to construct a satisfactory microscopic theory of the transition.

In his original paper[1] Mott suggested that magnetic insulators such as NiO should undergo a transition to a metallic phase at high pressures. However, experimentally it is found that the pressures required for such materials exceed that available in the laboratory.[2] Recently it has been found that the metal-to-insulator transition at $T \approx 170°K$ in V_2O_3 discovered by Föex[3] is part of a generalized phase diagram.[4-6] This generalized phase diagram is obtained by applying pressure and by the addition of small amounts of Cr_2O_3. It is found that there are three phases, a paramagnetic metallic phase M, a paramagnetic insulating phase I, and an antiferromagnetic insulating phase AF. All three phase transitions, AF-I, AF-M, and M-I, are first order. The M-I phase transition involves no change in long-range order since both phases are paramagnetic and have the corundum crystal structure. The M-I phase boundary terminates at a line of solid-solid critical points. The phase diagram has been interpreted as an example of a Mott transition.[4,5] The interested reader can find a full discussion of the experimental aspects of this work elsewhere.[4,6]

The theoretical studies of the Mott transition have centered on the Hubbard model Hamiltonian. In this Hamiltonian, introduced by Hubbard[7] and others,[8,9] a localized or Wannier representation is taken for the electronic wave functions and only two terms of the full Hamiltonian are retained. One term represents the Coulomb repulsion U when two electrons are on the same site and the other term the tunneling of electrons from one site to its neighbors. For a half-filled band in the atomic, or large U limit, the model is clearly insulating while for small U it is metallic. Thus the model Hamiltonian describes a metal-insulator transition as U is varied. The simplifying assumption made in this model may well make comparison with experiment impossible. However even within this model the possible phase diagrams for the metal-insulator transition have not been established. In Sec. 2 of this paper we discuss this problem in the light of the results of the Gutzwiller approach to the metallic phase near the metal-insulator transition.[10,11] In fact, we show that the Gutzwiller approach predicts a screening length that becomes infinite as the transition is approached without involving nested phases as in the model of Kohn.

In Sec. 3 we discuss the insulating phase in the atomic limit. The properties of an extra electron or hole in this limit were investigated recently by the present authors[13] by analyzing the problem in terms of walks on a lattice. It was shown that a reasonable approximation is obtained by summing all paths which are self-retracing with no closed loops. In this paper the problem will be formulated using Green's functions. It is found that the equivalent approximation is not obtained by truncating the hierarchy of Green's function equations but by approximating the equations by a simple recurrence relation.

2. SCREENING IN STRONGLY CORRELATED METALS

As the metal-insulator phase boundary is approached from the metallic side, the metal will become strongly correlated and the average number of doubled occupied sites will become small. An interesting approach to this problem was suggested some years ago by Gutzwiller.[10] He constructed a trial wave function by starting with the conventional Bloch state for noninteracting electrons and reducing the amplitude of all components in which ν atoms are doubly occupied by an amount η^ν where $0 < \eta < 1$. He calculated the needed matrix elements for the up spin electrons by assuming a random distribution for the down spin electrons, arguing that this procedure should be a good approximation to an optimally chosen generalization of his wave function. This led to an explicit and spin-symmetric expression for the energy as a function of η. This expression was then minimized with respect to η and the ground state energy obtained. Gutzwiller used his results to obtain a criterion for itinerant ferromagnetism. Recently, the present authors[11] have pointed out that Gutzwiller's approach gives a metal-insulator transition for a half-filled band.

The Hubbard Hamiltonian may be written as

$$H = \sum_{i\sigma j} t_{ij} C^+_{i\sigma} C_{j\sigma} + U \sum_i n_{i\uparrow} n_{i\downarrow} \qquad (1)$$

where U is the intra-atomic repulsive interaction and t_{ij} is the matrix element for tunneling between sites i and j. The operator $C_{i\sigma}^+$ creates an electron of spin σ in the Wannier function localized at site i and $n_{i\sigma} \equiv C_{i\sigma}^+ C_{i\sigma}$. In the Bloch representation the creation operator for an electron in state k and spin σ is given by $a_{k,\sigma} = 1/\sqrt{N} \sum_i C_{i\sigma}^+ \exp(i\mathbf{k} \cdot \mathbf{R}_i)$ where $\mathbf{R}_i$ is the position of the lattice site i. Thus the kinetic energy in the Bloch representation $\epsilon(\mathbf{k})$ is given by

$$\epsilon(\mathbf{k}) = \sum_j t_{ij} \exp[\mathbf{k} \cdot (\mathbf{R}_i - \mathbf{R}_j)] \qquad (2)$$

Following Gutzwiller[10] let us assume for simplicity that the average energy in the band is zero.

His approximate expression for the ground state energy per site is

$$<H>_N = 2\bar{m}q\bar{\epsilon} + \bar{\nu}U \qquad (3)$$

where $\bar{m}$ and $\bar{\nu}$ are the number of particles of one spin per site and the fraction of doubly occupied sites respectively and $\bar{\epsilon} = m^{-1}\sum_{k<k_F}\epsilon(\mathbf{k})$ is the average energy of the electrons without correlation. The quantity q is the discontinuity in the single particle occupation number $<n_k>$ at the Fermi surface. By minimizing the energy in Eq. (3) with respect to the parameter η, Gutzwiller derived the following equation relating η to $\bar{\nu}$ and q to $\bar{\nu}$

$$\eta = \frac{[\bar{\nu}(1 - 2\bar{m} + \bar{\nu})]^{1/2}}{(\bar{m} - \bar{\nu})} \qquad (4)$$

and

$$q = \frac{(\bar{m} - \bar{\nu})[(1 - 2\bar{m} + \bar{\nu})^{1/2} + \bar{\nu}^{1/2}]^2}{\bar{m}(1 - \bar{m})} \qquad (5)$$

For the special case of the half-filled band for which $\bar{m} = 1/2$ we find on minimizing the energy with respect to $\bar{\nu}$ (or equivalently with respect to η)

$$\bar{\nu} = \frac{1}{4}\left(1 - \frac{U}{U_0}\right) \qquad (6)$$

and

$$q = \left[1 - \left(\frac{U}{U_0}\right)^2\right] \quad (7)$$

and substituting in Eq. (3) we find for the expectation value of the energy[11]

$$<H>_N = \bar{\epsilon}\left(1 - \frac{U}{U_0}\right)^2 \quad (8)$$

where $U_0 = -8\bar{\epsilon}$. At the critical value of the interaction strength $U = U_0$, both the number of doubly occupied sites and the discontinuity in the single particle momentum occupation number at the Fermi surface go to zero. At the same time the value of the energy also approaches zero, the expectation value of the energy in the completely localized insulating ground state. Since some magnetically ordered localized state will have a lower energy than the completely localized ground state, a transition to a magnetically ordered state will presumably occur for a value of U less than U_0. Nevertheless, we can use this trial wave function to study the properties of a strongly correlated paramagnetic metallic state in which the fraction of doubly occupied sites is small. The present authors[13] showed that in this state both the effective mass and spin susceptibility were enhanced by a common factor

$$\frac{m^*}{m} = q^{-1} = \left[1 - \left(\frac{U}{U_0}\right)^2\right]^{-1} \quad (9)$$

and

$$\chi_s^{-1} = \frac{1 - (U/U_0)^2}{\rho(\epsilon_F)}\left[1 - [U\rho(\epsilon_F)]\left(\frac{1 + U/2U_0}{(1 + U/U_0)^2}\right)\right] \quad (10)$$

where $\rho(\epsilon_F)$ is the band structure density of states at the Fermi energy. Thus as U approaches U_0 both the susceptibility and the effective mass diverge in proportion to $[1 - (U/U_0)^2]^{-1}$.

It is also of interest to examine the nature of the screening in this strongly correlated metallic phase. The screening constant q_s^2 can be defined in terms of long wavelength limit of the static dielectric constant[14]

$$\lim_{q \to 0} \epsilon(q, 0) = 1 + \frac{q_s^2}{q^2}$$

It is directly related to the derivative of the chemical potential μ with respect to the number of electrons N

$$q_s^2 = \frac{4\pi e^2}{\Omega} \frac{dN}{d\mu} \qquad (11)$$

where Ω is the volume and e is the electronic charge. The screening constant without correlations q_{s0}^2 is proportional to the density of states

$$q_{s0}^2 = 8\pi e^2 \rho(\epsilon_F) \qquad (12)$$

To calculate the derivative in Eq. (11) we must evaluate the second derivative of the ground state energy with respect to the number of particles. Let $\bar{m} = 1/2 + \delta$ and expand the energy (Eq. (3)) in powers of δ

$$<H>_N = E_0(\nu) + \delta E_1(\nu) + \delta^2 E_2(\nu) \qquad (13)$$

The energy is minimized with respect to ν and expanding about the old equilibrium value ν_0 given by Eq. (6) we write

$$\nu = \nu_0 + A\delta \qquad (14)$$

Substituting in Eq. (13) and minimizing $A = -E_1'(\nu_0)/E_0''(\nu_0)$ where the primes denote derivatives with respect to ν. Upon substitution back in Eq. (13) we get the following expansion for the energy

$$<H>_N = E_0(\nu_0) + \delta E_1(\nu_0) + \delta^2 \left(E_2 - \frac{(E_1')^2}{2E_0''} \right) \qquad (15)$$

It remains to calculate the coefficients E_0, E_1, and E_2. To evaluate these coefficients we expand $\bar{\epsilon}$ and q to second order in δ

$$\bar{\epsilon} = \bar{\epsilon}_0 + \frac{\delta^2}{\rho(\epsilon_F)} \quad (16)$$

and from Eq. (5)

$$q(\delta) = 16\bar{v}\left(\frac{1}{2} - \bar{v}\right) + 8\delta(4\bar{v} - 1) + 4\delta^2\left(\frac{-1}{2\bar{v}} - 3 + 8\bar{v} - 16\bar{v}^2\right) \quad (17)$$

After some algebra we arrive at the result

$$\frac{q_s^2}{q_{s0}^2} = \left[1 - \frac{U^2}{U_0^2}\right]^{-1}\left[1 + \frac{\rho(\epsilon_F)U}{(1 - U/U_0)^2(1 + U/U_0)}\right]^{-1} \quad (18)$$

Thus as $U \to U_0$ we find

$$q_s^2 = \frac{8\pi e^2}{U_0}\left(1 - \frac{U}{U_0}\right) + 0\left(1 - \frac{U}{U_0}\right)^2 \quad (19)$$

Therefore as U approaches U_0 the screening constant is proportional to $\bar{v}$, the fraction of doubly occupied sites, and goes to zero. Equivalently the screening length becomes infinite. It is interesting to express this result in terms of the Landau Fermi Liquid theory. Using the relation[14]

$$\frac{q_s^2}{q_{s0}^2} = \frac{m^*/m}{1 + A_0} \quad (20)$$

where A_0 is the uniform average of the singlet part of the Landau f function and Eqs. (9) and (18), we find

$$A_0 = U\rho(\epsilon_F)\left(1 + \frac{U}{U_0}\right)^{-1}\left(1 - \frac{U}{U_0}\right)^{-2} \quad (21)$$

Thus, although the effective mass diverges as the transition is approached, A_0 diverges faster and as a result the screening constant goes to zero.

The Gutzwiller approach gives an appealing description of the metallic phase in which the Fermi surface always encloses the correct volume while the magnitude of the discontinuity in single particle occupation number at the Fermi surface goes to zero as the system approaches the metal-insulator phase boundary. In the strongly correlated metallic state there are two types of particle-hole excitations. One type corresponds to the fluctuations of the spin on the singly occupied sites. These excitations build up at low frequency as $U \to U_0$ and are responsible for the large effective mass and the large susceptibility. As we pass over into the insulating phase, these excitations become the spin waves of the magnetic insulator. The second class of excitations corresponds to the creation of doubly occupied and empty sites. The main weight of these excitations builds up at an excitation energy of order U as U approaches U_0. These are the excitations which carry the current and are responsible

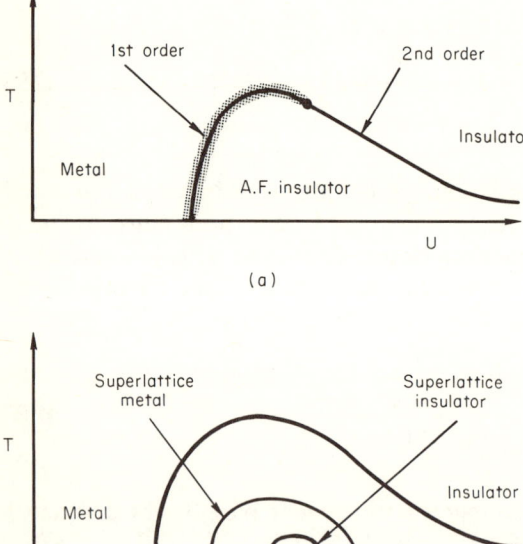

Fig. 1 *Two possible phase diagrams for the metal-insulator transition in a Hubbard model extended to include long-range interactions.*

for the screening. As we pass into the insulating phase, an energy gap develops for the creation of such an excitation leading to insulating behavior.

The detailed nature of the metal-insulator phase transition in the Hubbard model is still an open question. The Gutzwiller[10] approach has been applied only to ground state and to properties which are obtained by differentiating the ground state. Thus the very interesting question of the stability of the Gutzwiller wave function against antiferromagnetism has not been investigated. The present authors[11] have pointed out that the Stoner enhancement factor in χ_s in Eq. (10) is reduced by the correlations and therefore the possibility of itinerant ferromagnetism prior to the metal-insulator transition is less. They have further speculated that this may be true also for $\chi(\mathbf{Q})$ and thus for itinerant antiferromagnetism, and have suggested that a phase diagram of the form in Fig. 1a is possible. This would be in contrast to the form in Fig. 1b suggested by Kohn[12] on the basis of the Hartree-Fock theory for a model with long-range interactions at zero temperature. In this latter case the Fermi surface is successively truncated by the superlattice itinerant antiferromagnetic energy gaps.

3. A GREEN'S FUNCTION APPROACH TO THE ATOMIC LIMIT

For large values of the Coulomb interaction Hubbard[7] showed that the Hamiltonian (1) for a half-filled band describes an insulator with a Mott-Hubbard correlation energy gap which is $\sim U$ for large U. However, unlike the weak coupling limit where the limit $U \to 0$ is trivial, the atomic limit where t_{ij} is held fixed but $t_{ij}/U \to 0$ is a nontrivial problem. The atomic limit has been studied by a number of authors. Harris and Lange[15] have shown that the one particle spectral weight function has two bands separated in energy by U, and that the values of the first two nonzero moments for these bands are unchanged from that of the tight-binding band with total spectral weight (1/2). Nagaoka[16] has formulated the Hubbard model in this limit in terms of the number of possible paths on a lattice. The present authors[13] have studied the form of the density of states for a single hole (or electron) in a half-filled one-band Hubbard model in the atomic limit in detail using Nagaoka's formulation.[16] It is found that the bulk of the band for an additional electron or hole is reduced by approximately 25 or 20 percent depending on whether the spins are ordered antiferromagnetically or are disordered. There are, however, band tails which extend all the way to the full bandwidth of a particle without interactions. It was found also that a reasonable approximation to the bulk of the band was obtained by summing all paths which are self-retracing with no closed loops. We shall give here a Green's function derivation of this approximation. This is of interest for two reasons. First, it can be directly compared to the several Green's function decoupling schemes which have been proposed and second, it can hopefully be generalized more easily to the case of finite U.

Let us consider the atomic limit of the Hamiltonian in Eq. (1). For simplicity we will consider only the simple cubic lattice where the number of nearest neighbors $z = 6$ and we will take t_{ij} to be nonzero only when i and j are nearest neighbors. We will use the double-time Green's function technique of Zubarev[17] introduced to this problem by Hubbard.[7] In this technique one studies the equations of motion of a Green's function $\langle\!\langle A; B \rangle\!\rangle_E$ which is regular everywhere in the whole complex E plane except on the real axis.

$$\langle\!\langle A; B \rangle\!\rangle_E = \langle\!\langle A; B \rangle\!\rangle_E^{(+)} \quad \text{if} \quad \text{Im}(E) > 0$$
$$= \langle\!\langle A; B \rangle\!\rangle_E^{(-)} \quad \text{if} \quad \text{Im}(E) < 0 \qquad (22)$$

where (+) and (−) denote the Fourier transforms of the advanced and retarded Green's functions

$$\langle\!\langle A(\tau); B(\tau') \rangle\!\rangle^{(\pm)} = \mp i\theta\{\pm(\tau - \tau')\}\langle[A(\tau), B(\tau')]\rangle \qquad (23)$$

where $A(\tau) = \exp(iH\tau) A(0) \exp(-iH\tau)$; $[A, B] = AB - BA$ and $\theta(x)$ is the step function $\theta(x) = 1$ if $x > 0$ and $= 0$ otherwise. The angular brackets denote the trace over all states

$$\langle X \rangle = Z^{-1} Tr\{X \exp(H - \mu N)\}, Z = Tr\{\exp[-\beta(H - \mu N)]\} \qquad (24)$$

where β is reciprocal temperature and μ is the chemical potential.

The basic equation of motion satisfied by the Green's functions is

$$E\langle\!\langle A; B \rangle\!\rangle_E = \langle[A, B]\rangle + \langle\!\langle [A, H]; B \rangle\!\rangle_E \qquad (25)$$

We shall apply this technique to the study of the single particle, equal-site Green's function

$$G_{ii}^\sigma(E) = \langle\!\langle C_{i\sigma}; C_{i\sigma}^+ \rangle\!\rangle_E \qquad (26)$$

The basic commutators we shall need are

$$[C_{i\sigma}, H] = t \sum_j C_{j\sigma} + U n_{i,-\sigma} C_{i\sigma} \qquad (27)$$

$$[n_{i\sigma}, H] = t \sum_j (C_{i\sigma}^+ C_{j\sigma} - C_{j\sigma}^+ C_{i\sigma}) \qquad (28)$$

and

$$[S_{i,-\sigma}, H] = t \sum_j (C_{i,-\sigma}^+ C_{j\sigma} - C_{j,-\sigma}^+ C_{i\sigma}) \qquad (29)$$

where we have introduced the notation $S_{i,\pm\sigma} = C_{i,\pm\sigma}^+ C_{i,\pm\sigma}$ for the spin-flip operators on site i and $t_{ij} = t$ for i and j nearest neighbors.

The equation of motion for G_{ii}^σ is given by

$$E G_{ii}^\sigma(E) = 1 + t \sum_j \langle\langle C_{j\sigma}; C_{i\sigma}^+ \rangle\rangle_E + U \langle\langle n_{i,-\sigma} C_{i\sigma}; C_{i\sigma}^+ \rangle\rangle_E \qquad (30)$$

Now the last term on the right-hand side involves two destruction operators on the same site and is therefore $\sim 1/U$ for large U. However since it enters Eq. (30) multiplied by U we must evaluate this term to order U^{-1} and then take the limit $U \to \infty$. We obtain

$$E \langle\langle n_{i,-\sigma} C_{i\sigma}; C_{i\sigma}^+ \rangle\rangle_E = \langle n_{i,-\sigma} \rangle + U \langle\langle n_{i,-\sigma} C_{i\sigma}; C_{i\sigma}^+ \rangle\rangle_E$$

$$+ t \sum_j \langle\langle n_{i,-\sigma} C_{j\sigma}; C_{i\sigma}^+ \rangle\rangle_E$$

$$- t \sum_j \langle\langle S_{i,-\sigma} C_{j,-\sigma}; C_{i\sigma}^+ \rangle\rangle_E$$

$$- t \sum_j \langle\langle C_{j,-\sigma}^+ C_{i,-\sigma} C_{i\sigma}; C_{i\sigma}^+ \rangle\rangle_E \qquad (31)$$

where we have made use of the identity $n_{i,-\sigma}^2 = n_{i,-\sigma}$. The last term on the right-hand side is again of order $1/U$ and may be dropped. Solving Eq. (31) and substituting in Eq. (30) we arrive at the following equation for $G_{ii}^\sigma(E)$ in the limit $U \to \infty$

$$EG_{ii}^{\sigma}(E) = 1 - \langle n_{i,-\sigma} \rangle + t \sum_{j} \mathcal{G}_{ij}^{\sigma}(E) \qquad (32)$$

where $\mathcal{G}_{ij}^{\sigma}$ denotes the combination of terms

$$\mathcal{G}_{ij}^{\sigma}(E) = \langle\langle (1 - n_{i,-\sigma}) C_{j\sigma}, C_{i\sigma}^{+} \rangle\rangle_{E}$$
$$+ \langle\langle S_{i,-\sigma} C_{j,-\sigma} ; C_{i\sigma}^{+} \rangle\rangle_{E} \qquad (33)$$

Let us examine these terms in detail. In the atomic limit for a half-filled band a complete set of states for the system is obtained by specifying the spin of the electron on each site. There are two possibilities of interest which are shown in Fig. 2. We denote the spin states $\pm \sigma$ by ↑ and ↓ respectively. (Clearly the spin on site i must be up). The two possibilities correspond to states in which the spin on site j is (1) parallel or (2) antiparallel to that on site i. In the former case the first term in Eq. (33) corresponds to the creation of the hole on site j and its subsequent destruction on site i. The second case corresponds to creating a hole in the spin ↓ state on j simultaneously flipping the spin on site i and then subsequently destroying the hole on site i as illustrated in Fig. 2b. Thus the Green's function which enters here is not the usual one where we create a hole on site i and subsequently destroy (in the same spin state) it on j irrespective of spin configurations. Rather it is a Green's function in which the hole propagates through the lattice, disturbing the spins where necessary, to avoid doubly occupying any intermediate site.

In a similar fashion we can derive the equation of motion for $\mathcal{G}_{ij}^{\sigma}(E)$

$$E\mathcal{G}_{ij}^{\sigma}(E) = tG_{ii}^{\sigma} + t \sum_{k \neq i} \mathcal{G}_{ijk}^{\sigma}(E) \qquad (34)$$

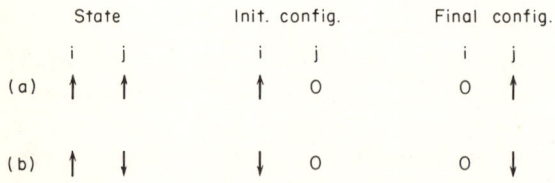

Fig. 2 *The initial and final configurations of the spins for the two possible spin states on the sites i and j.*

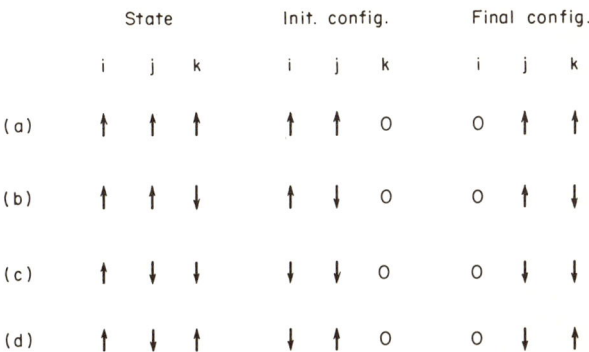

Fig. 3 *The initial and final configurations of the spins for the four possible spin states on sites i, j, and k.*

where k is any nearest neighbor of j but $k \neq i$. The Green's function $\mathcal{G}^\sigma_{ijk}(E)$ is defined as

$$\begin{aligned}
\mathcal{G}^\sigma_{ijk}(E) &= \langle\langle (1 - n_{i,-\sigma})(1 - n_{j,-\sigma}) C_{k\sigma}; C^+_{i\sigma} \rangle\rangle \\
&+ \langle\langle (1 - n_{i,-\sigma}) S_{j,-\sigma} C_{k,-\sigma}; C^+_{i\sigma} \rangle\rangle \\
&+ \langle\langle S_{i,-\sigma}(1 - n_{j\sigma}) C_{k,-\sigma}; C^+_{i\sigma} \rangle\rangle \\
&+ \langle\langle S_{j,+\sigma} S_{i,-\sigma} C_{k\sigma}; C^+_{i\sigma} \rangle\rangle
\end{aligned} \qquad (35)$$

The four terms correspond to the four possibilities illustrated in Fig. 3a–d for the motion of the hole through the lattice from site k to site i

We can in a similar fashion write down an equation for $\mathcal{G}^\sigma_{ijk}(E)$ which will couple to $\mathcal{G}^\sigma_{ijkl}(E)$ where l is any nearest neighbor site of k but $l \neq j$. The general equation we obtain is

$$E\mathcal{G}^\sigma_{i_0 i_1 \cdots i_n}(E) = t\mathcal{G}^\sigma_{i_0 i_1 \cdots i_{n-1}}(E) + t \sum_{i_{n+1} \neq i_{n-1}} \mathcal{G}^\sigma_{i_0 i_1 \cdots i_{n+1}} \qquad (36)$$

So far we have made no approximations and these equations are exact in the limit $U \to \infty$. The basic approximation, which corresponds to the summation in the earlier work of all self-retracing paths without closed loops, is to ignore any possibly coupling back to lower order Green's functions other than the coupling to the

11. Brinkman, W. F., and T. M. Rice: *Ibid.*, **B2**:4302 (1970).
12. Kohn, W.: *Phys. Rev. Lett.*, **19**:789 (1967).
13. Brinkman, W. F., and T. M. Rice: *Phys. Rev.*, **B2**:1324 (1970).
14. See, for example, P. Noziéres: "Theory of Interacting Fermi Systems," W. A. Benjamin, Inc., New York, 1963.
15. Harris, A. B., and R. V. Lange: *Phys. Rev.*, **157**:295 (1967).
16. Nagaoka, Y.: *Solid State Comm.*, **3**:409 (1965) and *Phys. Rev.*, **147**:392 (1966).
17. Zubarev, D. N.: *Usp. Fiz. Nauk.*, **71**:71 (1960); *Soviet Phys. Usp.*, **3**:320 (English translation.)

DISCUSSION on *Papers by D. Adler and T. M. Rice*

R. B. GRIFFITHS (to T. M. Rice): It is not clear to me what distinguishes a Mott transition from other types of insulator-metal transition. That is to say, there are transitions which occur under pressure in which there is an important change in crystal structure as well as a change in electrical conductivity, which one would presumably not call "Mott transitions." What distinguishes these from the transition in V_2O_3 and the transitions found when Cr is added to this compound?

T. M. RICE: There is a real difference between the metal-insulator transitions in the V_2O_3 system and a transition such as that observed in Ge under pressure discussed by David Adler in the previous talk. In Ge one has a simple first-order transition between a band semiconductor to a band metal with no magnetic phases involved. In the case of Ge the crystal phases of the semiconductor and metal are different but this need not be the case. For example, under pressure Bi and Yb change continuously from semimetals to semiconductors with no change in crystal structure. The theory of the excitonic insulator was developed to describe this situation and while the theory suggests the possibility of a magnetic super lattice excitonic phase very near the critical pressure, the basic transition is from a nonmagnetic band semiconductor to a nonmagnetic band metal. On the other hand Mott in 1949 considered the transition between the localized or Mott insulating phase, which is not describable within the framework of band theory, and the itinerant or band metallic phase. Such a transition may or may not involve a crystal symmetry change. In the case of Cr in V_2O_3 there is no change in crystal symmetry. The Mott insulating phase must be ordered magnetically at $0°K$ and in the V_2O_3 the transition observed at low temperatures is from an antiferromagnetic phase with a moment of $1.2\ \mu\beta$ per V ion to a paramagnetic metallic phase.

B. WIDOM: May I ask either of the speakers to tell us the present experimental status of two phenomena which in the past were thought perhaps to be metal-nonmetal transitions. The first is mercury, which as a high-density vapor has an electrical conductivity that is a rapidly varying function of the density. The second is solutions of alkali metals in liquid ammonia, where at one time it was suggested that there is a locus of conductivity anomalies that terminates at the critical consolute point.

D. ADLER (to B. Widom): The present situation is that neither transition appears to be a Mott transition, but both are primarily due to atomic rearrangements. The locus of conductivity anomalies correlates exactly with phase separation in the two systems. The analysis of the problem of the metal-ammonia solutions has been reviewed by M. H. Cohen and J. C. Thompson.[1] One of the reasons for the conclusion that the transition is not a Mott transition is that positron-annihilation experiments by P. G. Varlashkin and A. T. Stewart[2] show no change in the momentum distribution function through the transition. In the case of dense mercury vapors, Franck and Hensel[3] give strong evidence that the conductivity jump is primarily associated with atomic density changes. A Mott transition is apparently a quite elusive phenomenon.

T. M. RICE: In agreement with what David said, the transition in solutions of Li in ammonia is believed to be a mobility transition rather than a many body transition.

B. WIDOM: Is there a locus of sharp transitions, or just a range where the conductivity is changing gradually?

D. ADLER: Although a locus of transitions can be drawn, the actual conductivity changes occur gradually rather than sharply in the metal-ammonia systems and in the mercury vapor system above the critical point.

S. C. MOSS: At the metal-insulator transition, the fact that there is no crystal structure change is really irrelevant if, in fact, the vanadium atom position changes with respect to the three oxygens above and below it. Since the vanadium does not sit symmetrically in this site, it would seem that the entropy of the transition, much as W. N. Paul has suggested, could be taken up in an optical mode softening where the vanadiums and oxygens vibrate against each. Aram Monadian has indeed noticed such a softening in Ti_2O_3 where the transition, albeit reversed in high and low temperature phases is similar and this possibility should be seriously considered for V_2O_3.

T. M. RICE: There are changes in the ion positions across the metal-insulator phase boundary, but the vanadium-vanadium distances just pull apart. This is just what is expected in a simple picture of a Mott transition, which is somewhat surprising since we would expect hybridization with the oxygens to be important. However, Dernier has shown that the vanadium oxygen distances remain the same.

S. C. MOSS: How can the vanadium-oxygen distances remain constant if the vanadium shifts in the hole?

T. M. RICE: This is what happens. The corundum structure always stays the same. There is a certain amount of freedom in this structure, since there are 10 atoms per unit cell.

D. ADLER: Although all these results are consistent with a Mott transition, they are also consistent with an ordinary band transition. The argument that the insulating state has higher entropy than the metallic state and that this therefore eliminates the possibility of a band uncrossing, is incorrect if the phonon entropy changes significantly during the transition. Since the entropy change is quite small, and the decrease in entropy from a band semimetal to a band semiconductor is also small, only a minor change in phonon entropy can overturn your conclusion. There is a striking analogy between the high-temperature transition of V_2O_3 and the Ti_2O_3 transition. Mooradian and Raccah have shown that a 20 percent softening of a phonon mode occurs, despite the fact that the corundum symmetry is maintained. The c and a lattice parameters in Ti_2O_3 and V_2O_3 change in opposite directions in the 500-600°K range and the conductivity anomalies are similarly opposite.

T. M. RICE: While there is a certain resemblance between the high temperature anomaly in V_2O_3 and that observed in Ti_2O_3 in that both are continuous, there are two important differences between the two cases. First, in V_2O_3 one goes from metal to insulator with increasing temperature while the opposite is true for Ti_2O_3. Secondly, in Ti_2O_3 the magnetic susceptibility of the insulating phase is very much smaller than in the metallic phase while in V_2O_3 the susceptibility increases on going from the metal to insulator. Both these differences are strong evidence in favor of the interpretation of V_2O_3 as a Mott transition and not a band upcrossing. I should also like to point out that there is a continuous variation with Cr concentration for concentration greater than approximately 1 percent and of course Cr_2O_3 is believed to be a Mott insulator.

S. MOSS: Can't all the entropy of the transition be taken up by the electrons?

T. M. RICE: The change in entropy across the antiferromagnetic to insulating phase transition is roughly half the entropy that comes from $k \log (2S + 1)$ where $S = 1$, which is the maximum electronic entropy change possible.

S. MOSS: No, I mean the high-temperature transition.

T. M. RICE: That entropy change is very small, only 0.2 entropy units.

D. ADLER: That means only a small phonon entropy increase can more than compensate for the entropy decrease associated with a band uncrossing.

T. M. RICE: It is possible, but a band picture doesn't explain why Cr impurities do not bring about a metallic state.

D. ADLER: But we have also the added complication of a possible Anderson localization when Cr is added.

Ph. CHOQUARD: We are not easily going to exhaust the problems dealing with these kinds of transitions. For them, we have no order parameters, which is one reason why we have failed to solve these problems.

D. ADLER: There are a number of theoretical questions that still bother me. First, you claim the correlation energy is of the order of 5 eV as is the bandwidth, whereas the transition temperature is only 0.01 eV. This would appear to be an amazing accident, the transition temperature being only 1/500 of the other relevant energies.

T. M. RICE: Remember that we can pick up an order of magnitude from the fact that the transition temperature is proportional to $2zt^2/U$ where $2zt$ is the bandwidth. Since $2z$ is about 10, we pick up a factor of 10. Until one has a microscopic theory of the Mott transition it seems to me premature to discuss this question in detail.

D. ADLER: But it isn't clear what z would be used in the corundum structure, when only one C-axis nearest-neighbor exists. And even if $2z = 10$, we still must account for a factor of 50. Another point is that we are not sure how applicable the Hubbard approximation is to the Mott transition. In Mott's original paper, he invoked interatomic correlations and screening in order to argue that the nonmetal-metal transition is sharp, and these are all neglected in the Hubbard Hamiltonian. The Gutzwiller approximation is then a model of a model, and really cannot be trusted in explaining a real problem like V_2O_3. We are not even sure if the Hubbard Hamiltonian really has an insulator-metal transition in it if solved exactly. In one dimension, the exact solution of Lieb and Wu shows no transition.

T. M. RICE: As Overhauser first showed, one-dimensional models always have nested Fermi surfaces, and show no transition. I admit the Hubbard model may not apply to real cases like V_2O_3 and screening effects may be important, but until good calculations are performed we won't know the answer.

D. ADLER: I also have some reservations about the sacredness of Luttinger's theorem. After all, it depends on the convergence of a perturbation expansion.

T. M. RICE: Luttinger's theorem was shown to all orders of perturbation theory. Also, we have the added insight of experimental results. I know of no metal in which Luttinger's theorem breaks down, so why should we expect it to break down before the Mott transition?

D. ADLER: Isn't it peculiar that eight oxides of vanadium all undergo insulator-metal transitions and only one is magnetic? Why is the magnetism important only in V_2O_3?

T. M. RICE: I want to emphasize that the transition in V_2O_3 is different from the other transitions because it is antiferromagnetic at low temperatures.

D. ADLER: None of the others can be Mott transitions.

T. M. RICE: That is right. The others are not Mott transitions.

J. VILLAIN: For the Hubbard model, how many electrons per atom are necessary to obtain a Mott transition?

T. M. RICE: In a one-band Hubbard model, exactly one electron per atom is necessary. For a multi-band model, then one can have an integral number of electrons per atom larger than one.

G. ALEFELD: In respect to the nonequilibrium transitions, what is known about the change in magnetic properties, if you go from the insulating to the conducting state?

D. ADLER: Nothing is known about this because conduction takes place along a filamentary path in the interior of the material, and this filament has not yet been studied in the conducting state. In the chalcogenide glasses, of course, the atoms are all nonmagnetic. Experimental studies of filaments in the conducting state are extremely difficult.

REFERENCES:

1. Cohen, M. H., and J. C. Thompson: *Adv. Phys.*, **17**:857 (1968).
2. Varlashkin, P. G., and A. T. Stewart: *Phys. Rev.*, **148**:459 (1966).
3. Hensel, F., and E. U. Franck: *Rev. Mod. Phys.*, **40**:697 (1968).

RECENT DEVELOPMENTS OF THE LEE-YANG CIRCLE THEOREM AND ANALYTICITY OF THE FREE ENERGY

Masuo Suzuki[*]

Department of Physics
Cornell University
Ithaca, New York

ABSTRACT

The aim of this paper is to summarize recent work on the Lee-Yang theorem that all the zeros of the partition function of ferromagnetic models lie on a unit circle of the complex fugacity plane (i.e., they lie on the imaginary axis of the complex field plane). The original Lee-Yang theorem has been extended to the ferromagnetic Ising model with general spin. The generalized Ising model with many-body interaction is also discussed with applications to the fully anisotropic Heisenberg and ferroelectric models.

[*]On leave of absence from the Institute for Solid State Physics, University of Tokyo, Tokyo.

1. INTRODUCTION

In 1952, Lee and Yang[1] proved that all the zeros of the partition function lie on a unit circle in the complex fugacity plane for the ferromagnetic Ising model with spin one half. An important physical consequence is that the limiting free energy of such a system is analytic except in a vanishing field ($H = 0$), and that there occurs no phase transition for all (real) nonzero magnetic field H.

In detail, the partition function of the ferromagnetic Ising model is written as

$$\Phi(z, T) = Tr\, e^{-\beta \mathcal{H}} = \sum_k a_k(T) z^k$$

$$= A \prod_k (z - \exp i\theta_k) \qquad (1)$$

where z is the fugacity defined by

$$z = \exp\left(\frac{-m_B H}{kT}\right) \qquad (2)$$

with Bohr magneton m_B.

Now, this analyticity property is very useful for investigating critical behavior, in particular for studying critical exponents, as discussed by G. A. Baker, Jr. in this meeting already. Furthermore, Abe[2] and Suzuki[3] derived the scaling relations between critical exponents by assuming the asymptotic form of the distribution of zeros.

Thus, it seems very interesting to ask whether or not the Lee-Yang circle theorem is only a special feature of the Ising model with spin one half. Recently, we have found that the circle theorem is valid in quite general ferromagnetic systems. That is, the analyticity of the free energy has been established for several ferromagnetic systems.

Here, we summarize the recent results concerning this problem and explain briefly the essence of our proof in general cases.

2. EXTENSION TO THE ISING MODEL WITH HIGHER SPIN

Asano[4] and Suzuki[5] proved the circle theorem for the ferromagnetic Ising model with *spin 1 and 3/2*. The proof of it is based on a simple extension of a method used by Lee and Yang[1]. Griffiths[6] succeeded completely in proving the circle theorem for the ferromagnetic Ising model with *general spin*, by reducing the Ising model with general spin to that of spin one-half.

3. MONOMERS AND DIMERS

Quite recently, Heilmann and Lieb[7] proved the circle theorem for monomers and dimers systems, by applying a technique on the Sturm sequence to recurrence relations for the partition function. Kunz[8] gave a slightly different proof for this system.

4. EXTENSION TO THE FERROMAGNETIC HEISENBERG MODEL AND THE GENERAL ISING MODEL WITH MANY-BODY INTERACTIONS

Heilmann and Lieb[7] were able to prove the circle theorem for a ferromagnetically coupled Heisenberg spin system at high enough temperatures. Earlier Suzuki[9] had given a complementary proof for the Heisenberg model valid for low enough temperatures. In both of these cases the range of temperature (or T^{-1}) for which the proofs are valid shrinks to zero in the thermodynamic limit. However, numerical studies by Kawabata and Suzuki[10] for several small Heisenberg systems indicated that the circle theorem was probably valid for all T. Indeed very recently Asano[11] has considered the general Heisenberg Hamiltonian

$$\mathcal{H}_N = - \sum_{(ij)} [J_{ij}^x S_i^x S_j^x + J_{ij}^y S_i^y S_j^y + J_{ij}^z S_i^z S_j^z]$$

$$-2 \sum_{i=1}^{N} [H_i^x S_i^x + H_i^y S_i^y] - 2H^z \sum_{j=1}^{N} \mu_j S_j^z$$

$$= \sum_{(ij)} \mathcal{H}_{ij} + \sum_j \mathcal{H}_j - 2H^z \sum \mu_j S_j^z \quad (3)$$

in the uniform cylindrically symmetric case, $\mu_j \equiv 1, J_{ij}^x \equiv J_{ij}^y$ and $H_i^x \equiv H_i^y \equiv 0$. In this case, $M_N^z = m \sum_j S_j^z$ commutes with $\mathcal{H}_N$ and so the partition function $Z_N(z)$ is, as for an Ising model, still a polynomical in z and z^{-1}. Asano then has proved that all the zeros of $Z_N(z)$ lie on the unit circle $|z| = 1$ provided

$$J_{ij}^z \geq |J_{ij}^x| = |J_{ij}^y|. \tag{4}$$

Suzuki and Fisher[12] have extended the circle theorem to the nonuniform, fully anisotropic general spin Heisenberg model ($\mu_i \neq$ const., $J_{ij}^x \neq J_{ij}^y$, $H_i^x \neq H_i^y \neq 0$) under the "ferromagnetic" conditions

$$\mu_i \geq 0, J_{ij}^z \geq |J_{ij}^x|, \text{ and } J_{ij}^z \geq |J_{ij}^y| \tag{5}$$

We remark that in the fully anisotropic case the partition function is not merely a polynomial in z and z^{-1}. Thus, the circle theorem for the present case should be expressed[12] by the statement that *all zeros of the partition $Z_N(H^z)$ for the real Heisenberg Hamiltonian $\mathcal{H}_N$ defined in Eq. (3) lie on the imaginary H^z axis* $(\text{Re}\{H^z\} = 0)$ *under the condition of Eq. (5)*.

To prove the theorem for general S we may use the method of Suzuki[9] to reduce a general spin Hamiltonian to one for spin ½. Furthermore, the above theorem is easily extended[12] to pair interaction of the form

$$\mathcal{H}_{ij} = - \sum_\alpha \sum_\beta J_{ij}^{\alpha\beta} S_i^\alpha S_j^\beta \qquad (\alpha, \beta = x, y, z) \tag{6}$$

under the "ferromagnetic" condition

$$2J^{zz} \geq [(J^{xx} - J^{yy})^2 + (J^{xy} + J^{yx})^2]^{1/2}$$
$$+ [(J^{xx} + J^{yy})^2 + (J^{xy} - J^{yx})^2]^{1/2}$$
$$+ 2(|J^{xz}|^2 + |J^{yz}|^2)^{1/2} + 2(|J^{zx}|^2 + |J^{zy}|^2)^{1/2} \tag{7}$$

where we have assumed

$$J^{xx}, J^{yy}, J^{zz}, \text{ and } J^{xy}, J^{yx} : \text{real, but}$$
$$J^{xz}, J^{yz}, J^{zx}, J^{zy} : \text{pure imaginary} \tag{8}$$

The above results have been derived from the following general theorem recently developed by Suzuki and Fisher:[12]

RECENT DEVELOPMENTS OF LEE-YANG THEOREM

All the zeros of the partition function of the general Ising model with many-body interaction lie on a unit circle in the complex fugacity plane under the conditions that

(1) spin inversion symmetry

$$q^*(\sigma_1,\ldots,\sigma_k) = q(-\sigma_1,\ldots,-\sigma_k) \qquad (9)$$

where $\sigma_j = \pm 1$, and

$$q(\sigma_1,\ldots,\sigma_k) = \exp\{-\beta \mathcal{H}_{1,\ldots,k}(\sigma_1,\ldots,\sigma_k)\} \qquad (10)$$

is a partial Boltzman factor for the system and
(2) ferromagnetic condition

$$q(1,1,\ldots,1) \geq \frac{1}{4} \sum_{\sigma_1=\pm 1} \cdots \sum_{\sigma_k=\pm 1} q(\sigma_1,\ldots,\sigma_k) \qquad (11)$$

This general theorem yields the proof of the circle theorem for the fully anisotropic Heisenberg model above mentioned, by combining Trotter's formula[13,14]

$$\exp(A + B) = \lim_{n \to \infty} \left[\exp\left(\frac{1}{n}A\right) \exp\frac{1}{n}B\right]^n \qquad (12)$$

or by using simple formulas

$$\exp(A + B) = \lim_{n \to \infty} \left[\left(1 + \frac{1}{n}A\right)\left(1 + \frac{1}{n}B\right)\right]^n \qquad (13)$$

and more generally

$$\exp\left(\sum_{j=1}^m A_j\right) = \lim_{n \to \infty} \left[\prod_{j=1}^m \left(1 + \frac{1}{n}A_j\right)\right]^n \qquad (14)$$

Now the partition function is defined by

$$Z_N(H^z) = \text{Tr}\{\exp[-\beta \mathcal{H}_N(H^z)]\}$$

$$= \sum_{\sigma_1=\pm 1} \cdots \sum_{\sigma_N=\pm 1} \langle \sigma_1,\ldots,\sigma_N | \exp[-\beta \mathcal{H}_N(H^z)] | \sigma_1,\ldots,\sigma_N \rangle \qquad (15)$$

where we have used the representation of states, $|\sigma_1,\ldots,\sigma_N\rangle$, in which each S_i^z is diagonal with eigenvalues $1/2\sigma_i = \pm 1/2$. Then

$$Z_N(H^z) = \lim_{n\to\infty} \text{Tr}\left(1 - \frac{\beta}{n}\mathcal{H}_{12}\right)z^{\sigma_j} \cdot \left(1 - \frac{\beta}{n}\mathcal{H}_{13}\right)\cdots$$

$$= \lim_{n\to\infty} \sum_{\{\sigma_j=\pm 1\}} \cdots \sum \langle\{\sigma\}|1 - \frac{\beta}{n}\mathcal{H}_{12}|\{\sigma'\}\rangle$$

$$\times \langle\{\sigma'\}|1 - \frac{\beta}{n}\mathcal{H}_{13}|\{\sigma''\}\rangle \cdots \quad (16)$$

Here, using the notation $\epsilon = \beta/n$, we have

$$\langle \sigma_1, \sigma_2, \ldots | 1 - \epsilon \mathcal{H}_{ij} | \sigma'_1, \sigma'_2, \ldots \rangle$$

$$= \langle \sigma_i, \sigma_j | 1 - \epsilon \mathcal{H}_{ij} | \sigma'_i, \sigma'_j \rangle \prod_{k \neq i,j} \delta(\sigma_k, \sigma'_k) \quad (17)$$

Thus, the present system is equivalent to the Ising model with four-body interaction, in which a partial Boltzman factor is expressed by

$$q(\sigma_i, \sigma_j, \sigma'_i, \sigma'_j) = \langle \sigma_i, \sigma_j | 1 - \epsilon \mathcal{H}_{ij} | \sigma'_i, \sigma'_j \rangle \quad (18)$$

Therefore, the "ferromagnetic" condition (2) for the partial Boltzman factor $q(\sigma_i, \sigma_j, \sigma'_i, \sigma'_j)$ is easily found from the matrix

$$q(\sigma_i, \sigma_j, \sigma'_i, \sigma'_j) =$$

$$\begin{bmatrix} 1 + \epsilon J_{ij}^z & 0 & 0 & \epsilon(J_{ij}^x - J_{ij}^y) \\ 0 & 1 - \epsilon J_{ij}^z & \epsilon(J_{ij}^x + J_{ij}^y) & 0 \\ 0 & \epsilon(J_{ij}^x + J_{ij}^y) & 1 - \epsilon J_{ij}^z & 0 \\ \epsilon(J_{ij}^x - J_{ij}^y) & 0 & 0 & 1 + \epsilon J_{ij}^z \end{bmatrix} \quad (19)$$

That is, we obtain the ferromagnetic condition

$$|J^x + J^y| + |J^x - J^y| \leq 2J^z \quad (20)$$

which is simplified to the condition of Eq. (5). Here, we omit the proof of the circle theorem for the general Ising model with many-body interaction. (For details see Ref. 12).

5. EXTENSION TO FERROELECTRIC MODELS

Finally, we discuss lattice models of ice and hydrogen-bonded ferroelectrics[15] which have recently been the subject of significant numerical[16,17] and exact analysis. These models are easily transformed[12] to the general Ising model with four-body interactions (some of which should go to infinity). Thus, we can apply again our general theorem to these problems. The results are summarized as follows.[12]

5.1. KDP Model

In this case, the "ferromagnetic" condition (2) is equivalent to

$$T \leq T_0 \tag{21}$$

where T_0 is defined by

$$T_0 = \frac{\epsilon}{k_B \ln 2} = T_c \tag{22}$$

which happens to be identical with the exact critical or phase transition temperature[15,18] of the model. That is, for all temperatures below the critical temperature the zeros of the partition function of the KDP model lie on the imaginary axis in the complex electric field plane. These results have been confirmed numerically by Katsura, et al.[17] They have found numerically, however, *two-dimensional* distribution of zeros for all temperatures a little above the critical point. It seems to be connected with the peculiar singularity of the ferroelectric model.

5.2. Model with Double-ions

Sutherland[19] has recently discussed models in which "doubly-ionized" configurations are allowed (all four arrows inward or all four outward). The ferromagnetic condition (2) reduces to $T \leq T_1$, where T_1 should be the critical temperature when the original ferroelectric lattice is the plane square lattice, according to Sutherland's argument[19] (similar to that used originally by Kramers and Wannier for the Ising model).[20] Again we conclude that the zeros must lie on the imaginary axis in the complex electric field plane.[12]

ACKNOWLEDGMENTS

The author would like to express his sincere thanks to Professor M. E. Fisher for the hospitality at Cornell University. The support of the National Science Foundation is gratefully acknowledged.

REFERENCES

1. Lee, T. D., and C. N. Yang: *Phys. Rev.*, **87**:410 (1952).
2. Abe, R.: *Prog. Theor. Phys.*, **38**:72, 568 (1967).
3. Suzuki, M.: *Ibid.*, **38**:289, 1225 (1967).
4. Asano, T.: *Ibid.*, **40**:1328 (1968); *J. Phys. Soc. Japan*, **25**:1220 (1968).
5. Suzuki, M.: *J. Math. Phys.*, **9**:2064 (1968); *Prog. Theor. Phys.*, **40**:1246 (1968).
6. Griffiths, R. B.: *J. Math. Phys.*, **10**:1559 (1969).
7. Heilmann, O. J., and E. H. Lieb: *Phys. Rev. Lett.*, **24**:1412 (1970).
8. Kunz, I.: Private communication.
9. Suzuki, M.: *Prog. Theor. Phys.*, **41**:1438 (1969).
10. Kawabata, C., and M. Suzuki: *J. Phys. Soc. Japan*, **28**:16 (1970).
11. Asano, T.: Preprint and *Phys. Rev. Lett.*, **24**:1409 (1970).
12. Suzuki, M., and M. E. Fisher: *J. Math. Phys.* (in press).
13. Trotter, H. F.: *Proc. Am. Math. Soc.*, **10**:545 (1959).
14. Ginibre, J.: In lectures given at the Cargése Summer School in Statistical Mechanics, July 1969.
15. Slater, J. C.: *J. Chem. Phys.*, **9**:16 (1941); Takahashi, H.: *Proc. Phys.-Math. Soc. Japan*, **23**:1069 (1941).
16. Nagle, J. F.: *J. Math. Phys.*, **7**:1484, 1492 (1966).
17. Katsura, S., Y. Abe and K. Oukouchi: *J. Phys. Soc. Japan* (in press).
18. Lieb, E. H.: *Phys. Rev. Lett.*, **18**:692, 1046; **19**:108 (1967); Sutherland, B.: *Ibid.*, **19**:103 (1967).
19. Sutherland, B.: Preprint (1970).
20. Kramers, H. A., and G. H. Wannier: *Phys. Rev.*, **66**:252 (1941).

DISCUSSION on *Paper by M. Suzuki*

C. N. YANG: Do you know of other models that display two-dimensional distribution of zeros?

M. SUZUKI: Yes, Katsura, et al. have found numerically the two-dimensional distribution of zeros for various ferroelectric and antiferroelectric models, at least in sufficient high temperature. Furthermore, Kawabata, Ono, Karaki, Ikeda, and I have found that some of zeros deviate from the unit circle and from the negative real axis for the antiferromagnetic Ising model. I am not sure whether this distribution is two-dimensional or one-dimensional, because our studied lattices are too small ($4 \times 4, 5 \times 5, 4 \times 6$, and $3 \times 3 \times 3$).

P. CHOQUARD: The zeros being, by definition, independent of the magnetic field, do you think it plausible that we shall be able to obtain the solution in the nonzero magnetic field for some zero field known soluble model like the second Ising one?

M. SUZUKI: No, I think that it is too optimistic, because the solution in the presence of the magnetic field contains much more information than that in the absence of the field.

C. N. YANG: In answer to Professor Choquard's question perhaps it is useful to make the following remark:

In 1952, after we proved the circle theorem, Lee and I thought that surely with the knowledge of the position of the zeros, and with the exact solutions known for the $H = 0$ case for the two-dimensional Ising model, it should not be difficult to guess at the solutions for all $H \neq 0$. It was a great disappointment to me that I was too optimistic. I learned from this the lesson that the mere knowledge of the position of the cuts of an analytic function, together with many properties of the discontinuity across the cuts, are not necessarily enough to solve a dynamic problem. I believe many high-energy theorists today have not yet learned the same lesson.

G. A. BAKER: I am sure that everyone who has worked on these problems which you described finds the results to be of great beauty and a substantial accomplishment. I wonder if you have investigated the exchange model?

M. SUZUKI: Yes, I have tried it. I am convinced that the Lee-Yang theorem is also valid for the ferromagnetic exchange model from the observation for small finite lattices. Though I have not succeeded in general, I believe that I will be able to give a proof, at least, for spin 1 and spin 3/2.

Summary Agenda Discussion:
CRITICAL PHENOMENA IN ALLOYS, MAGNETS, AND SUPERCONDUCTORS

E. Montroll*
Physics and Astronomy Department
University of Rochester
Rochester, New York

R. E. Mills†
Department of Physics
University of Louisville
Louisville, Kentucky

The agenda for the final session began with a discussion of recent developments in the use of light scattering for the investigation of turbulence, and went on to discussions of models with periodicity, population growth, molecular field theory, diffusion constants, and the law of rectilinear diameters.

*Chairman.
†Secretary.

Montroll started a discussion of lambda points in turbulent flow with the observation that the transition from laminar to turbulent flow is one of the older phase transitions known. Reynolds' observations were done during the same period in which Andrews studied critical phenomena and Van der Waals developed his famous equation of state. Recently, Goldstein and Hagen[1] used light scattering techniques, like those which Benedek discussed earlier, to investigate water flowing in a pipe. They measured the Doppler shift in the scattering arising from the turbulence. In this case, at a critical Reynolds number of about 2300, the flow changed rapidly from laminar to turbulent. The line width of the scattered light gave a measure of the velocity distribution function. Since water itself isn't a very good scatterer, tiny polystyrene spheres were added to the flow. The size of these spheres can be very well determined, so application of Stokes' law to the line width data can be used to extract the viscosity or the diffusion constant.

An interesting question concerns how these spheres affect the flow themselves. The critical Reynolds number came out at the right place. Check experiments were done previously by Yeh and Cummins[2] by measuring the velocity profile of the laminar flow at different distances from the center of the tube. The Poiseuille law predicts a parabolic distribution in the speeds from the middle, and it was verified very well.

In the Goldstein-Hagen experiment, line widths were plotted versus the Reynolds number. There was a small line width when the flow was laminar. When the flow was turbulent, the line width was about ten times as great. In the vicinity of the transition, however, the line width rose very strongly as the Reynolds number was approached, giving a curve very like that of the lambda curve for the specific heat divergence. Further, although the line widths in the laminar region were indicative of a Gaussian distribution, in the turbulent region the line widths were much larger and clearly non-Gaussian. This, then, is the sort of experimental evidence which a proper theory of turbulence should try to fit.

Benedek discussed some experiments based on the scattering of light from small particles in both laminar and turbulent flows. He recalled that at the twenty-fifth annual meeting of the Division of Fluid Dynamics held at Lehigh University about a year ago, some very beautiful work was presented by a group at the University of Tennessee. They were able to take measurements in a flow and make detailed comparison with predictions of the Navier-Stokes equations. This left him with the feeling that solutions of the Navier-Stokes equations valid for very low Reynolds numbers could be tested in this, if no other, way. Also, a collaboration of people at Harwell, Malvern, and Kent had studied turbulent flow by looking at the self-beat spectrum of the scattered light.[3] This gave the velocity difference between the various particles. If the size of the region they examined was very small, compared to the correlation length, the velocities in the region would be correlated, and there would be no differences. Thus the line width of the scattered light was very small. But if they increased the size of the region which they probed, they got to sizes where the velocities would no longer be correlated. Thus by changing the size of their apertures, they were able to study the correlation range of the velocity fluctuations. One could also study anisotropic distributions by changing the direction of the light beam relative to the mean direction of the flow. With the heterodyne technique, it was also possible to study anisotropy. There was a correction needed due to the fact that the particles remain in the beam only a finite time. This gave an additional line width.

Benedek remarked that no one had then performed an experiment which studied density fluctuations in the turbulent flow itself. This would measure something different from the velocity correlation function, specifically correlations of the energy-momentum tensor through the pressure-pressure correlation function. This would require the use of extremely pure water. Since some of the experiments were conducted using the water at Oxford, it was not possible to remove all the suspended particles.

There is one other feature which can be explored in such experiments. By changing the scattering angle, one can probe fluctuations which range essentially from the size of the tube to the wavelength of the light. The line width of such fluctuations gives some information concerning the damping times of the fluctuations of the various Fourier components. This is important for the theory of turbulence, for this theory holds that energy dissipation takes place by macroscopic transfer of energy to smaller and smaller turbules, until finally viscosity takes over and gives microscopic dissipation. Variation of the scattering angle should thus make possible a study of these macroscopic transfers in turbulence. Benedek thought that this would be a feasible experiment.

Kadanoff observed that in the usual theory of turbulence, the fluctuations in a scalar quantity such as the density are very small, and so this might affect the observability of density variations in turbulent flow. Benedek replied that most theories of turbulent flow neglect the effect of density fluctuation. Of course, in a real fluid some density fluctuations do occur, and it is necessary to ask if the fluctuations are large enough to affect the hydrodynamic motion.

Benedek thought the answer to that question was no, but that the density fluctuations might still be large enough to use them to probe the flow using scattered light.

Onsager pointed out that velocity fluctuations in turbulence increase with the rate of flow. As the rate of dissipation increases, the microscale decreases. For both these reasons, the light scattering would be expected to increase as well, and one should not expect spectacular light scattering at the onset of steady flow.

Montroll then discussed an extrapolation scheme which might be useful to people who study systems containing impurities. It is desirable to get as much information as possible from a calculation, e.g., a perturbation calculation involving impurities. With one defect in a material, results to first order in the concentration on the effects of some of the properties of the material are easily obtained: With two defects and some averaging, one can get results to second order. With three defects, the calculation usually stops. The results at very high concentrations are also usually available in the same manner. The problem is then to find a suitable way of interpolating between them.

In the mathematical literature in the 1920s, there was a generalization of the Taylor expansion by a man named Lidstone. In the Taylor expansion about one point, one uses all the derivatives. In the Lidstone expansion, about two points, one uses only the even derivatives at either end. In the simplest form, one might try, with c being the concentration

$$f(c) = (1 - c)f(0) + cf(1)$$

In the Lidstone expansion, one has

$$f(c) = f(0)\Lambda_0(1 - c) + f(1)\Lambda_0(c)$$
$$+ f''(0)\Lambda_1(1 - c) + f''(1)\Lambda_1(c) + \cdots$$

One can show that the polynomials Λ_n satisfy a recursion formula

$$\Lambda_n'' = \Lambda_{n-1}$$

together with the boundary condition that the polynomials vanish at either end of the interval. The first two are

$$\Lambda_0(c) = c \qquad \Lambda_1(c) = \frac{c(c^2 - 1)}{3!}$$

The simplest Lidstone series is that for $\sinh(x)$

$$\sinh(x) = \sinh(1) \sum \Lambda_n(x)$$

Not all functions have such expansions, of course, because their second derivatives may vanish. The function $\sin \pi c$ is an example.

Onsager recalled a specialization of the Lagrange interpolation scheme. With it, one could also use the odd derivatives as well as the even ones. To use N derivatives, all that is needed is to construct the Lagrange interpolation, and then specialize it to some set of N points. This might work a little better, because if the even derivatives are available, the odd ones usually are too.

Much of the meeting had involved discussions of mathematical models, so Montroll discussed a group of solvable models and their characteristics. The models were all based on lattice networks. All the cosine functions which are common to such models come from the group-theoretical description of the translational invariances of the lattices. If one has a Laplace difference equation relating the value of a function at a lattice point to its values at nearest neighbor sites, one looks for periodic solutions and finds the associated eigenvalues. Or, one can look for a Green's function solution by putting in a delta function as a source term and then find the analog of the energy denominators. The latter, however, have the periodicity of the lattice.

In doing problems involving electrons in crystals, one does not usually find formulas like those at all, except in the tight-binding approximation. In the case of the two-dimensional Ising problem, there are two cosine functions in the solution. If it is a symmetric Ising problem, the cosines appear symmetrically. One of the tempting questions is: Why is it necessary to work so hard in the Ising case with Lie algebras or the spinor analysis, etc., to reach this point, when in the other cases the periodicity enters in an almost trivial way?

He discussed a few of the other soluble models. First, he considered lattice vibrations, where the displacement from equilibrium at one site is related to the displacements at neighboring sites. In doing the Ising problem, one never starts with something which is that local. In random walks, the probability that the walker is at a given point is expressed in terms of things at neighboring points at an earlier time. These formulations again lead to the Laplace difference equation and its typical solution.

A problem which is older than the Ising problem doesn't seem to get as much attention. Montroll had found a solution to it. It had to do with the electrons in a lattice. Bloch theory allows one to do calculations of the various dispersion curves, densities of states, etc., with all sorts of complicated models. But every time one teaches a course in solid state, one wishes for a nice three-dimensional model to discuss, something better than the Kronig-Penney model which would still have all the right qualitative features. On one end, you'd have the free particle, and on the other the tight-binding model. Then you could work things out in the middle. It turns out to be almost trivial if you put in the restriction that the electrons propagate only on wires. The crystal is thus characterized by the kind of network of wires which you show to freshmen students. In fact, Coulson had done this for graphite in a free electron model, with the electrons on wires in a hexagonal network.

It is necessary to take proper account of nodes, etc., by solving Schrödinger's equation along each line in the network. The wave function at one nodal point is thus related to the wave functions at the neighboring points. Once this is done, the problem is recast in the form known from the lattice vibration problem. Instead of having a difference equation for lattice displacements, however, one has a difference equation for the wave functions at neighboring points.

The right thing to do is to take a solvable one-dimensional potential, for example,[4]

$$\phi(x) = \text{sech}^2 \gamma x$$

This has one bound state, and by taking γ large, this becomes very deep. Thus one can approach the tight-binding case. As γ decreases, the binding decreases until at last it vanishes as γ does. The wave functions are elementary, but the eigenvalue equation involves a complicated transcendental function of the energy $f(E)$. At this point, one learns a lesson: If one insists on calculating energy levels by writing a formula with E equated to something, then one has set the problem up wrong. If you let the problem tell you what the transcendental function $f(E)$ is, you get the Laplace difference equation, and one is then back to the same well-known problem. Instead of energy levels, you find a discrete set of values of the function of E. In the case of Coulson's free electron model, the function is just $\cos(lk)$, where l is the lattice spacing and k^2 is proportional to E.

The model can also be done with more than one bound state. For example, use the Teller-Pöschel potential, which is also solvable in terms of hypergeometric functions. All this does is change the form of $f(E)$. Montroll stressed that one shouldn't insist on energy E as a variable, but rather one should first find a suitable variable which will give the Laplace difference equation.

He then asked what happens in the case of the Ising model. The one-dimensional Ising model is solvable in this and apparently all other schemes. The first function one would think of as characterizing something locally is the correlation function with a spin at the origin and a spin at point j. If one used the Callen identity[5] which Clapp mentioned earlier in the week and a second identity, one obtained just the analog of the one-dimensional Laplace equation. The correlation function $c(j)$ was related to its values at $j \pm 1$. He had started with a basic equation with the lattice periodicity in it. If one did this on a circle and let the circle get large, one got the usual hyperbolic tangents raised to powers.

Montroll pointed out that in the spherical model, the spin correlation function was just the Green's function for the Laplace equation, so if one were to use the correlation functions for the spherical model, he could write down a similar set of equations and proceed from there.

So, he asked, how would one do this in two dimensions? If this would go trivially, something would be wrong, because the Ising correlation functions in two dimensions are very complicated things. They have the right cosines in the denominators, but they are very complicated expressions. One would find that in order to use the Callen identity, one had to take the hyperbolic tangent of a sum of four spin variables. Then one had not just the sum of single spins; you end up with pairs, triples, and even quadruples in all of these. So, as Clapp mentioned earlier, one found too many complicated correlation functions.

To proceed, one first threw away all the odd ones, because they averaged to zero anyway. The correlation coefficient between the spins at the origin and the point c still involved the correlations between the origin and the points $c \pm 1$, but there were also fourth-order coefficients. The latter could be shifted to the right-hand side of the equation. The second-order functions could then be obtained in terms of the fourth-order ones using the Green's functions. This solution would then have the right geometrical features. Montroll remarked that he had been thinking about this for some time, but hadn't dared to do it because if the denominators came out right, he would have to spend the next few months on the three-dimensional problem. He asked if perhaps Clapp knew if the right terms obtained when this was done. Clapp replied that he had done something like that, but he couldn't compare it with what Montroll had suggested. The appearance of the fourth-order denominators was discouraging.

Gruber stated that the coefficient β_1 was given by

$$\beta_1 = \frac{1}{8}\{\tanh[2\beta(J_1 + J_2)] + \tanh[2\beta(J_1 - J_2)] + 2\tanh(2\beta J_1)\}$$

β_2 was then obtained by permutation of the indices 1 and 2. In any case one does not obtain the coefficients which appear in the exact solution of the free energy. Suzuki observed that Callen's identity

$$\langle \sigma_i \sigma_j \rangle = \langle \sigma_i \tanh \beta E_j \rangle$$

which is equivalent to the Fisher transformation, and an explicit polynomial expression for the spin function $\tanh(\beta E_j)$ have been used frequently in discussing dynamics of the kinetic Ising model. There the transition probability $W_j(\sigma_j)$ is assumed to be

$$\dot{W}_j(\sigma_j) = \frac{1}{2\tau}(1 - \sigma_j \tanh \beta E_j)$$

where the parameter τ indicated the relaxation time of a free spin which interacted with the heat bath. Thus it was possible to find several expressions for the above function $\tanh(\beta E_j)$ in two and three dimensions there.[6] Montroll concluded that it appeared that the spin correlation functions were not the right local variables. The question, then, was what are the right ones? He stated that even though the spin correlation functions don't give the Laplace difference equation, that with the right local (not thermodynamic) variables, one can wind up with the proper formulation. Kadanoff pointed out that the variables that Kaufmann used were the right variables since one does get a Laplace difference equation in these spinor variables. The problem then is that we do not understand the variables very well. Montroll agreed, but thought that was giving too much, that it made the work too hard. There are also graphs on lattices which give variables which are completely different, but which also work. He asked if there were variables to which one comes in a more direct way which will work.

Montroll then turned to a discussion of a particular nonlinear rate equation relevant to a study of population and ecology.

At the beginning of the Brussels school of statistics came Verhulst, who asked what the population would be. This was in the 1830s, at the beginning of King Leopold's reign. The King, who was very scholarly, had read Malthus's book and was concerned (they were even in those days) about the possibility that the population of Belgium would explode. Verhulst's response was that it would saturate sometime. The Malthus equation said that the rate at which the population would increase was proportional to the population, but Verhulst said that there must be a second term

$$\frac{dN}{dt} = \frac{kN(\theta - N)}{\theta}$$

which would give saturation. This equation can be solved in first order, and one gets curves which were later rediscovered by Pearl and Reed. They plotted populations of various countries, and it is remarkable how the populations agree with the various curves. Montroll went on to discuss "noise" in the solutions.

Volterra and Lotka, Volterra in a general way and Lotka in a two-body problem, considered the case of competing species. They didn't consider the saturation curve here, because their species

$$\frac{dN_j}{dt} = \frac{k_j N_j (\theta_j - N_j)}{\theta_j} + \frac{1}{\beta_j} \sum_{i,j} a_{ji} N_i N_j$$

were eating each other at such a rate that they never got near to saturation. They needed only the linear terms. But the sum

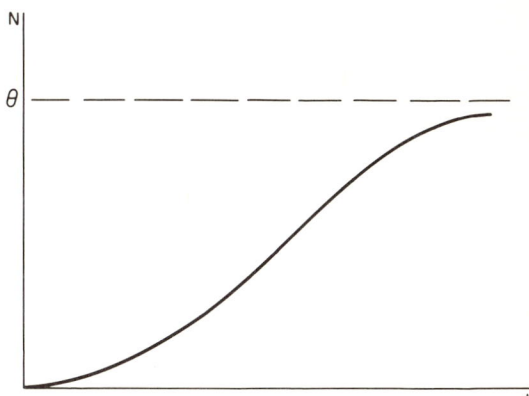

Fig. 1 *Saturation of population.*

included the binary collisions. If species j were to eat species i, then the population of j would increase as a result of the binary collisions. This was an antisymmetric situation: If one went to the growth of species i, a_{ij} would be the negative of a_{ji}. Our experience with Pfaffians tells us that if the derivatives are set to zero, the N_j cancel out. Ordinarily with quadratic terms, this wouldn't happen, but it was stated to be so in this case. This gave a linear equation for the equilibrium case. The condition for stability was that the determinant of the coefficients vanish. Since they are antisymmetric, this seemed correct, for it happens also for Pfaffians.

Montroll then went back to the two-body case, the so-called big fish–little fish problem. There was an equilibrium point. One would plot N_2 as a function of N_1 by eliminating time. There were periodic solutions. If one started with an initial condition very close to equilibrium, one got just sines and cosines, i.e., small oscillations. On the other hand, if one started far from equilibrium, the things called explosions or crashes in ecological systems were found. The little species would start growing, the big species then eat the little species and grow rapidly. There would almost be spikes when one started far from equilibrium. Thus in the very nonlinear range, the process would be one where the process went slowly, then there would be an enormous spike, followed by another slow period, followed by another spike, and so on. This was an interesting feature in itself.

A remarkable feature of the Volterra model is that it leads to a "constant of motion" of the population dynamics of the competing species. Let $\{q_i\}$ represent the set of steady-state populations of the various species, and let

$$v_i = \log \frac{N_i}{q_i}$$

Clearly, when $N_i = q_i(1 + \epsilon_i)$ and $|\epsilon_i|$ is small, then v_i is small. It can be shown from the Volterra equations that

$$G = \sum_i \beta_i q_i [-1 - v_i + \exp(v_i)] = \text{constant}$$

This expression is analogous to the Hamiltonian of a dynamical system. It is quadratic in the v_i when the v_i are small.

The existence of a constant of the motion suggested to Kerner that some of the ideas of statistical mechanics might be applied to the competition between species. The population distribution function might be expected to have the canonical ensemble form

$$\mathcal{P}(v_i) \propto \exp\left\{\beta_i q_i \frac{1 + v_i - \exp(v_i)}{g}\right\}$$

where g is a constant proportional to G. A detailed discussion of these problems will be given by Goel, Maitra, and Montroll in a forthcoming issue of *Reviews of Modern Physics*.

In the simplest development with competing species, the fact that the a_{ij} have positive and negative signs implied that this sum over i is essentially a random variable. If such a variable, called F_j, had zero mean, one could derive an equation which had the form of the equation for Brownian motion with a forcing term. Then a Fokker-Planck equation would also be obtained. After a few transformations, the result was an equation in the form of the Schrödinger equation with a Morse potential. One could make an interesting interpretation of all this in terms of the Morse potential: The early part of the population growth is like the motion of a free particle in a field. The particle is being accelerated, i.e., the population growth is accelerating. Finally it begins to compete with other things, i.e., it comes down to the first bound state in the Morse potential. Then it goes down from state to state until finally, in the saturation range, one sees a noisy population. The noisy population is thus like the zero-point motion in the Morse potential. (Incidentally, Montroll observed, if you grow fruit flies in a crowded cage, you find out that the population goes up and then it starts to fluctuate, i.e., there is noise in it.) The equilibrium distribution is then just what one would get with a canonical ensemble.

Montroll concluded with a remark about the network problem. Since there are a variety of kinds of network, if each particle is connected with the total, then the ecology is like statistical mechanics. If there were a loose end, though, the rest would be like a heat bath, and one got a periodic function in contrast to the periodic solution with noise. He felt one could do much more with the Volterra equation than he had exhibited, and the model would be interesting to consider for those who like nonlinear problems.

Kadanoff asked if the population spikes were not those usually connected with big fish–little fish situations. Montroll replied that they sometimes were and sometimes were not. There were some that looked periodic, the classical cases of the foxes and hares, and, of course, the big fish–little fish. The latter were observed by D'Anconna in fish catches in the Adriatic a number of years ago. He was the one who stimulated Volterra into setting up these equations. Most of the cases were noise, such as would be expected from the statistical mechanical description. The ecologists work hard to make cycles out of these, but they probably are not really there.

Clapp asked what would happen if a completely new species, such as a mutant virus strain, were injected into the population. Montroll remarked that the question was interesting and would probably be explored soon.

Several short discussions on different topics followed. The first of these started with a discussion by Kac on features of the molecular field theory. Kac was concerned with why molecular field theories, which are supposed to be rigorously valid only in the limit of weak long-range interactions, account well for experimental data in situations in which there is no reason to believe that the interactions are of such a nature. He was unable to answer this question, but, he pointed out, if one looks upon the mean-field approximation in a somewhat different way from the usual one, one does obtain quite good approximations (*away* from the critical point). As an example, he took the two-dimensional square net Ising model with nearest-neighbor interactions.

Kac observed that it was easily checked that the spectral properties of the transfer matrix are equivalent to those of the kernel

$$K(\mathbf{u}, \mathbf{v}) = \sum_{\mu} \exp\left(\nu \sum_{1}^{M} \mu_k \mu_{k+1}\right) \exp\left(-\frac{1}{4} \sum_{1}^{M} (u_k - 2\sqrt{\nu}\mu_k)^2\right)$$

$$\times \exp\left(-\frac{1}{4} \sum_{1}^{M} (v_k - 2\sqrt{\nu}\mu_k)^2\right)$$

where $\nu = J/kT$ and the summation is over the 2^M configurations

$$\boldsymbol{\mu} = (\pm 1, \pm 1, \ldots, \pm 1)$$

In particular (except for a trivial constant)

$$\frac{-\psi}{kT} = \lim_{M \to \infty} \frac{1}{M} \log \Lambda_1$$

where Λ_1 is the maximum eigenvalue of the kernel.

The modified mean-field approximation consisted in taking as the principal trial function

$$\phi = \exp\left(-\frac{1}{4} \sum_{1}^{M} (u_k - 2c\sqrt{\nu})^2\right)$$

and choosing c so as to maximize the Raleigh-Ritz ratio

$$\frac{\langle K\phi, \phi \rangle}{\langle \phi, \phi \rangle} = \exp(-Mc^2 \nu) \sum_\mu \exp\left(\nu \sum_1^M u_k u_{k+1} + \nu c \sum u_k\right)$$

$$\sim \left\{ e^\nu e^{-\nu c^2} \left[\cosh 2\nu c + \sqrt{\cosh^2 2\nu c - (1 - e^{-4\nu})}\right] \right\}^M$$

The equation for c which maximized the above expression was

$$\eta = \frac{2\nu \sinh \eta}{[\cosh^2 \eta - (1 - e^{-4\nu})]^{1/2}} \qquad (\eta \equiv 2\nu c)$$

which had the typical appearance of the mean-field equation.

For ν less than a certain ν_{crit}, $c = 0$, but for $\nu > \nu_{\text{crit}}$, $c\nu$ would be an increasing function of ν. For $\nu \gg 1$, $c \sim 1$.

With $c = 1$, the expression for the free energy agreed with the exact one up to and including $\exp(-8\nu)$, so that the error was of the order $\exp(-10\nu)$. Thus, at least the free energy was given accurately (almost up to the critical point!) by a mean-field-like approximation.

If one were to take the approximation seriously, one would obtain a displaced critical point and classical exponents at this approximate critical point.

Griffiths suggested that Kac's first approximation might be equivalent to a simple extension of the molecular field. He noted that the square Ising model consists of a series of linear chains in the horizontal direction, and supposed that one solved the linear chain problem exactly in a magnetic field. The vertical bonds were then treated as a self-consistent field in order to obtain an approximate solution of the two-dimensional Ising model. Griffiths suggested this might give the same answer as the first approximation discussed by Professor Kac. Kac was not certain, but thought it might be so. If one were to go back to the original transfer matrix, however, then it could be said, as a first guess for low temperatures, that the eigenvectors would all be concentrated in the ordered state. One wouldn't get nearly the same number of solutions. But if this were replaced by a completely disordered eigenvector, then one would get a rough estimate of the transition good to about 10 to 12 percent. With the anisotropic Ising model, the same procedure located the correct transition temperature to the leading order in ϵ.

Martin asked whether it would be possible to relate these comments to the so-called Ginsburg criterion. This states that if z is the number of neighbors contributing to the mean field, and $z \gg 1$, then the mean-field theory is approximately correct to the order

$$\left(\frac{1}{z}\right)^2 \left(\frac{T - T_C}{T_C}\right)$$

and critical behavior occurs closer to T_C. This criterion has been particularly helpful, for example, in understanding why superconductivity, where z is so large, the critical region is so small that it is only accessible where special tricks make the effective z considerably smaller. In experimental quantities it seems that the critical phenomena set in around the point determined by this criterion. Martin also asked about the logarithmic derivatives of free energy that are usually measured, i.e., susceptibilities and specific heats. He didn't understand how Kac's theory could give these quantities much better than the Ginsburg theory would support. In response, Kac stated that he was studying the case of four neighbors. In Brout's terminology, $z = 4$, and would suggest 1/16 as an error bound. This is much, much better. Nonetheless, it was necessary to stay away from narrow temperature ranges about T_C. After writing down the transfer matrix or kernel, one looked for the point where the one-hump eigenfunction changed into the two-hump eigenfunction. The mean field would be determined by the point where the approximation was centered. Then one got an unbelievably good approximation to the free energy. The T_C was incorrectly predicted, however, and the predicted exponents were wrong because they were used in expanding about the incorrect T_C.

Kadanoff inquired if Kac expected a similar accuracy for an application to the three-dimensional case. Kac replied that he couldn't tell, because he didn't have the exact solution. He had done that because a much nicer kernel came for the case where one had next-nearest-neighbors interactions of about the same strength. The Brout approximation would be much better because one then had eight neighbors. There, Kac thought, though no one had done the calculations, even the simplest guess would be very good. He pointed out that he was not discussing experiments, and that all he meant to say was that if one didn't include ordinary mean-field theory the way it was in all the books, but rather in a slightly modified way retaining a mean-field flavor, then one would get fantastic accuracy in the free energy doing almost no work at all. The accuracy of the magnetization is also very good. He hadn't tested the correlation functions, so he couldn't tell about them.

Kouvel asked if Kac would care to tell in which direction (and to what degree) did the Curie point in his mean-field approximation differ from the exact value. Kac answered that it was always higher. That was a general situation. Griffiths had given a demonstration of that. Griffiths said that had only been proven for a simple model. Kac felt however that it would more generally be true. Stanley discussed Griffiths' conjecture that Kac's approximation might correspond to a model in which *intra*-chain interactions are taken into account exactly, while the *inter*-chain interactions are approximated by the molecular field approximation. He believed this conjecture to be false for the following two reasons:

1. The cases $d = 2$ and $d = 3$ would be identical in the "chain model" (except for a change in z, the coordination number), while he believed they would be different in Kac's approximation.

2. Kaplan and he had studied the chain model in some detail, and he had not found such striking agreement with the exact result as Kac described for his approximation.

McCoy asked if Kac had applied this method to Dyson's long-range Ising models, and what predictions he might have for the critical temperature. Kac had discussed this point in his earlier lecture. It was a standard step he took whenever he wrote a transfer matrix or kernel, and he believed the estimates of T_C, etc., were quite good.

Onsager commented that this method, with perhaps a refinement, was applied by Kramers and Wannier. Comparison of their curve for the specific heat with his exact curve show that the two compare well, to within perhaps the thicknesses of the curves. Montroll added that the only tricky thing was that there was a sharp break in it, i.e., it was continuous but with a corner, whereas the true one didn't have the corner. Onsager continued that in the Rayleigh-Ritz approximation used by Kramers and Wannier, they used one more parameter below T_C, and none above. Consequently, they got a slightly better free energy below T_C than above, and a positive error in T_C.

Alefeld posed an additional question to which he thought the theoreticians might have an answer: The shortcomings of the mean-field theory were often stated as ignoring interaction of fluctuations. If there were a system, such as he described two days ago, namely hydrogen in metals, for which all 10^{23} modes of fluctuations are suppressed (on the basis of an argument due to Cahn) except for very few modes for fluctuations which have the wavelength of the size of the sample, what would happen to the mean-field theory? One would be inclined to ignore in a free-energy expression the significance of the few critical modes compared to the consisted of a gas of magnons. In doing this, Heller used neutron-scattering data to obtain values for the magnon velocities and mean free paths. For the purposes of a crude estimate an average was taken of the contributions from magnons of different wave-vectors. In a comment, Ferrell pointed out that a more accurate computation should also include the variation with q of the Boltzmann factor. In any case the result of this estimation for the diffusion constant D was again of the order of magnitude of 10^{-3} cm^2/sec at $T/T_c = 0.95$.

Lastly, Heller discussed what could be deduced if the principle of dynamical scaling were assumed to apply to both parts of the relaxation function. Then one would have $D(T) \propto \kappa^{-1/2}$ or roughly $D(T) \propto (T_c - T)^{-1/3}$ below T_c. This was not inconsistent with the scattering data though it *was* inconsistent with the temperature dependence predicted in the magnon thermal conductivity estimation). As another application of the scaling assumption, Heller obtained an estimate of $\tau_S(T)$ by taking κ to be the wave-vector for which the two time constants describing the relaxation merged. For MnF$_2$ this gave

$$\tau_S^{-1}(T) \approx 0.1 \text{ meV } ^\circ\text{K}^{-1}(T_c - T)$$

below T_c. Heller considered this result to be physically reasonable in the light of the measured behavior of the staggered relaxation rate above T_c, namely

$$\Gamma_{\text{stag}}(T) = 0.031 \text{ meV } ^\circ\text{K}^{-1}(T - T_c)$$

It should be made clear, however, that the available experimental data are not good enough to decide whether dynamical scaling provides a quantitatively accurate account of the relaxation behavior below T_c.

Ferrell noted that it would be interesting to see how much the average magnon velocity would be changed if a Boltzmann weighting factor were included.

The chairman then introduced the last topic which began with a discussion by Widom on the law of rectilinear diameters. Widom discussed further a point discussed earlier in connection with Cooper's paper, having to do with symmetry and analyticity in liquid-vapor equilibrium in the neighborhood of the critical point. He discussed this in the context of the penetrable sphere model which Rowlinson and he recently discussed.[8] First he recalled the special features of the model.

In the penetrable sphere model, an interaction is defined between molecules in a one-component fluid as follows: Around each of the molecules in some configuration, spheres were circumscribed, all of a

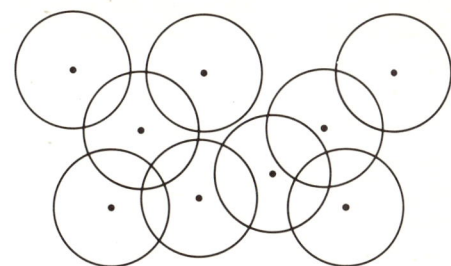

Fig. 2 *The interpenetrating sphere model.*

size. Some of these were isolated, but many were overlapping. The total volume covered by this complex of interpenetrating spheres was called W. If the overlap were considerable, W would be small. If the overlap were small, the covered volume W would coincide with N, the number of spheres, times v_0, the volume of one of them. The difference between these expressed the energy of interaction as a function of the configuration

$$E = \frac{e}{v_0}(W - Nv_0)$$

Here e was an energy parameter.

A symmetry of this model was the exact analog, even in this continuum model, of the hole-particle symmetry of the lattice gas. In that case, one could say that the coexistence curve is symmetrical about the density of half filling. On that line of symmetry, the

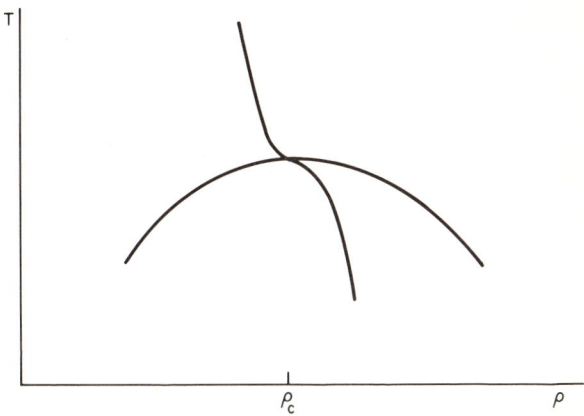

Fig. 3 *The line of symmetry in the sphere model.*

chemical potential would be an exactly known, simple function of T. Here in the $T - \rho$ plane there was again a line of symmetry, but it was no longer a line of constant density. The line in fact had a singularity in it that reflected the singularity in C_V at T_C. But again, in this case, the chemical potential was an exactly known, simple function of the temperature. This line of symmetry, as in the lattice gas model, coincided in the coexistence region with the diameter of the coexistence curve. Since this line of symmetry had a singularity at T_C, the law of rectilinear diameters had to be replaced by

$$\rho_l + \rho_g - 2\rho_c \sim (T_c - T)^{1-\alpha}$$

This suggested questions which Widom thought were of a rather fundamental character, having to do with the nature of liquid-vapor equilibrium. This singularity for one thing was weak, and so this result, like the law of rectilinear diameters, still chose the density as a variable which was somehow unique, and distinguished from any nonlinear function of the density.

Widom didn't know who first made the remark, but he had learned it from Griffiths, who had dropped a hint on one occasion. If one didn't have a classical coexistence curve, i.e., the coexistence curve itself is of algebraic degree not equal to two, then if one transformed the diagram from the density to any nonlinear function of the density, say the volume, one would automatically introduce into that new diameter in the new representation a singularity which would be just 2β. In the original law of rectilinear diameters, which Widom believed to be false, and in his model, the singularity was much weaker than 2β. Therefore these said that the density played a unique role in describing the symmetry between liquid and vapor.

Widom commented also on the analyticity of the chemical potential. Just as in the lattice gas model, the chemical potential in his model on the whole line of symmetry was an analytic function of T, analytic even at T_C. In fact if z were the activity, and v_0 were the volume of an isolated sphere, then on the line

$$v_0 z = \left(\frac{e}{kT}\right) \exp\left(-\frac{e}{kT}\right)$$

Thus in both the lattice gas model and the penetrable sphere model the chemical potential would be an analytic function of T.

Widom reviewed the experimental situation. The data were not all that accurate, but they had been examined by Sengers, Vicentini-Missoni, and Green at the Bureau of Standards. They found that the data bore out this analyticity in the two-phase region. It was by no means obvious that this should be the case. It was certainly not a thermodynamic necessity. If so, then, only the pressure, and not the chemical potential, contributed the whole of the specific heat singularity. Yang had mentioned this a couple of years earlier. Widom asked if analyticity, then, were a law of nature?

If this were so, it would give us a line of symmetry in a liquid-vapor system without any particular reference to any one model. Thus one could establish in the two-phase region as accurately as possible what analytic function of T the chemical potential actually would be. If that function were continued into the one-phase region, then what would the locus be in the latter region such that the chemical potential depends on T in the same manner?

Widom gave a geometrical interpretation of the symmetry in his model. In the lattice gas, the statement was that when liquid and vapor were in equilibrium with each other, the density of holes in the liquid would be equal to the density of particles in the vapor.

In the penetrable sphere model, he noted the following feature:

If one looked at a very dilute fluid in this model, one saw practically isolated spheres. He called the total exposed area of these spheres A. Then in the dilute case, A was roughly proportional to the number of molecules, and so there would be a total amount of exposed area per unit volume. In the very dense fluid, there would be a great deal of overlap, and the exposed area was the area of almost isolated holes. The criterion for equilibrium in the penetrating sphere model was that the exposed area per unit volume of the particles in the gas would be equal to the exposed area per unit volume of the holes in the liquid

$$\left(\frac{A}{v}\right)_g = \left(\frac{A}{v}\right)_l$$

But this quantity (A/v) then played the role of a thermodynamic potential. A plot of A/v vs. W/v gave the phase diagram as shown in Fig. 4. Two phases were in equilibrium for a single value of A/v, but different values of W/v. In this way, he described the phase equilibrium in purely geometrical terms without mentioning a thermodynamic variable. Of course, thermodynamics is in back of it. The degree of the coexistence curve near the critical point was

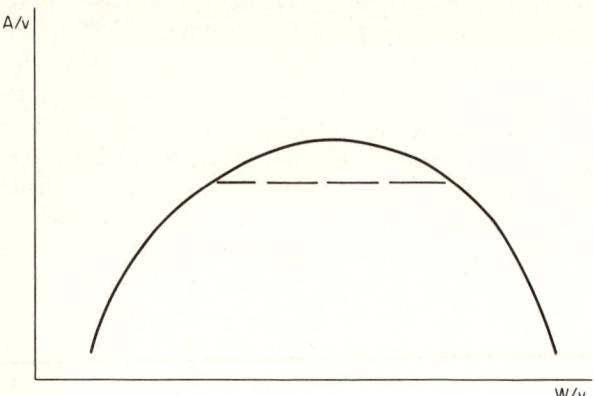

Fig. 4 *The general phase diagram in the sphere model.*

$\beta/(1 - \alpha)$. In the geometrical language, one had a renormalized coexistence curve. Ferrell asked what combination of ρ and ϕ, when plotted against T, gave a rectilinear diameter? Widom answered that if one were to plot $\rho - (\epsilon/kT)(1 - \rho - \phi)$, where ρ was dimensionless and where ϕ was a dimensionless energy density (related to W), the line of symmetry would become a straight line. Widom pointed out that N. David Mermin formulated a lattice analog of the penetrable sphere model. On the lattice, he covered lattice sites with molecules such that neighboring molecules overlapped. The interaction energy was the same as before, and no repulsion was included. Thus an arbitrary number of molecules could coexist on a site. Mermin showed this model was exactly equivalent to the original Ising model by the use of a simple transformation. However, the transformation was such that the original symmetry was lost, and what was a vertical line then became an oblique diameter. The new diameter was still analytic, however, unlike in Mermin's later models.[9] The critical indices were the same as those of the original lattice gas.

Josephson then asked what was known about Ising models in which the conditions for the Yang-Lee theorem were violated, so that one could get an asymmetric critical point. Griffiths answered that as long as one used only pair interactions in the Ising model (in addition to the magnetic field) and as long as the critical temperature is at $H = 0$, one had to expect the usual Ising ferromagnet symmetries and the analogous symmetries in the lattice gas. One could expect that with a suitable choice of interactions (including, of necessity, some antiferromagnetic interactions), one might obtain the situation shown in Fig. 5. At one of these critical points, one would

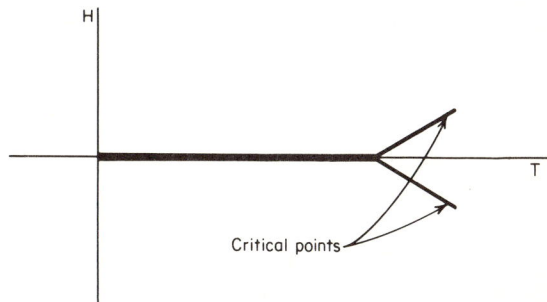

Fig. 5 *Hypothetical magnetic phase boundaries.*

expect the aforementioned symmetry to be absent. On the other hand, the investigation of singularities at such a point by numerical (series) methods would be an order of magnitude more difficult than in the case where the critical point is at $H = 0$ because, for example, one did not know from the outset at what values of H the critical point would occur. Griffiths wondered if the series experts could be persuaded to investigate such difficult problems.

Rushbrooke raised a question concerning the Lee-Yang theorem. As he understood this, if one were to regard the free energy as a function of the magnetic field H, and the temperature T, the Lee-Yang theorem proved that at a fixed T, F could not be singular as a function of H, except possibly at $H = 0$. He asked what could be said about the behavior of F as a function of T at fixed H. Griffiths replied that the situation with respect to analyticity in T for the Ising ferromagnet was as follows: For sufficiently large H (depending on T), one could demonstrate analyticity in both H and T.[10] The Lee-Yang circle theorem would give analyticity on H for $\text{Re}(H) > 0$ for fixed T. These results could be combined by a purely mathematical argument on functions of two complex variables to show that the free energy would be analytic in both T and H for $\text{Re}(H) > 0$ and some region in the complex T variable which included the real T axis.[11] For the Heisenberg model one could also show analyticity for H large enough, and the recent work of Asana (and the extensions by Suzuki and Fisher) provided the corresponding circle theorem. Thus the Lebowitz and Penrose argument could again be used to show that the Heisenberg ferromagnet had a free energy analytic in both H and T for $\text{Re}(H) > 0$.

Montroll concluded the meeting with the following remarks:

Part of my responsibility is to summarize the present status, and perhaps to foresee a bit of the future. There is a famous precedent for this. In the 1890s, Lord Kelvin was invited to Baltimore to give the Baltimore lectures. He was paid the sum of $2,000 to give a series of three or four lectures, which was fantastic at the time. He concluded his remarks with the comment that everything in physics was fine, and that only a few of the more important variables needed to be measured to a few more significant figures. He noted that there were three small clouds on the horizon, however. One had to do with the ether: it didn't seem to be understood completely. A second had to do with the specific heat of hydrogen gas: the degrees of freedom seemed not to have been counted properly. The third had to do with heat sources in the Earth: his calculations of the temperature and the age of the Earth differed from what the geologists thought it should be.

It might be suitable here to follow Lord Kelvin's example and suggest that one might measure the thermodynamic quantities with a bit more accuracy, and perhaps a bit closer to the critical point. There might be three little clouds on the horizon, however. One has to do with the three-dimensional Ising model (which will probably continue to be a little cloud). A second has to do with the relations between laws of force, lattice structures, and critical exponents: there might yet be something about this which we don't fully understand. The third has to do with nonlinear processes; turbulence is one such problem, dynamical scaling another. When we have these points cleaned up, the whole subject should become clear.

REFERENCES

1. Goldstein, R. J., and W. F. Hagen: *Phys. Fluids,* **10**:1349 (1969).
2. Yeh, Y., and H. Cummings: *Appl. Phys. Lett.,* **4**:176 (1964).
3. Bourke, P. J., et al.: *J. Phys.,* **A3**:216 (1970).
4. Montroll, E. W.: *J. Math. Phys.,* **11**:635 (1970).
5. Callen, H. B.: *Phys. Lett.,* **4**:161 (1963).
6. Yahata, H., and M. Suzuki: *J. Phys. Soc. Japan,* **27**:1421 (1969).
7. Cahn, J. W., and R. Kikuchi: *J. Chem. Phys. Solids,* **20**:94 (1961); **23**:137 (1962); **27**:1305 (1966).
8. Widom, B., and J. S. Rowlinson: *J. Chem. Phys.,* **52**:1670 (1970).
9. Mermin, N. D.: *Phys. Rev. Lett.,* **26**:168, 959 (1971).
10. Galavotti, G., and S. Miracle-Sole: *Commun. Math. Phys.,* **7**:274 (1968).
11. Lebowitz, J. L., and O. Penrose: *Ibid.,* **11**:99 (1969).

INDEX

INDEX

Activation rate:
  phase slips in superconductors, 453
  prefactor, 452–456
    experimental, 452, 454, 456–466
    Langer-Ambegaokar, 453
    McCumber-Halperin, 454
Adler, D., paper by, 576–590
  discussion of: energies in metal-insulator transitions, 611
    Luttinger's theorem, 612
    Mott transition: and band crossing, 610
    in metal-NH_3 solutions, 609
$AgAuZn_2$, 246
Al, resistivity transition in, 455, 456
Al-Ag alloys, 331
Al-Zn alloys, 250, 284, 286
Alefeld, G., paper by, 399–362
  discussion of: coalescence of triple and critical points, 406
    diffusion of elastic dipoles, 361
    elastic interactions, 475
    exponents for H-containing metal, 372
    mean-field theory, 637
    range of forces in liquid crystals, 475

Alefeld, G., paper by, discussion of (*Cont.*):
    solid-liquid crystal transition, 406
    use of continuum theory, 362
Alloys:
  Al-Ag, 331
  Al-Zn, 250
  Au-Pd, 330
  binary, 245–247
  close-packed, 253, 254
  Co-Fe, 244, 245, 250
  Cu-Al, 321, 329
  Cu-Au, 327
  Cu-Ni, 333
  Cu-Pt, 329
  Fe-Mo, 331, 332
  loose-packed, 246, 254
  Ni-Pd, 329
  $\underline{Pd}$-Ag, 352
  $\underline{Pd}$Fe, 437–449
  Pd_3Fe, 438 ff.
  $\underline{Sn}$In, 457 ff.
  ternary, 246
Ammonium halide, 255
Amplification factor, 55, 263–265, 274
Antiferroelastic order, 359

650 INDEX

Antiferromagnet(s), 382-384
  magnon damping, 433
  neutron scattering, 483-488
  NMR linewidths, 483-488
  relaxation in isotropic, 423-436
  uniaxial, 382, 483-488
Atomic misfit parameter, 273
Au-Ni, 271-274, 284, 286-288
Au-Pd alloys, 330
p-Azoxyanisole (PAA), 403

Baker, G. A., Jr., paper by, 221-229
  discussion of: beta-brass data and the Ising model, 297
    C_p for rigid Ising model, 255
    effect of lattice changes on exponents, 478
    location of spinodal curve, 64
    random exchange energies, 153
    scaling of Ising theory, 86
    smoothness postulate and analyticity, 390
    weak spin-lattice coupling, 123
Bardham, P., coauthor of paper, 309-337
Barmatz, M., paper by, 541-559
  discussion of: Kawasaki theory and ultrasonic experiments, 564
    Landau-Khalatnikov relaxation mechanism, 564
BaVS$_3$, electrical conductivity of, 570
Benedek, G. B., paper by, 503-527
  discussion of: anomaly in thermal diffusivity for Xe, 562
    interpretation of dynamic scaling, 564
    Kawasaki's linewidth formula, 563
    location of spinodal curve, 421
    optical homodyning in study of turbulent flow, 624
Berge, P., coauthor of paper, 529-540
Beta-brass (β-CuZn), 245, 248, 255, 289-298

Beta-brass (β-CuZn) (*Cont.*):
  Debye-Waller factor, 290, 293, 297, 298
  Ising model, 290, 296
  long-range order, 289, 290, 295
Bethe's hypothesis, 14-16, 18
Bi, resistivity transition in, 455
Binary alloys, 245-247
  compressibility, 246, 250, 251
  critical temperature, 249, 250
  expansion coefficient, 251
  first-order transition, 245
  heat capacity, 246, 250
  pair-correlation function, 247-249
  short-range order, 248, 271
  x-ray scattering, 248
Binary critical systems, 41-65
Binary fluid mixture, 489, 497-500, 529-540
Binary solid solutions, 251, 252, 257 ff
  critical solution temperature, 251
  H_2 in metals, 252, 255
  short-range order, 277
Bose-Einstein condensation of concentration fluctuations, 277-280
Bose fluid, 186, 281
Bose gas model, scaling equation, 415, 419
Brillouin scattering, 507, 508
Brinkman, W. F., coauthor of paper, 593-608
Buttner, H., secretary of agenda discussion, 471-480
  discussion of transition rates in superconductors, 473, 474

C_7F_{14}-C_7H_{14}, 44-46
Cahn, J. W., paper by, 41-65
  discussion of: accuracy of molecular field theories, 637, 638
    location of spinodal curve, 64
Callen identity, 365, 367, 630
Cannell, D., coauthor of paper, 503-527
CdS, electrical conductivity in, 569
Chalcogenide glasses, 588

Chang, Do-Ren, coauthor of paper, 105–123
Chipman, D. R., coauthor of paper, 289–298
Choquard, Ph., discussion of:
  lack of order parameter in metal-insulator transitions, 611
  possibility of finite-field solutions, 621
Clapp, P. C., paper by, 299–308
  discussion of: Cowley theory of long- and short-range order, 365
  diffusion of elastic dipoles, 361
  elastic energies in pairwise interactions, 370
  precursors in binary alloy transitions, 405
  ranges of correlations and forces, 475
  transitions in loose-packed alloys, 254
Clustering, 257
  systems, 277, 281, 283
CO_2, relaxation phenomena in, 561, 564
Co-Fe alloys, 244, 245, 250
Coexistence curve, 70, 83, 84, 414, 510, 643
  measurements along, 510, 516
  scaling on, 84, 414 ff.
  symmetry on, 84, 643
Cohen, J. B., paper by 309–337
  discussion of: Fourier transformation of atomic distributions, 336
  kind of ensembles in computer simulations, 337
Coherency, 59, 60, 62
Comley, J. B., coauthor of paper, 437–449
Computer modeling of local order, 310, 315 ff.
Concentration:
  fluctuations, 497, 531–539
  variation, 258
Configuration entropy, 259, 269

Configuration entropy (Cont.):
  in Ising lattices, 300
Connective constant, 92
Cook, H. E., paper by, 257–275
  discussion of: elastic constants: in liquid crystals, 475
  of Au-Ni, 287
  elastic energies in pairwise interactions, 370
  Fourier transformation of atomic distributions, 336
  limit of Kawasaki equations at high temperatures, 500
  metastable gap in H-containing metals, 373
Cooper, M. J., paper by, 77–87
  discussion of: corrections to scaling equations away from T_c, 421
  magnetic scaling equation, 86
  parametric representation and smoothness postulate, 390
Correlation:
  dynamic, 143
  function: binary solutions, 497
  concentration fluctuations, 487
  high T, 25–29
  low T, 29–37
  order parameter, 192
  scaling of, 69
  spin-pair, 93, 97–99, 100–102
  spin-spin, 13, 149–151, 386–388
  time, 495
  two-particle, 181
  static, 142
Cowley, J. M., discussion of Wilkins' theory of scattering intensity, 368–370
$CrBr_3$, 410, 411
Critical exponents (indices):
  dependence on: impurities, 478
  isotropy of interaction, 206
  lattice dimensionality, 208
  lattice spacing, 478
  nonuniformity of exchange factors, 207
  spin quantum number, 204

652 INDEX

Critical exponents (indices), dependence on (*Cont.*):
  symmetry of ground state, 208
  magnetic, 440
Critical field:
  electric, 136
  magnetic, 383
Critical line, 106, 378, 384
  in antiferromagnets, 383
  in $FeCl_2$, 384
  ferromagnet under pressure, 383
  in He^4, 106
Critical point:
  dependence on parameters, 71, 377–391
  tangency plane, 79
Critical surfaces, 263, 264
Critical wavelength, 264
Cr_2O_3, electrical conductivity of, 581
CrS, electrical conductivity of, 570
Cu-Al alloys, 321, 329
Cu-Au alloys, 327
Cu_3Au, 315, 318, 323–325, 253, 303–307
Cu-Ni alloys, 333–335
Cu-Pt alloys, 329
CuZn, 243, 244, 248–250, 373
Cyclohexane-anilene, light scattering from, 529–540

Deuterium:
  in Nb, 349
  in Pd, 355, 356
  in Ta, 349
  in V, 349
Diameter:
  phase boundary, 84, 85
  rectilinear, 71, 84
Diffuse scattering intensity, Fourier coefficients, 311 ff., 364
Diffusion coefficient, 41, 54, 263
Diffusivity, 41, 48, 50
  thermal: in SF_6, 520
  in Xe, 514
Dimensionality:
  effect on critical exponents, 208

Dimensionality (*Cont.*):
  effect on critical temperature, 213, 215
Domb, C., paper by, 89–103
  discussion of: beta-brass exponents and the Ising model, 297
  coexistence curve for spherical model, 76
  convergence of Rushbrooke's calculation, 477
  corrections to scaling in the Ising model, 86
  exponents for $\underline{Pd}Fe$, 448
  homogeneous function in equation of state, 103
  lack of connection between Kouvel's and Rushbrooke's work, 477
  liquid crystals and Ising systems with impurities, 474
  low-temperature series for a random system, 164
  magnetic analog of liquid crystals, 405
  Ornstein-Zernike pair-correlation function, 103
  series expansions: for ferroelectric models, 175
  for multiple correlations near T_c, 308
  transformations of spin-½ systems, 218
  utility of linear parameter representation for Ising systems, 477
Droplet picture, 103
Dynamic scaling:
  basic hypothesis, 507
  in binary solutions, 495, 497–500
  in ferromagnets, 492–495
  in pure fluids, 506 ff.
  laws, 424, 479, 561–565

Elastic constants in liquid crystals, 475, 476
Elastic modulus, 347, 348
  effective, 273

Elastic stability criteria, violation of, 348, 351
Elasticity, microscopic theory of, 258
Electric polarization, 137
Electrical conductivity:
    as function of composition, 572, 573
    as function of pressure, 569, 570
    as function of temperature, 570, 571
Electronic phase transitions, 567 ff.
    nonequilibrium transitions, 582–589
Electrons, screening of, 595–600
Energy gradient, 280, 282, 283
    coefficients, 266, 267, 270, 274
Equation of state:
    analyticity, 413
    asymptotic, 409 ff.
    fluid, 504, 505
    homogeneity, 103, 190
    magnetic, 74, 438 ff.
    parametric, 73–87, 412–420
    scaling, 407–421

Falk, G., paper by, 135–137
FeAl, 245
Fe_3Al, 243, 245, 246, 248, 249
$FeCl_2$, 384
FeF_2, 283
$FeLiO_2$, 335
Fe-Mo, alloys, 331, 332
Fe_3O_4, electrical conductivity of, 570, 571, 578
Ferroelasticity, 339–362
    analogs with electric systems, 340–346
Ferroelectric(s), 69, 72
    critical: states, 135–137
        temperature, 136
    Lee-Yang theorem, 619
    planar models, 165–175
    specific heat, 137
    transitions, 130, 135
Ferrell, R., chairman of agenda discussion, 471–480
    analyticity of thermodynamic functions, 390

Ferrell, R. (Cont.):
    determination of linewidth coefficient, 559
    dynamic effect in isotropic ferromagnets, 479, 480
    fluctuations in superconductors, 468
    logarithmic factor in ferromagnet linewidth, 436
    mode decoupling, 501
    physics of mode decoupling, 564
    theory for thermal conductivity, 559
Ferromagnetism:
    compared with liquid crystals, 395 ff.
    critical: index inequalities, 221 ff.
        dynamic scaling in isotropic, 479, 480
    itinerant, 179–185
    Lee-Yang theorem, 613 ff.
    mean-field, 69, 72
    PdFe alloys, 437–449
    under pressure, 379–382
    with random impurities, 145–164
    relaxation, 423–429
    scaling equation, 73, 407–421
FeS, electrical conductivity of, 570, 578
$FeSi_2$, electrical conductivity of, 570, 571
Fe_3Si, 245
$FeSiO_4$, electrical conductivity of, 582
Fick's law, 263
First-order transitions:
    in binary alloys, 245
    in liquid crystals, 394 ff.
    order-disorder, 245
Fluctuations:
    composition, 50, 55
    concentration, 497, 531–539
    critical, 489–502
        frequency spectrum of, 494
        slowing-down of, 490, 492, 505
    in liquid crystals, 402–404
    kinetic equations for, 490–492
    magnetic, 471, 472
    relaxation of, 530

Fluctuations (*Cont.*):
  spin, contribution to NMR linewidth, 483
  in superconductors, 472 ff.
  thermal, 48
Fluids, pure:
  light scattering from, 503–527
    in SF_6, 515–525
    in Xe, 510–515
de Fontaine, D., papers by, 257–275, 277–288
  discussion of: elastic energies in pairwise interactions, 371, 372
    liquid crystal molecules as elastic dipoles, 476
    long-range forces and correlations, 475
    phase boundaries, 373
Free energy:
  analyticity of, 613–620
  of binary solid solutions, 278
  smoothness postulate, 377

GaAs, electrical conductivity of, 569
Gap indices, 222
GaSb, electrical conductivity of, 570
Gd, 251
  magnetism in, 411
Ge, electrical conductivity of, 569, 570, 573, 582
Gehlen, P. C., secretary of agenda discussion, 363–374
Gibb's conditions of critical phases, 129, 130
Giglio, M., coauthor of paper, 503–527
Ginzberg-Landau theory:
  activation rate prefactor, 454–456
  barrier energy in, 453
  fourth-order corrections, 456
Glasser, M. L., paper by, 165–175
  discussion of: analytic properties of ferroelectric models, 174
    Mossbauer effect in liquid crystals, 406
Goldstone modes, 177
  normalization conditions, 185

Gorsky effect, 349, 352, 354, 357
Gragg, J. E., coauthor of paper, 309–337
Graph theory, 157–159
Gravity effect, 542, 549–551
  in He^4, 550
Griffiths, R. B., paper by 377–391
  discussion of: analytic properties of ferroelectric models, 175
    asymptotic scaling, 75
    critical indices for very dilute samples, 449
    Lee-Yang theorem, 644
    mean-field theories of metastability, 133
    Mott and metal-insulator transitions, 608
    pure and impure superconductors, 469
    Rushbrooke's impurity calculation, 478
    series for low-temperature susceptibility, 164
    shifts of mean-field critical temperatures, 636
    smoothness postulate, 390
Gruber, C.:
  discussion of higher-order correlation functions, 39
  discussion of Ising coefficients, 630
Guttman, L., paper by, 243–256
  discussion of: AB ordering in loose-packed alloys, 255
    diffuse scattering at high temperatures, 288
    dilutent in liquid crystals, 406
    values of correlation function exponents, 218

H-Pd, 372
Harmonic concentration waves, 261
He^3-He^4 mixtures, 386
Heisenberg model:
  anisotropic, 225, 229
  classical, 209
  ferromagnet, 229

INDEX

Heisenberg model, ferromagnet (*Cont.*):
   dynamic scaling, 492–495, 497
   Lee-Yang theorem, 615
   randomly dilute, 155–164
   variation of critical exponents, 204 ff.
Heisenberg-Ising model, 221, 223, 225
Helium-4:
   critical point measurements, 541–559
   gravity effect in, 550
   liquid, 105–170
     compressibility, 115, 118
     entropy, 117
     specific heat, 113, 115, 551–552
     thermal expansion coefficient, 114, 117–120
   sound dispersion, 557–559
   transport properties, 557–559
   velocity of sound, 545, 548, 552–557
Heller, P., paper by, 483–488
   discussion of: measurements of diffusion factor, 639–640
   susceptibilities and smoothness postulate, 389
Hubbard model, 594 ff.
   atomic limit, 610–607
Hydrogen in metals, 339–362
   in Nb, 349, 352, 353, 358
   in Pd, 352, 355, 356
   in Ta, 349, 352, 358
   in V, 349

Ice condition, 165
Impurities:
   frozen-in, 147
   mobile, 147
   random, 145–154
InAs, electrical conductivity of, 569
Inequalities on critical indices, 221–229
Infinite-order phase transition, 169
InP, electrical conductivity of, 569
InSb, electrical conductivity of, 569
Insulator-metal transitions, 567 ff.
   mechanisms for, 574–576

Integrated intensity, 280, 293
   condition, 282
Interface, 59, 62
Ising ferromagnets, 386–388
Ising kinetic, 231
Ising lattices:
   cluster probabilities in, 299 ff.
   configurational entropy, 300
   long-range order, 300
   order-disorder transformation, 299
   short-range order, 300
Ising model:
   1D, 32–37
   2D, 3–12, 74, 89, 91, 191, 209, 222, 223, 243, 248, 252, 440, 445, 477
     nearest-neighbor interactions, 25–32
     random impurities, 145–154
     scaling: equation of state, 412, 414, 419
     test of, 141
   3D, 73, 75, 86, 89, 91, 209, 297, 440, 444, 445, 477
     in beta-brass, 373
     Lee-Yang theorem, 615
     scaling equation of state, 412, 414, 419
Ising systems, 186
   randomly dilute, 155–164
Isochore, critical, 85, 513, 514, 519, 520, 524, 545
Isotherm:
   critical 82, 85, 440, 443, 516, 518, 545
   flat parts of, 24, 25, 36
   loops in, 42
   pressure-density, 42
   Van der Waals, 42

Jona-Lasinio, G., paper by, 189–201
   discussion of renormalization group results, 200–201
Josephson, B. D., paper by, 73–76
   discussion of: asymptotic scaling, 76

656   INDEX

Josephson, B. D., paper by, discussion of (*Cont.*)
   Lee-Yang theorem, 644
Josephson effect, influence of fluctuations on, 472

Kac, M., paper by, 23–29
   discussion of: continuity for vanishing interaction, 21
   factorable potentials, use of, 39
   free energy for random walks, 103
   higher-order correlation functions, 39
   nonergodicity of XY model, 239
   Ornstein-Zernike results, 39
   pair-correlation functions, 39
   spectrum gap and phase transitions, 38
Kadanoff, L. P., chairman of agenda discussion, 561–565
   convergence of mode-mode coupling, 501
   correlation function exponents, 560
   critical exponents, effect of impurities, 478
   mean-field theory and liquid crystals, 475
   origins of smoothness postulate, 391
   properties of SF_6 and Xe, 563
   scaling functions, 86
   summary of status of mode-mode coupling, 564–565
Kawasaki, K., paper by, 489–502
   discussion of: convergence of perturbation theory, 502
      mean-field theory and mode-mode coupling, 501
Kinetic equation, 490, 497
   for critical fluctuations, 490–492
Kouvel, J. S., paper by, 437–449
   discussion of: critical exponents and interaction ranges, 448, 449

Lambda:
   line: density change on, 109
      entropy on, 109
      in $FeCl_2$, 384
      in metamagnets, 384
      lattice changes on, 110
      specific heat on, 110
      thermodynamic relations near, 108
   transition, 69, 71, 378
      in $FeCl_2$, 384
      in laminar-turbulent flow, 624–626
      in liquid helium, 111
      in metamagnet, 378, 384
      secondary variables in, 105–120
Laminar flow, 624–626
Landau theory:
   of fluctuations, 262, 277, 278
   of liquid crystals, 395–397
   of metamagnets, 385
Lastovka, J. B., coauthor of paper, 503–527
Lattice changes, effect on exponents, 478
Lattice constant, 91
Lee-Yang theorem, 223, 613–620, 644–645
Light scattering:
   cyclohexane-anilene, 529–540
   pure fluids, 503–527
   SF_6, 515–525
   Xe, 510–515
Linear response theory, 423
Linear scaling function, 413 ff., 476, 477, 542, 546–549
   gravity effect in He^4, 549–551
Liquid crystals, 393–406
   critical behavior, 400–402
   discussion of, 474–476
   elastic constants in, 475, 476
   fluctuations in, 402–404
   and Ising system with impurities, 474
   Landau theory, 395–397
   mean-field theory, 474, 475
Litster, J. D., paper by, 393–406
   discussion of: elastic constants in liquid crystals, 475

Litster, J. D., paper by, discussion of (*Cont.*):
   mean-field theory in liquid crystals, 474
   utility of linear scaling function, 476
   value of exponent η, 560
Local order, 309 ff.
London theory, 280–283
Long-wavelength relations, 267, 271
Luttinger's theorem, 612

McCoy, B. M., paper by, 145–154
   discussion of: four-particle correlation functions, 307
   mean-field theory for liquid crystals, 405
   random-exchange energies, 153
Magnetic systems:
  discussion of, 476–480
  dynamic scaling in, 479–480
Magnetization:
  2D Ising model, 14
  as a parameter, 126
  longitudinal, 425, 431
  staggered, 430
Magnon damping rate near T_c, 424
Marshall, W., discussion by:
  of Debye-Waller factors, 297
  of mean-field theory, 372
  of origin of energy anisotropies, 372
Martin, P., rapporteur's introduction, 141–144
  discussion of:
    convergence of perturbation theory, 501
    liquid crystals and Ising systems with impurities, 406
    phase transitions in two dimensions, 38
    pressure and the order of transitions, 255
MBBA (*see* p-methoxybenzylidene p-n-butylanilene)
Mean-field theory (MFT):
  in alloys, 368

Mean-field theory (MFT) (*Cont.*):
  discussion of accuracy, 633–638
  for ferroelastics, 346–349
  in liquid crystals, 477
  for metamagnets, 385
  scaling equation of state, 413
Meissner effect, influence of fluctuations on, 472
Memory switching, 588, 589
Metal-insulator transition, 576–590, 594 ff.
  discussion of, 608–612
Metal-NH_3 solutions, 609
Metamagnets, 384–386
Metastability, 41–53
Metastable states, 125, 126, 128
p-Methoxy benzylidene p-n-butylanilene (MBBA), 393, 397 ff.
  critical fluctuations in, 400–403
  magnetic birefringence in, 397–400
Mills, R. E., paper by, 177–188
Miscibility gap, 286
Mixtures:
  binary liquid, 489, 497–500, 529–540
  fermion and boson, 20
Mode decoupling, 501, 564
Mode-mode coupling, 423, 501, 502, 535, 537, 557, 561–565
  convergence of, 501
Molecular field theory (*see* Mean field theory)
Moments, giant magnetic, 446
Montroll, E., chairman of summary agenda, 623–646
  ecology, populations, and statistical mechanics, 630–633
  lambda points in turbulent flow, 624
  Lidstone series, 626
  solutions of models on periodic lattices, 627–628
Moss, S. C., chairman of agenda discussion, 363–374
  exponents and nonstoichiometry, 297
  long-range potentials, 448

Moss, S. C. (*Cont.*):
  metal-insulator transitions and mode softening, 609
  unmixing temperature in Au-Ni alloys, 287
  x-ray critical opalescence, 288
Mott transitions, 576 ff., 593–608
  discussion of, 608–612
  possible occurance, 578, 609, 610

NbO_2, electrical conductivity, 570, 571
Nearest-neighbor interaction:
  in alloys, 364
  in magnetic systems, 156, 207
Néel temperature, 382
Nematic-isotropic transition, 393
  in MBBA, 397 ff.
Neutron scattering
  and Debye-Waller factors, 297
  in FeF_2, 483 ff.
Ni, magnetism in, 410, 411, 440
Ni-Pd alloys, 329
NiO, 594
  electrical conductivity, 581
NiS, electrical conductivity, 570, 571, 578
Ni_3V, 306
NMR linewidth:
  contribution from spin fluctuations, 483
  in FeF_2, 483 ff.
Nonequilibrium transitions, 582–589
  mechanisms for, 584–587
Nucleation, 43, 46–49, 52
  in C_7F_{14}-C_7H_{14}, 49
  droplet theory, 46, 49, 50
  in water, 48, 50

One-dimensional problems, 14–20
Onsager, L.:
  autobiographical sketch, xix
  discussion of paper by, 3–12
    behavior of correlation function through T_c, 390
    Landau criterion for continuous transition, 255

Onsager, L., discussion of paper by (*Cont.*):
  opalescence in laminar-turbulent transition, 626
  spinodal extrapolation and impurities, 405
  three- and four-particle correlation functions, 308
Opalescence, critical, 61
  in liquid crystals, 403
  in turbulent-laminar transition, 626
Optical mixing spectroscopy, 509, 529, 530
Order:
  Cowley theory, 365
  -disorder transformation, 275
  local, computer modeling, 310, 315 ff.
  long-range, 365
  short range, 336, 365
Ordering kinetics, 257 ff.
Ordering systems, 281, 283
Ornstein-Zernike, 26, 100, 103, 499, 560

Padé approximants, 64, 224, 295
Pair-interaction parameters in alloys, 364, 365, 368, 369
Pair-probability, conditional, 310 ff.
Paraelasticity, 346–349
Paramenters, dependence of critical points on, 71, 377–391
  thermodynamic, 378
Parametric representation, 40, 78, 412–420
Penetration depth, 460
Percolation, 156
Pippard relations, 384
Polygon, self-avoiding, 91
Potential surface, thermodynamic, 77
Pressure effects:
  on ferromagnets, 379–382
  order of transition, 255
Principal eigenfunction, 31, 35
Psuedoparticle, 279, 280

INDEX

Random impurities, 145–154
Random lattice, 156 ff.
Random walks, 89 ff.
Range, magnetic interaction, 445
Rayleigh scattering, 505, 508
Rectilinear diameters, 70, 640 ff.
Relaxation process in magnets, 423
Relaxation time:
  in ergodic systems, 235
  in linear response, 232
  in nonergodic systems, 236–239
  spin-wave, 427, 433–436
Renormalization group, 186, 189–201
  functional equation, 193
Resistance, electrical, near T_c, 425, 455
Resistivity transition:
  in Al, 455, 456
  in Bi, 455
  in Sn, 456–466
Rice, O. K., paper by, 105–123
  discussion of: analogy of magnetic and fluid thermodynamic functions, 390
    transitions in NH_4-halide crystals, 255
    weak coupling near the lambda line, 126
Rice, T. M., paper by, 593–608
  discussion of: Luttinger's theorem, 612
    metal-insulator transitions and mode softening, 609
    Mott and other metal-insulator transitions, 608
    Mott transitions and band crossings, 610
    orders of magnitude of energy contributions, 611
Rotons, 113
Rushbrooke G. S., paper by, 155–164
  discussion of: exponents for H(½) system, 217
    his results and Kouvel's data, 478
    Lee-Yang theorem, 646

S-matrix, 20
Sb_2Se_3, electrical conductivity of, 572, 577
Scaling:
  correlation functions, 69, 71, 97–99
  and dimensionality, 144
  dynamic (see Dynamic scaling)
  equation of state, 69, 70, 407–421, 441
  homogeneity, 142
  hypothesis, 73, 93, 407, 409
  laws, 73, 186, 409
  linear function, 413 ff.
  modifications beyond simple, 84, 420
  and renormalization group, 193–201
  strong, 97, 144
  thermodynamic 69, 70
  universality, 143, 144
  weak, 97, 98, 143, 144
Se, electrical conductivity of, 569
Secondary variables:
  effect on critical point, 69, 71
  lambda transition in liquid helium, 105–120
  pressure, 106
  volume, 106
Self-avoiding polygon, 91
Self-avoiding walk, 89
Series, high temperature, 212
SF_6:
  discussion of transport properties, 561–563
  light scattering from, 515–525
  Rayleigh linewidth anomaly, 520–525
Short-range order, 266
  Warren, 261, 270, 364
Si, electrical conductivity of, 569, 570, 573
Sievert, P. R., secretary, agenda discussion, 561–565
Slowing down:
  critical, 231–240
  kinetic, 233
  of fluctuations in liquid crystals, 397
  thermodynamic, 233

Smoothness postulate, 378 ff.
Snoeck effect, 349
Snoeck relaxation, 349, 352, 354
Solutions, binary, 495
    correlation functions, 497–500
Sound:
    dispersion of, 557–559
    speed in He^4, 545, 548, 552–557
Spectrum:
    of operators, 32, 33
    of scattered light, 505, 530
Spherical model, 209
Spin waves:
    damping near T_c, 425
    long wavelength, 427, 434–436
Spinodal 41, 43, 53–58, 283, 284, 413, 421
    decomposition, 56, 59, 257, 262, 264, 266, 272, 274, 275
Staggered field, 382
    on antiferromagnets, 382
    on ferromagnets, 388
    on metamagnets, 385
Stanley, H. E., paper by, 203–219
    discussion of: Callen identity, 367
        gap parameter, 218
        Rudermann-Kittel interaction, 370
        susceptibility series, 218–219
Stenschke, H., paper by, 125–133
    discussion of: magnetic moments in ordered and random state, 478
Stephen, M. J., discussion by:
    of liquid crystals and Ising systems with impurities, 474
    of second-order transitions in liquid crystals, 474
Superconductors, 451–469
    dimensionality, 453
    discussion of, 473–474
    fluctuation temperature range, 452
    quantum phase slips, 353, 464–467
    resistivity transition: in Al, 455, 456
        in Bi, 455
        in Sn, 456, 466
    thermodynamic fluctuations, 452 ff., 471, 472

Superheated liquid, 47
Supersaturations, 50, 52
Susceptibility:
    electric, 137
    ferromagnetic, 131, 132
    longitudinal staggered, in FeF_2, 486
    parallel, by neutron scattering, 485
    paramagnetic, 131, 132
    series for, 156, 157–159, 215, 218
Suzuki, M., paper by, 231–240, 613–620
    discussion of: anisotropic Heisenberg ferromagnets, 229
        critical exponents and concentration, 163
        models with planar distributions of zeroes, 620
        relaxation times, 240
        singularities of ferroelectric models, 174
        solution of finite-field problems, 621
Symmetry:
    of coexistence curve, 83
    dynamic rearrangement of, 187
Syozi model, 155, 163

Tangent plane, critical, 79
$Te_{81}Ge_{15}As_4$, memory switching in, 589
Ternary alloys, 246
Thomas, H., discussion by:
    of ferroelectric critical exponents, 137
    of magnetization without long-range order, 38
    of nucleation in gas-liquid transition, 64
Ti_xO_y:
    electrical conductivity in, 570, 571, 578, 579, 582, 610
    Mott transition in, 610
Transfer matrix:
    2D Ising model, 4–6, 32
    ferroelectric models, 167

Transport properties:
  in critical region, 489–502
  of cyclohexane-anilene, 531–539
  of fluids, 505
  in He^4, 557–559
  of SF_6, 515 ff.
  of Xe, 510 ff.
Tricritical points, 378, 384, 385
Turbulent flow, 624–626

Ultrasonic measurements in He^4, 543–545
Universality hypothesis, 208 ff.

V_xO_y:
  electrical conductivity in, 570, 571, 578–582, 610
  Mott transition in, 610
Van Hove, L., banquet address, xxv
Vertex consistency, 185
Vertex functions, 195
  and critical indices, 196
  and correlation length, 197
Vicentini-Missoni, M., paper by, 407–421
  discussion of: spinodal line, 421
  utility of the linear model, 477
Villain, J., paper by, 423–436
  discussion of: agreement with dynamic scaling, 436
  relaxation time, 240
Viscosity, shear, 498
  anomaly in, 499
Volochine, B., paper by, 529–540

Walk:
  random, 91
  self-avoiding, 89–103
Walker, C. B., paper by, 289–298
  discussion of: Debye-Waller factors, 297, 298
Warburton, R. J., paper by, 451–469
  discussion of: prefactor for resistivity transition, 469
    pure and dirty superconductors, 469

Webb, W. W., coauthor of paper, 451–469
Widom, B., rapporteur's introduction, 69–72
  discussion of: coexistence curve for H-Pd alloys, 372
  exponent renormalization, 448
  long-range forces versus correlations, 475
  metal-insulator transitions, 609
  Ornstein-Zernike theory below T_c, 39
  rectilinear diameters, 640–644
  scaling law, 448
  solubility of bose-fermi liquid mixtures, 20
Wiener-Hopf method, 11

Xe, 403, 510, 542, 551
  light scattering from, 510–515
  discussion of transport properties, 561–563
X-ray:
  diffracted intensity, 364
  scattering in binary alloys, 248, 289–298
XY model, linear, 231

Yang, C. N., paper by, 13–21
  discussion of: computer simulation of ordering, 337
  continuity of interaction parameter, 21
  correlation functions for boson systems, 20
  effect of varying interaction strength, 21
  ice condition in ferroelectric models, 175
  solubility of finite-field problems, 621
  zeroes of grand partition function, 38

ZnS, electrical conductivity of, 569